D0712652

ATMOSPHERIC DIFFUSION
Third Edition

ELLIS HORWOOD SERIES IN ENVIRONMENTAL SCIENCE

Series Editor: **R. S. Scorer,** Imperial College of Science and Technology, University of London

A series concerned with nature's mechanisms – how earth and the species which inhabit it fit together into a dynamic whole, and the means by which evolution has taught them to survive. We are *not* primarily concerned to exploit the environment to human advantage, although that may happen as a result of understanding it. We are interested in the basic nature of the physical world, the special forms that it takes on earth, the style of life species which exploit special aspects of nature as well as the details of the environment itself.

POLLUTION CONTROL AND CONSERVATION
Edited by M. KOVACS, Hungarian Academy of Sciences, Vacratot, Hungary

ATMOSPHERIC DIFFUSION 3rd Edition
F. PASQUILL and F. B. SMITH, Meteorological Office, Bracknell, Berkshire

LEAD IN MAN AND THE ENVIRONMENT
J. M. RATCLIFFE, National Institute for Occupational Safety and Health, Cincinnati, Ohio

THE PHYSICAL ENVIRONMENT
B. K. RIDLEY, Department of Physics, University of Essex

CLOUDS, NATURAL AND ARTIFICIAL
R. S. SCORER, Imperial College of Science and Technology

ENVIRONMENTAL ECONOMICS Analysis and Policy
R. WATERTON, School of Economics and Accounting, Leicester Polytechnic

ATMOSPHERIC DIFFUSION

Third Edition

F. PASQUILL, D.Sc., F.R.S.
and
F. B. SMITH, Ph.D.
Senior Principal Scientific Officer
Meteorological Office, Bracknell, Berkshire

ELLIS HORWOOD LIMITED
Publishers · Chichester

Halsted Press: a division of
JOHN WILEY & SONS
New York · Chichester · Brisbane · Ontario

First published in 1983 by

ELLIS HORWOOD LIMITED
Market Cross House, Cooper Street, Chichester, West Sussex, PO19 1EB, England

The publisher's colophon is reproduced from James Gillison's drawing of the ancient Market Cross, Chichester.

Distributors:

Australia, New Zealand, South-east Asia:
Jacaranda-Wiley Ltd., Jacaranda Press,
JOHN WILEY & SONS INC.,
G.P.O. Box 859, Brisbane, Queensland 40001, Australia

Canada:
JOHN WILEY & SONS CANADA LIMITED
22 Worcester Road, Rexdale, Ontario, Canada.

Europe, Africa:
JOHN WILEY & SONS LIMITED
Baffins Lane, Chichester, West Sussex, England.

North and South America and the rest of the world:
Halsted Press: a division of
JOHN WILEY & SONS
605 Third Avenue, New York, N.Y. 10016, U.S.A.

British Library Cataloguing in Publication Data
Pasquill, F.
Atmospheric diffusion. – 3rd ed.
1. Atmospheric turbulence 2. Diffusion
3. Air – Pollution
I. Title II. Smith, F. B.
551.5'153 QC882

Library of Congress Card No. 83-219

ISBN 0-85312-426-4 (Ellis Horwood Ltd., Publishers – Library Edn.)
ISBN 0-85312-587-2 (Ellis Horwood Ltd., Publishers – Student Edn.)
ISBN 0-470-27404-2 (Halsted Press)

Typeset in Press Roman by Ellis Horwood Ltd.
Printed in Great Britain by R. J. Acford, Chichester.

Table of Contents

Preface 7

Notation and Symbols 9

1 Introduction and Brief Outline 13

2 Turbulence in the Atmospheric Boundary Layer 20
- 2.1 Essentials in the statistical description of turbulent flow 20
- 2.2 The mean-flow properties of the boundary layer 35
- 2.3 Thermal stratification and the maintainence of turbulent kinetic energy 49
- 2.4 Spectral representation – its significance and limitations 54
- 2.5 The small-scale structure of atmospheric turbulence 59
- 2.6 The statistics of boundary layer turbulence 67
- 2.7 Lagrangian properties 80

3 Theoretical Treatments of the Diffusion of Material 88
- 3.1 Eddy diffusivity and the gradient transfer approach 88
- 3.2 Solutions of the equations of diffusion 94
- 3.3 Similarity theory applied to diffusion 105
- 3.4 Statistical theory of dispersion from a continuous source 115
- 3.5 Applications and extensions of Taylor's theory 123
- 3.6 Stochastic modelling of diffusion 133
- 3.7 The expanding cluster 149
- 3.8 Diffusion of falling particles 158
- 3.9 The fluctuation of concentration 165
- 3.10 The effect of wind shear on horizontal spread 169

4 Experimental Studies of the Basic Features of Atmospheric Diffusion . . . 179
- 4.1 Principles of technique and analysis 179
- 4.2 The form of the cross-wind distribution at short range from a maintained point source 186

4.3 The magnitude of the cross-wind spread from a maintained point source . . . 189
4.4 The explicit relation between cross-wind spread and the statistics of the wind fluctuations . . . 198
4.5 Vertical diffusion from a near-surface release of passive particles . . . 204
4.6 Generalizations in terms of theories of vertical diffusion . . . 208
4.7 Dispersion from an elevated source . . . 215
4.8 The growth of clusters of particles . . . 220

5 The Distribution of Windborne Material from Real Sources . . . 233
5.1 The rise of hot effluent . . . 233
5.2 Deposition of airborne material . . . 246
5.3 The distribution of gaseous effluent from industrial stacks . . . 264
5.4 The effects of surface features . . . 283
5.5 Pollution in urban areas . . . 298

6 The Estimation of Local Diffusion and Air Pollution from Meteorological Data . . . 309
6.1 Qualitative features of diffusion and their variation with weather conditions . . . 310
6.2 The special meteorological factors in diffusion calculations . . . 313
6.3 Dispersion formulae for the local distribution of concentration from surface releases . . . 319
6.4 Estimation of vertical diffusion, from a surface release, through the depth of the boundary layer . . . 326
6.5 Allowance for elevation of source and for plume rise . . . 333
6.6 Presentation of dispersion estimates in terms of routine meteorological data . . . 335
6.7 Mathematical modelling for multiple sources . . . 341
6.8 Estimates using routine meteorological data and historical data on pollution . . . 351
6.9 Limitations and uncertainties in meteorological estimates of pollution . . . 357

7 Dispersion over distances dominated by Mesoscale and Synoptic Scale Motions . . . 359
7.1 The atmosphere on a large scale . . . 359
7.2 Dispersion into the whole atmosphere . . . 362
7.3 Mesoscale dispersion . . . 366
7.4 Long-range transport and the acid rain problem . . . 378
7.5 Long-range transport models . . . 380

References . . . 400

Index . . . 431

Preface to the Third Edition

In this new edition of *Atmospheric Diffusion* we have taken the opportunity of including several new developments that have come into prominence in the last decade. During the writing of the second edition the windborne transfer of pollution to areas remote from its origin had just become an international issue, and has since attracted sufficient scientific activity to warrant the collection of the relevant aspects of large-scale, long-range dispersion in a separate chapter. Otherwise the chapter structure of the former edition has been retained, with the content brought up to date in accordance with the continued improvement in the description of turbulence in the atmospheric boundary layer, advances in the basic theory of dispersion, and the search for more effective modelling procedures for the prediction of pollution levels.

As in the earlier editions we do not claim this latest survey to be encyclopaedic. Our aim is rather to be selective in providing coverage for which definitive progress seem to have been made

We are indebted to the following authorities for permission to include additional diagrams: Pergamon Press (Figs. 4.7 and 6.7), American Meteorological Society (Fig.s 2.13, 2.14, 2.16, 2.17 and 3.8), Royal Meteorological Society (Figs. 3.7, 4.17, 7.1, 7.2 and 7.4), and Mrs. Jane Blonder for the preparation of several other diagrams.

Bracknell, Berks
November 1982

Notation and Symbols

Symbols are defined on first introduction in the text, but for convenient subsequent reference a summary of those which are conventionally accepted or most frequently used is collected here. Equation or Figure numbers in which certain specialized quantities are defined are also indicated. Where there are alternative meanings identification will be obvious from the context. Occasional different usage, and isolated usage of additional symbols not included below, are explained as necessary in the text.

A subscript notation is used to indicate, for example, axis of reference (x, y, z), velocity component involved (u, v, w), duration of sampling or release (τ), or of averaging (s). A zero subscript is used to denote a ground-level value or an initial value. A subscript g denotes a geostrophic value.

An overbar refers to a mean value (usually with respect to time).

Primes usually denote departures from a mean value.

- $a(t)$ particle acceleration
- b numerical constant in the relation $\mathrm{d}\bar{Z}/\mathrm{d}t = bu_*$
- B dimensionless Stanton number
- c_p specific heat of air at constant pressure
- CIC crosswind integrated concentration
- D structure function – Eq. (2.2)
  - or divergence term in turbulent energy balance
  - or normalized dispersion
  - or rate of deposition of material per unit area
- E total turbulent kinetic energy per unit mass of air
  - or collection efficiency
  - or as subscript referring to Eulerian
- $E(\kappa)$ three-dimensional spectrum function – Eq. (2.62)
- f Coriolis parameter
  - or longitudinal correlation function – Fig. 2.11
  - or reduced frequency $nz/\bar{u}$

F turbulent flux
or stack parameter $\equiv gQ_H/\pi\rho c_p T_0$

$F(n)$ or $F(\kappa)$ normalized one-dimensional spectral density in terms of frequency n or wavenumber $\kappa, \equiv S(n)/\sigma^2$ or $S(\kappa)/\sigma^2$ – Eq. (2.10) *et seq.*

g acceleration due to gravity
or as subscript indicating geostrophic value
or transverse correlation function – Fig. 2.11

h mixing depth

h_s height of source

h_c and h_s are sometimes used to denote mixing depths in convective and stable conditions respectively
or height of base of overhead stable layer

H height of source or of plume centre-line
or vertical heat flux

i intensity of turbulence $\equiv \sigma_u/\bar{u}$ etc.

k von Kármán's constant – Eq. (2.51)

K eddy diffusivity – Eq. (2.44)

L Monin–Obukhov length: $-\left(\dfrac{u_*^3}{H}\right)\left(\dfrac{\rho c_p T}{kg}\right)$

l mixing length
or integral length-scale of turbulence – Eq. (2.6)

n frequency, usually in cycles/sec

n_m frequency at which $nS(n)$ is a maximum

p pressure
or a deposition coefficient $\equiv v_d/\bar{u}$

P probability density
or a stability parameter

q source density, rate of mass emission from unit area

Q source strength, rate of emission (point source) per unit length (line source)

Q_H rate of output of heat from a chimney

$R(x), R(t), R(\xi)$ correlation coefficient with reference to spatial separation (x) or time-lag (t or, for Lagrangian reference, ξ) – Eq. (2.3) *et seq.*

R_f flux Richardson number $\equiv -\dfrac{gH}{\rho c_p T u_*^2 (\mathrm{d}u/\mathrm{d}z)}$

Ri Richardson number $\equiv \dfrac{g}{T}\dfrac{\mathrm{d}\theta/\mathrm{d}z}{(\mathrm{d}u/\mathrm{d}z)^2}$

s averaging time
or parameter in exponential form of vertical distribution of concentration (Eq. 3.25)

or exponent in σ_z, x power law relation (Eq. (6.22) and Table 6.VI)

$S(n)$, $S(\kappa)$ spectral density with reference to frequency or wave-number, defined by $\int_0^\infty S(n)\,dn = \sigma^2$ etc.

t time

or time scale of turbulence, then usually with subscript s, or E or L, – Eq. (2.7)

T absolute temperature

or time of travel

u, v, w velocity components along axes x, y, z

u_* friction velocity $\equiv (\tau_0/\rho)^{1/2}$

v_d deposition velocity $\equiv D/\chi$

v_s settling velocity of particle

V wind velocity vector

x, y, z rectangular coordinates, x usually along mean wind and z vertical

y separation of pair of particles

z_0 roughness length

z_s height of source (alternative to H or h_s)

X, Y, Z displacements of particles, usually from mean position thereof, or (Z) from ground in case of ground-level source

Y_0 half-width of plume, defined by concentrations one-tenth of centre-line value

Z_0 vertical dimension of plume

α turning of wind direction with height

β ratio of Lagrangian and Eulerian time-scales

γ Euler's constant or a measure of kurtosis

Γ dry adiabatic lapse rate

Gamma function

ϵ rate of dissipation of turbulent kinetic energy per unit mass of air, also a randomly selected number from a Gaussian distribution with zero mean and unit standard deviation

θ potential temperature

or wind direction

or angular crosswind spread of plume $\equiv 2\tan^{-1}(Y_0/x)$

Θ moments of a probability distribution

κ wavenumber, cycles per unit length unless otherwise stated

λ wavelength $\equiv 1/\kappa$

λ_m $\equiv \bar{u}/n_{\bar{m}}$

Λ washout coefficient

ν kinematic viscosity

ξ time-lag in Lagrangian sense

ρ air density

σ^2 variance
τ horizontal shearing stress
or duration of sampling release
or time-scale (alternative to t_s etc.)
ϕ inclination of wind
or Monin–Obukhov universal function – Eq. (2.42)
χ concentration (mass of material per unit volume of air)
ψ an angular deviation
or $\equiv d\bar{v}/dz$.

1

Introduction and brief outline

Effective dispersion of gaseous or finely divided material released into the atmosphere near the ground depends on natural mixing processes on a variety of scales. In the main, this mixing is a direct consequence of turbulent and convective motions generated in the *boundary layer* itself. This is the layer containing typically some ten per cent of the overlying mass of air, in which the flow properties are determined partly by the aerodynamic friction of the underlying surface (as in classical boundary layer aerodynamics on a laboratory scale) but also to an important extent by the density stratification of the air which results from differences in temperature of the surface and the air. These differences in temperature arise over land primarily in the course of the daily cycle of radiative heating and cooling of the ground, but they may also arise from the overflowing of air from warmer or cooler regions of the earth.

The layer is often referred to alternatively as the *mixing* layer. It is also sometimes called the *Ekman* layer, in recognition of a systematic turning of the direction of motion with distance from the boundary, analogous to that discussed by Ekman for wind-driven ocean currents. In the aerodynamic sense the boundary layer is simply the layer from which momentum is extracted and transferred downward to overcome surface friction. In contrast to the classical notion of a purely frictional boundary layer, the effective mixing layer of the atmosphere has important variations in depth as well as in mixing efficiency, arising from the control on vertical transport by the density stratification. Thus by day over land a well-defined top to the mixing layer is frequently set by an overhead layer with stable density stratification, beginning at a height typically in the range $\frac{1}{2}$ to 2 km, in which turbulent motion is suppressed. This sort of stable stratification is a typical feature of the free atmosphere away from the ground. At night, however, especially when the sky is clear and the wind light, the stable layer formed next to the ground by radiative cooling reduces or even completely suppresses the turbulence and mixing is confined to a shallower layer than in the daytime. In contrast there are special circumstances, both day and night, when convection extends over much deeper layers of the atmosphere in the formation of cumulus clouds and showers.

The systematic study of vertical mixing in the lower atmosphere effectively began with G. I. Taylor's (1915) examination of the redistribution of heat in a current over relatively cold sea. Taylor (1927) also provided the first direct measurements of the turbulent velocities in the horizontal by using the widths of the traces produced by conventional wind speed and direction recorders. More refined measurements of the range of fluctuation of the vertical component of the eddy motion, as well as the crosswind component, were then made with the aid of a more responsive vane mounted in a universal bearing (the bi-directional vane). These relatively simple measurements were extended soon afterwards by Scrase (1930) and Best (1935), in studies which reveal the marked dependence on the thermal stratification of the air and also the existence of a very wide spectrum of frequencies in the generally irregular fluctuation.

About the same time as Scrase's fine-scale investigation of atmospheric turbulence an investigation on a rather larger scale, using routine wind recorders specially adapted to give open-scale time-lapse traces, was undertaken at Cardington, Bedfordshire, in connection with the effects of wind gusts on airships. This classical study, reported by Giblett and others (1932), is particularly noteworthy in providing a qualitative classification of eddies into four main types according to the thermal stratification of the air. It also included the first steps in the auto-correlation and cross-correlation analysis of the turbulent fluctuations in the natural wind.

The evolution from these pioneering stages to the now familiar pattern of detailed observations and elaborate statistical analyses of atmospheric turbulence did not follow rapidly. Progress was inhibited at first by the difficulty of recording the data in sufficient detail and then by the enormous labour involved in carrying out the numerical analysis on primitive calculating machines. With the removal of much of these difficulties by modern recording and computing techniques there was a rapid advance after 1950. A good deal of progress has been made in the statistical description of atmospheric turbulence in terms of its intensity, scale and spectral properties, and the earlier bias towards the more accessible lowest regions of the boundary layer is being reduced with the introduction of improved techniques and facilities for measurements at higher levels. This statistical description of turbulence and of the associated vertical distributions of mean wind speed and temperature are an essential background to any discussion of turbulent mixing, and a general outline of these features is provided in several sections in Chapter 2.

In interpreting and using the data on atmospheric turbulence it is important to keep in mind the *ensemble* average nature of any supposedly *normal* or *standard* value of a property such as intensity, scale or spectral distribution. At best such values represent the most likely value, whereas that experienced at a particular time and place may be considerably different. Furthermore, such data on their own reveal little or nothing about the actual mechanism of turbulent mixing.

The complexities and variability of the turbulent motions in the atmosphere have a direct bearing on the essentially patchy nature of the distribution of windborne material. In this respect the relative dimensions of the dispersive motions and of the volume of air over which the material has been spread at any instant are very important. Also, it is important to distinguish between the dispersive effects on material injected into the atmosphere in a steady continuous stream and the effects on a virtually instantaneous release of material.

The growth of the volume over which a given amount of suspended material is spread has been conventionally regarded as a result of an exchange process analogous to molecular diffusion, but on a much larger scale, with parcels of air replacing molecules. In reality the growth arises from a process of distortion, stretching and convolution, whereby a compact 'blob' or 'puff' of material which continues to occupy basically the same volume of fluid (we are here disregarding for the moment the actual molecular process) is distributed in an irregular way over a volume which is much larger owing to the effective enclosure of 'clean' air. Clearly such a process alone could not lead to any reduction of the density of suspended material in the strictest sense, i.e. where material existed it would locally have the original density or concentration. However, apart from the diluting action of molecular diffusion, the probability of encountering material at all will progressively have a less concentrated distribution in space and the corresponding average concentration over the larger volume containing the distorted blob will be less.

The continuous stream or 'plume' of effluent stretching from an industrial chimney may be thought of as a succession of elementary sections which behave somewhat like individual puffs. It is to be noted immediately that the amount of material contained in an element of plume of given length parallel to the wind will be inversely proportional to the wind speed. This direct dilution by the wind appears universally in theoretical formulations of the continuous point source, to the effect that concentration at a given distance is inversely proportional to wind speed. The crosswind and vertical spreads of a section of plume – which represent the remaining two dimensions of the volume over which a given amount of material is distributed – grow under the action of the small-scale distorting processes, and in this respect the dispersion of a plume is similar in two dimensions to that occurring in three dimensions in the case of a single puff. An important difference however is that the trajectories of successive sections of plume are not identical but are irregularly displaced by the larger-scale fluctuations in the flow. The result is a progressive broadening of the crosswind front over which material is spread at a given distance downwind of the source, and the same process is also effective in the vertical if the plume is elevated clear of the ground. Thus the average concentration produced downwind of a point source not only diminishes with distance from the source but also with the time of exposure. It is important to realize that this property of *time-mean* concentration is essentially a consequence of the existence of dispersive motions on a scale larger than the plume cross-section itself.

The formal representation of the dispersive effects is considered in Chapter 3. Historically the earliest efforts were preoccupied largely with analytical solutions of diffusion equations incorporating the gradient-transfer hypothesis, which implies an exchange by motions on a small scale relative to the volume over which the material is spread. Initially the eddy diffusivity K, analogous to the molecular diffusivity D, was taken to be a constant, but this was soon superseded by more realistic but simple power law functions of height. With finite-difference techniques and high-speed computing facilities, solutions may now be obtained with far more complicated shapes of K profile than could be handled by analytical methods, and the main limitation is now the actual specification of K profiles which are realistic in absolute magnitude as well as shape. However, from a physical standpoint the whole concept of gradient diffusion seems acceptable only for the *vertical* spread from a *ground-level* source, since only in this context is it reasonable to assume the dispersive motions never to be large compared with the vertical spread, a condition which is enforced by the restrictive effect of the boundary on the effective scales of vertical motion. A clear demonstration of this condition is provided by the absence of any detachment of a ground-based non-buoyant plume from the ground.

For the *time-mean* crosswind distribution from a continuous source, and also for the vertical distribution when the plume is elevated, the statistical treatment initiated by G. I. Taylor provides the most rational approach, subject only to there being a satisfactory approximation to homogeneity in the relevant conditions of turbulence. In its basic representation the method is entirely kinematical but the development into convenient practical form necessarily introduces the difficult physical problem of the relation between the Lagrangian (moving particle) and Eulerian (fixed point) descriptions of a field of turbulence. Given an acceptable transformation between the two systems the turbulent spread can then be derived from conventional fixed-point data on turbulence. Useful progress has been possible along these lines and the method has also been extended with some success to the rather more difficult problem of the expansion of clusters of particles, i.e. the 'puff' as distinct from the time-mean aspect. In several respects the properties of the atmospheric boundary layer contain sufficient departures from homogeneity to render invalid the original analytic form of the Taylor theory, but extension of the whole approach has recently been opened up by a combination of random-walk theory with high-speed numerical techniques and the new analytic short cuts provided by the integral equation. A complementary approach which has provided many satisfactory generalizations is the application of dimensional analysis as elaborated into its modern form of similarity theory. This approach has provided useful steps in relation to the puff problem, the vertical spread from a continuous source at ground level, and the dispersion in buoyancy-driven mixing.

There are other theoretical aspects of dispersion going beyond the direct effects of turbulent mixing on a passively suspended material. Particles may be large enough to undergo significant gravitational settling and the separation of the trajectories of the material particles and the fluid elements then needs to be taken into account. Also in the sheared flow which is characteristic of the lower part of the boundary layer there is the possibility of an additional *indirect* horizontal spread caused by the interaction between direct turbulent spread in the vertical and the systematic variation of wind velocity with height. These aspects are given attention in Chapter 3 from both the gradient-transfer and statistical approaches.

The testing and development of the theoretical background has from the beginning relied substantially on careful full-scale experiments. In Chapter 4 the data which have provided significant steps are assembled and put into perspective, and a general summing-up of the present stage of understanding and generalizations is offered. The foundations were laid in the 1920s in the field trials carried out by the Chemical Defence Establishment at Porton. Using controlled sources of smoke in selected near-ideal conditions of airflow over open downland on Salisbury Plain a basic specification of the crosswind and vertical spread was obtained for a range of a few hundred metres from the source. The principal results were ultimately brought out and interpreted theoretically in the work of O. G. Sutton. These early tracer experiments have since been extended both in range and in elaborateness, in long-term programmes both at Porton and in several places in the U.S.A.

Interpretations of tracer experiments in an absolute sense, i.e. as regards the absolute magnitude of tracer concentrations from a source of given strength are rarely straightforward, owing to uncertainty in the conservative nature of practical tracers. Because of this the most reliable conclusions come from relative concentrations and the magnitude of crosswind and vertical spread which these denote. Enough data of this kind have now been accumulated to make clear the extent to which short-range dispersion is predictable, for ideal sources and terrain, given a measurement or climatological estimate of the appropriate flow properties in the lower part of the boundary layer.

Real cases of dispersion of windborne material include a whole complex of processes additional to the dispersion of a passive conservative material as assumed in all considerations so far, and these are reviewed in Chapter 5. The very important effluent from power stations and heating plants rises significantly above the chimney exit as a consequence of its high temperature and buoyancy relative to the outside air. For the sort of exit diameters and vertical velocities usually involved it is evident that at first the effluent plume responds to the natural wind only by being bent over from the vertical and quickly taking on the horizontal speed of the wind. The early progressive growth of the plume cross-section and the resultant decrease in its buoyancy are a result of entrainment of air in a process generated by the relative vertical motion of air and volume, and it is only in due course that the natural mixing action of the air exerts any controlling influence. Much effort has gone into both theoretical and observational studies of plume rise. There are many rival formulae from which to choose and it is important to have the essential basis and limitations brought out. Effluent plumes also undergo depletion by various processes – sedimentation of large particles, uptake of small particles and gases at the ground and washout by rain.

Much information is now available from surveys which have been made of the actual distribution of effluent concentration, especially for sulphur dioxide, downwind of individual stacks. These provide a valuable practical basis in examining the extent to which simple models of dispersion provide a satisfactory basis for generalization and hence for predicting the pollution levels from new power stations and industrial plant. The use of these simple models, based as they are on studies of airflow, terrain and source conditions which are so often a considerable idealization of those encountered in practice, is open to severe criticism unless their limitations are fully realized and their predictions are adopted with adequate qualification. It is essential to have a realistic view of the complications arising from various features, such as the aerodynamic effects of buildings both in an individual sense and collectively in an urban area, the control exerted by natural topography and the influence of the 'heat-island' created by a modern city.

For the most convincing assessment of the pollution levels created by specified sources there may often seem to be no substitute for a properly planned monitoring survey in which either the actual pollutant or a simulant is measured in the real conditions of source and terrain, and in the particular conditions of weather and wind direction which are deemed to be most important. Such surveys are, however, likely to be laborious and expensive, as well as prolonged if a wide range of weather conditions needs to be covered. It has accordingly become a common practice to consult meteorologists for theoretical estimates of the likely concentration levels from specified sources, either in lieu of a survey or as part of the planning for it. There is also increasing interest in the provision of forecasts and warnings of the build-up of pollution levels in adverse weather situations. It therefore seems important to put the developments and experience outlined in Chapters 2–5 into the most convenient form for assimilation and use by meteorologists in practice. An attempt on these lines has been made in the sixth chapter. This chapter has been written in a fairly self-contained form, in which there is some repetition and restatement of material from the earlier chapters, but this has been kept to the minimum essential for effective use without intensive study of those chapters.

Although the more obvious problems of air pollution are of a local nature, it is inevitable that any material which is not rapidly removed from the atmosphere by the processes already mentioned will be transported over virtually unlimited distances, albeit with continuing dilution, until spread over the whole global atmosphere. Long-range transport by the atmosphere on a continental or hemispheric scale has long been recognized in respect of radioactive debris from the testing of nuclear weapons and on the natural plane in respect of spores, pollens, seeds and insects. Now the process is attracting renewed interest as regards possible ecological effects of industrial effluent remote from the source and conjectures about long-term influences on weather and climate. Full appreciation of these processes demands more critical examination of the physical actions of the atmosphere on scales greater than those generally considered in the earlier chapters. These aspects, and especially the progress in modelling the travel of pollution on an international scale, are considered in the seventh and final chapter.

2

Turbulence in the atmospheric boundary layer

2.1 ESSENTIALS IN THE STATISTICAL DESCRIPTION OF TURBULENT FLOW

For a comprehensive treatment of turbulent flows several texts are now available. Here we need consider only those aspects which are particularly relevant to atmospheric flow and to the theoretical and experimental study of the dispersion problems which are the main concern of this book.

The quantitative specification of turbulent fluctuations depends on the principle that the whole motion can be resolved into a fluctuating component superimposed on a general mean *flow*. In practice there are complications in the interpretation of the term *mean*, to which there will be further reference, but taking initially the simplest view that there is a basic flow of constant uniform velocity then instantaneous components of velocity may be defined in a rectangular system of coordinates as follows:

$$\begin{aligned} u &= \bar{u} + u' && \text{along the } x \text{ axis} \\ v &= \bar{v} + v' && \text{along the } y \text{ axis} \\ w &= \bar{w} + w, && \text{along the } z \text{ axis} \end{aligned} \tag{2.1}$$

For atmospheric flow the x and y axes are conventionally set in a horizontal plane and the z axis vertical; the quantities with overbars refer to the mean velocity and those with primes are the instantaneous fluctuations therefrom, variously referred to as turbulent or eddy velocities. Usually the system is simplified by taking the x axis in the direction of the mean flow ($\bar{v}=0$), and by assuming the mean flow to be horizontal ($\bar{w}=0$).

Physically the velocity components refer to an *element* of air passing through a specified point, and in the *Eulerian* system the velocities are in principle specified at all positions in the field of flow at a given instant. In addition to this *Eulerian-space* description fluid dynamicists sometimes use an *Eulerian-time* description in which the velocities at a point in a coordinate system moving with the mean flow are specified as a function of time. Both of these systems involve a large number of different elements of air. A third system known as

Lagrangian, is concerned with the variations in time of the velocity of a particular element, which is of course continuously changing its position. This system is obviously directly involved in the description and analysis of the diffusion of particles.

Although analysts of atmospheric turbulence may sometimes be concerned with a practical approximation to a Langrangian system, such as that provided by a floating balloon, usually they are concerned with the fluctuation recorded by a fixed instrument responding more or less rapidly to the relative motion of the air. The description so provided does not fall into any of the categories of the previous paragraph, but it is customarily regarded as equivalent to the Eulerian-space description. This equivalence is based on the hypothesis, originally due to G. I. Taylor, that the sequence of variations at a fixed point is statistically the same as if the spatial pattern of velocities were suddenly frozen and swept past the measuring instrument with speed $\bar{u}$.

Whichever system is involved definition of the mean value is the first step, and the next step is a statistical specification of the fluctuations about this mean. This requirement is most simply met by the *second moment* (about the mean), i.e. the mean-square-deviation or variance ($\overline{u'^2}$ etc.), or the square-root thereof, i.e. the standard deviation. Such quantities however contain no information about the scale in time or space on which the contributory fluctuations occur, and it would be possible to have identical variances associated with very different scales.

An indication of spatial scale is contained in the properties of the relative velocity at two points. If these are close together the velocity difference will reflect only the fluctuations on a correspondingly small scale, and those on a larger scale will become apparent only as the separation is increased. A statistical indication is provided by the variance of the velocity difference as a function of separation. Thus, if positions 1 and 2 are taken on a line parallel to the x-axis, with separation x,

$$D(x) = \overline{(u_1 - u_2)^2} \tag{2.2}$$

This quantity was originally used in Russian work on turbulence and was termed the *structure function.* Substituting from Eq. (2.1) and expanding, and assuming variances to be independent of position (i.e. a homogeneous field)

$$D(x) = 2(\overline{u'^2} - \overline{u'_1 u'_2}) \tag{2.3}$$

The second term in the bracket is the familiar *covariance* $C(x)$, which on division by the variance gives the *correlation coefficient* $R(x)$.

Corresponding *time*-functions. applicable to the Eulerian-time, Langrangian and fixed-point systems, are defined in terms of velocities at instants 1 and 2 separated by an interval (or lag) t, i.e. the *auto-covariance*.

$$C(t) = \overline{u'_1 u'_2} \tag{2.4}$$

and the *auto-correlation coefficient*

$$R(t) = \overline{u'_1 u'_2}/\overline{u'^2} \tag{2.5}$$

Stationary conditions are implied, in that the above functions are regarded as dependent only on *time lag* and not on the actual time at which the sequence of values begins.

By definition the correlation functions are unity at zero separation or lag. To the extent that they have values greater than zero for large separation or time-lag the fluctuation pattern is of large scale in space or time. A convenient overall representation is provided by the *integral scales*

$$l_s = \int_0^\infty R(x)\,\mathrm{d}x \tag{2.6}$$

$$t_s = \int_0^\infty R(t)\,\mathrm{d}t \tag{2.7}$$

Similar relations to the above Eq. (2.3) to (2.5) may be written for the other components of eddy velocity and, in the case of the space functions, for the other coordinates.

When the time-correlation refers to the fixed-point of measurement the Taylor hypothesis implies.

$$R(t) = R(x) \quad \text{when } x = \bar{u}t \tag{2.8}$$

$$\bar{u}t_s = l_s \tag{2.9}$$

However, note that while these last two relations may be applied for any component of eddy velocity they apply only to variations along the x-direction, i.e. the direction of the mean flow.

Regarding the overall variation as composed of a *spectrum* of fluctuations it is clear that a large integral time-scale will be associated with large periodic times or low frequencies, and the relative contributions from various frequencies are reflected in the shape of the correlation function. Mathematically the two are related by the following inverse Fourier transforms,

$$F(n) = 4\int_0^\infty R(t)\cos 2\pi nt\,\mathrm{d}t \tag{2.10}$$

$$R(t) = \int_0^\infty F(n)\cos 2\pi nt\,\mathrm{d}n \tag{2.11}$$

The quantity $F(n)$ is the *energy spectrum* or *power spectrum* function, or *spectral density*, representing the fractional contribution to the total variance from frequencies between n and $n + dn$, i.e.

$$\int_0^\infty F(n)\,dn = 1 \tag{2.12}$$

Because of its reference to the proportionate rather than the absolute magnitude of contribution to the variance $F(n)$ is often described as the *normalized* power spectrum function. A corresponding absolute spectrum function $S(n)$, equal to $\overline{u'^2}F(n)$, is defined by

$$\int_0^\infty S(n)\,dn = \overline{u'^2} \tag{2.13}$$

and the above Fourier transforms may be rewritten with $F(n)$ replaced by $S(n)$ and $R(t)$ replaced by $C(t)$.

This introduction to the spectrum of turbulence in terms of frequency is particularly appropriate in the sense that turbulent fluctuations are usually measured as variations in time rather than in space. Moreover, in the statistical theory of diffusion there is necessarily a reference to a frequency spectrum in a Lagrangian sense. However, the basic theory of the structure of turbulence is most directly and logically developed in an Eulerian-space framework, and it will be necessary to make reference to this at various stages. In this respect there are relations equivalent to (2.10) and (2.11) with $R(t)$ replaced by $R(x)$ and $F(n)$ by $F(\kappa)$, a *wavenumber* power spectrum function. Also, as in the case of the correlation functions, the frequency spectrum and the wavenumber spectrum along the x axis (the mean flow direction) are usually assumed to be related by the Taylor hypothesis in the sense that

$$\bar{u}S(n) = S(\kappa) \quad \text{when } n = \bar{u}\kappa \tag{2.14}$$

The significance of the spectrum function in relation to the integral scale should be noted. From Eq. (2.10)

$$\underset{n\to 0}{F(n)} = 4\int_0^\infty R(t)\,dt = 4t_s \tag{2.15}$$

and correspondingly

$$\underset{\kappa\to 0}{F(\kappa)} = 4l_s \tag{2.16}$$

In other words the integral scale is determined by the very slow or very large-scale fluctuations. Consequently the evaluation of the true scale is often difficult or even impossible in practice.

By definition $F(\kappa)$ is a *one-dimensional* property describing fluctuations along a line, and is to be distinguished from a three-dimensional spectrum function (see 2.4). The finite limit in $F(\kappa)$ as $\kappa \to 0$ does not imply turbulent energy actually at zero wavenumber, but represents the projected effect of 'waves' with finite wavenumber and with the normals to their 'crests' inclined to the reference line.

The statistical effects of finite sampling and averaging

We now consider the analysis of fluctuations which contain a very wide range of frequencies. If the variation is considered only over some finite duration of *sampling* τ (see Fig. 2.1), the effects of the slow variations will be partially excluded. In this case the magnitude of the turbulent component, say u', is expressed precisely as in Eq. (2.1), except that now the mean value is defined as

$$\bar{u}_\tau = \frac{1}{\tau}\int_{t'-\tau/2}^{t'+\tau/2} u \, dt \tag{2.17}$$

and then $u = \bar{u}_\tau + u'$ (2.18)

The mean velocity is thus recognized as a purely arbitrary quantity, dependent on the sampling duration, τ, and time of origin, t', of the sample of fluctuations.

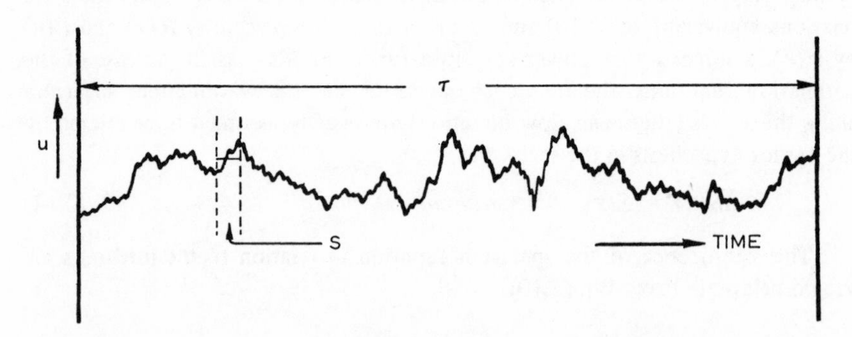

Fig. 2.1 – Sampling duration (τ) and averaging time (s) in the analysis of velocity fluctuations.

At the other end of the scale some of the fine structure of the variation will be smoothed out, since any individual observation must represent an average over some period s. Even if this averaging effect is not imposed deliberately in the process of 'reading' the variation, it will still exist as a consequence of the inertia of the measuring instrument. The effects of *sampling duration* τ and *averaging time* s on the apparent statistical properties are thus of great importance.

Considering the whole variation as a composition of sinusoidal components let the variation associated with a frequency n be represented by

$$y = a \sin 2\pi nt \tag{2.19}$$

The process of averaging over time interval s amounts to replacing y in the above by

$$\bar{y} = \frac{1}{s}\int_{t'-s/2}^{t'+s/2} y \,\mathrm{d}t = \frac{a}{\pi ns} \sin \pi ns \sin 2\pi nt' \tag{2.20}$$

and the original sinusoidal variation is replaced by a variation of the same frequency but with an amplitude which is reduced to $\sin \pi ns/\pi ns$ of the original value. This means that the variance contributed by the frequency n is reduced to $\sin^2 \pi ns/(\pi ns)^2$ of its original value. The form of this reduction factor is shown in Fig. 2.2.

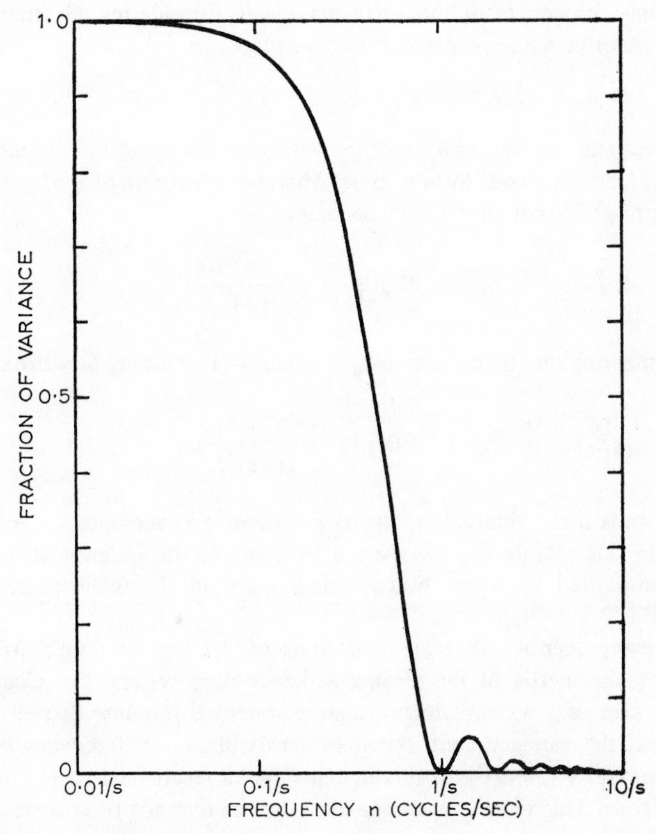

Fig. 2.2 – Effect of averaging, over a time s, on the variance of a sinusoidal fluctuation of frequency n.

Generalizing to the whole spectrum, the variance $\sigma^2_{\infty,s}$, obtained by averaging over time interval s and sampling for infinite time, is related to the whole variance $\sigma^2_{\infty,0}$, obtained with infinitesimal averaging time, by the equation

$$\sigma^2_{\infty,s} = \sigma^2_{\infty,0} \int_0^\infty F(n) \frac{\sin^2 \pi n s}{(\pi n s)^2} \, \mathrm{d}n \tag{2.21}$$

In effect the original spectrum is cut off by a weighting function of the shape shown in Fig. 2.2. As s is increased, and the curve is displaced to lower frequencies, more of the spectrum is cut off and $\sigma^2_{\infty,s}$ is reduced.

The complementary effect of sampling over finite time is easily derived. Consider an indefinitely long series of equally-spaced values representing departures (with infinitesimal averaging time) from the long-term mean. The series is then subdivided into sequential samples of length τ. For each sample the mean square departure from the overall mean is simply the sum of the particular variance and the square of the particular mean. Considering all samples, and using the subscript notation above, it follows that

$$\sigma^2_{\infty,0} = \sigma^2_{\infty,\tau} + [\sigma^2_{\tau,0}]_\infty \tag{2.22}$$

where the square bracket with subscript ∞ implies averaging the variances from consecutive periods τ over infinite time. Then by substituting for $\sigma^2_{\infty,\tau}$ the form equivalent to (2.21) equation (2.22) becomes

$$[\sigma^2_{\tau,0}]_\infty = \sigma^2_{\infty,0} - \sigma^2_{\infty,0} \int_0^\infty F(n) \frac{\sin^2 \pi n \tau}{(\pi n \tau)^2} \, \mathrm{d}n \tag{2.23}$$

and, remembering that by definition $\int_0^\infty F(n)\,\mathrm{d}n = 1$, this may be written

$$[\sigma^2_{\tau,0}]_\infty = \sigma^2_{\infty,0} \int_0^\infty F(n) \left(1 - \frac{\sin^2 \pi n \tau}{(\pi n \tau)^2}\right) \mathrm{d}n \tag{2.24}$$

In other words if the fluctuating quantity is observed over time τ, the variance apparent in this sample will *on average* be equal to the variance of the whole spectrum modified by a function which has a shape complementary to that shown in Fig. 2.2.

As already mentioned, some smoothing of the rapid fluctuations will be imposed by the inertia of the sensing and recording system. For example, in the simple case of a system which has an exponential response $(\exp -t/\alpha)$ to a step change, the variance contributed by an oscillation of frequency n will be reduced to $1/[1+(2\pi n\alpha)^2]$ of its original value, a result familiar in alternating current theory. This filter is less steep in its cut-off than the filter corresponding to simple linear averaging.

In summary, for practical purposes the essential effects of the simple numerical and instrumental filters are as listed in Table 2.I below. In round figures periodic times between τ and $10s$ (or 30α) are satisfactorily included, but those $> 10\tau$ and $< s$ (or $< \alpha$) are effectively excluded.

Table 2.I – Summary of separate filtering effects of sampling over time τ, averaging over time s, and of instrument with exponential time-constant α.

Reduction of spectral density to	Frequency at which reduced spectral density applies as a result of		
	Limited sampling	Averaging	Instrument inertia
<5%	$<0.125/\tau$	$>0.81/s$	$>0.69/\alpha$
50%	$0.44/\tau$	$0.44/s$	$0.159/\alpha$
>95%	$>0.81/\tau$	$<0.125/s$	$<0.037/\alpha$

When s is sufficiently large the weighting function excludes all but the very lowest frequencies, for which [see Eq. (2.15)] $F(n) \to 4t_s$, effectively independently of n. Accordingly, in Eq. (2.21), $F(n)$ may be taken outside the integral sign and the remaining integral evaluated to give

$$\sigma^2_{\infty,s}s = 2\sigma^2_{\infty,0}t_s \qquad s \to \infty \tag{2.25}$$

i.e. the reduced variance tends asymptotically to a variation with $1/s$. It may be noted here that Eq. (2.25) is also the limiting form of the following relations,

$$\frac{\mathrm{d}(\sigma^2_{\infty,s}s^2)}{\mathrm{d}s} = 2\sigma^2_{\infty,0}\int_0^s R(t)\,\mathrm{d}t \tag{2.26}$$

$$\sigma^2_{\infty,T}T^2 = 2\sigma^2_{\infty,0}\int_0^T\int_0^s R(t)\,\mathrm{d}t\,\mathrm{d}s \tag{2.27}$$

As will be seen in Chapter 3 these relations were first obtained by G. I. Taylor as expressions for the mean square displacement of a particle after travelling for a specified time with randomly fluctuating velocity (Eq. (3.59), in which $[X^2]$ is equivalent to $\sigma^2_{\infty,T}T^2$ above). It is clear however that Taylor's analysis is in no way confined to particle velocity, nor even to velocity *per se*, but is in fact applicable to any so-called stationary random function of time, say $p(t')$. The expansion of the quantity $\mathrm{d}[\overline{p_s}^2 s^2]/\mathrm{d}s$ into a form equivalent to Eq. (2.26) follows from the definition of time-mean value as in Eq. (2.17) and from the rule for differentiating an integral. Thus

$$\frac{\mathrm{d}[\bar{p}_s^2 s^2]}{\mathrm{d}s} = 2\ \bar{p}_s s \frac{\mathrm{d}(\bar{p}_s s)}{\mathrm{d}s} = 2\int_0^s \overline{p(t')p(s)}\,\mathrm{d}t' \tag{2.28}$$

On writing $t' = s - t$ and substituting the definition of $R(t)$ we obtain Eq. (2.26). The identity of the alternative expressions (2.27) and (2.21) for $\sigma^2_{\infty,T}$ may also be demonstrated by substituting in Eq. (2.27) the Fourier transform (2.11) for $R(t)$.

So far the effects of finite averaging and sampling times have been considered separately. The precise effect of the two in combination depends on the order in which the averaging and sampling operations are carried out. Basically there are two alternative procedures.

(a) Take samples of length τ and for each sample derive averages over sub-interval s. These sub-intervals must not overlap, otherwise the data near the end of a sample will be used with less weight than those near the middle. Consequently τ/s must be a whole number.

(b) Derive averages over intervals of length s and from the smoothed series so formed take samples of length τ. In this system the averages may be taken in an overlapping manner, and indeed this is preferable in that the distortion of the high-frequency end of the spectrum by 'aliasing' is minimized (see later).

The important difference in the two procedures lies in the effective sampling lengths, procedure (a) being similar to procedure (b) with τ replaced by $\tau - s$. Using procedure (a), with consecutive sampling intervals τ, we may apply Eq. (2.22) with the infinitesimal averaging time replaced by s, i.e.

$$[\sigma^2_{\tau,s}]_a = \sigma^2_{\infty,s} - \sigma^2_{\infty,\tau} = [\sigma^2_{\tau,0}] - [\sigma^2_{s,0}]$$

$$= \sigma^2_{\infty,0}\int_0^\infty F(n)\left(\frac{\sin^2\pi ns}{(\pi ns)^2} - \frac{\sin^2\pi n\tau}{(\pi n\tau)^2}\right)\mathrm{d}n \tag{2.29}$$

Thus the resultant effect of (a) is equivalent to *subtracting* the separate weighting functions for averaging over s and τ. Using procedure (b) the analysis is much more complicated. The derivation is given in full by F. B. Smith (1962a) starting with the expression

$$\lim_{\Lambda\to\infty}\frac{1}{2\Lambda}\int_{-\Lambda}^{+\Lambda}\frac{1}{\tau}\int_{T-\tau/2}^{T+\tau/2}\Bigg[\frac{1}{s}\int_{t-s/2}^{t+s/2} u(\xi)\,d\xi\ -$$

$$-\frac{1}{\tau}\int_{T-\tau/2}^{T+\tau/2}\frac{1}{s}\int_{r-s/2}^{r+s/2} u(\zeta)\,\mathrm{d}\zeta\,\mathrm{d}r\Bigg]^2\,\mathrm{d}t\,\mathrm{d}T$$

(the notation differs from that of the original paper in respect of sampling duration and averaging time). The expansion of the integral in terms of covariance functions and the substitution of the Fourier transform expression for the covariance leads ultimately to

$$[\sigma^2_{\tau,s}]_b = \sigma^2_{\infty,0} \int_0^\infty F(n) \frac{\sin^2 \pi n s}{(\pi n s)^2} \left(1 - \frac{\sin^2 \pi n \tau}{(\pi n \tau)^2}\right) dn \qquad (2.30)$$

In this case the resultant effect is equivalent to applying the *product* of the separate weighting functions. The difference between the two forms for $\sigma^2_{\tau,s}$ becomes on rearrangement

$$[\sigma^2_{\tau,s}]_b - [\sigma^2_{\tau,s}]_a = \sigma^2_{\infty,0} \int_0^\infty F(n) \frac{\sin^2 \pi n \tau}{(\pi n \tau)^2} \left(1 - \frac{\sin^2 \pi n s}{(\pi n s)^2}\right) dn \quad (2.31)$$

i.e. precisely the variance which would be obtained by following procedure (b) with averaging time τ and sampling time s. Its magnitude decreases as τ/s increases and for a spectrum form $nF(n)$ constant it amounts to 16 per cent of $[\sigma^2_{\tau,s}]_b$ when τ/s is 2 but only 0.8 per cent when τ/s is 6 (see Smith, 1962).

Spectrum analysis of atmospheric turbulence

There are two important complications which arise in the analysis of atmospheric turbulence as distinct from turbulence in pipes and wind tunnels. The first is that the atmospheric spectrum extends up to the large scales involved in the disturbances of the general circulation. Consequently the statistical properties depend considerably on the sampling duration, and in many practical examples the variance and autocorrelation do not reach their respective constant or zero limiting values. The second complication is that because of the effects of variable terrain, of diurnal heating and nocturnal cooling of the ground, and of the contunually changing large-scale pattern of airflow, turbulence in the atmosphere is neither *homogeneous* nor *stationary*, i.e. the statistical properties depend also on the particular *place* and *time* at which the observations are made. This feature is troublesome not only in the obvious sense that extensive observations are required before any representative description can be assembled, but also because the major developments in the theory of time-series analysis rely on the simplifying assumption that conditions are stationary. This is particularly so, for example, in the case of the Fourier-transform relation between the autocorrelation and spectrum function. In this and other related aspects there are difficulties which have yet to be clarified and which are beyond the scope of this book. In the meantime, as will be seen, the Fourier-transform method is used with various adjustments designed to eliminate gross error.

The methods by which a series of observations can be analysed to give a spectrum may be considered under five headings:

(a) Classical harmonic analysis. (See for example Brooks and Carruthers, 1953.)
(b) Numerical filter techniques.
(c) The use of electrical filters.
(d) The application of the Fourier-transform to the correlogram of the series.
(e) The Fast Fourier-transform.

(a) gives *periodogram* ordinates for $N/2$ values of frequency, where N is the total number of observations. In practice adjacent ordinates are found to differ widely, and despite the already considerable numerical effort a reproducible smoothed spectrum is apparently not obtained without time-consuming smoothing procedures.

One of the simplest forms of (b) consists of applying different averaging times to the data. From Fig. 2.2 this is equivalent (in electrical terms) to *low-pass* filtering with a cut-off at frequencies which are inversely proportional to the averaging time. The difference in the variances obtained with two different averaging times s_1 and s_2 corresponds to a *spectral window*, the shape of which is obtained by subtracting the transmissions (i.e. the ordinates of Fig. 2.2) for the larger value of s from those for the smaller value. For approximate purposes the total transmission through this window may be regarded as equivalent to 100 per cent transmission over the frequency range $0.44/s_1$ to $0.44/s_2$ (the frequencies for 50 per cent transmission through the separate filters). The approximation depends on the precise shape of the spectrum and in any case requires that s_1/s_2 be reasonably large, preferably 6 or more. Otherwise, as just discussed, the indicated variance will tend to be in error on the low side to an important extent. A version of the technique was used at an early stage by Panofsky and McCormick (1952). The complementary method, using different durations of sampling (i.e. *high-pass* filtering) has been employed by F. B. Smith (1961), the difference in variance for sampling durations τ_1 and τ_2 being assigned to the frequency band $0.44/\tau_1$ to $0.44/\tau_2$. Departure of the filter shapes from an ideally sharp cut-off means that the spectrum obtained will be a distorted version, but certain broad features may be usefully and conveniently examined by these simple techniques. A 'sharpening' of the numerical filter to give a closer approach to a definite cut-off at a given frequency can be introduced by using suitably weighted averages, but the numerical work is then greatly increased and the procedure has apparently not been used in evaluating turbulence spectra.

Method (c), in which a continuous electrical signal representing the variable is fed into an electrical filter, is especially attractive in its avoidance of laborious extraction and computing processes. However, for detailed resolution of the spectrum narrow band-pass filters with sharp cut-off characteristics are required. In this case the electrical equipment is elaborate and intricate, especially when the low frequencies of atmospheric turbulence are concerned, and the method has not yet been widely adopted in this connection. Examples of the use of this analogue computing technique have been described by MacCready (1953) and Businger and Suomi (1958), and it is noteworthy that in a later discussion Businger (1959) refers to the need frequently to test the accuracy of the results obtained. If the requirement for sharp cut-off is relaxed, much simpler electrical filters may be used, yielding a resolution similar to that obtained by the simple numerical filter technique outlined above.

An application in this direction was designed by Jones and Pasquill (1959), based on the demonstration (Fig. 2.3) that the 'filtering' effect of sampling for a specified duration can be reproduced to a good approximation by a simple two-stage capacitor-resistor filter. Since in Eq. (2.24) $\sigma^2_{T,0}/\sigma^2_{\infty,0}$ falls to 0.5 for $n\tau = 0.44$, it follows that a high-pass electrical filter of adequately similar shape will give an output corresponding to a sampling duration of $0.44/n'$, where n' is the frequency for 50 per cent power transmission, or in the present case 1.78 CR sec, where C is the capacity in microfarads and R the resistance on megohms. The second stage is the conversion of this filtered output (which automatically has zero mean value), by rectification and then by smoothing with a low-pass filter, into an output which represents a smoothed mean deviation, and which can be recorded at a relatively slow chart speed. With the assumption of a Gaussian distribution of the component fluctuations, the scale can be adjusted to give readings of standard deviation directly. This system of recording can of course be adopted for any variable which can be arranged to give a linear electrical output. Practical details and tests against a numerical evaluation of the standard deviation of wind direction fluctuations are contained in the original paper, and the further development into a multi-band-pass system is described by Jones (1963).

Until recently the method which has been used most widely is (d), especially with the modern automatic digital computing methods which facilitate the otherwise laborious process of deriving the autocorrelogram for large numbers of lags. Moreover, there is the advantage (over (a)) that a relatively smoothed spectrum is obtained. One of the difficulties encountered in this method is that when very slow variations are contained in the velocity trace the autocorrelogram may retain a positive value over the range of lags employed. A formula for eliminating this effect has been given by Webb (1955). If the correlogram of the whole variation (fluctuation plus slow trend) is denoted by $R_q(t)$, and tends to a steady value R_q' at large t, the correlogram of the fluctuation alone is

$$R_s(t) \cong \frac{R_q(t) - R'_q}{1 - R'_q} \tag{2.32}$$

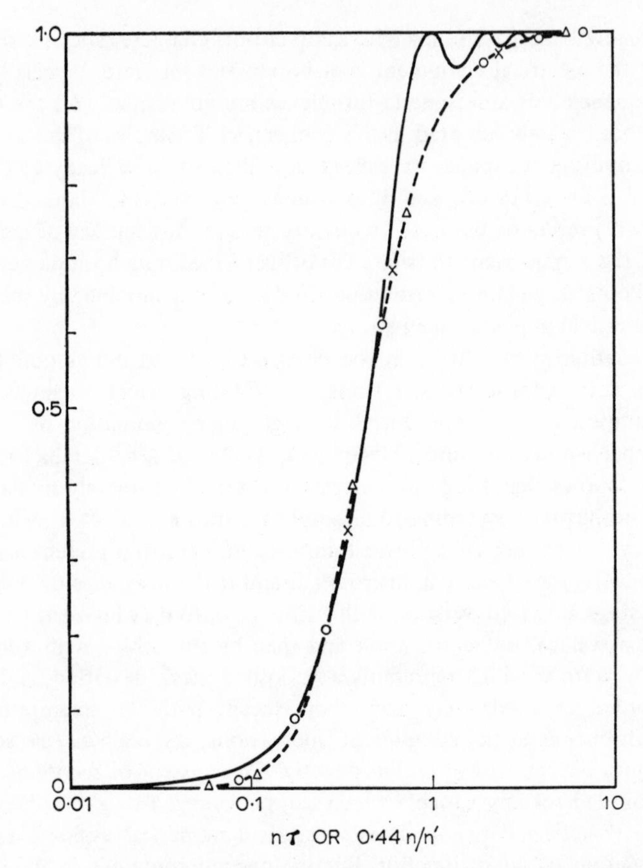

Fig. 2.3 – Electrical filter corresponding approximately to a finite duration of sampling.

——— $f = 1 - \sin^2 \pi n\tau/(\pi n\tau)^2$ where n is frequency (cycles/sec) and τ is sampling-duration (sec).

– – – – $f = [1 + 1/(\omega^2 C^2 R^2)]^{-2}$ where $\omega = 2\pi n$, C and R are the capcity (farads) and resistance (ohms) of a two-stage high-pass filter, for which the frequency for 50 per cent power transmission is $n' = 0.247/CR$. The points are measured values for CR 101 (○), 16.9 (×) and 2.8 (△) sec. (Jones and Pasquill 1959).

The formula applies to the case of a linear trend, or more generally, to any slow variation uncorrelated with the relatively high-frequency fluctuation which it is required to analyse. The validity of applying it when the correlogram has reached a value which is not constant but merely diminishing very slowly, as may well occur with a wide continuous spectrum, has not been made clear. If it is applied in such a case it is presumably at least necessary that the spectrum from $R_s(t)$ should not be evaluated for frequencies lower than the reciprocal of the maximum lag.

The most comprehensive practical procedure which has yet been offered, for the Fourier-transform treatment of time-series of the type encountered in atmospheric turbulence measurements, is that developed by Tukey for the analysis of noise problems in electrical communications (Tukey 1950, Blackman and Tukey 1958). Summaries of the procedure in the present context have been given by Panofsky and McCormick (1954) and R. A. Jones (1957). The method is intended to provide a realistic analysis of a finite series of discrete observations such as would be obtained by reading a velocity record at prescribed intervals. With any set of N discrete observations at interval δt products are formed from pairs with $m+1$ intervals of lag. The cosine-transform equation is then evaluated in series form with n replaced by $h/2m\ \delta t$, $h=0, 1, 2 \ldots, m$ and time-lag t by $k\ \delta t$, $k=0, 1, 2, \ldots, m$. This provides spectral estimates averaged over frequency bands of width $\frac{1}{2}m\ \delta t$ except at the high-frequency and low-frequency ends when the width is $\frac{1}{4}m\ \delta t$. Experience shows that further smoothing is desirable, and application of a particular 3-term weighted average is recommended to provide estimates of spectral density which are effectively averages over the overlapping frequency bands shown in Fig. 2.4.

Fig. 2.4 – Frequency-bands over which estimates of average spectral density are obtained from observations at intervals δt, following the procedure recommended by Tukey.

The choice of number of lags is important. For resolution of the spectrum into narrow bands m should be as large as possible, but if it is too large the computational work may be prohibitive and, more important probably, the accuracy of the estimate decreases. Tukey suggests that m should be small enough in relation to N to make the number of degrees of freedom

$$f = \frac{2(N-m/4)}{m} \tag{2.33}$$

satisfactorily large, and the values which have been commonly used are $6<m<30$ and $f>30$. In round figures, if the number of lags is equal to 1/20 of the N observations in a sample of duration τ, the spectrum is evaluated in $[(N/20)-2]$ bands over the frequency range $10/\tau$ to $\frac{1}{2}\ \delta t$, and $f \cong 40$. Assuming that the original data X_i have a Gaussian distribution the spectral estimates from different samples will have χ^2/f distribution, and in the last example ($f = 40$) the 90 per cent confidence limits of the estimates will be 70 and 140 per cent of the values computed.

The analysis amounts to subjecting the spectrum to band-pass filters which are imperfect in the sense of not having definite cut-offs, and an additional error is thereby introduced when the spectrum changes rapidly with frequency. To remedy this situation Tukey has suggested a *pre-whitening* procedure, by which the original time series is transformed to one not varying rapidly with frequency. For the case of a spectrum in which energy increases with decreasing frequency this procedure amounts to emphasizing the high-frequency fluctuations, and in practice is effected by the following linear transformation of X_i to a new series Y_i

$$Y_i = X_i - bX_{i-1} \tag{2.34}$$

with b usually 0.75. The spectrum of the transformed series is evaluated as before, and a reverse transformation is then applied to give the estimates appropriate to the original spectrum.

Another source of error, known as *aliasing,* arises from the fact that with observations at discrete intervals the variations associated with a high-frequency oscillation effectively appear at a lower frequency (see Fig. 2.5). This is reduced by making δt as small as possible, consistent with the averaging time of the observations, and by discarding the high-frequency end of the spectrum. It is also customary to apply a statistical correction for the effect of the finite averaging time s. From the previous discussion leading to Eq. (2.21) it follows that this is achieved by multiplying the spectral estimates by $(\pi ns)^2/\sin^2 \pi ns$.

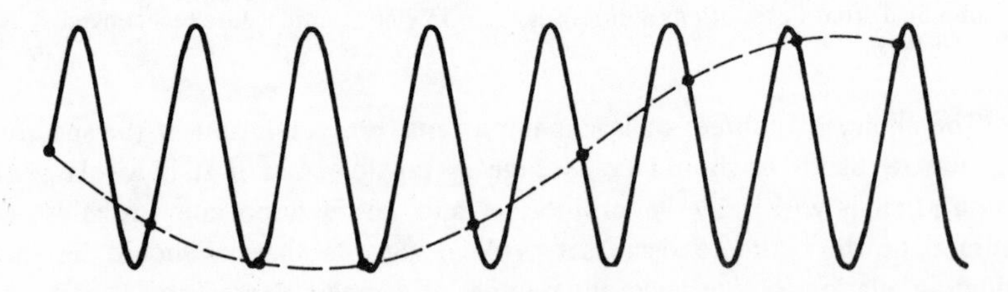

Fig. 2.5 – The *aliasing* of a high-frequency oscillation as a result of using discrete observations.

The Fast Fourier Transform (F.F.T.) is essentially a more economically organized version of classical harmonic analysis stemming from a paper by Cooley and Tukey (1965). A full account of the method in the context of spectral analysis of atmospheric turbulence and a comparison with the cosine-transform technique will be found in the paper by Rayment (1970).

A simple appreciation of the essential advantage of the F.F.T. is evident from consideration of the summation series giving the amplitudes or coefficients of the Fourier series. With N the number of discrete values at intervals δt, the

Fourier coefficients $C(k\,\delta n)$, $k = 0, 1, \ldots, N-1$, are evaluated at frequency intervals δn $(= 1/N\,\delta t)$ from

$$C(k\,\delta n) = \frac{1}{N}\sum_{j=0}^{N-1} x(j)W^{jk} \qquad W = \exp(2\pi i/N) \tag{2.35}$$

Evaluation of (2.35) in the original way required about $N^2/2$ operations of addition and multiplication to produce the maximum of $N/2$ coefficients. However, in carrying out this operation the various different values of x will have been multiplied separately by the W^{jk} terms. Many of these W^{jk} terms are identical or merely of opposite sign, so if the values of x are first sorted to bring together those multiplied by the same W^{jk} a large saving in arithmetic is made.

The relative effort required in applying the Blackman–Tukey cosine-transform system and the F.F.T. depends on the frequency resolution demanded, greater efficiency being achieved with the former for relatively low resolution and with the latter for relatively high resolution. The F.F.T. also has the advantage that the Fourier coefficients obtained for two or more different series may also be used conveniently to evaluate cospectra, which could not be evaluated from the autocorrelations but would require cross-correlations to be derived specially for the purpose. In general the F.F.T. now seems to have largely superseded the cosine-transform method in spectral analysis of atmospheric turbulence.

2.2 THE MEAN-FLOW PROPERTIES OF THE BOUNDARY LAYER

The physical nature of the atmospheric boundary layer

The boundary layer, or mixing layer, is the lowest layer of the atmosphere under direct influence of the underlying surface. Its character and depth vary in an evolutionary way in response to changes in the physical nature of the surface, the surface fluxes of sensible and latent heat, as well as changes in the 'external' synoptic field. The layer extends upwards from the surface to where all turbulent flux-divergences resulting from surface action have fallen to zero (or virtually so). Sometimes the boundary layer is capped by a well-marked inversion, generated either synoptically or by the evolutionary growth processes within the layer. Across the inversion there is usually some wind shear, a decrease in humidity and a drop in the turbulent fluxes to virtually zero. On other occasions, however, no clearly marked top exists and the turbulent fluxes decrease only very gradually with increasing height.

In the presence of convective cloud, significant vertical fluxes may continue upwards from the boundary layer proper into the clouds whilst being very small outside at the same height. Such a situation is an example of a 'break-down' process in which the boundary layer can no longer be considered as a storage buffer-zone into which heat, moisture, momentum and pollution are fed from the surface and accumulated, but the zone is leaking its 'contents' into a much deeper layer of the whole atmopshere. Other boundary-layer break-down processes occur at synoptic fronts and in flow over mountain ranges.

Typically the height of the boundary layer evolves continuously in response to spatial and temporal changes in surface conditions, varying considerably between 400 and 2000 metres deep by day and between a few tens of metres to about 400 metres deep by night. Since the flux of momentum, or shearing stress, is by definition significant within this layer, it enters into the simple balance of forces between the pressure gradient force and the coriolis force, causing vertical shears of the horizontal wind in both magnitude and direction. The wind is usually backed relative to the geostrophic, causing a partial flow down the pressure gradient and gaining momentum lost by surface friction. The turning of the wind relative to the geostrophic direction, though quite variable, influenced as it is by local topography and by meso- and synoptic-scale accelerations, amounts at the surface to some 10° over the sea, 20° over the land in unstable conditions and 30° or more in stable conditions, and decreases with height.

In so far as a parcel of air experiences changes in the synoptic pressure field, in surface roughness and fluxes on a time-scale of several hours at least, the boundary layer can transmit changes in heat input, etc., throughout its depth reasonably adequately and remain in quasi-equilibrium. Unfortunately many changes in surface parameters, particularly as met with at coastlines or in sharp mountainous regions, occur on a shorter time-scale and pose very special problems of adjustment.

Classically most theoretical interest has centred on the neutral boundary layer since the problem is simplified by the exclusion of buoyancy forces, so the character of the layer is dominated by the flux of momentum to the surface. However, except in strong winds, exceeding 12 m s^{-1}, and overcast conditions, a truly neutral boundary layer hardly ever exists even near dawn or sunset. Should quasi-neutral conditions exist, the depth of the boundary layer, although not normally well defined by a capping inversion, varies most directly with wind speed (see Table 2.II).

Table 2II – Typical depths of the boundary layer in quasi-neutral flow.

Geostrophic wind speed (m s^{-1})	2	5	10	20
Depth of boundary layer (m)	200	450	900	1680

Note. These values refer to typical mixed agricultural countryside, $z_0 \cong 20$ cm. Over a much smoother surface, say with $z_0 \cong 0.01$, the depths would be approximately half the above.

In neutral boundary layers there are two important sources of turbulent energy. The first arises from the basic instability of the wind speed gradients found in the lowest few tens of metres which are caused by the mechnanical drag of the underlying surface. The second arises from the turning of the wind with height and the velocity inflection point associated with this. At this level there is a maximum in the vorticity field and displacements of air that cross this level experience forces that enhance these movements rather than damp them out. These motions (described more fully by Brown (1974)) are known as Ekman instabilities and are theoretically capable of forming large rolls filling much of the boundary layer. However, the total wind-direction shear across actual neutral atmospheric boundary layers is only some 10°–20° resulting in rather slow roll-growth-rates and it usually takes another source of energy, buoyancy forces caused by surface heating, to produce significant motions.

Over land, convectively unstable boundary layers are said to exist when additional turbulence is generated by buoyancy forces arising mainly from the input of sensible heat into the air at the ground. In the absence of significant amounts of water (vapour and liquid), latent heat and condensation effects are of relatively little importance and furthermore radiation sources and sinks within the layer may usually be neglected. These boundary layers are normally in a state of evolution since the depth responds to the input of heat which is directly related to the diurnal cycle of incoming radiation.

The generation of turbulence by buoyancy forces may be greatest in a horizontally averaged sense in the lowest part of the boundary layer. However, observations show that large rolls filling the boundary layer tend to form over rather homogeneous terrain which may have their origin either in the Ekman instabilities referred to earlier (but strongly enhanced by the upward flux of heat) or may have developed purely from thermal convection becoming increasingly organized with height. Local thermally-driven generation of turbulence is at its greatest in the relatively narrow upward moving portions of these rolls somewhere near the middle of the boundary layer. The broader and slower moving downward areas are relatively much less turbulent. These rolls have their axes orientated almost along the wind, backed a few degrees from the geostrophic and moving slowly in time across the ground towards lower pressure.

In the presence of a large urban heat island or a few dominant hills or mountains, lee eddies are generated which can be similar in size to the thermally driven rolls. When these occur, radar studies demonstrate that the lee eddies may be reinforced and persist a long way downstream. Furthermore the field of rolls may be locked on to the feature (the city or the hill), minimizing the horizontal drift of the rolls towards low pressure.

Over more heterogeneous terrain the resulting complex of interacting wakes can largely destroy the order of the rolls and strong convective upflows will be more transient and haphazard.

In very unstable conditions the rolls are replaced by cellular-type motions which tend to be more efficient at transporting heat upwards. They have a much more three-dimensional character than the rolls. The nature of the thermals which drive these larger-scale motions has been described by M. J. Manton (1977).

When significant condensation occurs within the boundary layer the dynamics and the overall heat balance are much more complex, especially if precipitation occurs, and relatively little is known about the structure of such boundary layers. Nevertheless it is likely that the cloud reduces the sensible and latent heat fluxes from the ground surface, that the top of the cloud is cooled by radiation and that as a result the stability of the boundary layer tends to that of a well-mixed near-neutral layer.

Over the sea, water vapour plays a much greater role by enhancing the much smaller buoyancy forces and in determining the overall stability of the boundary layer than is normal over land even when condensation does not occur. The depth of the layer and its advective warming rate are also more dependent on the evaporation rate from the sea and the latent heat released through condensation and precipitation processes where these exist. Diurnal variations in surface fluxes are, however, much less and boundary layer modifications usually take place much more smoothly.

On occasions the sensible heat flux and the latent heat flux may be in opposite directions with the result that the cooling of a warm dry airstream passing over a cold sea may not be inhibited by stabilization of the surface layer as might happen in the absence of evaporation. Over the sea a rather well-mixed turbulent layer can be maintained by the virtual balance of opposing buoyancy forces. However, sometimes in this situation shallow sea fog or low stratus can form, capped by a temperature inversion created by radiational cooling at the cloud top. Downward convection can then gradually develop making itself manifest by the appearance of structure within the cloud of fog.

Stable layers present many challenging problems. Their generation and development over land depend on a balance between radiational cooling of the surface, internal radiational flux divergences (particularly when condensation occurs) and the effects of turbulent mixing. Nocturnal boundary layers typically develop by the cooling of the surface and adjacent layers when vertical turbulence has to do work against the resulting buoyancy forces. In extreme (but not uncommon) situations this results in an increase in the Richardson number at each height until a critical level is reached at one height of the order of 10–40 metres when turbulence is largely suppressed. The air below is then starved of momentum flux from above and although the flow then turns even further down the pressure gradient the velocity tends to fall and turbulence dies out leaving a quiescent stable layer. Subsequently velocity gradients tighten until the layer becomes dynamically unstable and a turbulent breakdown again occurs which may penetrate down to the surface, mixing the chilled air of the layer through a greater depth and ultimately leading to the re-establishment of an even deeper stable layer. Turbulence is, in these situations, highly intermittent in character. Above the level reached by these bursts, but within the old day-time boundary layer, an imbalance of forces persists between the coriolis force and the pressure gradient which accelerates the air. In consequence a super-geostrophic jet has been observed to form over a time-scale of several hours, the height of which is typically some 100–200 metres above ground.

Gravity waves commonly occur in stable layers and are caused by the effect of upstream surface obstacles. In an ideal airstream without velocity shear and with a constant increase of potential temperature θ with height, the wavelength is determined by the layer's natural frequency $N = \left(\frac{g}{T}\frac{\mathrm{d}\theta}{\mathrm{d}z}\right)^{-\frac{1}{2}}$ called the Brunt–Vaisala frequency.

In real stable surface boundary layers N can vary considerably especially in the first few metres. However above this, N usually varies more slowly and acoustic sounder records often show evidence of occasional large dominant waves with a frequency close to N given by the bulk θ-variation. This is so even though in theory the waves must depend to some degree on wind shear and other parameters.

Gravity waves invariably transfer momentum through the vertical and in so far as they are subject to dissipation can transfer heat, moisture and pollution (if present). Such fluxes should be measured directly and not inferred indirectly as sometimes can be done for normal turbulence.

It is useful to note the customary statement of the balance of forces acting horizontally on the air. With axes fixed in the earth and following the notation of the previous section.

$$\rho\frac{\mathrm{d}\bar{u}}{\mathrm{d}t} - \rho f\bar{v} = -\frac{\partial p}{\partial x} + \frac{\partial}{\partial z}(\tau_{zx}) \tag{2.36}$$

$$\rho\frac{\mathrm{d}\bar{v}}{\mathrm{d}t} + \rho f\bar{u} = -\frac{\partial p}{\partial y} + \frac{\partial}{\partial z}(\tau_{zy}) \tag{2.37}$$

Here ρ is air density, p pressure, τ_{zx} and τ_{zy} the components of the horizontal shearing stress (or in other words the vertical fluxes of u-momentum, $\overline{\rho w'u'}$, and v-momentum, $\overline{\rho w'v'}$, respectively) and f the Coriolis parameter (equal to $2\omega \sin \phi$ where ω is the angular velocity of rotation of the earth and ϕ is latitude). The term containing this parameter is an apparent deviating force which is purely a consequence of the reference to axes in the earth. (For fuller detail of these equations and their development from the classical Navier–Stokes equations see, for example, Sutton 1953, pp. 40, 61, 70.)

Outside the boundary layer τ and hence $\partial\tau/\partial z$ are by definition zero, and with the additional conditions of uniformity and steadiness ($\mathrm{d}/\mathrm{d}t$ zero) there is a balance between the pressure gradient and Coriolis terms which yields the familiar relations for the components of the geostrophic wind, u_g and v_g

$$\rho f u_g = -\frac{\partial p}{\partial y}, \quad \rho f v_g = \frac{\partial p}{\partial x} \tag{2.38}$$

Within the boundary layer the resultant horizontal stress decreases upwards from the value corresponding to the surface drag to zero at the top of the layer. Adjacent to the surface a *surface boundary layer* may be defined within which the stress has fallen of only slightly and may be regarded as effectively constant. This is usually called the *constant-stress* layer, though *surface-stress* layer would be more appropriate, and is perhaps some tens of metres deep.

A confirmatory estimate of the approximate depth of the surface-stress layer may be made as follows. Integrating Eq. (2.36), with $d\bar{u}/dt = 0$, and omitting the subscripts on τ for convenience

$$\tau(z) - \tau(0) = f \int_0^z \rho(v_g - \bar{v})\, dz \tag{2.39}$$

Taking the x-axis along the surface wind direction, $\tau(0)$ is the resultant stress at the surface and for small enough z (such that the wind has not turned appreciably from its direction at the surface) $\tau(z)$ is the resultant stress at height z. Similarly $\bar{v}$ may be neglected. Also $v_g = V_g \sin \alpha$ (approximately $V_g \alpha$) where V_g is the resultant geostrophic wind speed and α is the total turning of direction in the friction layer. With these approximations

$$\frac{\tau(0) - \tau(z)}{\tau(0)} \cong \frac{\rho f z V_g \alpha}{\tau(0)} \tag{2.40}$$

Observations over land give values near 10^{-3} for $\tau(0)/\rho V_g^2$, the *geostrophic drag coefficient*, and 0.3 radians for α, and taking $V_g = 10^3$ cm/sec, and $f = 10^{-4}$ as appropriate for middle latitudes,

$$\frac{\tau(0) - \tau(z)}{\tau(0)} \cong 3 \times 10^{-5} z \tag{2.41}$$

indicating that the proportionate fall in τ is less than 10 per cent for heights up to about 30 m.

The properties of the surface-stress layer

From the assumption of stress invariant with height important results follow for the relations between vertical flux and vertical gradient. For detailed discussions reference should be made to the texts by Priestley, Lumley and Panofsky and Monin and Yaglom. Here we shall concentrate on the approach, now widely accepted, which originates in the *similarity* arguments of Monin and Obukhov.

At the basis of this approach is the hypothesis that for any transferable property, the distribution of which is homogeneous in space and stationary in time, the vertical flux/profile relation is determined uniquely by the parameters $z, \rho, g/T$, u_*, $H/c_p\rho$ where in addition to symbols already defined or familiar u_* is the *friction velocity* represented by $(\tau/\rho)^{1/2}$ and H is the vertical heat flux carried by turbulence (assumed constant with height as in the case of momentum). Dimensional analysis, along lines which have been usefully summarized by Calder (1966), leads to a flux/gradient relation which may be expressed in the following general form:

$$\frac{d\bar{S}}{dz} = \frac{S_*}{kL} g_s \left(\frac{z}{L}\right) = \frac{S_*}{kz} \phi_s \left(\frac{z}{L}\right) \tag{2.42}$$

Here $\bar{S}$ is the mean amount of the property in unit mass of air, and S_* is a scale with the dimensions of S defined by the vertical flux F_s (i.e. rate of vertical transfer per unit horizontal area) and the parameters ρ and u_* in the form $F_s/\rho u_*$. The parameter L is a length scale formed from the foregoing physical quantities as follows:

$$L = -\frac{\rho c_p T u_*^3}{kgH} \tag{2.43}$$

With upward heat flux taken as positive, L is negative. g_s and ϕ_s are universal functions, which may be different for different properties. The dimensionless constant k is added for subsequent convenience. (Note that the quantity g/T appears only in L and there the important point is that it appears in association with H. With no heat flux, and hence implicitly no buoyancy forces, L and g/T do not appear.)

An effective diffusivity, or *eddy diffusivity*, K_s, is defined by the ratio of flux to gradient in the form

$$K_s = -F_s/\rho \frac{d\bar{S}}{dz} \tag{2.44}$$

(see 3.1 for further discussion of this *gradient-transfer relation*) and from (2.42)

$$K_s = ku_*z/\phi_s \tag{2.45}$$

Thus, for momentum, with $S = \bar{u}, F_s = -\rho u_*^2$

$$\frac{d\bar{u}}{dz} = \frac{u_*}{kz} \phi_M \left(\frac{z}{L}\right) \tag{2.46}$$

$$K_M = ku_*z/\phi_M \tag{2.47}$$

and on integrating (2.46) with the condition $\bar{u} = 0$ at $z = z_0$

$$\bar{u}(z) = \frac{u_*}{k}\left[f_M\left(\frac{z}{L}\right) - f_M\left(\frac{z_0}{L}\right)\right] \tag{2.48}$$

Similar relations may be written for heat and water vapour. In the former case, because of the compressible nature of the air, the appropriate gradient is that of potential temperature $\frac{d\theta}{dz}$, or alternatively $dT/dz + \Gamma$ where Γ is the dry adiabatic lapse rate. These temperature gradients reflect the gradient of air density, positive or negative values designating respectively stable or unstable states with respect to vertical air motion.

For neutral conditions the buoyancy parameter g/T and the heat flux H must vanish from the relations, and, as could have been deduced from the beginning on the hypothesis that the determining parameters must be u_* and z,

$$\frac{d\bar{u}}{dz} = \frac{u_*}{kz} \tag{2.49}$$

$$K_M = ku_*z \tag{2.50}$$

$$\bar{u}(z) = \frac{u_*}{k} \ln \frac{z}{z_0} \tag{2.51}$$

(In terms of Eq. 2.46 ϕ_M has been set equal to unity in neutral conditions ($z/L \to 0$), the whole numerical constant then being represented in k.)

The logarithmic form of wind profile in Eq, (2.51) has been generally established for neutral flow in the surface-stress layer. The quantity z_0, formally the height at which $\bar{u}$ goes to zero, is interpreted as a roughness length characterizing the surface. Values of z_0 derived from wind profiles range from 10^{-4} to about 1 m respectively for surfaces ranging from ice sheets to forested or city areas (for additional details, see 6.2). Because of the effect of individual roughness elements on the flow the foregoing relations are restricted to heights clear of these elements, i.e. to $z/z_0 \gg 1$. Parameters corresponding to z_0 appear in the profile relations for temperature and humidity, but generally these differ in magnitude from z_0 (for a full discussion of these aspects see Brutsaert, 1982).

In the above equation k is recognizable as von Kármán's constant familiar in the development of the *mixing-length* concept and the experimental establishment of the relation in Eq. (2.51) for flow in rough pipes and over rought plates. The early laboratory studies undertaken in this connection gave $k = 0.4$. A more recent laboratory study of a turbulent Ekman layer in a rotating apparatus (Caldwell, van Atta and Helland 1972) has substantially confirmed this. In the lower atmosphere direct measurements of u_*, using the drag-plate techniques initiated by P. A. Sheppard (1947), are also largely in agreement on a value near 0.4. From the latest drag-plate measurements to be reported, Pruitt, Morgan and Lourence (1973) advocate a value of 0.42. Eddy correlation measurements (giving $\overline{w'u'}$) are a preferable basis for k, in view of inevitable doubts about the precise representativeness of a drag-plate as an undisturbed element of a natural surface. However eddy correlation measurements of $\overline{u'w'}$ in the atmosphere have proved particularly troublesome in other respects and are not completely in agreement about the value of k. An analysis by Dyer and Hicks (1970) of Australian date referred to in more detail below requires $k = 0.41$, whereas measurements in the U.S.A. reported by Businger, Wyngaard, Izumi and Bradley (1971) are claimed to support a lower value of 0.35. Added to this, theoretical arguments are now being advanced (Tennekes, 1973) for a dependence of k on the surface Rossby number, V_g/fz_0, the significant parameter which emerges in treatments of the boundary layer as a whole. Tennekes's discussion contains the claim for a lower bound to the value of k (near 0.33) as $V_g/fz_0 \to \infty$ (the limit approached over a very smooth surface), with larger values than 0.33 only at small V_g/fz_0 (i.e. over rough surfaces). However, a recent discussion by Wieringa (1980) has raised questions about the observational support for 0.35. Moreover, in a reanalysis of international measurements made in 1976 Dyer and Bradley (1982) have reaffirmed the support for the earlier value of 0.4.

In the absence of a theoretical derivation the forms of the functions ϕ and f have to be specified empirically. In the present context by far the greater interest is in respect of ϕ, since this immediately and directly conveys the magnitude of the effect of thermal stratification on the eddy diffusivity. The overall evidence from observations is that for all three properties – momentum, heat and water vapour – ϕ decreases in unstable stratification (L – ve) and increase in stable stratifications. In unstable conditions the forms were first comprehensively prescribed from field studies undertaken by the Australian C.S.I.R.O. Division of Meteorological Physics, on sites and in conditions which closely approach the ideal in uniformity and stationarity. The profile measurements were made at heights up to 16 m and the resulting range of $-z/L$ was 0.005 to 5.0. Some analyses of the results are summarized in Figs. 2.6 and 2.7. Note especially the demonstration of equality in the magnitudes for heat and water vapour and the different (closer to unity) magnitudes for momentum. At the largest values of $(-z/L)$ both ϕ_H and ϕ_W fall to roughly 0.1 whereas ϕ_M falls to roughly 0.5. Note also that these particular results in unstable conditions indicate a close approach to $\phi_H = \phi_M{}^2$, which according to definition implies an equality in the quantities z/L and the Richardson No. defined in 2.3.

Measurement of the vertical fluxes of heat and momentum (and to a lesser extent water vapour) have continued and these have included several internationally collaborative field studies aimed at more firmly establishing the behaviour of the Monin–Obukhov functions. In the event these studies have shown discrepancies which remain to be resolved and there is a continuing discussion about the possible sources of these discrepancies, for example experimental error, non-conformity to ideal fetch and exposure and inadequacy of sampling duration. A useful review has been provided by Yaglom (1977), including a comprehensive table and graphical summary of the various analytic forms of $\phi(z/L)$ proposed as representative of the *average* trend indicated by particular sets of data. According to this, with k assumed equal to 0.4 $\phi_M(0)$ has average

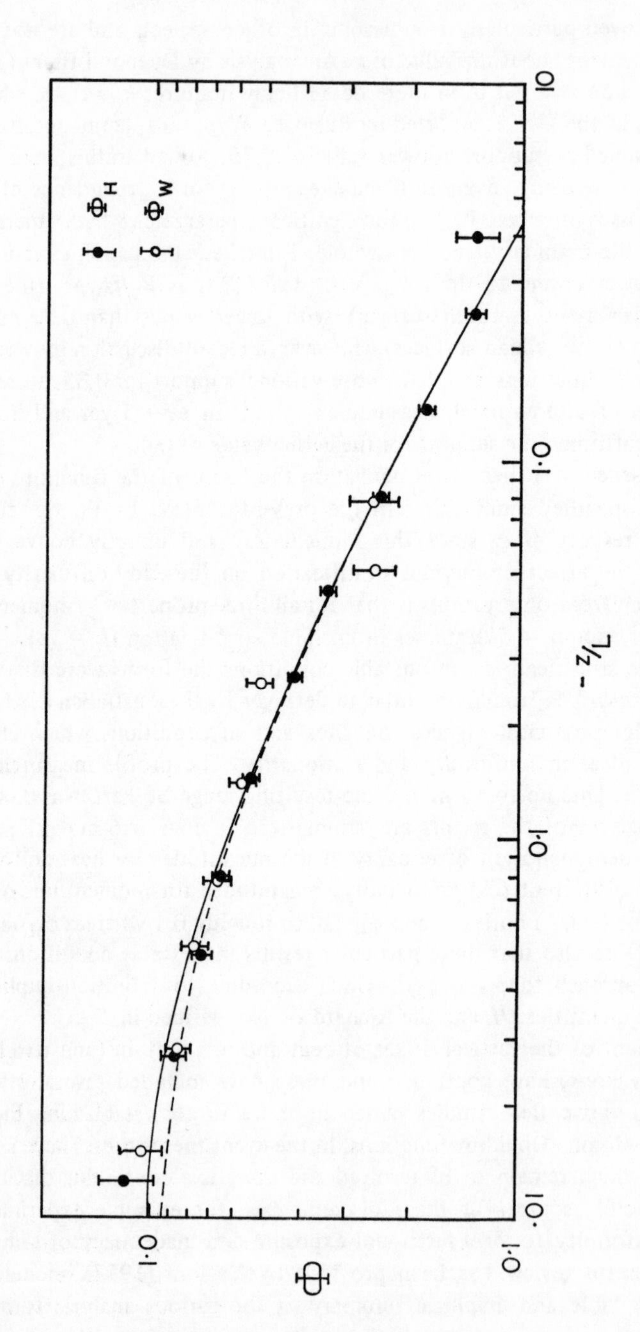

Fig. 2.6 – Geometric means values of ϕ_H (•) and ϕ_W (○), the Monin-Obukhov functions for heat and water vapour, from several field expeditions in Australia (Dyer, 1967). The broken line is an empirical expression $\phi = (1 - 15\, z/L)^{-0.55}$.

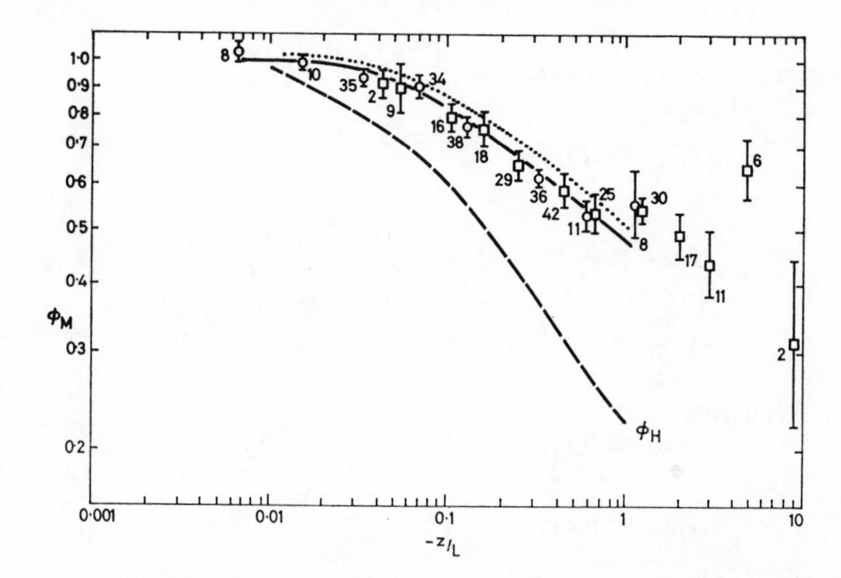

Fig. 2.7 – Geometric mean values of ϕ_M as a function of $-z/L$ (after Dyer and Hicks, 1970). The dotted line represents ϕ_M determined from shape function analysis (Swinbank and Dyer 1967) and the dashed line is the earlier determination of ϕ_H (Dyer 1967). o Wangara expedition 1967, □ Gurley 1970.

values ranging from 0.93 to 1.14 (which is equivalent to giving k a possible range of 0.43 to 0.35), while $\phi_H(0)$ has the range 0.84 to 0.98. For many practical purposes these uncertainties may be unimportant but unfortunately in non-neutral conditions they become considerably wider, reaching for example a factor of 1.5 for ϕ_H in unstable conditions and a factor of about 3 for ϕ_M in stable conditions. Corresponding uncertainties have therefore to be admitted in respect of the effective eddy diffusivities as defined in Eq. (2.45).

The most widely used of the proposed analytic forms of ϕ are due to Webb (1970) for stable flow (with $k = 0.41$), and Businger *et al.* (1971), (see also Businger, 1973), the latter based on the 'Kansas' flux-profile measurements and incorporating the lower value (0.35) for k. The most recently advocated set incorporating $k = 0.4$ is that of Dyer and Bradley (1982) for unstable conditions. The relations for the quantity k/ϕ ($\equiv k/u_*z$) are summarized below and a numerical comparison is given in Table 2.III.

$$\text{Webb} \quad k/\phi_M = k/\phi_H = 0.41\left(1 + 5.2\frac{z}{L}\right)^{-1} \qquad 1 > \frac{z}{L} > 0 \tag{2.52}$$

Businger

$$k/\phi_M = 0.35\left(1 - 15\frac{z}{L}\right)^{\frac{1}{4}} \qquad -2.2 < \frac{z}{L} < 0$$

$$k/\phi_M = 0.35\left(1 + 4.7\frac{z}{L}\right)^{-1} \qquad 1.2 > \frac{z}{L} > 0$$

$$k/\phi_H = (0.35/0.74)\left(1 - 9\frac{z}{L}\right)^{\frac{1}{2}} \qquad -2.2 < \frac{z}{L} < 0 \qquad (2.53)$$

$$k/\phi_H = 0.35\left(0.74 + 4.7\frac{z}{L}\right)^{-1} \qquad 1.2 > \frac{z}{L} > 0$$

Dyer and Bradley

$$k/\phi_M = 0.4\left(1 - 28\frac{z}{L}\right)^{\frac{1}{4}}$$

$$\qquad -4 < \frac{z}{L} < 0 \qquad (2.54)$$

$$k/\phi_H = 0.4\left(1 - 14\frac{z}{L}\right)^{\frac{1}{2}}$$

Table 2.III – Comparison of the Businger *et al.* and Dyer and Bradley values of k/ϕ $(= K/u_*z)$ in neutral and unstable conditions.

	$\frac{z}{L}$	0	−1	−4
k/ϕ_M	Businger *et al.*	0.35	0.70	0.98
	Dyer and Bradley	0.4	0.93	1.30
k/ϕ_H	Businger *et al.*	0.47	1.50	2.88
	Dyer and Bradley	0.4	1.55	3.02

The planetary boundary layer as a whole

The classical approach to formulating the wind profile through the whole depth of the boundary layer was to solve Eqs. (2.36) and (2.37) with the stress components expressed according to the gradient-transfer relation. This has yielded results that are mostly of doubtful value in one of the primary requirements in the present context, namely the deduction of the effective K profile above the surface layer. In the first place experience indicates that the form of the wind profile in the upper layers is not sensitive to the assumed form and magnitude of K assumed in those layers, as the major part of the momentum transfer occurs at lower levels. Secondly it is now widely appreciated that the vertical transfer above the surface-stress layer in typical daytime conditions is often dominated by organised convective motions and is then not satisfactorily represented by the gradient-transfer relation.

Even in near-neutral flow it is not obvious that the upper-level K values can be derived from the wind profile with the accuracy possible in the surface layer, remembering the various complications noted in the introduction to this section. This uncertainty is well illustrated in Fig. 2.8 which shows K profiles reported by Clarke (1970). These were derived directly from the wind profiles, without making any assumptions in advance about the form of K, by applying the 'geostrophic departure' method represented in Eq. (2.39). The wind data were from the Leipzig pilot balloon study first analysed in the present context by Lettau (1950) and from a comprehensive boundary layer study in Australia. It is thought that the differences may be linked with the effects of baroclinicity, which in Clarke's neutral cases tended to bring an increase of geostrophic wind with height.

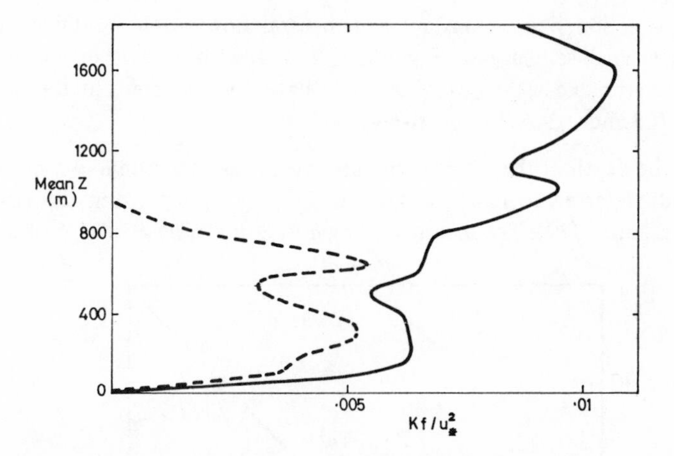

Fig. 2.8 – Profiles of normalized eddy viscosity in near-neutral conditions (after Clarke, 1970). The full line is the profile derived by Clarke from wind profile observations in Australia. The broken line is the 'Leipzig' profile.

The organization of observational data and the development of functional relationships for the boundary layer as a whole have been greatly facilitated by extension of the 'scaling' considerations already noted in respect of the surface layer (for a comprehensive discussion see Tennekes, 1982). To the scales z_0 already introduced as specially relevant to the low-level profiles must also be added the length V_g/f, which is specially relevant to the boundary layer as a whole and is implicitly related to the horizontal scale 0(100 km) over which the wind speed can adjust to a new pressure gradient. From the latter and z_0 is formed the non-dimensional surface Rossby number V_g/fz_0, the parameter to which boundary layer properties are expected to be uniquely related in the absence of buoyancy effects.

Several of the consequences of these scaling considerations are of particular interest to our present considerations.

(a) 'Matching' of the similarity relations for the wind profile at small and large z leads to the result that in neutral flow there must be a layer $z_0 \ll z \ll u_*/f$ in which $(z/u_*)/(du/dz)$ is constant. Thus the logarithmic wind profile emerges in a way that is not necessarily confined to the surface-stress layer, and in reality the experience is that this form is often observed up to about 100 m.

(b) In thermally stratified conditions, with the scale L also included, the depth h of the whole layer is expected to be universally in accordance with

$$hf/u_* = F(u_*/fL) \tag{2.55}$$

The function F is a constant a in neutral flow, while in stable flow various analyses suggest $F = c\ (u_*/fL)^{-\frac{1}{2}}$ with c another constant. Rather variable estimates of a and c have been offered, in the range 0.2–0.3 and 0.2–0.7 respectively.

Data on the depth of the stable boundary layer (from the Minnesota boundary layer study referred to in more detail in 2.6) are shown in Fig. 2.9 (from Smith and Blackall, 1979). The formula giving h in terms of $u_*^2\ H^{-\frac{1}{2}}$ corresponds

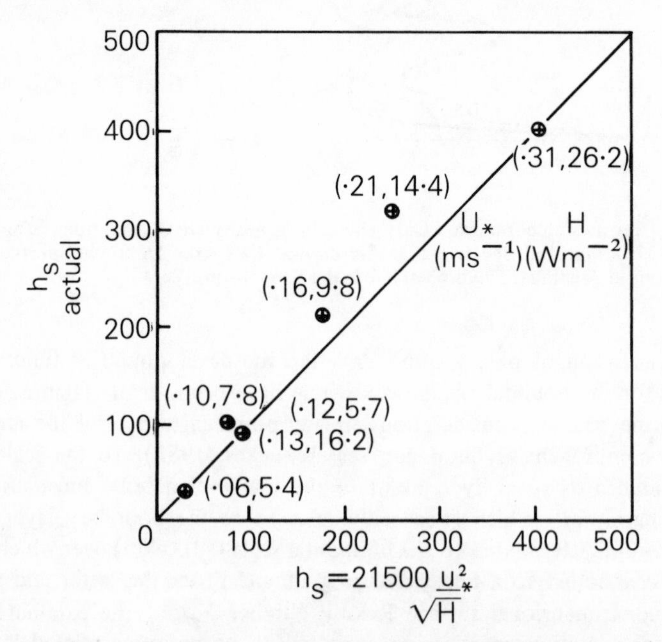

Fig. 2.9 – Data from the Minnesota Experiment confirming the formula $h = 21500\ u_*^2 H^{\frac{1}{2}}$ for the depth of the stable mixed layer.

to the foregoing form for F with a coefficient $c = 0.7$. According to the theoretical modelling studies such a relatively high value is apparently appropriate to the early stages of development of the stable layer but c is progressively decreased with approach of a steady state.

For the daytime unstable boundary layer several theories have been developed to represent the evolutionary process in which the turbulent flux of heat is redistributed into a deepening layer with a dry adiabatic lapse rate. A simple formula for the growth of h (Carson, 1973) is

$$\mathrm{d}h^2/\mathrm{d}t = 2(1 + 2A)H(0, t)/\rho\, c_p\, \gamma \tag{2.56}$$

where $H(0, t)$ is the surface heat flux, $AH(0, t)$ (with A typically 0.2–0.5) is the heat brought down into the mixed layer by mixing at the interface with the stable layer above and γ is the gradient of potential temperature in that stable layer. Carson's complete form includes the effect of general subsidence of the air mass, and Deardorff (1976) and others have included the effects of cloud cover. A nomogram giving h according to these theories is included in 6.2.

2.3 THERMAL STRATIFICATION AND THE MAINTENANCE OF TURBULENT KINETIC ENERGY

In turbulent shear flow turbulent energy is supplied by the working of the horizontal shearing stresses. When the flow is thermally stratified, as is usually the case in the atmospheric boundary layer, buoyancy forces operate, and according to their direction, either augment or suppress the turbulent energy. The magnitudes of these effects can be derived in an elementary way as follows.

For the action of a horizontal shearing stress (τ), consider a thin horizontal layer of thickness h, over which the velocity gradient $(\mathrm{d}\bar{u}/\mathrm{d}z)$ is presumed constant. The shearing force exerted by the air at the upper boundary plane on that at the lower boundary plane is τ per unit horizontal area, i.e. τ/h per unit volume of air in the layer. The relative velocity of the two planes is $h(\mathrm{d}\bar{u}/\mathrm{d}z)$ and so the rate of working of the shearing force is simply $\tau(\mathrm{d}\bar{u}/\mathrm{d}z)$ per unit volume of air, or $(\tau/\rho)(\mathrm{d}\bar{u}/\mathrm{d}z)$ per unit mass of air. In the thermally stratified flow the vertical migration of elements of fluid results in the appearance of the turbulent fluctuations of temperature. Let the fluctuation in the temperature from the mean (T) at any level be T'. With mean density ρ at that level this is equivalent to a density difference from the environment of $\rho T'/T$. The corresponding buoyancy force on unit volume is $g\rho T'/T$, and if the instantaneous vertical component of velocity is w' the instantaneous rate of working is $\rho g w' T'/T$. Averaging over all vertical motions the mean rate of working is $\rho g \overline{w'T'}/T$ per unit volume, or $g\overline{w'T'}/T$ per unit mass of air. The quantity $\rho c_p \overline{w'T'}$ is the vertical flux of sensible heat (H) and so the rate of working may be alternatively written as $gH/\rho c_p T$. If the heat flux is upwards (w' and T' positively correlated), as is the case with an unstable thermal stratification near the ground, there is an addition to the energy of the vertical turbulent motion at the rate $gH/\rho c_p T$ per unit mass. Conversely, if the stratification is stable, and the heat flux downward, turbulent energy is extracted at this rate.

Consider now the typical relative magnitudes and variations of these contributions to turbulent kinetic energy. In the surface-stress layer near the ground we have, from Eq. (2.46)

$$\frac{\tau}{\rho}\frac{d\bar{u}}{dz} = \frac{u_*^3}{kz}\phi_M\left(\frac{z}{L}\right) \tag{2.57}$$

It is immediately obvious that this mechanical contribution to the turbulent energy, as it is usually termed, falls off with height rapidly. The variation is a simple inverse relation in neutral conditions [$\phi_M(z/L) = 1$] or (approximately) even in stratified conditions for small enough values of z. However, with increasing height in stratified conditions the variation becomes significantly more or less rapid than the inverse relation according as the heat flux is upward or downward, $\phi(z/L)$ then respectively decreasing or increasing with $|z/L|$. There must always be a decrease in the mechanical contribution as a consequence of the decrease of τ with height and this no doubt becomes an important effect at heights above the surface-stress layer. The positive or negative buoyancy contribution, being proportional to the vertical heat flux, H, is effectively constant in the surface-flux layer, but this will ultimately diminish as H decreases with height. In this connection notice that the heat flux and hence the buoyancy contribution may even change sign at some height – e.g. in the convective stirring of the atmosphere below a stable layer.

The ratio of these buoyancy and mechanical terms (the Flux Richardson number R_f) is identical with $z/L\phi_M$. With ϕ_M decreasing with $|z/L|$ in unstable conditions, as indicated in 2.2, $|R_f|$ increases rapidly with height, becoming 10 when $|z/L| \cong 4$. In the very unstable conditions represented by $L = -5$ this condition applies at a height of about 20 m. Note that with the maximum heat fluxes experienced in middle latitudes such a value of L would occur only in very light winds (see 6.2), but even in stronger winds the buoyancy term may be expected ultimately to become dominant at a sufficiently great height in a convective boundary layer, though the foregoing characteristics of the surface-stress layer would not then apply.

The Richardson number and related stability parameters

From the simplest standpoint the increase or decrease of the resultant turbulent energy may be considered to be determined primarily by the relative magnitude of the two terms considered above. Thus in an unstable (or even neutral) stratification the intensity of turbulence would be expected to increase, though clearly some other factors operate to prevent this continuing indefinitely. Conversely, when in stable stratifications the buoyancy term becomes the larger of the two, then the intensity of turbulence must decrease and may even be reduced to zero. This simple standpoint was adopted by Richardson (1920) as valid for the case of initially very slight turbulence, on the grounds that in such conditions other factors could be assumed negligible.

In Richardson's original analysis the quantities τ and H were replaced by the gradient expressions of the form Eq. (2.44). The rate of change of mean total turbulent kinetic energy per unit mass of air ($\bar{E}$) may then be written

$$\frac{\partial \bar{E}}{\partial t} = K_M \left(\frac{d\bar{u}}{dz}\right)^2 - \frac{g}{T} K_H \left(\frac{\partial T}{\partial z} + \Gamma\right)$$

$$= K_H \left(\frac{d\bar{u}}{dz}\right)^2 \left[\frac{K_M}{K_H} - \frac{g}{T} \frac{(\partial T/\partial z + \Gamma)}{(d\bar{u}/dz)^2}\right] \tag{2.58}$$

the second (dimensionless) quantity in the bracket being the Richardson number Ri. Assuming $K_M = K_H$ Richardson was thus able to conclude that a slightly turbulent motion would remain turbulent if $Ri < 1$ and would be suppressed if $Ri > 1$.

The whole question of the suppression (or initiation) of turbulent activity is much more complex than represented in the foregoing argument. There is now considerable theoretical support and experimental evidence for initiation of turbulence in a stably stratified fluid at a much lower value of Ri, indeed at $Ri < \frac{1}{4}$. For suppression of turbulence early observations in the atmosphere suggested values over the wide range 0.4 to 1.0, and controversy over the precise value continues, though with a tendency to return to a value nearer 1.0, as originally and simply estimated.

Quite apart from consideration of its critical value for suppression of turbulence the Richardson number has been widely adopted as a stability parameter generally characterizing the effects of both unstable and stable stratifications on the form of the wind profile, the magnitudes of the components of turbulence and the rates of eddy transfer and diffusion. A disadvanrage of this parameter is that it is a function of height (to a first approximation it is proportional to height, since both the wind and temperature gradients are nearly inversely proportional to height). At a height of 1 m over grassland in middle latitudes Ri is typically in the range ± 0.1 and values significantly outside this range at that height require a combination of particularly intense surface heating or cooling with very light winds.

Reference has already been made to the Flux Richardson number R_f and with Ri and z/L we thus have three stability parameters related as follows:

$$R_f = \frac{K_H}{K_M} Ri = \frac{z}{L\phi_M} \tag{2.59}$$

Of these parameters z/L is now widely preferred as the most basic for the surface-stress layer, L being independent of height, whereas generally Ri and R_f vary with height in a non-linear way. In many studies, however, only the gradients of wind velocity and temperature are specified and the Richardson number then continues to provide the most convenient description. Note that all three parameters, Ri, R_f, and z/L tend to approximate equality at small magnitudes, i.e. near-neutral conditions, while in stratified conditions useful empirical relations between Ri and z/L are now available. In unstable conditions, as noted in the previous section, approximate equality of Ri and z/L appears to extend over much of the range of instability normally encountered, while in stable conditions the log-linear law implies.

$$Ri = \frac{z}{L}\left(1 + \alpha\frac{z}{L}\right)^{-1} \tag{2.60}$$

α being close to 5 according to the analysis by Webb (1970).

The turbulent kinetic energy balance in more detail

The complete mathematical expression of the turbulent energy balance may be formed from the equations of motion of a viscous compressible fluid. If the instantaneous values of velocity, temperature and pressure are replaced by mean values plus instantaneous eddy fluctuations therefrom [as in Eq. (2.1)], and averages are taken, equations for the mean motion are obtained. These are similar in form to the original equations but now contain additional terms representing the eddy fluxes of momentum and heat. Subtraction of the original and averaged equations gives the equations for the eddy fluctuations of velocity and temperature and these may be manipulated to give the mean square fluctuation. This type of analysis was first followed through completely by Calder (1949a). A more recent discussion of the process and results is given by Lumley and Panofsky (1964). Their discussion brings out the following main features.

(1) Energy is extracted from the mean motion by the working of the shearing stresses and appears as turbulent energy.
(2) By the action of the buoyancy forces potential energy is converted into kinetic energy in the w-component, or vice versa, according to the direction of the heat flux.
(3) Viscous forces act to transfer energy from one component to another but also to give a net dissipation into heat.
(4) Fluctuating pressure forces also act to transfer energy between components and in addition cause transfer from place to place.
(5) There are eddy fluxes of turbulent kinetic energy, divergences in which imply local rates of change in total turbulent energy.

For horizontal mean flow which is also homogeneous in the sense that there is no horizontal transfer of turbulent energy by either the mean or eddying components the complete balance is expressed by the equation

$$\frac{\partial \bar{E}}{\partial t} = \left(u_*^{\ 2} \frac{d\bar{u}}{dz} + v_*^{\ 2} \frac{d\bar{v}}{dz} \right) + \overline{w'T'} \frac{g}{T} - \epsilon - D \tag{2.61}$$

where $\bar{E}$ is the mean total turbulent kinetic energy per unit mass $\frac{1}{2}(u'^2 + v'^2 + w'^2)$, $v_*^{\ 2} = \overline{v'w'}$, analogous to $u_*^{\ 2}$, ϵ is the rate of viscous dissipation, and $D = \partial(\overline{p'w'}/\rho + \overline{w'E})/\partial z$, with E instantaneous turbulent energy and p' the eddy fluctuation of pressure. The divergence term, as well as the buoyancy term, may take on either sign. It is often assumed that conditions are stationary, hence $d\bar{E}/dt = 0$, and this is no doubt an acceptable simplification in many circumstances. Even with this simplification, however, there remains a rather complex balance, in which the relative importance of the terms undoubtedly varies substantially. In neutral flow the general impression is that the mechanical production and dissipation are the dominant terms and should be found essentially in balance if the distorting effects of changing terrain are avoided. In stable flow near the ground the energy budget terms are not reliably estimated but the broad expectation is of a rough balance between mechanical production and loss by viscous dissipation and working against buoyancy forces.

When the conditions are unstable there now seems little doubt that the vertical divergence of the vertical flux of turbulent kinetic energy may be significant, typically decreasing from a positive value (i.e. a vertical export) near the ground, ultimately to a negative value (a vertical import) in the upper part of the layer. This implies that at some intermediate height there should be a negligible divergence term, leaving the balance essentially between the buoyancy production and dissipation, the mechanical production also possibly having fallen to a negligible value. However the details of the precise balances cannot be said to be firmly settled for all circumstances, as is evident for example in the detailed estimates constructed by Wyngaard and Coté (1971) from full observations of fluxes and profiles in the first 30 m above an extensive site in Kansas, U.S.A. The production and dissipation terms and the vertical flux divergence term show a growing net imbalance, as instability increases, which the authors are unable to explain, except speculatively as being due to the unmeasured pressure fluctuation term. Further sophisticated instrumental measurements and critical appraisal are evidently required to clarify the position in general.

In addition to their direct implications regarding the intensity of turbulence, the foregoing considerations of the balance of turbulent energy also have application in the problem of estimating the eddy viscosity, as discussed in 2.5.

2.4 SPECTRAL REPRESENTATION – ITS SIGNIFICANCE AND LIMITATIONS

The procedures for obtaining the spectrum of a fluctuating quantity have been outlined in 2.1. Before summarizing the data on atmospheric turbulence it will be useful to consider in a little more detail the interpretation of spectra. For the purposes of illustration, and to fix ideas, the normalized frequency spectrum corresponding to an autocorrelogram of simple exponential form is shown in Fig. 2.10. This is not the best representation of the shapes of the spectra of atmospheric turbulence, but it is close enough for the immediately following considerations, and the mathematical form is convenient. Most real spectra do show, after averaging or smoothing, a more or less continuous decrease of spectral density with increasing frequency. In view of the wide range of frequency usually involved there is convenience in plotting against $\ln n$. If instead of $F(n)$ the dimensionless product $nF(n)$ is plotted as the ordinate there is the further advantage that the area under the spectrum curve between specified frequency limits still represents the fraction of the total variance, since $nF(n)\,d(\ln n) = F(n)\,dn$. Also, this method of plotting produces a peak in the graph, at a frequency n_m (or wavelength $\lambda_m = u/n_m$) which is related to the integral time-scale, and which can be used as a more practical specification of the spectrum scale.

In practice the range of frequencies over which a spectrum is evaluated is limited at one end by the response of the measuring system and at the other

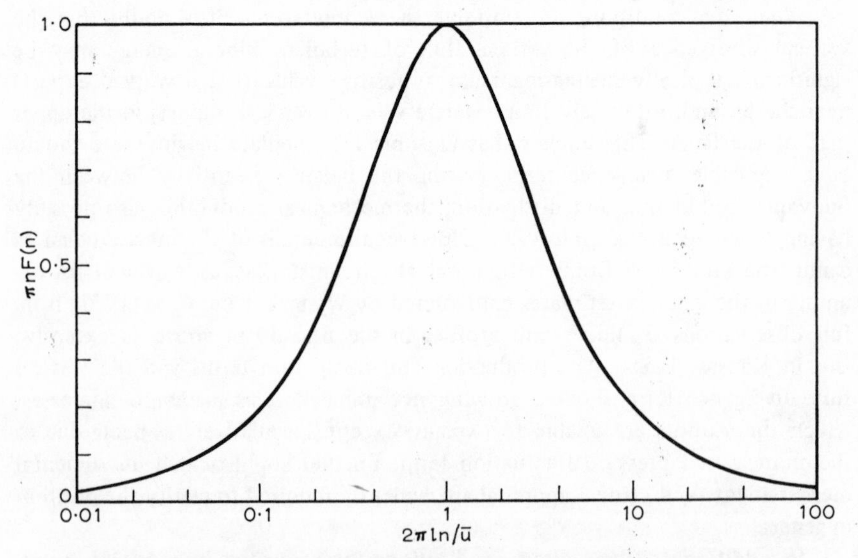

Fig. 2.10 – Normalized frequency spectrum corresponding to an auto-correlogram of exponential form. The curve is $nF(n) = 2/[\pi(a + 1/a)]$ where $a = 2\pi nl/\bar{u}$ with n in cycles/sec and corresponds to Eq. (2.69) with $2\pi n/\bar{u} = \kappa$.

by the duration of record available, as discussed in 2.1. As pointed out there the latter limitation means that the integral time-scale is not usually well determined. Also, the spectral densities evaluated from a single record exhibit considerable irregularities, consequently it is customary to compound individual spectra obtained in nominally similar conditions. These average spectra usually display much smoother characteristics, confirming the expectation that the irregularities of individual spectra are really a result of imperfect sampling of a velocity field which is always patchy to some degree, rather than an indication of genuine peaks in the spectrum.

Wavenumber spectra are virtually never obtained directly in the strict sense of using simultaneous measurements in space. Instead they are evaluated from time-lapse measurements with a single instrument, using Taylor's hypothesis that if the mean speed $\bar{u}$ of the air relative to the instrument is large compared with the turbulent components of velocity the spatial characteristics of the field of turbulence will not have changed significantly over the time lags involved. Consequently, for some range of time lag t may be replaced by $x/\bar{u}$, and t_s by $l_s/\bar{u}$, and the correlogram may be interpreted as a space correlogram *along* the mean wind direction. Correspondingly the spectrum may be interpreted as a wavenumber spectrum *along* the mean wind direction, with n replaced by $\bar{u}\kappa$. The same principle applies, even more effectively, when the instrument is carried at high speed on an aeroplane, in which case the relative air speed V is used in the above transformations in place of $\bar{u}$, and the reference line is the track relative to axes moving with the mean wind.

The available data and evidence for Taylor's hypothesis have been reviewed by Lumley and Panofsky (1964). We should note in particular the evidence from spectra obtained from concurrent measurements with instruments mounted on a tower (or captive balloon) and carried on an aeroplane in level flight at the same altitude. The first comparison of 'fixed-point' and 'aeroplane' data, for the vertical component at a height of 91 m, showed a broad similarity in shape and position of the $nS(n)$ peak when the spectra were plotted against equivalent wavenumber, i.e. apparent frequency divided by relative airspeed (Gifford, 1956). Further and more detailed comparisons have been made in U.S.A. by Lappe, Davidson and Notess (all three components) and in England by Mrs. A. Burns (vertical component only) and the spectra resulting from these have been collected in a brief summary by Pasquill (1963). For equivalent wavenumbers greater than about 5×10^{-3} m^{-1} no significant difference between 'fixed-point' and 'aeroplane' spectral densities was apparent, but over a range of low wavenumbers the 'fixed-point' values were more often lower than the 'aeroplane' values.

For a much lower height (2 m) Panofsky, Cramer and Rao (1958) have compared correlations obtained from instruments arranged along the mean wind direction. When plotted against equivalent separation ($\bar{u}t$) the autocorrelograms mostly agreed excellently with the space correlograms for spacings up to 90 m – the maximum considered. However, two of the four space correlograms for the v-component showed an increase at large separation which is not reflected in the autocorrelograms at equivalent times. In these cases the true wavenumber spectra would presumably show a greater proportion of energy at low wavenumber than would be evident from the frequency spectra, a discrepancy which is in the same sense as noted above for 'fixed point' and 'aeroplane' spectra. One possible reason for this sort of discrepancy is the existence of flow variations which are induced by and fixed relative to terrain irregularities, but which are not apparent in the 'fixed point' measurements. Another possible reason to be considered in the case of sheared flow is that the speed with which the eddy is convected may not be the local mean wind speed, especially as far as low-frequency variations are concerned. To explain on this basis the tendency for 'fixed-point' spectra to give relatively low energy densities at low wavenumbers would require the appropriate convecting speed to be *lower* than the local mean (since the spectral densities generally decrease with increasing frequency or wavenumber), whereas with wind speed increasing with height one might have expected the reverse. These features still require clarification and the most that can be said at the present stage is that the Taylor hypothesis appears to be acceptable with confidence for the relatively high-frequency sections of spectra. However, as the integral scale is reflected in the low-frequency contribution the equivalence of l_s and $\bar{u}t_s$ is in some doubt, and this is an important reservation to be borne in mind in subsequent discussions.

So far in these considerations there has been no recognition of the import of the essentially three-dimensional nature of turbulence. The spectrum which is derived in practice on a frequency basis and subsequently converted to a wavenumber basis is necessarily a one-dimensional representation – i.e. it describes the velocity fluctuations which would be apparent *along a line.* The effect of this is most easily appreciated by considering first a two-dimensional system of waves of constant wavelength λ with the displacements in, say, the x, z plane and the crests parallel to the y-axis. As long as the amplitudes of the waves are observed in the x-, z-plane they will be associated with the correct wavelength λ, but if observed in any other plane through the z-axis they will apparently be associated with a wavelength greater than λ. In the one-dimensional view of turbulent fluctuations there will be contributions from 'waves' with all possible orientations with respect to the reference line, with the result that part of the energy which is really associated with wavenumber $1/\lambda$ will appear to be associated with smaller wavenumbers. For a correct association with wavenumber it is necessary to define a three-dimensional spectrum function $E(\kappa)$ to represent the contribution to the speed variations (irrespective of direction) of various wavenumbers (irrespective of orientation). In fluid mechanics it is conventional to define the function so that its integral with respect to wavenumber is the total turbulent kinetic energy per unit mass of air, i.e. $\frac{1}{2}(\overline{u'^2} + \overline{v'^2} + \overline{w'^2})$, but, as already noted, in meteorological applications it has become customary to define in terms of the variances, omitting the factor of $\frac{1}{2}$ and this convention will be followed here, so that in the case of $E(\kappa)$

$$\int_0^\infty E(\kappa)\,\mathrm{d}\kappa = \overline{u'^2} + \overline{v'^2} + \overline{w'^2} \tag{2.62}$$

The three-dimensional and one-dimensional functions are related in a way which must satisfy continuity (incompressibility being assumed as usual in problems of the present type). Explicit relations have been derived (see Batchelor (1953) or Lumley and Panofsky (1964)) only for the simple condition of *isotropic* turbulence, for which it is assumed that, in addition to homogeneity, all statistical properties are unchanged by a rotation or reflection of the coordinate axes. An obvious consequence of this assumption is

$$\overline{u'^2} = \overline{v'^2} = \overline{w'^2} \tag{2.63}$$

It also follows that in isotropic turbulence the correlation and spectrum properties are completely defined by the two one-dimensional representations shown in Fig. 2.11. This is an aspect of representation which is applicable to any component of velocity and the terms *longitudinal* and *transverse* are not to be confused

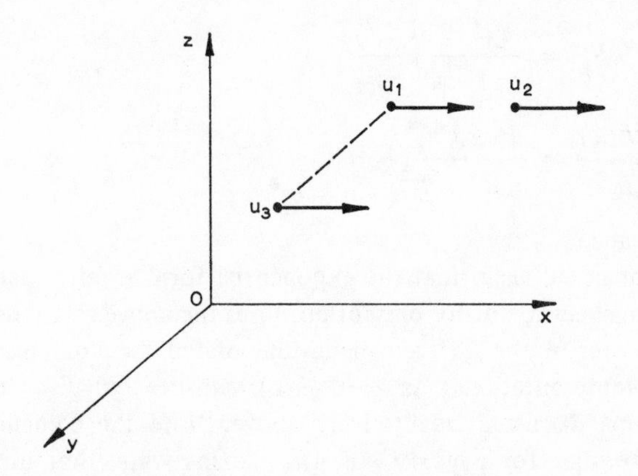

Fig. 2.11 – Longitudinal correlation (u_1 and u_2) and transverse correlation (u_1 and u_3).

with the directions of the velocity components with respect to the mean flow direction. Using the symbols f and g to denote the longitudinal and transverse correlation functions, and S_f and S_g the corresponding absolute spectrum functions, the relations are

$$g = f + \frac{r}{2}\frac{\partial f}{\partial r} \tag{2.64}$$

when r is spatial separation

$$S_g(\kappa) = \tfrac{1}{2}S_f(\kappa) + \frac{\kappa}{2}\frac{dS_f(\kappa)}{d\kappa} \tag{2.65}$$

$$E(\kappa) = \kappa^3 \frac{\partial}{\partial\kappa}\left(\frac{1}{\kappa}\frac{\partial S_f(\kappa)}{\partial\kappa}\right) \tag{2.66}$$

Using for illustration the simple exponential form as representative of the *longitudinal* correlation, i.e.

$$f(x) = \exp(-x/l_s) \tag{2.67}$$

then in isotropic turbulence

$$g(x) = \left(1 - \frac{x}{2l_s}\right)\exp\left(-\frac{x}{l_s}\right) \tag{2.68}$$

and the spectrum functions are

$$\kappa F_f = \frac{2\kappa l_s}{\pi(1 + \kappa^2 l_s^2)} \tag{2.69}$$

$$\kappa F_g = \frac{\kappa l_s(1 + 3\kappa^2 l_s^2)}{\pi(1 + \kappa^2 l_s^2)^2} \tag{2.70}$$

$$\frac{\kappa E(\kappa)}{u'^2} = \frac{16\kappa l_s \kappa^4 l_s^4}{\pi(1 + \kappa^2 l_s^2)^3} \tag{2.71}$$

with κ in radians m^{-1}.

It is emphasized again that the exponential form is not a particularly good fit to Eulerian spectra, but for present purposes the above relations bring out the important matter of the different behaviour of the three-dimensional and one-dimensional representations. As $\kappa \to 0$ $E(\kappa)$ vanishes but $F(\kappa)$ remains finite, for the reasons discussed qualitatively above. Note the coincidence that the radian wavenumber for peak κF_f is $1/l_s$ radians while that for peak $E(\kappa)$ is $\sqrt{2}/l_s$ – in other words the scale defined by the peak in the κF_f function is not very different from that defined by peak spectral density in a three-dimensional representation.

The distinction between the three-dimensional and one-dimensional forms, especially regarding the finite limit of the latter as $\kappa \to 0$ and the qualitative difference in the longitudinal and transverse forms, is necessary irrespective of the precise forms of $E(\kappa)$ and $F(\kappa)$. In considering the spatial significance of the 'fixed-point' frequency spectra obtained in practice, that for the u-component is implicitly longitudinal whereas those for v and w are both transverse and all refer to the fluctuations along the direction of the mean flow. However, if the instrument is carried at high speed on an aeroplane the spectrum is predominantly longitudinal for the head-on component, whatever the orientation with respect to the mean flow direction, and transverse for the vertical and normal components. Furthermore, flights at high speed across the flow provide an approximation to cross-flow correlations and spectra. Regarding the estimation of scale and the difficulty which arises from the limitation of the low-frequency end of the spectrum, an apparently attractive alternative is provided by the peak in the $\kappa F(\kappa)$ function, and this is widely used in practice. However, in so far as this is used to estimate the integral scale l_s the advantage is largely illusory, since the precise relation between l_s and the wavenumber for peak $\kappa F(\kappa)$ is dependent on the low-frequency shape of the spectrum. Finally, the distinction between the longitudinal and transverse forms is a consequence of the vector nature of the property, and does not arise in the case of a scalar property such as temperature.

2.5 THE SMALL-SCALE STRUCTURE OF ATMOSPHERIC TURBULENCE

When plotted in the customary fashion in terms of the product $nS(n)$ the smoothed or averaged spectra of atmospheric turbulence typically have a single rather flat peak, roughly similar to a limited central section of the theoretical spectrum represented in Fig. 2.10. On the high-frequency side of this peak the spectra show an approximation to a simple power-law relation of the type $S(n) \propto n^{-p}$, with the exponent p necessarily greater than unity. The properties of this high-frequency section of spectra have received a great deal of attention, especially in relation to the theory that in general turbulence tends to have a universal small-scale structure, and it is appropriate to begin the discussion of atmospheric turbulence spectra with this aspect.

The idea that the spectrum of atmopsheric eddies extends over a wide range of sizes, and that the kinetic energy of the turbulence is handed *down* the scale of eddies ultimately to be dissipated as heat by viscous action, was implicit in L. F. Richardson's discussion of the diffusive action of turbulence. The development of the idea in its present formal and quantitative terms began however with A. N. Kolmogorov (1941). The theory is that all turbulent motions, irrespective of their origins and subject only to the condition that the energy thereof is fed in on a scale sufficiently large relative to that on which it is dissipated, have a universal *locally isotropic* structure on a small scale. This means that structure functions for *small* separation, or energy densities associated with *large* wavenumbers, are independent of rotation or reflection of the coordinate axes and, for the latter properties, the relations in Eqs. (2.65)–(2.66) are deemed to apply. It also follows that in this range of wavenumbers there is no contribution to the shearing stress and heat flux.

The general physical picture, represented schematically in Fig. 2.12, is as follows (note that the discussion applies strictly to three-dimensional turbulence

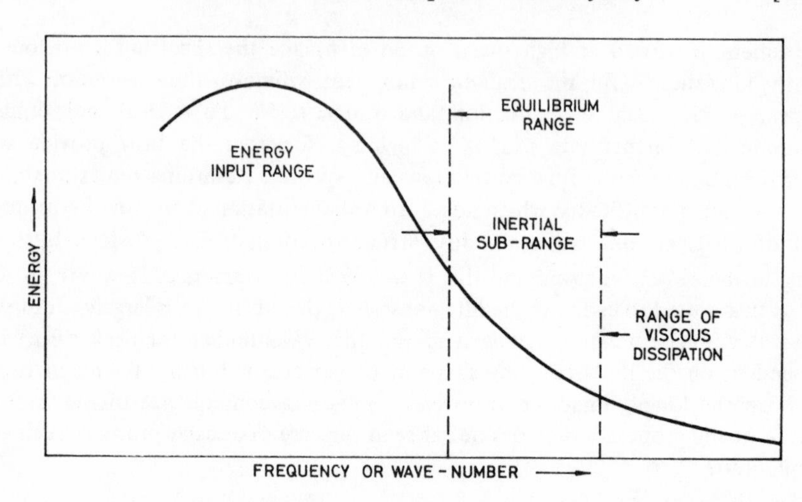

Fig. 2.12 – Schematic representation of energy spectrum of turbulence.

and not to the essentially two-dimensional form which may exist in highly stable flow). It is supposed that the turbulent energy is fed into the system over a limited range of wavenumbers according to the scale of the feeding mechanisms involved. At this stage it must also be supposed that the effect is generally anisotropic, in that the feed is directly into the horizontal components from the working of the horizontal shearing stresses and into or out of the vertical component by the action of the buoyancy forces. Two other mechanisms then come into play – a sharing of energy between the different components by the action of viscous stresses and pressure forces – and a removal of energy from low wavenumbers to high wavenumbers by the mean and fluctuating gradients of velocity. Kolmogorov's idea was that in this process there would be a progressive decoupling of the structure from the original anisotropic form and a consequent tendency to local isotropy. Once this stage is reached the properties must be conditioned firstly by the rate at which the energy is handed down, which because of the small viscous dissipation by the relatively large eddies of the range is equal to the energy originally fed in and ultimately dissipated at rate ϵ, and secondly by the viscosity ν governing this rate of dissipation. Accordingly, the properties on a small enough scale should be functions only of the two parameters ϵ and ν. Forming length and velocity scales from these parameters, respectively $(\nu^3/\epsilon)^{1/4}$ and $(\nu\epsilon)^{1/4}$, it follows on dimensional grounds that the three-dimensional spectral density $E(\kappa)$ must be proportional to the product of the velocity scale squared and the length scale, and a universal function of the only dimensionless combination $\kappa(\nu^3/\epsilon)^{1/4}$, i.e.

$$E(\kappa) = (\nu^5\epsilon)^{1/4}\, F(\kappa\nu^{3/4}\epsilon^{-1/4}) \tag{2.72}$$

where the proportionality factor has been incorporated in the function F. The length $(\nu^3/\epsilon)^{1/4}$ is known as the Kolmogorov microscale and is regarded as defining the order of wavenumber above which viscous action begins effectively to annihilate the turbulence. From measurements in the planetary boundary layer this microscale has a magnitude of order 1 mm. If there is a large separation between the energy feeding and dissipating ranges of wavenumbers Kolmogorov further suggested that there could be a subrange of relatively low wavenumbers in which neither the original feeding nor the ultimate dissipation is important, and in which therefore only the *inertial transfer of energy* by the distorting action of the *fluctuating* velocity gradients is relevant. In this case, the spectral properties should be independent of ν, and on applying this condition to Eq. (2.72) there follows the well-known *inertial subrange* law

$$E(\kappa) = \alpha\epsilon^{2/3}\,\kappa^{-5/3} \tag{2.73}$$

where α is a dimensionless constant to be determined from observation.

Using Eqs. (2.65)–(2.66), applicable to isotropic turbulence, it follows that the one-dimensional spectra in the inertial subrange must have the same power law dependence on ϵ and κ, i.e.

$$S(\kappa) = C\epsilon^{2/3}\kappa^{-5/3} \tag{2.74}$$

but with the constant C different according as longitudinal or transverse description is involved. With the previous notation C_g is equal to $\frac{4}{3}C_f$. Corresponding laws can be formulated in terms of the structure function defined in Eq. 2.2), viz.

$$D(r) \propto \epsilon^{2/3}r^{2/3} \tag{2.75}$$

with r the separation distance. The proportionality constants are again different for the longitudinal and transverse representation and are almost exactly four times the C's in Eq. (2.74) when, as is intended above, wavenumber is in radians per unit length. Note that Eq. (2.75) can be expressed in terms of covariance or correlation functions, using Eq. (2.3). Also note that when the reference line is along the mean wind direction all the relations can be converted into frequency or time-lag forms appropriate to 'fixed-point' fluctuations, by applying Taylor's hypothesis.

The conditions for the existence of local isotropy and of an inertial subrange have been discussed in general by Lumley and Panofsky (1964) and, in particular relation to measurements discussed below, by Pond, Stewart and Burling (1963). Essentially it is argued that for local isotropy the time-scale characterizing the distorting action of the turbulence must be small relative to that characterizing the feeding mechanisms. From this it is predicted that in unstratified sheared flow near the ground only wavenumbers large compared with $1/z$ can be expected to be isotropic, and of these only those which are very small compared with the reciprocal of the Kolmogorov scale can be expected to be within an inertial subrange.

Observational evidence for the inertial subrange law

In the observational tests which have been made of the applicability of the inertial subrange relation in the atmosphere by far the greatest attention has been given to the *shape* of the spectrum. As already mentioned frequency spectra for the atmosphere show a close approach to a power law in n at high frequency, and through Taylor's hypothesis imply the same power law over an equivalent range of wavenumber, and the point at issue is the agreement of the observed exponent with the theoretical value of $-\frac{5}{3}$. On the whole the earlier data cannot be said to be decisive. A summary by Pasquill (1963) of the estimates then available from eleven different sources of turbulence measurements lists values of the exponent in the range -1.2 to -2.1. In some of the cases no attempt was made to derive a best-fitting exponent and the results were used merely to demonstrate that in a logarithmic plot the data points fell reasonably close to a line with a slope of $-\frac{5}{3}$. It is noteworthy that at heights of a few metres the power-law relation evidently extended to equivalent wavelengths several times the height, while at heights of 50–500 m the corresponding wavelength limit was roughly 200 m. The extent to which the deviations from the minus-five-thirds power relation are truly a demonstration against the inertial subrange law, rather than a result of inadequate sampling of a patchy field of turbulence, has not been made clear. However, in one of the low-level cases for which agreement with a minus-five thirds relation was claimed for equivalent wavelengths up to several times the height, examination of the spectral contribution to the shearing stress and of the relative magnitudes of the longitudinal and transverse correlations indicated that the turbulence could not have been isotropic for such large wavelengths (Taylor, 1955).

The first really critical observational tests of the theory of local isotropy were provided by R. W. Stewart and his co-workers. Those reported by Pond, Stewart and Burling (1963) refer to the u-component at a height of 2 m above tidal water with waves approximately 0.3 m high. Usable data were obtained for three periods, two of duration 30 min and one of duration 15 min, and the analysis was made on a wavenumber basis by applying Taylor's hypothesis. The results are especially remarkable in two respects – first in that the spectra extended to high enough wavenumber to permit direct evaluation of ϵ from the expression $15\nu \int_0^\infty \kappa^2 S(\kappa)\, d\kappa$ – second in that they could be compared with previous measurements in the water, for which the values of ϵ were smaller by a factor of 10^2–10^3. For the air measurements the logarithmic plots of $S(\kappa)$ against κ show an impressively close fit to a minus-five-thirds relation over the radian wavenumber range $5 \times 10^{-3} - 1$. However, as the authors point out, closer scrutiny reveals a tendency for points in the lower wavenumber range 10^{-2} to 10^{-1} to fall below the extension of the minus-five-thirds power relation through the higher wavenumber range 10^{-1} to 1, by a factor of about 2 in $S(\kappa)$. It may be significant in this connection that the authors estimate, on the grounds referred to previously, that the region of local isotropy should be confined to wavenumbers $> 10^{-1}$ (i.e. to equivalent wavelengths less than the height above the boundary).

In the foregoing measurements of the u-component there was some doubt about the precise absolute calibration of the hot wire anemometer, and the calibration finally adopted was one which in addition to satisfying a somewhat subjective estimate of the *mean* wind speed also placed the spectral data in exact conformity with the measurements in water by Grant, Stewart and Moilliet (1962), when the function $S(\kappa)/(\epsilon\nu^5)^{1/4}$ was plotted against κ/κ_s, where κ_s is the reciprocal of the Kolmogorov length scale. The composite plot on a logarithmic basis follows a minus-five-thirds law closely over the range $10^{-4} < \kappa/\kappa_s < 10^{-1}$. A more rapid fall of spectral density follows at higher values of κ/κ_s, implying that viscous dissipation of turbulent energy becomes significant at equivalent wavelengths about ten times the Kolmogorov scale.

The general consistency in the foregoing results, for states of turbulence with widely different magnitudes of ϵ, is strong support for the idea of local isotropy, and inspires confidence in the estimates of the Kolmogorov constant C which follow. Fitting their results to this equation over the radian wavenumber range 10^{-1} to 1, Pond *et al.* derive values of C which together with Grant *et al.*'s estimates for water cover an overall range of about 0.35 to 0.60. There is no sign of any systematic variation with ϵ and the overall mean value is 0.46. In reviewing this estimate and other recent data Lumley and Panofsky (1964) concluded that the magnitude of C could be confidently taken to be between 0.45 and 0.50. Note that this applies when κ is measured in radians – if it is in cycles, which is conventional in spectra of atmospheric turbulence, the foregoing value of the constant must be divided by $(2\pi)^{2/3}$ or 3.4. Also, the foregoing value refers to the longitudinal form of spectrum and for the transverse form must be multiplied by $\frac{4}{3}$. Thus, substituting $n = \bar{u}\kappa$ (according to Taylor's hypothesis) in Eq. 2.74), and taking $C = 0.5$, the expressions for the 'fixed-point' frequency spectra of the three components of turbulence in the inertial subrange are

$$S_u(n) = 0.15\bar{u}^{2/3}\epsilon^{2/3}n^{-5/3} \tag{2.76}$$

$$S_v(n) = S_w(n) = 0.2\bar{u}^{2/3}\epsilon^{2/3}n^{-5/3} \tag{2.77}$$

n being measured in cycles/sec.

Subsequent evidence has not been entirely consistent. Measurements at a height of about 100 m, using a hot-wire anemometer carried on a light aircraft, have been reported by Sheih, Tennekes and Lumley (1971). While these show an impressively extensive minus-five-thirds range followed by a dissipation range, the Kolmogorov constant derived from the latter is 30 per cent higher than that obtained by Stewart and his colleagues. On the other hand Wyngaard and Coté (1971) and Kaimal, Wyngaard, Izumi and Coté (1972) report estimates much closer to 0.5, from hot-wire and sonic measurements in the first 30 m of the atmosphere.

Most recently Antonia, Chambers and Satyprakash (1981) have presented new data, from both laboratory and atmospheric measurements, which favour the higher value of C. The atmospheric data are from Bradley, Antonia and Chambers' (1981) measurements of turbulence at a height of 4 m in near-neutral and slightly unstable conditions. The spectral estimates derived therefrom give $C = 0.65$. Antonia *et al.* make the point that values near 0.55 may be obtained indirectly by using the structure-function data provided by the same turbulence measurements, but emphasize that it would be inappropriate to use the lower value in the spectral density equations.

There is also conflict about the evidence for local isotropy. The point has often been made that for the u-component the minus-five-thirds form appears to extend to much greater equivalent wavelengths (in relation to height) than might be expected from the likely scale of the energy feed. To this have now been added several indications against isotropy in regions where it would be expected – i.e. for wavelengths small compared with the height. In particular the later measurements of Stewart and his co-workers have failed to confirm the expected relation between the longitudinal form of spectrum represented by the frequency spectrum for the u-component and the transverse form represented by the w-component. Data presented by Stewart (1969) show w and u spectral densities which are equal within a few per cent over the whole mutual minus-five-thirds region, instead of in the ratio 1.33. Over the same range the uw co-spectrum has values which cannot be regarded as negligible, as required in the theory. In addition Sheih *et al.*'s data do not conclusively support local isotropy, the ratio of the w and u spectral densities being widely scattered over the range 0.8 to 5, about a mean value undoubtedly *more* than 1.33, in contrast to Stewart's data. This is in conflict with conclusions reached by Panofsky (1969) that the required u/w ratio for isotropy has been verified, albeit at much smaller equivalent wavelengths than those to which the u-component obeys the minus-five-thirds law, and also with the evidence recently reported by Wyngaard and Coté (1971). For the present therefore, although there is considerable encouragement for the practical use of the inertial subrange law, some reservations must continue to be held about its precise validity in real conditions of turbulent flow and about the precise magnitude of C.

The combination of the inertial subrange law with turbulent energy balance considerations

The inertial subrange relation is of great importance on its own in that it provides a means of estimating ϵ, now generally regarded as a fundamental parameter of turbulence, from reasonably practicable measurements of turbulent fluctuations. An indication of the general behaviour of ϵ is also provided by simplified forms of the balance equation for turbulent kinetic energy. In combination the inertial subrange law and these balance equations are particularly useful in predicting the level of high-frequency turbulence in terms of the mean parameters τ and H, or conversely in predicting τ and K_m from measurements of the high-frequency fluctuations.

From Eq. (2.61), neglecting the divergence of the vertical flux of energy and the variation of v with height and assuming steady conditions,

$$\epsilon = u_*^2 \frac{d\bar{u}}{dz} + \frac{gH}{\rho c_p T} = u_*^2 \frac{d\bar{u}}{dz}(1-R_f) \tag{2.78}$$

and, for the surface-stress layer, substitution of the similarity relation for du/dz and the definition of L gives

$$\epsilon = \frac{u_*^3}{kz}\left(\phi_M - \frac{z}{L}\right) \tag{2.79}$$

Certain useful predictions follow immediately. In the simple case of effectively neutral conditions, when $H = 0$,

$$\epsilon = u_*^2 \frac{d\bar{u}}{dz} \tag{2.80}$$

and within the surface-stress layer with $z/L = 0$ and $\phi_M = 1$

$$\epsilon = u_*^3/kz \tag{2.81}$$

Thus, for given surface roughness and in neutral conditions in the surface-stress layer

$$\epsilon \propto \bar{u}_1^3/z \tag{2.82}$$

where $\bar{u}_1$ is the mean wind speed at some reference height z_1. On the other hand, in unstable conditions the buoyancy production term is large and at sufficiently large z may dwarf the mechanical term. The reduction of Eq. (2.78) to

$$\epsilon = gH/\rho c_p T \tag{2.83}$$

cannot be taken as a necessary consequence in these conditions, especially in view of the unknown importance of the energy flux divergence, but it may be a rough approximation applicable over a limited range of height, over which ϵ may therefore be expected to vary little with either wind speed or height.

If the general behaviour of ϵ is as outlined, substitution of Eq. (2.77) in the foregoing equations suggests that the high-frequency spectral density may be expected to vary with wind speed and height as follows:

$$S(n)_z \propto \bar{u}_z^{2/3}\bar{u}_1^2 z^{-2/3} \tag{2.84}$$

in the surface-stress layer in near-neutral conditions and

$$S(n) \propto \bar{u}_z^{2/3} \tag{2.85}$$

for given heat flux in convective conditions. The variation as Eq. (2.84), implying an overall variation with $\bar{u}^{8/3}$ at given z, has been given some observational support from low-level data for the u and v components of turbulence, as summarized by Lumley and Panofsky (1964). Some evidence for the contrasting relatively slight variation with wind speed in convective conditions has been found in daytime measurements of the fluctuation of wind inclination at heights of 300 and 1200 m (Pasquill, 1967).

In neutral conditions substitution in Eq. (2.80) of the relation for eddy viscosity K_M gives

$$K_M = \frac{\epsilon}{(d\bar{u}/dz)^2} \tag{2.86}$$

Hence an alternative way of estimating K_M is available from measurements of turbulence in the inertial subrange in conjunction with measurements of the vertical gradient of wind speed. A test of this method was made by Pasquill (1963a) using measurements of the high-frequency fluctuation of wind inclination at heights up to 16 m in near-neutral conditions to obtain values of K_M which were then in effect compared with the values derived from the wind profile using Eqs. (2.50) and (2.51). Taking $k = 0.4$ it was found that exact agreement required the *transverse* spectral constant to be 0.18, which is to be compared with the value of 0.2 since advocated. Note that although this test was made in the surface-stress layer the validity of Eq. (2.86) as such is not restricted to this layer. In conditions other than neutral, however, when R_f in Eq. (2.78) cannot be neglected, the method requires further data or assumptions concerning the heat flux. Estimates of the latter may for example be made from measurements of temperature fluctuation at high frequency, using the corresponding balance equation for mean-square temperature fluctuation. However, it should be recalled that the diffusion terms in the balance equations are of uncertain importance in highly stratified conditions, and insufficient data of a critical nature is available for any firm generalizations yet to be made about the full potential of the method.

The profile of ϵ in the boundary layer

Summaries of the variation of ϵ with height in the boundary layer have been given by Priestley (1959), Ball (1961) and Zilitinkevitch, Laikhtman and Monin (1967). All display a general reduction with height, especially at low levels, and here the evidence for a fall-off as $1/z$ is fairly strong. There is, however, a spread in the data which becomes very pronounced at the greater heights. For example, near $z = 1000$ m, the Russian review gives values ranging from 10^{-1} to 10^{-3} cm^2/sec^3. There seems little doubt that this wide variability is a consequence of a complex joint effect of wind speed and stability.

A useful further step in establishing some order in the behaviour of ϵ over the whole depth of the boundary layer has been provided by a preliminary study at Cardington, England, reported by Readings and Rayment (1969) and Rayment (1973), in which estimates of ϵ were derived from measurements of the high-frequency fluctuation of wind inclination. These were made with an instrument carried on the cable of a captive balloon, usually in the form of short samples (5 min) at different heights in succession as the balloon ascended or descended in stages. Representative individual vertical profiles could not be obtained in this way and it was necessary to eliminate the considerable random variations in the short samples by combining the profiles into two classes, according to concurrent records of net incoming radiation. In this way essentially non-convective conditions were separated from convective conditions. From sequences of short samples at a given height it was found that the values of ϵ conformed closely to a log-normal distribution, in accordance with the hypothesis stated by Gurvic and Yaglom (1967) in extension of Kolomogorov's original description of small-scale turbulence. The composite profiles were therefore formed by geometric averaging of the observations. They showed markedly different fall-off with height over the range 37–900 m, the reduction factors being roughly 3 and 20 respectively in the convective and non-convective conditions.

The progress to a more detailed generalization about the ϵ profile will depend on the accumulation of the more representative data from longer samples collected simultaneously at several heights. Such profiles will be particularly useful in the present context in providing a better basis for evaluating the form of K in terms of ϵ and the spectrum scale λ_m.

2.6 THE STATISTICS OF BOUNDARY LAYER TURBULENCE

The principal major experiment undertaken so far in the study of the turbulent structure of the whole depth of the boundary layer over land is the Minnesota Experiment (Kaimal *et al.*, 1976), in which turbulence probes were used mounted on the cable of a large captive balloon. This was preceded some years earlier by the Kansas Experiment (Haugen *et al.*, 1971, Kaimal *et al.*, 1972), a detailed study of the lowest 30 metres of the boundary layer, and by other low-level turbulence measurements reported extensively in the literature. The two quoted experiments gave data in good accord in these lowest layers. Later studies at Ashchurch in Southern England (Caughey and Palmer, 1979) supplemented the Minnesota results by providing data just below the top of the boundary layer in unstable conditions, where the turbulence scales decrease in response to the restraining capping inversion. The latest review of the progress to date is that by Caughey (1982).

The spectral parameters λ_m and σ

In unstable and near-neutral conditions, the wavelength λ_m corresponding to the peak in the $nF(n)$ spectrum shows an almost linear growth for the vertical velocity component w up to about one-tenth of the boundary layer height h and then approaches a virtually constant value of about $1.5h$ for much of the mid-section before falling rather rapidly just below the capping inversion (see Figure 2.13). It should be emphasized that these are average variations but that individual data values exhibit considerable scatter.

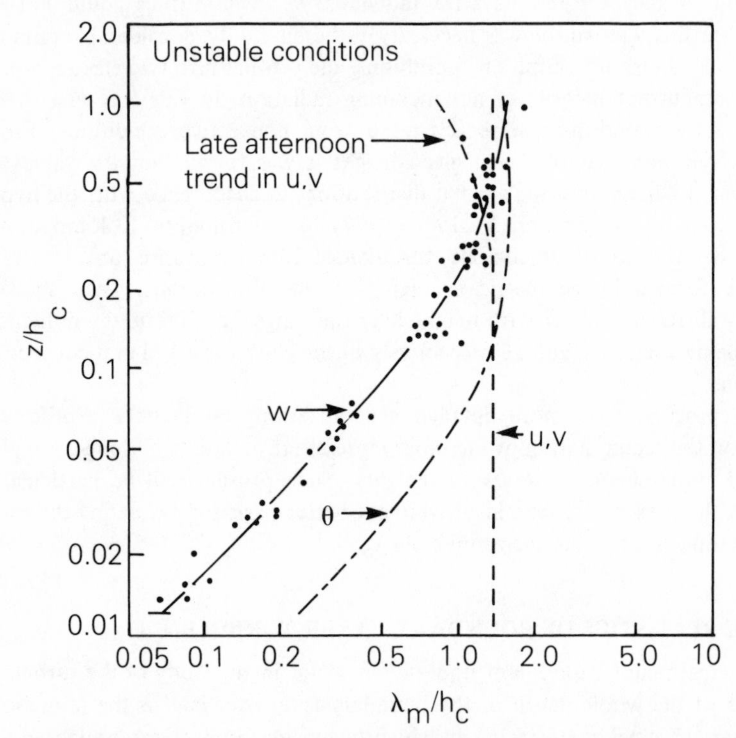

Fig. 2.13 – The behaviour of λ_m/h with z/h during the Minnesota Experiment λ_m is the wavelength of the peak in the $\kappa F(\kappa)$ vertical velocity spectrum (Kaimal *et al.*, 1976).

In stable conditions determination of λ_m is made more difficult by the presence of gravity waves, as mentioned earlier in the general description, and this contamination may be present in the Minnesota data as illustrated in Figure 2.14.

Empirical equations representing these data in unstable conditions are as follows:

$$\lambda_{m_w} = \begin{cases} 6z/(3 - 2z/|L|) & \text{for } z < |L| \\ 5.9z & \text{for } z < 0.1h \\ 1.8z_i\left(1 - \exp\left(-\dfrac{4z}{h}\right) - 0.0003\exp\left(\dfrac{8z}{h}\right)\right) & \text{for } 0.1h < z < h \end{cases} \tag{2.87a}$$

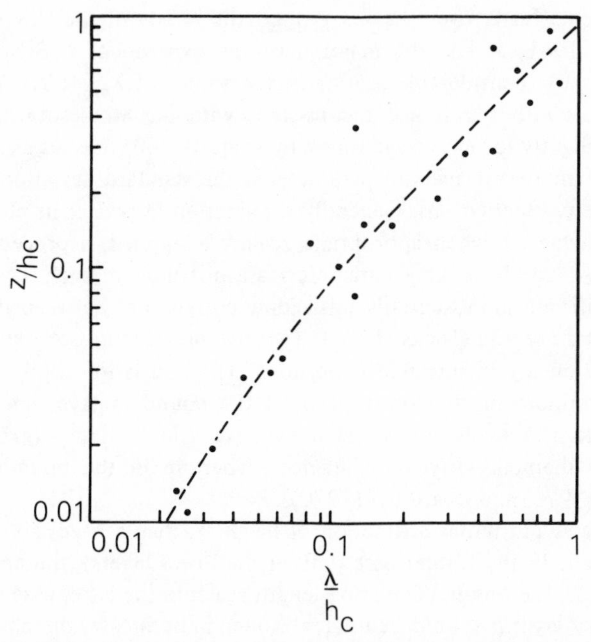

Fig. 2.14 – Minnesota data on the length-scale λ_m in stable conditions (Caughey *et al.*, 1979).

On the stable side:

$$\lambda_{m\,w} = \begin{cases} z/(0.5 + z/L) & \text{for } z \leqslant L/2 \text{ (Wamser and Muller, 1977)} \\ z & \text{for } L/2 \leqslant z \leqslant h \end{cases} \tag{2.87b}$$

The other components, u and v, have values of λ_m which are much more invariant with height. In unstable conditions $\lambda_m(u,\ v) = 1.3h$ according to the Minnesota results. In stable conditions on the other hand (with possible contamination from non-dispersive undulations)

$$\lambda_{m_u} \cong 2h\left(\frac{z}{h}\right)^{\frac{1}{2}} \tag{2.88a}$$

$$\lambda_{m_v} \cong 0.7h\left(\frac{z}{h}\right)^{\frac{1}{2}} \tag{2.88b}$$

fit the data reasonably well above about $0.02h$, although a lot of scatter is evident, due perhaps to the influence of the ratio h/L which is a relevant scaling parameter. These stable values apply only to very level terrain; in the presence of slopes or hills intermittent down-slope motions of relatively cold air will have significant influence. In neutral conditions the Kansas data (lowest 20 m) show

$$\lambda_{m_v} \cong 5z, \quad \lambda_{m_u} \cong 15z \tag{2.88c}$$

Measurements made over the sea around the coasts of the U.K. (Nicholls and Readings, 1981) and in the major maritime experiment JASIN (Nicholls *et al.*, 1982) show considerable scatter in the values of $\lambda_{m_w}(z/z_i)$. While the variation of λ_m with z/h is not inconsistent with the Minnesota data in an average sense, slightly lower average values by some 10–40% are suggested.

The other important spectral parameter is the standard deviation σ of the turbulent velocity fluctuations. Generally a reduction in scatter in plots of the data may be achieved when appropriate scaling is applied, a procedure fully acceptable provided the scaling parameters are not used in such a way as to introduce significant and essentially misleading correlations between the resulting ordinate and abscissa (Hicks, 1978). Two scaling velocities contend for the standard deviation σ_w in unstable conditions. The first is the friction velocity u_*, most appropriate in the lower parts of the boundary layer and in near-neutral conditions; the second is the convective velocity $w_* = (gHh/\rho c_p T)^{\frac{1}{3}}$ appropriate to thermally-driven turbulence higher up in the boundary layer (see Tennekes (1970) and Deardorff (1970)).

If the data is plotted as a function of height z, then z needs to be correspondingly scaled. In the former case (within the lower layers), the appropriate scaling for z is L, the Monin–Obukhov length scale. In the latter case the depth of the boundary layer h should be used, although in both cases there is a danger of introducing the false correlations mentioned above. Fig. 2.15 shows the two plots. The data all come from the Minnesota Experiment. Interestingly the scatter in plot (a) is not greatly worse at large z than that in plot (b) where the scaling should be most effective. The degree of scatter can be reduced further by plotting σ_w/u_* against both z/L and z/h, (or z/h and h/L), according to Smith and Blackall (1979). The lines drawn on Fig. 2.15 are close fits to the data and have the following analytic forms:

$$\frac{\sigma_w}{u_*} = 1.2\left[1 + 20\frac{z}{|L|} + 0.8\left(\frac{z}{|L|}\right)^2\right]^{\frac{1}{6}} \tag{2.89a}$$

$$\frac{\sigma_w}{w_*} = 0.745\left(1 + 0.255\left(\frac{z}{h}\right)^{-\frac{1}{2}}\right)^{-\frac{2}{3}} \tag{2.89b}$$

satisfying the prediction that σ_w behaves like $z^{\frac{1}{3}}$ in the free convection layer. This prediction follows from the hypothesis that the only relevant scaling parameters are z and H. Note that the plots do not include any of the Ashchurch data concentrated just below the capping inversion which seem to indicate a reduction in σ_w as z approaches h (see Caughey and Palmer, 1979).

In the surface stress layer, equation (2.89a) seems numerically consistent with the data of Wyngaard, Coté and Izumi (1971) and with the equation due to Panofsky *et al.* (1977)

$$\frac{\sigma_w}{u_*} = 1.3\left(1 + 3\frac{z}{|L|}\right)^{\frac{1}{3}} \qquad \text{for } z < 6\,|L| \tag{2.90}$$

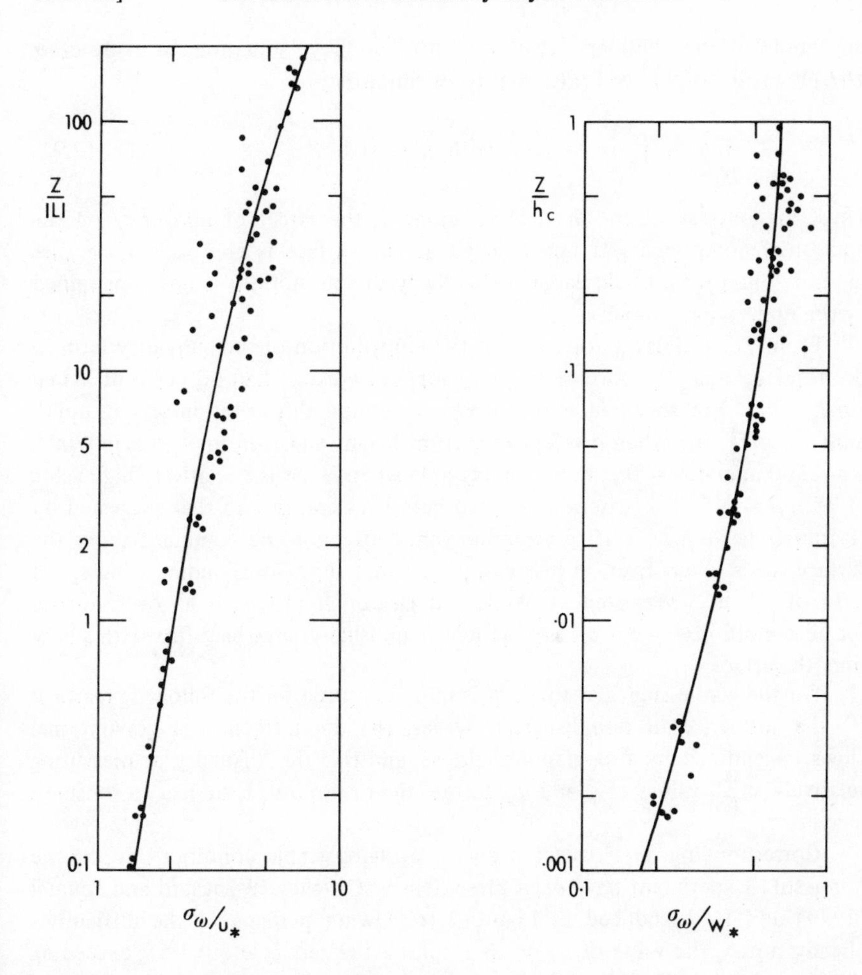

Fig. 2.15 – Minnesota data on the vertical velocity standard deviation σ_w plotted in two different ways; the first (a) is in terms of Monin–Obukhov scaling, the second (b) in terms of convective layer scaling.

For the u and v components, σ_u and σ_v are numerically indistinguishable. Panofsky *et al.* (1977), using surface layer data, concluded that

$$\sigma_{u,v} = u_* \left(12 + \frac{h}{2|L|}\right)^{\frac{1}{3}} \tag{2.90a}$$

provides a good fit. The Minnesota data are given by Caughey and Palmer (1979) and are also consistent with $\sigma_{u,v}$ being invariant with height throughout the whole of the boundary layer and with Panofsky's equation. In many cases $h/|L|$ is sufficiently large for (2.90a) to be simplified to

$$\frac{\sigma_{u,v}}{w_*} = \left(\frac{k}{2}\right)^{\frac{1}{3}} = 0.58 \text{ with } k = 0.4 \tag{2.91}$$

Thus, in contrast to the vertical component, the effect of buoyancy on the horizontal components is not reduced as the surface is approached, because the horizontal scale of the large convectively driven motions is not constrained by the underlying ground.

From Eq. (2.90a) it follows that the contribution of the buoyancy term to the total σ_v near the surface is dominant, say greater than 90 per cent, when $-h/L \geqslant 65$. For specified h and surface roughness this criterion sets an upper limit to $u/H^{\frac{1}{3}}$, 0.5 when u refers to a 10 m height and is in ms^{-1}, H is in Wm^{-2}, $h = 1500$ m and $z_0 = 0.2$ m. So for example when H has the relatively high value of 350, $u \leqslant 3.5$. The criterion adopted here is in contrast to that suggested by Deardorff i.e. $-h/L \geqslant 10$ for the buoyancy effects to be dominant above the surface-stress layer. Even so it is easily seen that the corresponding wind speed limit of 12 ms^{-1} suggested by Willis and Deardorff (1976) is an overestimate for all conditions except a combination of unusually large heat flux with a very smooth surface.

On the stable side the situation is more confused for the following reasons: (i) the influence of non-dispersive waves, (ii) the influence of gravitational flows engendered by topographical slopes, and (iii) the difficulty of measuring relatively small values of σ and u_*, so that their ratio σ/u_* is subject to considerable error.

Corresponding data collected during evolving stable conditions during the Minnesota Experiment have been presented by Caughey, Wyngaard and Kaimal (1979) and is reproduced in Figure 2.16. Owing perhaps to the difficulties already noted, the value of σ_w/u_* as z approaches zero is about 1.55, exceeding the expected 1.3. Overall a fair fit to the data (but forcing a fit with a value of 1.3 at $z = 0$) is

$$\frac{\sigma_w}{u_*} = 1.3\left(1 - \frac{z}{h}\right)^{\frac{1}{2}} \tag{2.92}$$

where h is the height of the stable boundary layer.

Caughey *et al.* (1979) also present data on σ_u and σ_v in stable conditions. An empirical fit to their data is given by

$$\sigma_{u,v}^2 = \begin{cases} 6\,u_*^2\left(1 - 3\dfrac{z}{h} + 2\left(\dfrac{z}{h}\right)^2\right) & \text{for } z < 0.2h \\ 3.75\,u_*^2\left(1 - \dfrac{z}{h}\right) & \text{for } 0.2h < z < h \end{cases} \tag{2.93}$$

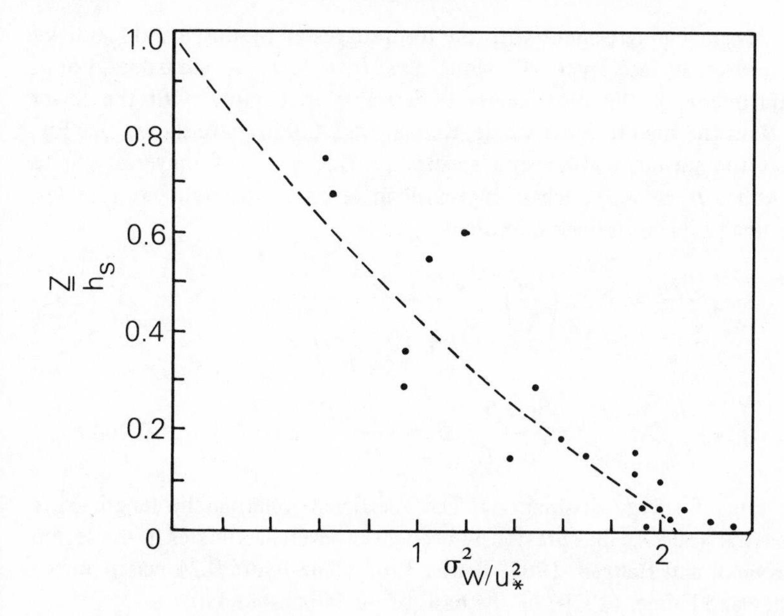

Fig. 2.16 – Minnesota data on σ_w in stable conditions. (Caughey *et al.*, 1979).

Many analytical forms for the spectra of the turbulent velocity components have been suggested. Ideally these forms should satisfy the following criteria:

(i) conformation with the inertial subrange form $F(n) = B\,(\lambda_m n/u)^{-5/3}$ on the high-frequency side of the peak u/λ_m; in unstable conditions B is about 0.48 for w and v, and about 0.36 for u; in stable conditions B is about 0.36 for all the components;
(ii) the area under the analytical spectrum should equal the variance σ^2 of the velocity component;
(iii) the $nF(n)$ should peak at $n = u/\lambda_m$;
(iv) the height of the peak spectral density should agree with the observed values (including their variation with height);
(v) the behaviour on the low frequency side should fit the observed behaviour reasonably well (e.g. the energy density at $n = 0.1\,n_m$ should be approximately 0.35 of the peak density) as this is important in diffusion applications.

A spectral form based on the Minnesota data in unstable conditions and virtually satisfying the first four of the criteria above is

$$\frac{nF(n)}{\sigma^2} = \frac{0.535\,N}{1 + 1.1\,N^{5/3}} \tag{2.94}$$

where $N = \dfrac{\lambda_m n}{u}$

This is in reasonable agreement with the form suggested by Busch and Panofsky (1968) based on surface layer data alone. Best fit is to the w and v data. For u, some adjustment of the coefficients is desirable to comply with the lower value of B on the high frequency side. Kaimal *et al.* (1976) have shown (see Fig. 2.17) that the unstable Minnesota spectra provide a set of universal curves varying with z/h, or λ_m/h when expressed in terms of different coordinates. Transforming (2.94) into these coordinates, we get:

$$\frac{nF(n)}{w_*^2\,\psi^{2/3}} = 0.22\left(\frac{\lambda_m}{h}\right)^{5/3} \frac{f}{1 + 1.1\left(\dfrac{\lambda_m f}{h}\right)^{5/3}} \tag{2.95}$$

where $\quad f = \dfrac{h}{\lambda_m} N,\ w_*^3 = \dfrac{gH_0 h}{\rho c_p T},\ \psi = \dfrac{\epsilon\,\rho c_p T}{gH_0}$ and $\epsilon = \sigma_w{}^3/0.264\,\lambda_m$

(ϵ is the rate of energy dissipation). The coefficient relating the length σ_w^3/ϵ to the spectral scale λ_m has previously been given several estimates in the region of 0.3 (Kaimal and Haugen, 1967, Hanna 1968). The figure 0.26 was reported by Caughey and Palmer (1979) on the basis of the Minnesota data.

The separation of the curves in Fig. 2.17 can be explained by the variation of λ_m with height described earlier in this section. However, equations (2.94) and (2.95) do not satisfy the last of the five criteria, and a better fit overall (except perhaps at high frequency) is obtained by the form suggested by Pasquill and Butler (1964).

$$\frac{nF(n)}{\sigma^2} = \frac{N}{(1 + 1.5N)^{5/3}} \tag{2.96}$$

which can be transformed to

$$\frac{nF(n)}{w_*^2\,\psi^{2/3}} = 0.41\left(\frac{\lambda_m}{h}\right)^{5/3} \frac{f}{\left(1 + 1.5\,\dfrac{\lambda_m f}{h}\right)^{5/3}} \tag{2.97}$$

These latter forms imply that the integral length-scale l_E (defined as $\frac{1}{4}(\lim_{n\to 0} F(n))$) is equal to 0.25 λ_m, in contrast to 0.134 λ_m from Eq. (2.94). The spectral data on the w-component give values of l_E/λ_m scattered widely over the range 0.1–0.7, providing slightly better support perhaps for Eq. (2.96) as regards the low-frequency part of the spectrum.

On the stable side Kaimal (1973) suggests an analytical form for the spectra differing from (2.94) only in the magnitude of the coefficients

$$\frac{nF(n)}{\sigma^2} = \frac{0.63N}{1 + 1.55N^{5/3}} \tag{2.98}$$

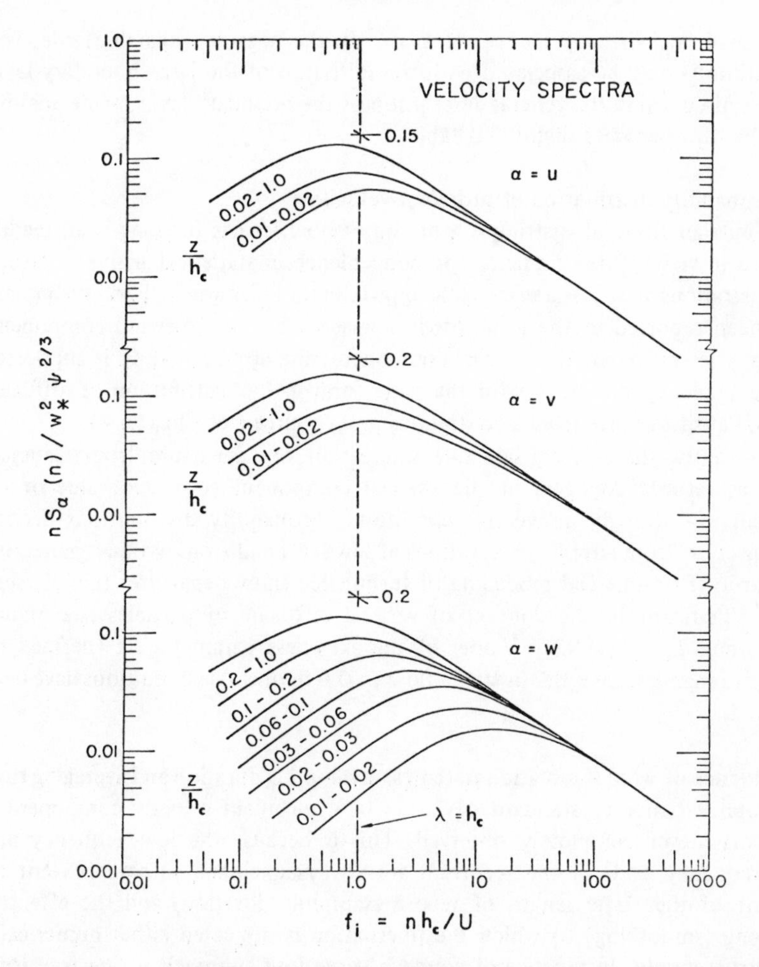

Fig. 2.17 – Universal curves for the velocity spectra expressed in mixed-layer similarity coordinates (Kaimal *et al.*, 1976).

The coefficients are to some degree dependent on the frequency bands over which the variance σ^2 has been computed, particularly the lower frequency limit in the 'spectral gap' region (see later in this section) and the upper limit set by the data sampling rate. Consequently it is not clear whether the differences between the quoted coefficients in stable and unstable conditions are real or not. However Kaimal *et al.* (1972) observed an interesting property in the u and v low-level spectra derived from the Kansas data which appears to be real. With frequency scaled in terms of u/z the spectra showed a discontinuity across neutral stability when plotted for specified values of z/L ranging from $+2.0$ to -2.0. Both sets of spectra showed a rather sudden increase in λ_m and low-frequency energy as z/L decreased from small positive to small negative values, with λ_m virtually independent of z on the negative (unstable) side. This discontinuity may be associated with the initiation of the large boundary-layer vortices discussed in the general description of the boundary layer, made possible once the flow becomes slightly unstable.

The probability distribution of turbulent velocities

The Gaussian form of distribution of eddy velocities has for long been readily adopted in view of its simplicity and convenience in statistical analyses. Several demonstrations that it is a reasonable approximation for atmospheric turbulence have been reported in the past. Also, in respect of the crosswind component, and to a lesser extent the vertical component, the approximation is supported by the tendency on average for the same form in the distribution of diffusing material at short range from a continuous point source (see Chapter 4).

Currently there is considerable interest in the departure from Gaussian form, as regards skewness of the vertical component (non-zero value of $\overline{w^3}$) especially in strongly convective conditions. Probability distributions derived by Lenschow from aircraft observations at low level and from w-values generated in Deardorff's numerical modelling of turbulence show departures that are seen to be significant in the context of vertical diffusion of particles (see Hanna (1982) and Lamb (1982)). Values of the skewness parameter Sk (defined by $\overline{w^3}/\sigma_w{}^3$) ranging from -0.2 in stable flow to 0.6 in unstable conditions have been reported.

Dependence of wind fluctuation statistics on sampling duration and averaging time

The total variance or standard deviation of a turbulent velocity component is in theory never completely observed. This is because the low-frequency and high-frequency ends of the spectrum are always excluded to some extent on account of the finite length of record (sampling duration) and the effective averaging (smoothing) to which the fluctuation is subjected either numerically or instrumentally. In practice of course a very close approach to the true total may be achieved if the sampling duration τ and averaging times are respectively long enough and short enough in comparison with the characteristic time-scale of the spectrum (as represented say by $1/n_m$). Except for the vertical component very near the ground when the fluctuations are concentrated at relatively high frequency, the second of the foregoing requirements is not difficult to achieve, though it may be necessary to use instruments with better response than those used in routine wind measurements. Adequate sampling duration may however be difficult to realize within the period over which conditions remain steady, especially for the horizontal components and for the vertical component at greater heights, since in both cases the time-scales are relatively large.

Assuming that the averaging time is virtually zero the effect of sampling duration is as in Eq. (2.24). If the sampling duration is so short that the contributions to the variance are confined to the high-frequency section of the spectrum, over which a simple power-law fall-off of spectral density applies, the result may be expressed in a simple form as follows. Representing the normalized spectrum by

$$F(n) = An^{-p}, \quad n > n_0 \tag{2.99}$$

we may substitute this for the *whole* spectrum in Eq. (2.24) provided $\tau < 0.1/n_0$ say. Hence

$$\begin{aligned}
\frac{\sigma^2_{\tau,0}}{\sigma^2_{\infty,0}} &= \int_0^\infty F(n)\left(1 - \frac{\sin^2 x}{x^2}\right) \mathrm{d}n \\
&= \int_0^\infty An^{-p}\left(1 - \frac{\sin^2 x}{x^2}\right) \mathrm{d}n, \quad \text{if } n_0 < \frac{0.1}{\tau} \\
&= A(\pi\tau)^{p-1} \int_0^\infty x^{-p}\left(1 - \frac{\sin^2 x}{x^2}\right) \mathrm{d}x
\end{aligned} \tag{2.100}$$

where $x = \pi n \tau$. The integral is not dependent on τ *per se* and provided convergence of this integral and of $\int F(n)\,\mathrm{d}n$ are satisfied, which requires $1 < p < 3$, it follows that

$$\sigma_{\tau,0} \propto \tau^{(p-1)/2} \tag{2.101}$$

It has previously been noted that spectra of atmospheric turbulence typically have a high-frequency section with a power-law variation as in Eq. (2.99), with p within the range specified above. More particularly, in the inertial subrange $p = \frac{5}{3}$ and then $\sigma_{\tau,0}$ increases as the cube-root of the sampling duration.

For spectra of the overall shape described in the previous section, in which the rise of $F(n)$ with decreasing frequency ultimately falls short of the simple power-law variation, it is obvious that the growth of σ must ultimately be slower than $\tau^{(p-1)/2}$, and hence slower than $\tau^{1/3}$ if inertial subrange conditions are assumed. The actual overall variation of σ of course depends on the precise shape of the spectrum. The only complete form of spectrum for which an analytical solution of Eq. (2.24) has been provided is that corresponding to an exponential longitudinal correlogram. For the transverse form of frequency spectrum corresponding to Eq. (2.70), the result (see Smith, 1961) is

$$\frac{\sigma^2_{\tau,0}}{\sigma^2_{\infty,0}} = 1 - \frac{2t_s}{\tau}\left[1 - \exp\left(-\frac{\tau}{2t_s}\right)\right] \tag{2.102}$$

in which t_s is the time-scale of the transverse spectrum, i.e. one-half the time-scale of the longitudinal spectrum or correlogram. This form of spectrum does not fit observed spectra very well. The equations which provide a better fit, Eqs. (2.94) and (2.96), do not permit analytic integration of Eq. (2.24) but results are available from graphical integration, using Eq. (2.96), and values of $\sigma_{\tau,0}/\sigma_{\infty,0}$ are shown in Table 2.IV with corresponding values from Eq. (2.102). The two forms of spectra give the same value of $\sigma_{\tau,0}/\sigma_{\infty,0}$ (0.76) at $n_m\tau \cong 0.5$, which is essentially a consequence of the nearly asymmetrical shape of the $nF(n)$ versus ln n forms and the placing of the low-frequency cut-off near $n = 0.44/\tau$ (see Fig. 2.3). The transverse spectrum corresponding to an exponential correlogram gives only slightly higher $\sigma_{\tau,0}/\sigma_{\infty,0}$ than Eq. (2.96) for $n_m\tau = 0.5$, but as $n_m\tau$ decreases from 0.5 there is a substantial fall away from the Eq. (2.102) values as a consequence of the different high-frequency forms, which behave as n^{-2} and $n^{-5/3}$ for spectra as in Eq. (2.70) and Eq. (2.96) respectively. This means that for short sampling times, when Eq. (2.101) tends to apply, $\sigma_{\tau,0}$ tends to behave as $\tau^{\frac{1}{2}}$ and $\tau^{\frac{1}{3}}$ respectively.

Table 2.IV – Effect of sampling duration τ on the standard deviation σ of a fluctuating property.

(a) $\sigma_\tau/\sigma_\infty$ *as a function of* $n_m\tau$

$n_m\tau$	0.025	0.05	0.1	0.25	0.5	1.0	2.5
Calculated $\sigma_\tau/\sigma_\infty$ for spectrum as in:							
Eq. (2.70)	0.228	0.317	0.433	0.622	0.764	0.878	0.952
Eq. (2.96)	0.36	0.44	0.53	0.67	0.76	0.85	0.92

(b) $\sigma_{u\tau}/\sigma_{1000\ \mathrm{m}}$ *as a function of* $u\tau$

$u\tau(m)$	10	30	100	300	1000
$\sigma_{u\tau}/\sigma_{1000\ \mathrm{m}}$					
from θ data in: Fig. 2.18	0.42	0.59	0.74	0.88	1.00
Eq. (2.70)	0.26	0.43	0.69	0.90	1.00
Eq. (2.96)	0.41	0.56	0.75	0.90	1.00
with $u/n_m = \lambda_m = 330$ m					

Notes: (i) Infinitesimal averaging time is assumed.
(ii) Eq. (2.70) is the transverse form of spectrum corresponding to a longitudinal exponential correlogram, Eq. (2.96) is an empirical form satisfying data and conditions as discussed earlier.
(iii) The value of 330 m for λ_m was obtained by fitting the calculated values from Eq. (2.96) to the observed value of $\sigma_{1000}/\sigma_{100}$.
(iv) Although a better fit with Eq. (2.70) could be achieved by adopting a larger value of λ_m this would be rather pointless as the spectral shapes observed do not fit that equation as well as Eq. (2.96).

The variation of $\sigma_{\tau,0}$ with sampling duration is of particular interest in the case of the v-component (i.e. the wind direction θ fluctuation) since this immediately determines the initial crosswind spread from a continuous point source of diffusing material (see 4.4 and 6.3). Systematic recording of approximate values of σ_θ (σ_v/u) may be conveniently achieved with the electrical filter technique described in 2.1. An analysis of about 1000 hr of such records, for a height of 16 m over open downland at Porton, England, has been made by Smith and Abbott (1961). The basic data consisted of hourly averages of the standard deviations of wind direction for the standard sampling durations of 5, 30 and 180 sec. For near-neutral conditions, as specified by

$$-0.01 \leqslant \Delta T_{7.1-1.2}/\bar{u}^2 \leqslant +0.01$$

where the numerator is the temperature in °F at a height of 7.1 m minus that at 1.2 m, and $\bar{u}$ is the mean wind speed (at 15.5 m) in m/sec, the variation of standard deviation σ_θ with wind speed and sampling duration, τ, is displayed in terms of the product $\bar{u}\tau$ in Fig. 2.18. The points are mean values for 1m/sec ranges of windspeed, the lower limits of the ranges being entered above the points, while the figures below the points represent the numbers of hourly averages included. In each case the standard deviation of the hourly averages is represented by the vertical line extending above or below the point. For wind speeds greater than 4 m/sec the groups of points of different sampling durations appear to follow a simple curvilinear relation with the length $\bar{u}\tau$. Such a relation would be consistent with a field of turbulence in which (a) the shape of the wavenumber spectrum is invariant with wind speed, (b) the total intensity of turbulence, $\sqrt{(\overline{v'^2})}/\bar{u}$, is independent of wind speed, and (c) the time-variation at a fixed point is statistically equivalent to the space-variation, with the transformation $x = \bar{u}t$.

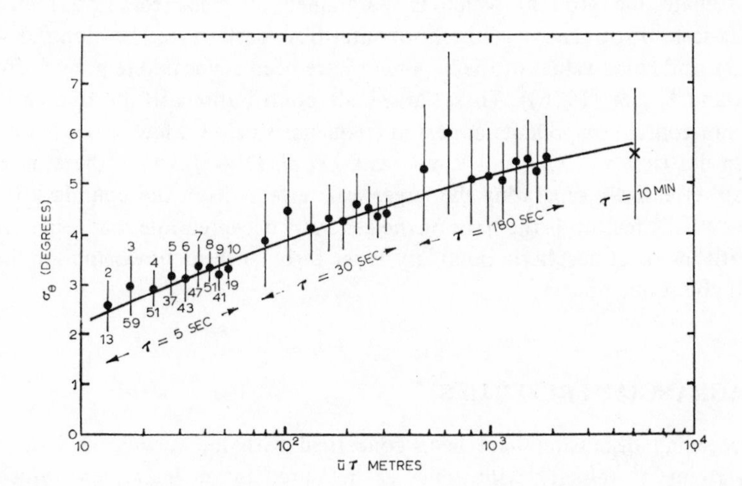

Fig. 2.18 – The standard deviation (σ_θ) of wind direction at a height of 16 m over open grassland, in neutral conditions of stability, as a function of sampling-duration τ and wind speed $\bar{u}$. • Porton, England. × O'Neill, Nebraska.

The single point for a sampling duration of 10 min was obtained from the observations at a height of 12 m at O'Neill (Lettau and Davidson, 1957, Vol. 2, Tables 5.2a), and is an average of eighteen individual values for similar near-neutral conditions. This value conforms reasonably closely to the trend of the Porton data, suggesting that the curve drawn in can be regarded as an approach to a universal empirical relation between σ_θ and $u\tau$ for an open grassland site, for wind speeds between 4 and 10 m/sec and sampling durations up to 10 min.

As demonstrated at (b) in Table 2.IV the shape of the $\sigma_\theta(u\tau)$ relation is fitted by calculated values from the graphical integration of Eq. (2.24) with a spectrum form as in Eq. (2.96) and $\lambda_m = 330$ m. It has already been noted that the Kansas data (Kaimal *et al.*, 1972) show a very wide spread of λ_m for near-neutral conditions. On the stable side $\lambda_m = 5z$, i.e. 80 m for the height (16 m) at which the measurements in Fig. 2.18 were made. However, on the unstable side λ_m is roughly 1000 m irrespective of height, most probably reflecting the depth of the boundary layer. The ensemble value of 300 m indicated by the foregoing analysis of $\sigma_\theta(\tau)$ data in near-neutral conditions is therefore quite consistent with the range of values indicated by spectral data.

It should be emphasized that the foregoing results refer to sampling durations of only a few minutes, whereas much longer durations are of practical concern in assessments of pollution levels. The corresponding σ_θ data then reflect lower-frequency fluctuations which are not directly a consequence of the boundary-layer generation of turbulence. As a result the fairly decisive termination of σ_θ growth as in Fig. 2.18 and Table 2.IV will not generally be observed. A widely used 'rule of thumb' is $\sigma_\theta \propto \tau^{\frac{1}{2}}$ (see Slade, 1968), but even this may underestimate the growth, which is maintained by meso-scale and then by synoptic-scale disturbances of the horizontal flow pattern (see Section 7.5 and Fig. 7.9), and larger values of the exponent have been advocated (e.g. by Gifford (1975) and Moore (1976)). These large-scale contributions to the fluctuations of the horizontal components appear at frequencies below a few cycles per hour (see van der Hoven (1957) and Vinnichenko *et al.* (1965)) where there usually is a 'gap' effectively separating the larger-scale effects from the boundary layer processes. This feature is the basis of the generally accepted rule that representative statistics of atmospheric boundary layer turbulence require sampling durations of about one hour.

2.7 LAGRANGIAN PROPERTIES

The preceding discussion have been concerned with the turbulence indicated by variations of velocity with time, as measured by an instrument which is usually fixed, but which may also be carried on a moving platform. For both cases the measurements will refer to a continually changing sample of air. In general this is the only type of measurement which is feasible, and it is in such terms that the effects of diffusion must be described in practice. However, in a qualitative way, it is obvious that the diffusion of a cloud of airborne material will depend on the *development* of the turbulence affecting particular elements of the cloud as they are carried along. In other words interest centres on the variations of velocity (with time) which would be observed in following the motion of the elements – the system described as Lagrangian – and it will be seen later that formal expressions of diffusion all involve the Lagrangian correlogram or spectrum. Because of the continuity of diffusion processes, down the scale of motion to the ultimate molecular agitation, the view is sometimes expressed that the term *Lagrangian velocity* has no obvious meaning in the case of a fluid. except presumably when referred to a single molecule. In practice this difficulty is evaded by thinking either in terms of an element of fluid so small that its own diffusive spread is negligible compared with its translation under the action of the larger-scale turbulence, or in terms of a solid particle of negligible buoyancy. Whatever formal view is taken, it is evident that the observation of such velocities presents considerable difficulties.

A casual examination of the behaviour of balloons or small puffs of smoke in the lower atmosphere shows that there is often a striking difference between the rapidity of turbulent fluctuation evident in their motion and that which is evident in the customary measurement of wind direction at a fixed point. The parcel of air identified by the balloon or smoke puff appears often to preserve its direction of travel with little change over times substantially longer than those over which the 'fixed-point' measurements display appreciable variations of direction. On the 'frozen eddy' hypothesis, i.e. the assumption that the spatial pattern of turbulence is swept along virtually unchanged at the speed of the mean flow, it has already been noted that the integral time-scale of the fluctuations evident at a fixed point is $l/\bar{u}$, where l is the integral length-scale. On the same hypothesis, particles may be considered to move through the fixed eddy pattern at a speed which may be statistically represented by say σ_u, and so the Lagrangian integral time-scale may be expected to be related closely to l/σ_u. Accordingly, the ratio of the two time-scales, t_L/t_E, should be of the order of $\bar{u}/\sigma_u$, i.e. the reciprocal of the intensity of turbulence, which in the atmosphere is typically of order 10.

Detailed theoretical examination of the relation between Lagrangian and Eulerian fluctuations

Several attempts have been made to provide more sophisticated representations of the relation between Lagrangian and Eulerian fluctuations, specifically with reference to the ratio of the integral time-scales appropriate to the Lagrangian and 'fixed-point' systems of reference. Basically the problem is formidable, and useful results have been achieved only at the expense of rather intuitive hypotheses and broad assumptions, the precise implications and limitations of which are not always very clear. The more straightforward analyses depend on one or other of three main principles:

(a) the 'frozen eddy' hypothesis already referred to, which may be developed in more detail,
(b) the conjecture, first specifically stated by Corrsin (1959), that after sufficiently long migration times particles may be considered to have velocities which are unbiased samples of the turbulent velocities at their positions in an Eulerian reference frame,
(c) the recognition of the difference in Eulerian and Lagrangian spectral characteristics which is implied by the 'inertial subrange' concept.

The application of the conjecture referred to in (b) above can be most easily appreciated by considering the behaviour of particles released from the origin of a system of coordinates moving with the mean flow. After each particle has travelled for time ξ a number of them will be found to have reached a point defined by the displacement vector r, and their contribution $\Sigma^N u(t)u(t+\xi)$ to the Lagrangian autocovariance will be, in Eulerian space-time terms, $\Sigma^N u(0,t)$ $u(r, t+\xi)$. The particle velocities involved in this summation will all carry a bias appropriate to their common displacement from the origin, but the assumption is that this bias diminishes as ξ increases and may be neglected at sufficiently large ξ. In this case the velocities of the sub-ensemble of particles defined by common displacement are virtually statistically identical with the *total* velocity characteristics appropriate to position r and time ξ in the Eulerian space-time sense. Consequently we may write as an approximation for large ξ,

$$\Sigma^N u(t)u(t+\xi) \cong N\mathcal{R}(r, \xi) \tag{2.103}$$

or integrating over all particles, i.e. over all positions r,

$$R(\xi) \cong \int_0^\infty \mathcal{R}(r, \xi)P(r, \xi)\,dr \tag{2.104}$$

where P is the probability density function describing the distribution of the particles and $\mathcal{R}(r, \xi)$ is the two-point two-time Eulerian correlation function.

The solution of Eq. (2.104) can be closed by assuming P to be Gaussian, with a variance which is related to $R(\xi)$ as in Eq. (3.59), and by adopting a tractable reasonable form for $\mathcal{R}(r, \xi)$. This process has been followed through in detail independently by Saffman (1963) and Philip (1967), assuming isotropic turbulence, to give t_L/t_E as a function of i. For small values of i the relations tend to $t_L/t_E \propto 1/i$, with numerical coefficients 0.8 (Saffman) and 0.35 (Philip). According to Philip the difference is a consequence of the different *transverse* forms which are implied by the respective analytical forms adopted for $\mathcal{R}(r, \xi)$.

The third approach was first applied by Corrsin (1963) with the assumption that the Eulerian and Lagrangian spectra may be *completely* represented by their inertial subrange forms. The original argument was in terms of the three-dimensional wavenumber spectrum. In terms of the one-dimensional frequency spectrum conventionally measured at a fixed point it may be restated as follows

$$\begin{aligned} S_E(n) &= Cu^{2/3}\epsilon^{2/3}n^{-5/3} \qquad & n \geqslant n_E \\ &= 0 & n < n_E \end{aligned} \tag{2.105}$$

$$\begin{aligned} S_L(n) &= B\epsilon n^{-2} \qquad & n \geqslant n_L \\ &= 0 & n < n_L \end{aligned} \tag{2.106}$$

where $\int_0^\infty S_E(n)\,dn = \sigma_E{}^2$ etc. Eq. (2.105) is already familiar. Eq. (2.106) follows from applying similar dimensional considerations to the *frequency* properties, which are the basic representation in the Lagrangian system, in contrast to *wavenumber* for the Eulerian system. Integration of the two forms for $S(n)$ gives $\sigma_E{}^2$ and $\sigma_L{}^2$, which are assumed equal, and elimination of ϵ then gives

$$\frac{n_E}{n_L} = \left(\frac{3}{2}\right)^{3/2} \frac{C^{3/2}}{B} \frac{1}{i} \tag{2.107}$$

in which n_E and n_L may be regarded as the inverse of the characteristic timescales.

The argument may be extended (Pasquill, 1968) to more realistic forms of spectra which have finite spectral density at zero frequency and reduce to the inertial subrange forms at large n. For example, the expression corresponding to Eq. (2.96)

$$nS_E(n) = \frac{\sigma_E{}^2 N_E}{(1+\frac{3}{2}N_E)^{5/3}}, \quad N_E = \frac{n_E}{n_{mE}} = 4t_E n_E \tag{2.108}$$

may be taken as a good representation of fixed-point spectra of the vertical component. A suggested corresponding form for the Lagrangian spectrum is

$$nS_L(n) = \frac{\sigma_L{}^2 N_L}{(1+N_L)^2}, \quad N_L = \frac{n_L}{n_{mL}} = 4t_L n_L \tag{2.109}$$

Equating the above to their inertial subrange forms at large n, eliminating ϵ, and taking $\sigma_E = \sigma_L$, it follows that

$$\frac{t_L}{t_E} = 2.76 \frac{C^{3/2}}{B} \frac{1}{i} \tag{2.110}$$

which differs from Corrsin's result only in the numerical coefficient.

The magnitude of the universal constant in the Eulerian inertial subrange form is now well established, but no estimates have yet been given for B. However, an estimate in the form of the quantity $C^{3/2}/B$ would be provided directly by simultaneous measurements of $S(n)$ in the two systems, at appropriate frequencies, since

$$\frac{C^{3/2}}{B} = \frac{[nS_E(n)]^{3/2}}{\bar{u}nS_L(n)} \tag{2.111}$$

A few such measurements are available from the work carried out by Angell (1964) using tetroons, to be discussed in more detail in this section. From average values of $nS(n)$ over the frequency range 0.2–2.0 cycles/sec, in which the tetroon data were not inconsistent with the n^{-2} law, the quantity in Eq. (2.111) has been estimated to be roughly 0.23 (see Pasquill (1968) for details). Using this figure the coefficient in the linear relation between t_L/t_E and $1/i$ is 0.4 in Eq. (2.107) and 0.6 in Eq. (2.110).

To sum up, the various theoretical estimates of the ratio $\beta = t_L/t_E$ all lead to the form βi = constant, the numerical values of the constant ranging from 0.35 to 0.8.

Direct measurement of Lagrangian fluctuations

Various workers have tried to approach a satisfactory measurement by observing the trajectories of *markers* floating more or less truly in the air. An early published account by Edinger (1952) contains an example of the form of the Lagrangian correlation coefiicient associated with small-scale turbulence. The technique used here was to release soap bubbles from a generator carried aloft by a captive balloon, and to record their motion by a high magnification cine-camera vertically below on the ground, the bubbles being identifiable in sunny conditions by the reflection of the sun from their surfaces. However, the techniques which have been most productive so far are those using horizontal balloon flights on a medium or large scale, and air trajectories computed from the synoptic charts used in meteorological practice.

Following a preliminary study (1953) with balloons, Gifford (1955) has reported further experiments which are particularly noteworthy in that they were accompanied by fixed-point measurements on a tower (at Brookhaven National Laboratory, Long Island, New York), so as to provide a direct comparison between Lagrangian and Eulerian time-spectra. The principle of using so-called *neutral* balloons, i.e. balloons adjusted in weight so that they truly float, is well known, as is also the difficulty of achieving this condition in practice. In this case the balloons were first inflated to an approximately neutral state, and brought to final equilibrium by a length of string tied to the neck. This adjustment was carried out in the tower elevator, from which the balloons were then released, readings being taken at 10-sec intervals and used to compute average vertical velocities over such intervals. The resulting serial values of average vertical component were then used to compute spectra by the technique advocated by Tukey, over frequencies from about 5 to just over 100 cycles/hr.

Despite the usual scatter in the spectral estimates, Gifford's results contain the general impression that the Lagrangian spectra are broadly similar in shape to the Eulerian spectra, but displaced from them toward lower frequencies. The results did not include any estimates of integral time-scales, but estimates were made of the frequencies n_{mE} and n_{mL} at which the smoothed spectra showed maxima in $nS(n)$. As noted in 2.4 this frequency is inversely related to the integral time-scale, with a coefficient which depends on the shape of the spectrum. If the Lagrangian and Eulerian spectral shapes are not very different the ratio n_{mE}/n_{mL} may be taken as an approximation to t_L/t_E. Gifford's results provide four estimates of this ratio, ranging from 1.7 to 4.0 and averaging approximately 3.

Extension of Gifford's pioneering measurements has been provided by Angell (1964) in a programme of measurements at Cardington in England and by Hanna's (1981a) comprehensive study in the daytime mixed layer over the plains east of Boulder, Colorado. Angell used *tetroons* to give quasi-Lagrangian observations and the tethered balloon system described by Jones and Butler (1958) to give 'fixed-point' observations at the same level for comparison. Tetroons are tetrahedral shaped inextensible balloons, which are filled to excess pressure and therefore maintain constant volume. The balloon flight is accordingly maintained at approximately constant air density and hence at approximately constant level (see Angell and Pack (1960) for further details). In the Cardington measurements the tetroons were filled to fly at about 700 m above ground and were tracked by radar over distances ranging from 6 to 26 km. From each trajectory vertical velocities were deduced, in the form of averages over 1-min periods, and used to compute spectra by the Tukey technique referred to in 2.1. Corresponding vertical velocities and spectra were deduced from the wind inclination measurements made at the same time from the tethered balloon. Hanna used both neutral balloons and tetroons, the former to provide data for all three components of turbulence, and the latter for the v-component only, the Eulerian (fixed point) data being obtained with tower-based instruments.

Angell's 'Lagrangian' and 'fixed point' spectra show the same broad features as Gifford's earlier observations at a lower level. Again the final comparison was made in terms of the frequency for peak $nS(n)$. In view of the irregularities in the spectra reproduced in Angell's paper it is obvious that the estimates of n_{mE} and n_{mL} have some subjective element. The resulting values of the frequency ratio are shown plotted against $1/i$ in Fig. 2.19. There is considerable scatter but nevertheless a discernible trend for increase with $1/i$. Lines corresponding to the extreme range of the theoretical linear relations are also shown and it is seen that these encompass roughly half the data points. Hanna's more numerous estimates of β show even more scatter than that in Fig. 2.19, with the major cluster of points referring to values of $1/i$ less than 5 and the best fitting linear relation $\beta i = 0.68$ (with i referring to the particular component), though the few cases at large $1/i$ are fitted better by 0.4. An overall summary of Hanna's t_L,

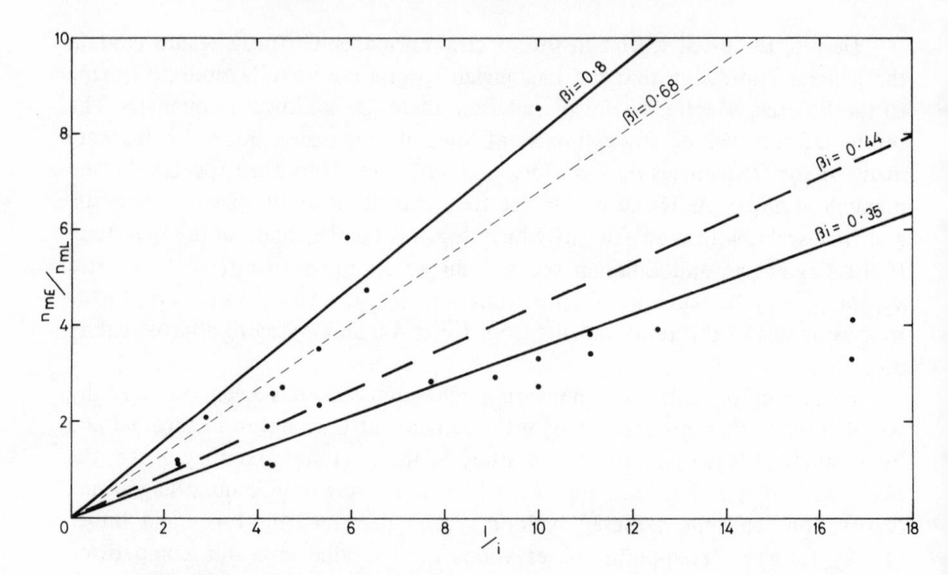

Fig. 2.19 – Estimates of the Lagrangian/Eulerian scale ratio from measurements of Lagrangian velocity fluctuations and from theory.

The individual points are values of n_{mE}/n_{mL} ($\approxeq \beta$) from 'tetroon' and 'fixed point' spectra of the vertical component (after Angell, 1964). The fine dashed line ($\beta i = 0.68$) is Hanna's (1981a) fit to further tetroon and neutral balloon measurements, including u and v components as well as w.

Lines labelled 0.35 and 0.8 represent the extremes of the theoretical relations referred to in the text.

$\beta i = 0.44$ is an early estimate based on assumed identity of the Monin–Obukhov and Taylor statistical forms of K, and early estimates of l_E/z in the neutral surface stress layer. This has been revised to $\beta i \approxeq 0.6$ (see Table 3.I). $\beta i = 0.68$ is supported by Hunt and Weber's (1979) statistical treatment of vertical spread in the neutral surface stress layer (see 3.5).

t_E and β is given in Table 2.V, and this shows encouraging agreement in the average values for the three components as well as demonstrating consistency in estimates obtained from the correlograms and the spectra.

The observations and theoretical consideration summarized above provide a fairly convincing specification of the general magnitude of the Lagrangian/Eulerian scale ratio, at least for levels well clear of the ground, but the detail of the variability of the ratio has yet to be satisfactorily explained. Certainly this detail does not appear to be completely represented by the predicted inverse variation with the intensity of turbulence, and the extent to which this is a reflection of imperfections in the measurements and their interpretations, or of limitations in the theoretical analyses, is not immediately clear, though the continuing attempts at improved determination of β from the tetroon technique show no sign of a reduction in the variability.

Table 2.V – Average values of t_L, t_E (sec) and β from Hanna's (1981a) study in the daytime mixed layer, (a) from autocorrelograms, taking the time at which $R = 1/e$, (b) from spectra taking the time-scales as one-sixth of the peak-periods of $nF(n)$

	From correlograms			From spectra		
	Balloon t_L	Tower t_E	β	Balloon t_L	Tower t_E	β
Pibal u	67	45	1.5	90	42	2.1
Pibal v	78	54	1.4	87	43	2.0
Pibal w	82	52	1.6	80	57	1.4
Tetroon v	88	50	1.8	62	35	1.8
Average	79	50	1.6	80	44	1.8

Individual balloons tracked over times in the range 10–50 min.

Several laboratory studies of Lagrangian properties have been reported – the latest being that by Snyder and Lumley (1971), who briefly review earlier laboratory work and then describe a wind tunnel study using four different types of particle, ranging from hollow glass spheres with terminal velocities of 1.7 cm/sec to copper spheres with terminal velocities of 48 cm/sec. The response of the particles to a grid-generated turbulence was observed photographically and corrections applied for the downstream decay of the turbulence in order to arrive at velocity correlations appropriate to stationary homogeneous turbulence. Snyder and Lumley's results show that the integral time-scale for the particle velocities falls off with particle time-constant or terminal velocity, with a fairly clear indication of an asymptote to $\beta = 1/i$ at very small terminal velocity. Thus the value of βi is somewhat higher than those summarized above from either the theoretical analyses or the atmospheric observations, and this is a discrepancy remaining to be explained. On the other hand a simplifying feature of their results is the striking demonstration of similarity in shape of the particle auto-correlation and the Eulerian spatial correlation, despite the dimensional implication of a difference in the spectral shapes, as in Eqs. (2.105) and (2.106).

Further discussion of this topic will be found in Chapters 3 and 4, where the additional implications of observations of the spread of particles are considered, though it must be admitted that at present these seem only to add to the irregularities already noted.

3

Theoretical treatments of the diffusion of material

Theoretical analyses of the diffusion of material in turbulent flow have developed along three main lines: the *gradient transfer* approach, the *statistical theory* of turbulent velocity fluctuations, and *similarity* considerations. In the first of these a particular physical model of mixing is implied, for which the classical background and mathematical development up to about 1950 have been fully surveyed by Sutton (1953). The statistical theory is essentially a kinematic approach in which the behaviour of marked elements of the turbulent fluid is described in terms of given statistical properties of the motion, though as already seen in Chapter 2 there are important physical problems arising in the relation between the directly relevant Lagrangian fluctuations and the 'fixed-point' fluctuations of velocity which are commonly observed. Historically this treatment started almost as early as the transfer theory but the full emergence of its application came later. In similarity theory the controlling physical parameters are postulated and laws relating the diffusion to these parameters are then derived on a dimensional basis. Numerical rather than analytic processes are now being undertaken more freely, and this has made possible some new theoretical approaches, notably the use of higher-moment conservation equations so as to avoid the original form of gradient-transfer assumption, and extension of the Lagrangian statistical method by numerical simulation of particle-dispersion in random-walk models.

3.1 EDDY DIFFUSIVITY AND THE GRADIENT-TRANSFER APPROACH

In the gradient-transfer approach it is assumed that turbulence causes a net movement of material down the gradient of material concentration, at a rate which is proportional to the magnitude of the gradient. The proportionality factor is of course analogous to the coefficients of viscosity or conductivity in the familiar laws for the transfer of momentum or heat in laminar flow. Generally, we write

$$F_s = -A\frac{\partial \bar{S}}{\partial n} = -\bar{\rho}K\frac{\partial \bar{S}}{\partial n} \tag{3.1}$$

where F_s is the eddy flux, i.e. the rate of eddy transfer per unit area across a fixed surface, and $\partial\overline{S}/\partial n$ is the gradient of the property ($\overline{S}$ being the mean quantity per unit mass of air). The negative sign is consistent with *down-gradient* flux. This definition in terms of an austausch coefficient A or an eddy diffusivity K has already been introduced for the case of momentum transfer in 2.2. Turbulent transfer following such a law is referred to as a *simple diffusion process* (Sutton 1953).

Representing the instantaneous value in terms of a mean and an eddy fluctuation therefrom, as in Eq. (2.1), the overall vertical transport through a horizontal plane is

$$\overline{\rho w S} = \overline{\rho w}\overline{S} + \overline{\rho w' S'} \tag{3.2}$$

(upward velocities and transports being conventionally taken as positive). The first term, arising from any mean vertical motion, is customarily assumed to be zero at low heights over level uniform terrain, leaving only the eddy flux [F_s in Eq. (3.1)].

The simplest possible basis for Eq. (3.1) is the direct analogy with transfer of momentum or heat by molecular agitation, or with the diffusion of a solute in a still solvent, which was treated by Fick as following a law identical with that for the conduction of heat. This carries the implication that the diffusivity is a constant, and diffusion processes in which this is so are described as *Fickian*. The analogy with molecular action is given a mechanistic basis in the classical *mixing length* theory, in which a discrete mass of fluid is supposed to leave some level, bearing the mean property $\overline{S}$ at that level and retaining it for some characteristic vertical distance, before mixing and becoming once again indistinguishable from its *mean* surroundings. Accordingly, the quantity S' in Eq. (3.2) is $-l(\partial\overline{S}/\partial z)$, where l is the vertical distance travelled by the mass of fluid since it was last representative of the mean conditions at some level. Substitution in the eddy flux term of Eq. (3.2) gives an expression equivalent to Eq. (3.1), with

$$A_s/\rho \cong K_s \cong \overline{w'l} \tag{3.3}$$

Recalling that the kinetic theory of gases leads to the expression $\frac{1}{3}cd$ for the kinematic viscosity, where c is average molecular velocity and d the mean free path, it is seen that eddy transfer is now similarly represented, with w' and l the eddy counterparts of c and d.

There is of course a considerable element of vagueness in the whole idea and statistical representation of a mixing length, and the formulation in Eq. (3.3) may be considered equally acceptable simply on dimensional grounds, postulating that the kinematic eddy diffusivity, of dimensions L^2T^{-1}, must be determined by the product of a characteristic eddy velocity and a characteristic length scale.

The further classical development of the mixing length theory is concerned specifically with transfer of momentum, in which case S' becomes

u' say. The additional assumption of proportionality in the w' and u' fluctuations, and the convenient absorption of the proportionality coefficient in the already vague specification of l, leads to the well-known expression for the Reynolds stress

$$\frac{\tau}{\rho} \equiv -\overline{u'w'} \cong l^2 \left|\frac{d\bar{u}}{dz}\right| \frac{d\bar{u}}{dz} \qquad (3.4)$$

the separation of the $d\bar{u}/dz$ terms satisfying the requirement that τ changes sign with $d\bar{u}/dz$. Assumptions of suitable forms for l and of *constancy of* τ with height allow Eq. (3.4) to be integrated to give the form of the wind profile.

The familiar forms

$$l = kz \qquad (3.5)$$

$$l = \frac{k(d\bar{u}/dz)}{d^2\bar{u}/dz^2} \qquad (3.6)$$

both lead to the logarithmic wind profile [Eq. (2.51)] and implicitly to $K_M = ku_*z$ [Eq. (2.50)]. The justification for these results is now more satisfactorily provided either by the similarity argument of 2.2 or as a direct application of laboratory laws for the wind profile over a surface of specified aerodynamic roughness (see Calder 1949, for a full exposé of this latter approach). Also, the similarity approach contains allowance for the effects of thermal stratification, through the function of Eq. (2.46).

An independent derivation of the flux-gradient relation for momentum transfer has been claimed by Monin (1965) on the basis of the so-called Friedman–Keller equations. These are a family of equations which are obtained by suitably operating on the Navier–Stokes equations and the corresponding heat equations, to give the rate of change of the variances and covariances of the turbulent fluctuations of velocity and temperature (i.e. they are a more complete and detailed set of the sort of equations given by Calder or Lumley and Panofsky for the total turbulent energy). For stationary and neutral conditions, and with some further simplifications which include some unfamiliar semi-empirical relations, Monin is able to achieve second-order closure of the equations and thus derive expressions for the stress terms without adopting the first-order closure assumption in Eq. (3.1). Thus for the transfer of u-component momentum

$$-\overline{u'w'} = \frac{l\sigma_w^2}{c_2 b} \frac{d\bar{u}}{dz} \qquad (3.7)$$

which is of the same form as Eq. (3.1) and implies

$$K_M = \frac{l\sigma_w^2}{c_2 b} \qquad (3.8)$$

In the above $b^2 = \sigma_u^2 + \sigma_v^2 + \sigma_w^2$ [$= 2\bar{E}$ in the notation of Eq (2.58)], l is referred to as the scale of turbulence but at small z is evidently the same as the l in Eq. (3.5), and c_2 is a numerical constant to be determined empirically. This independent demonstration of the flux-gradient relation from the basic equations is an important step in rationalizing the concept of diffusivity. However, the significance of the apparent dependence of a *vertical* diffusivity on the *horizontal* components of turbulences is not clear. Again, although Monin's analysis is not restricted to the surface-stress layer the explicit interpretation of l outside this layer is not immediately obvious.

An alternative method of deriving K_M without necessarily imposing restriction to the constant stress layer is provided by the combination of turbulent energy balance and inertial subrange laws, as outlined in 2.5.

In the present context the usefulness of the foregoing specifications of K_M depends on the further step of assuming identity in the eddy diffusivities for momentum and suspended material. Yet another method of specifying the required diffusivity, which avoids identification with momentum transfer and is instead directly concerned with the kinematics of suspended particles, is provided by the statistical theory considered in detail later in this chapter. Anticipating the ultimate result, in *homogeneous* turbulence the spread of a cloud of particles tends to a limit (at long time of travel) which may be formally represented by a diffusivity. This diffusivity is the product of the variance of the eddy velocity and the Lagrangian integral time-scale. By simple rearrangement, and introducing the ratio of Lagrangian and Eulerian scales this can be reduced to an explicit version of the 'eddy velocity times length scale' formulation of K.

The foregoing main forms of K are collected in Table 3.I with explanatory notes and further details.

For convenience in obtaining analytical solutions of the equations of diffusion referred to in the next section K and $\bar{u}$ have to be taken either constant, which is obviously erroneous, or as tractable functions of height. The most widely used form is the simple power law system, $K \propto z^n$, $\bar{u} \propto z^m$, and in this case identification with K_m and $\bar{u}$ in the constant stress layer means that $n = 1 - m$, the relationship then being known as Schmidt's *conjugate power law*. Unfortunately these mutually consistent forms of K and $\bar{u}$ are not supported by any of the preceding arguments, and their use is justified only in so far as they can be argued to represent an acceptable approximation to the true variations [see Calder's (1949) and van Ulden's (1978) treatments summarized in 3.2 and Chaudhry and Meroney's (1973) treatment in 3.3]. More realistic forms of K and $\bar{u}$ present greater difficulty analytically. In particular, the ideal neutral form $\bar{u} \propto \ln z$ is analytically intractable except in some special considerations (see 3.3). More complex variations of K, which may not even be monotonic, make analytical solutions impossible or at best achievable only in some series form. Fortunately the development of numerical solution techniques, in either analogue or digital form, means that in principle virtually any form of K may be

Table 3.I – Explicit forms of K for vertical transfer

	$\dfrac{ku_* z}{\phi_M}$	$\dfrac{\epsilon}{(1-R_f)(d\bar{u}/dz)^2}$	$\sigma_w{}^2 t_L = \beta i_w \sigma_w l_E$	$\dfrac{\sigma_w{}^2 l}{c_2 b}$
Basis	Similarity theory	Turbulent kinetic energy (T.K.E.) balance	Statistical theory ($\beta = t_L/t_E$)	Friedman–Keller equations
Relevant equations in text	2.51, 2.53–2.56	2.53, 2.96	3.69, 2.9	2.53, 3.7
Conditions	Surface-stress region	Divergence of vertical flux of T.K.E. negligible	Homogeneous turbulence	Neutral flow Divergence of vertical flux of T.K.E. negligible Other empirical assumptions
Region of applicability	Near-surface layer	Any height subject to holding of zero divergence of T.K.E.	All heights except where sharply layered or sheared	All heights subject to holding of zero divergence of T.K.E.
Data and other assumptions required for practical application	Profile of mean wind and temperature	Profile of mean wind High frequency fluctuations of wind and temperature Kolomogorov law	Intensity and scale of vertical component of turbulence Lagrangian/Eulerian scale ratio	Total energy and scale of turbulence

Notes

(i) The statistical theory form may be rearranged in various ways by substituting relations between the properties u_*, σ_w, l_E, λ_m and ϵ. Thus, with $k = 0.4$ and current estimates (see 2.6) of the turbulence parameter ratios ($\sigma_w/u_* = 1.3$ and $\lambda_m/z = 2$ in neutral conditions, $\lambda_m/l_E \cong 4$) identification with $K = ku_* z$ requires $\beta i \cong 0.6$ (a somewhat larger value than that originally suggested, 0.44, by an earlier estimate of l_E/z).

(ii) Taking $\sigma_w/u_* = 1.25$, $\sigma_u/u_* = 2.5$ and $\sigma_v/u_* = 2.3$ (so as to satisfy the condition implicit in Monin's (1965) analysis, that if $l = kz$ and $K = ku_* z$ near the ground the turbulent components must be related in the form $\sigma_w{}^2(\sigma_u{}^2 - \sigma_v{}^2) = u_*{}^4$), then $c_2 = 0.43$. The value given used, though obviously only at the sacrifice of the neat closed formulae which result from the simplest analytical solutions.

Whatever formal mathematical or numerical system is adopted the ultimate value of the approach rests on the physical validity of the whole concept of diffusivity. Lacking this, the prediction of concentration distributions and the generalizations from observed distributions which are so provided reduce to nothing more than rather arbitrary formula-fitting. Two questions in particular have to be critically considered. The first is whether or not the flux of suspended material can really be described as in Eq. (3.1), having regard to the geometry of the diffusing cloud; the second is whether or not the diffusivity can for convenience be identified with that for some other property, especially momentum.

It was noted in advance in the qualitative discussion of Chapter 1, that one important practical case, namely the continuous release of material, is not strictly representable as a simple diffusion process, and is more logically recognized as a scattering or dispersive process to be treated by statistical (kinematical) methods. Moreover, even if the interest is in the growth of a given volume of suspended material, it is evident that this volume will be acted on by a whole spectrum of turbulent fluctuations. Clearly only those fluctuations which are spatially small compared with the existing distribution of material can be expected to exert an action representable in mixing length terms, bearing in mind the implication that the transferring action of an element of fluid is confined to distances over which there is at least a statistical uniformity of material gradient. The fluctuations which are on a scale similar to or greater than that of the material distribution itself will exert actions ranging over convolution, systematic distortion and bodily movement of the volume. These effects obviously cannot be represented as a simple diffusion process except in the most superficial formal way. As will be evident from later discussions, experience so far shows that a real approach to the conditions of a simple diffusion process for suspended material has been convincingly demonstrated in only two circumstances in practice. Both are concerned with the vertical spread of material in flow conditions when the vertical scale of the turbulent motion is limited either by a stable density gradient or by the restrictive effect of proximity to the ground. On a more formal plane, some interesting reflections on the validity of the gradient-transfer relation are provided by the random walk considerations and the principle of limited velocity of diffusion discussed in the next section.

On the matter of the relation between the turbulent diffusivities for various properties, it was first suggested by Reynolds that identity should be assumed for momentum and heat, and it is an obvious generalization of the *Reynolds analogy* to include diffusivity for suspended material in the identity. There are, however, several aspects in which this assumption should be questioned. In the first place fluid momentum can be transferred by the action of pressure forces, a mechanism for which there is no parallel in the transfer of heat or gaseous material. This could be relevant in vertical transfer in a very stable environment, when masses of fluid may tend to oscillate vertically, and in so doing transfer some momentum by pressure forces, but without mixing the gaseous properties. It is also relevant to the process at a rough boundary, in which case momentum is primarily transferred by pressure forces on the roughness elements, whereas any transfer of gaseous material would depend solely on molecular processes (see Chapter 5 for further discussion of deposition processes).

There is also the possibility of a discriminatory process in the vertical transfer of properties in the presence of buoyancy forces, in the way suggested by Priestley and Swinbank (1947). On the classical mixing length theory the temperature fluctuation T' is $-l(\partial T/\partial z + \Gamma)$. However, upward motion is more likely to set in for fluid elements which have a temperature *higher* than the mean at their level, and conversely for downward motion. If this initial excess or deficit is T'' then the temperature fluctuation ultimately produced by the vertical motion becomes $T'' \; -l(\partial T/\partial z + \Gamma)$ and an addition is made to the *upward* flux of heat whatever the sign of $(\partial T/\partial z + \Gamma)$. Thus the effective diffusivity may be more or less than that exceeded on the original mixing length theory according as conditions are unstable or stable.

There has been much controversy on the magnitude of this discriminatory effect and on the implied enhancement of K_H relative to the K for other properties. As discussed in 2.2 it is now generally accepted that K_H and K_M differ in the sense expected in unstable conditions near the ground. There is also good evidence that K_v (the diffusivity for water vapour) is similar to K_H rather than to K_M. On the Priestley–Swinbank argument this requires that the fluctuations of temperature and vapour pressure be positively correlated. This seems likely for moist ground at least, in that the vapour pressure near the ground will be largely determined by the temperature. It does not follow, however, that gaseous or particulate material injected locally into the atmosphere will have such a correlation, and indeed it is difficult to envisage how this could arise naturally. If the material is released at a temperature significantly different from the ambient air temperature there will tend to be a bodily displacement relative to the air. This may have consequences on the subsequent diffusion of the material, but in a different way from that implied above, i.e. by inducing internal motion and entrainment into the volume of gas. However, as regards the diffusivity which is associated with the natural turbulence in the air there appears to be no reason to expect a discriminatory action of buoyancy as discussed above. Correspondingly there is no clear reason for equating the material diffusivity to K_H or K_v rather than to K_M, though empirical evidence in favour of K_H will be noted later.

3.2 SOLUTIONS OF THE EQUATIONS OF DIFFUSION

The differential equation which has been the starting point of most mathematical treatments of diffusion from sources is a generalization of the classical equation for conduction of heat in a solid and is essentially a statement of conservation of the suspended material. Denoting the local concentration by χ units of mass per unit volume of fluid (which is assumed incompressible)

$$\frac{\partial \chi}{\partial t} = -\left[\frac{\partial(u\chi)}{\partial x} + \frac{\partial(v\chi)}{\partial y} + \frac{\partial(w\chi)}{\partial z}\right] \tag{3.10}$$

Writing u, v, w and χ as the sum of a mean and eddy fluctuation therefrom, expanding, averaging and rearranging.

$$\frac{\partial \bar{\chi}}{\partial t} + \bar{u}\frac{\partial \bar{\chi}}{\partial x} + \bar{v}\frac{\partial \bar{\chi}}{\partial y} + \bar{w}\frac{\partial \bar{\chi}}{\partial z} = -\left[\frac{\partial(\overline{u'\chi'})}{\partial x} + \frac{\partial(\overline{v'\chi'})}{\partial y} + \frac{\partial(\overline{w'\chi'})}{\partial z}\right] \tag{3.11}$$

Replacing the eddy flux terms by the simplest gradient-transfer forms the equation becomes

$$\frac{d\bar{\chi}}{dt} = \frac{\partial}{\partial x}\left(K_x \frac{\partial \chi}{\partial x}\right) + \frac{\partial}{\partial y}\left(K_y \frac{\partial \chi}{\partial y}\right) + \frac{\partial}{\partial z}\left(K_z \frac{\partial \chi}{\partial z}\right) \tag{3.12}$$

(but see (3.155) for a more complete representation of three-dimensional transfer). It will be noted that Eq. (3.12) allows for differences in the eddy diffusivities in the component directions, i.e. for anisotropic diffusion, and also for spatial variation of these diffusivities. If the Ks are constant, independent of x, y or z, the simplified equation and the type of diffusion implied are Fickian. For a full discussion of the solutions of the Fickian equation, and a statement of the resulting expressions for the distribution downwind of sources of matter generated at a point or along a line, see Sutton (1953), Chapters 4 and 8. The essential feature of these expressions is that the distribution of the suspended material, with respect to distance from the centre of the 'puff' in the case of instantaneous generation at a point, or from a line or plane through the point or line source, is of Gaussian form with variances (defined as at the end of 4.1)

$$\sigma_x^2 = 2K_x t = 2K_x x/u \text{ etc.} \tag{3.13}$$

in the absence of boundaries. Comparison of the expressions with experimental data on diffusion in the atmosphere has from the beginning consistently shown that the equivalent values of K vary systematically with the time of travel, with position, and with scale of the diffusion process. This has led to much preoccupation with the solving of Eq. (3.12) with the K's variable.

The two-dimensional form of the equation

As discussed in the previous section the gradient-transfer formulation may be expected to be the most successful when the diffusive action of the turbulence is effectively confined to scales small relative to the volume occupied by the suspended material. Bearing in mind the scaling of the vertical component with respect to height (see 2.6) it is a reasonable hypothesis that for vertical spread this condition is approached when the material is injected at ground level. To put the matter in the simplest terms, the base of the plume of material from a source on the ground cannot be lifted clear of the ground in a way analogous to the sideways meandering, and the redistribution vertically is essentially a spreading under the action of turbulent motions on a scale small relative to the vertical depth within which the material is contained at any instant. Such is clearly not the case for time-mean lateral dispersion from a continuous point source (or for vertical dispersion if the source is elevated) and this aspect of dispersion is more appropriately treated by the statistical methods of 3.4 and 3.5.

Accordingly, the most promising application is the case of an infinite crosswind line source emitting at a constant rate at ground level. In Eq. (3.12) the left-hand side reduces to $\bar{u}(\partial\chi/\partial x)$ ($\bar{v}$ and $\bar{w}$ being zero, and a steady state, i.e. $\partial\chi/\partial t = 0$, being assumed). The second term on the right-hand side of Eq. (3.12) is zero, and neglecting the first term the equation finally becomes

$$\bar{u}\frac{\partial\chi}{\partial x} = \frac{\partial}{\partial z}\left(K_z\frac{\partial\chi}{\partial z}\right) \tag{3.14}$$

For the lower atmosphere, in adiabatic conditions, it has been seen that the wind velocity varies with the logarithm of the height, but such a variation proves intractable in the manipulation of Eq. (3.14), and progress has been made only by adopting a power-law form of wind profile. The significant mathematical step was provided by Roberts [unpublished, see Sutton (1953)] for the case when

$$K_z(z) = K_1\left(\frac{z}{z_1}\right)^n, \quad \bar{u}(z) = \bar{u}_1\left(\frac{z}{z_1}\right)^m \tag{3.15}$$

where $\bar{u}_1$ and K_1 are the values of $\bar{u}$ and K_z at a fixed reference height z_1. The boundary conditions are

$$\chi \to 0 \text{ as } x, z \to \infty \tag{3.16}$$

$$\chi \to \infty \text{ at } x = z = 0 \tag{3.17}$$

$$K_z\frac{\partial\chi}{\partial z} \to 0 \text{ as } z \to 0, x > 0 \tag{3.18}$$

and with Q the rate of emission per unit crosswind length

$$\int_0^\infty u\chi(x, z)\, dz = Q \quad \text{for all } x > 0 \tag{3.19}$$

The solution, valid for $r = m - n + 2 > 0$, is

$$\chi(x, z) = \frac{Qr}{z_1\bar{u}_1\Gamma(s)}\left[\frac{z_1^2\bar{u}_1}{r^2K_1x}\right]^s \exp\left[-\frac{z_1^{2-r}\bar{u}_1z^r}{r^2K_1x}\right] \tag{3.20}$$

where $s = (m+1)/r$. Note that when $m = n = 0$ the solution reduces to the Fickian form. Also, use of the momentum transfer analogy ($K = K_M$) with shearing stress constant with height implies $n = 1 - m$ (Schmidt's conjugate-power-law).

An attractive composition of this mathematical solution with the physical laws for the wind profile and the shearing stress in the lower atmosphere in adiabatic conditions has been developed by Calder (1949), using power forms approximating to the logarithmic forms, viz.

$$u/u_* = q(u_* z/v)^\alpha \tag{3.21}$$

and

$$u/u_* = q'(z/z_0)^\alpha \tag{3.22}$$

respectively for smooth and rough flow. For a unified mathematical treatment Calder represents these in the general forms

$$\tau_0 = \epsilon \rho u^{2\beta}(\delta/z)^{2\alpha\beta} \tag{3.23}$$

$$K(z) = (\epsilon/\alpha)\delta^{2\alpha\beta} u_1{}^{2\beta-1} z_1{}^{\alpha(1-2\beta)} z^{(1-\alpha)} \tag{3.24}$$

where, for smooth flow

$$\beta = 1/(1+\alpha), \quad \delta = v, \quad \epsilon = (1/q)^{2/(1+\alpha)}$$

for rough flow

$$\beta = 1, \quad \delta = z_0, \quad \epsilon = 1/q'^2$$

In practice the basic parameters are determined from wind profile observations over the anticipated height-range of diffusion, the appropriate fitting to the power forms being accomplished by first obtaining the effective value of α [u varies as z^α in Eq. (3.23)], and evaluating u_* and z_0 from the logarithmic forms of the wind profile. Substitution of α, u_* and z_0 in Eqs. (3.21) and (3.22) then gives q and q'. The full solution is given by substituting in (3.20) $m = \alpha$, $n = 1 - \alpha$ and K_1 as in Eq. (3.24) with z equal to z_1. With $\alpha = 0.187$ as recommended by Calder for a surface with $z_0 = 3$ cm, $\chi \propto x^{-0.86}$ and the height of cloud (arbitrarily the height at which $\chi(z)/\chi(0)$ is some chosen fraction) is proportional to $x^{0.73}$.

The solution in Eq. (3.20) may be rearranged into a more compact form, in terms of the magnitude of the vertical spread σ_z or $\bar{Z}$ (defined in 4.1) and the mean horizontal velocity u_p of the particles at distance x (see 3.3). The version given by van Ulden (1978) is

$$\chi(x, z) = (A/\bar{Z}u_p)\exp-(Bz/\bar{Z})^s \tag{3.25}$$

where $A = s\,\Gamma(2/s)/(\Gamma(1/s))^2$ and $B = \Gamma(2/s)/\Gamma(1/s)$

and $s = m - n + 2$ (different from that in Eq. (3.20)).

As we note later (see 3.3 and Table 3.II) van Ulden used this solution to derive $d\bar{Z}/dx$ in terms of $\bar{Z}$, $K(\bar{Z})$ and $u(\bar{Z})$, and then extended Calder's approach to the thermally-stratified surface-stress layer, assuming $K = K_H$ and adopting Monin–Obukhov forms for the wind and diffusivity profiles. Integration was then completed with certain approximations to provide $\bar{Z}/z_0$ as a function of x/z_0 and z_0/L for substitution in Eq. (3.25) (see 4.6 for the tests of these and related theories against measurements of dispersion).

A solution of Eq. (3.14), for wind constant with height ($m = 0$), and K varying linearly with height ($n = 1$), was stated by Bosanquet and Pearson (1936), and an explicit version of this, with K as in Eq. (2.50), has been put forward by Calder (1952) as a useful approximation. The result is

$$\chi = \frac{Q}{ku_*x} \exp\left(\frac{-uz}{ku_*x}\right) \tag{3.26}$$

and as Calder has demonstrated this gives an agreement with observations almost as good as that obtained with the more elaborate solution in his 1949 paper.

It may be noted that a detailed analysis of the two-dimensional equation of diffusion, though in the different context of the problem of evaporation from a plane, free-liquid surface, has been given by W. G. L. Sutton (1943), again for the conjugate-power-law condition, i.e. with $n = 1 - m$ in Eq. (3.15). Deacon (1949) and Rounds (1955) have given further discussion, adopting the form for K with $n \geqslant 1$. Solutions are given by Rounds for an arbitrary initial distribution of material, and these are then applied to several types of source. In particular, solutions are given in graphical form for the concentration at ground-level from a crosswind line source at arbitrary height, both for the previous condition when an increase of K with height is supposed to continue indefinitely, and also for the case when upward diffusion is suddenly halted by an overhead stable layer, by imposing the condition that the flux $K(\partial\chi/dz)$ is zero at this level (see also reference to Round's solution in 3.8).

In a more general context of the point-source problem F. B. Smith (1957a) has also treated the two-dimensional equation, for the case of an elevated line source, with the conjugate-power-laws for wind and diffusivity. Smith's analysis includes a proof of the so-called reciprocal theorem, namely that the concentration distribution at ground level due to a source at height H is identical with the horizontal distribution at height H due to a source at ground level. Later work by Smith (1957b and c) has dealt further with the vertical diffusion in a layer of finite depth h bounded by the ground and the base of a stable region in the atmosphere, using forms of K as follows, and assuming constant $\bar{u}$:

$$\begin{aligned}
&\text{(i)} && K = \text{constant} \\
&\text{(ii)} && K \propto z^{\alpha}, \quad 0 \leqslant z \leqslant h \\
&\text{(iii)} && K \propto (h-z)^{\alpha}, \quad 0 \leqslant z \leqslant h \\
&\text{(iv)} && K \propto z(h-z), \quad 0 \leqslant z \leqslant h \\
&\text{(v)} && K \propto z, \quad 0 \leqslant z \leqslant h/2 \\
& && \quad\ \propto (h-z), \quad h/2 \leqslant z \leqslant h
\end{aligned} \tag{3.27}$$

Forms (iv) and (v) in particular are possibly acceptable approximations to the real vertical profile of diffusivity in the atmospheric boundary layer. Adoption of (iv), in the explicit form $K = 0.4u_*z(1-z/h)$, has been followed up by Nieuwstadt (1980b). Note that in (v) the maximum of K occurs at the mid-height of the finite layer. The solutions are in the form of rapidly convergent series which can usually be evaluated adequately with a small number of terms.

Smith's treatment is also particularly notable in containing the first attempt to incorporate an additional transport mechanism specifically representing the effect of thermally-driven convection. An idealized model was assumed, in which bubbles or streams of air rise from the ground to the top of the layer, and are replaced by a uniform subsidence throughout the layer. In the first attempt (1957b) the 'thermals' were considered to be completely isolated from the surrounding atmosphere, but later (1957c) entrainment from the environment into the ascending currents was introduced.

The distribution of material by the convection process was represented mathematically by a series of sources which subside from the top of the layer, and by an appropriate reduction in the source strength assigned to the cloud spreading upward from ground-level in the usual way. For the diffusion from the sources K was assumed to have the forms

$$\left.\begin{array}{ll} K = a(h-z) & \text{(1957b)} \\ K = bz^{-\alpha} & \text{(1957c)} \end{array}\right\} \quad 0 \leqslant z \leqslant h \qquad (3.28)$$

where a and b are constants.

For further details of Smith's solutions reference should be made to the original papers. The treatments are so far purely formal in that the magnitudes of K, and of the convective circulations, have yet to be suitably assigned in relation to the dynamical conditions of atmospheric convection. They do however provide a mathematical framework in readiness for this step, and it is of interest to note, for example, the general concentration distribution for a crosswind line source at the ground, with $K = a(h-z)$ and λ, the compensating velocity of subsidence (generally assumed to be βa) taken to be $0.5a$. The distributions in non-convective and convective conditions are shown graphically in Fig. 3.1, in the form of isopleths of $\chi\bar{u}h/Q$ against height and distance in the non-dimensional terms z/h and $ax/\bar{u}h$ ($Q/\bar{u}h$ is the final uniform concentration). The second graph shows a more rapid attainment of uniform concentration, and a vertical profile of concentration with maxima at the ground and the top of the finite layer, both of which features are a consequence of the particular convective model assumed.

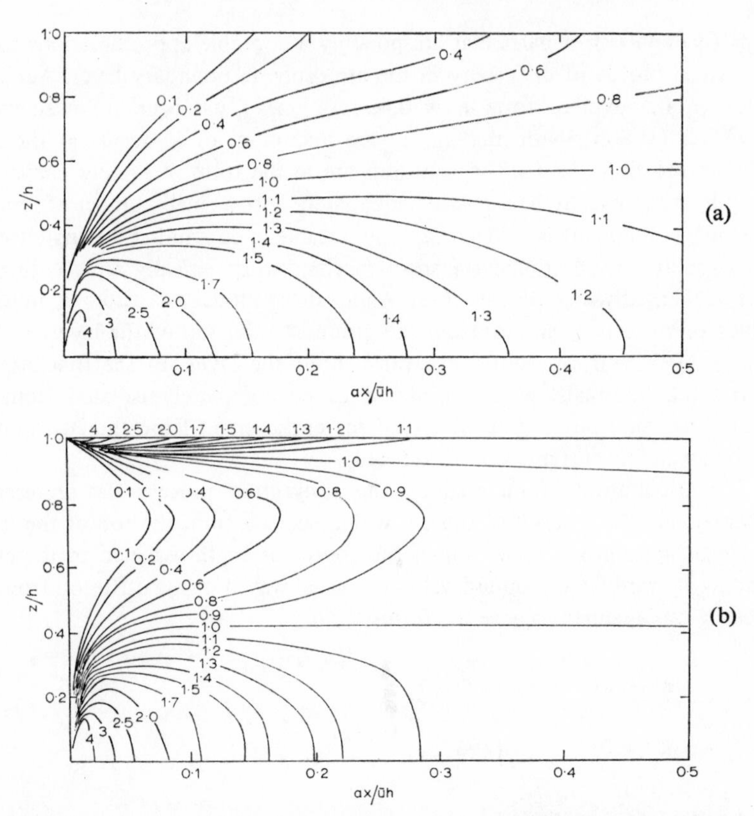

Fig. 3.1 – Isopleths of the concentration parameter $\chi uh/Q$ for a cross-wind line source at ground level, in an atmosphere with $K=a(h-z)$, for $0 \leqslant z \leqslant h$ and wind constant with height. (a) without convection; (b) with convection, $\lambda = 0.5a$. (Smith 1957b).

The limited velocity of propagation of material and the hyperbolic equation of diffusion

The preceding discussions have all been in terms of a differential equation which, once the gradient-transfer relation is introduced, is of parabolic form. The Gaussian or more complex exponential distributions derived therefrom are obviously unrealistic in the sense that they imply an unlimited rate of movement of matter outward from a source. A review of this point and of an alternative *hyperbolic* form of equation which avoids the difficulty has been given by Monin (1959) and more generally by Monin and Yaglom (1965).

The essential point is brought out in an unpublished analysis by F. B. Smith (1973a). A *random walk* model is assumed in which marked elements of fluid move in discrete steps, at constant velocity w^* for time τ. At the end of each step an element starts again but with equal probability of moving up or down. Considering the flux F at any given level and the conservation of marked elements (concentration χ) over an infinitesimal adjoining layer, the following equations are found to be satisfied as an approximation,

$$\frac{\partial \chi}{\partial t} = \frac{\partial F}{\partial z}, \quad \tau \frac{\partial F}{\partial t} + F = -w^{*2} \tau \frac{\partial \chi}{\partial z} \tag{3.29}$$

This is a hyperbolic set of equations of the same form as those discussed by Monin. Note that the first is simply the equation of conservation for one-dimensional diffusion, while the second reduces to the gradient-transfer relation when $\tau \to 0$ but $w^{*2}\tau$ remains finite. However this second equation shows that in general the gradient-transfer relation does not hold exactly if we adopt the realistic assumption of a finite limit to w^*.

Monin gives a solution to Eq. (3.29) for the special case of neutral conditions and neglecting the variation of wind with height. The quantity w^* is identified as the maximum vertical velocity of the diffusing smoke and both it and the quantity corresponding to τ are expressed on similarity grounds. The solution is

$$\chi(x, z) = \frac{Q}{k u_* x} \frac{(1 - \bar{u}z/\lambda u_* x)^{\epsilon - 1}}{(1 + \bar{u}z/\lambda u_* z)^{\epsilon + 1}}, \quad 0 \leqslant z \leqslant \frac{\lambda u_* x}{\bar{u}} \tag{3.30}$$

where $\epsilon = \lambda/2K \geqslant 1$ and λ is w^*/u_*. For $z = 0$ the expression is identical with that obtained by Calder, Eq. (3.26), and a comparison of the vertical profiles is shown in Fig. 3.2 for $\epsilon = 1.25$, an empirical value deduced by Monin from

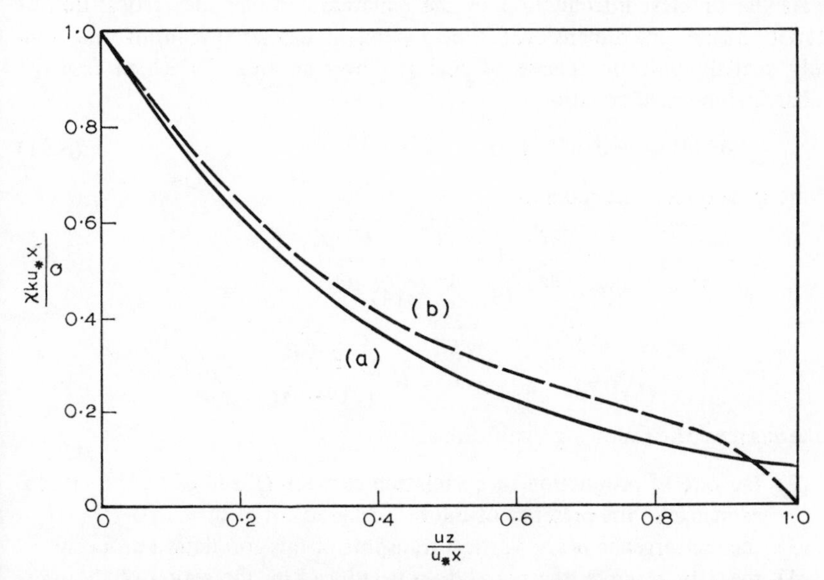

Fig. 3.2 – Vertical profiles of concentration according to (a) Calder's solution, Eq. (3.26) (b) Monin's solution, Eq. (3.30) with $\epsilon = 1.25$, $k = 0.4$.

examination of data on smoke plumes, and $k = 0.4$. As would be expected the discrepancy between the two becomes large only near the 'top' of the cloud, where Monin's profile, in accordance with the concept of finite propagation velocity, suddenly falls to zero concentration. Eq. (3.30) means that the concentration profiles are of similar shape at different distances, i.e. they can be expressed universally in terms of z/x, and Monin concludes that this is also approximately true for any stratification and for wind velocity varying with height.

The second-order closure solution of the diffusion equation

The doubts about the validity of the simple gradient-transfer hypothesis have led to interest in the use of the full and much more complicated equation which may be constructed for the eddy flux of material (e.g. $\overline{w'C'}$ in the vertical). This *second-moment* equation is one of a whole family derivable by writing the conservation equations for the components of wind velocity and for the scalar properties θ (potential temperature) and passive material concentration C, applying the Reynolds averaging rule ($C = \bar{C} + C'$ etc.) so as to obtain equations for dC'/dt and dw'/dt, then multiplying by w' and C' respectively to give two equations which in combination give $d\overline{w'C'}/dt$. The corresponding equations for the variances of the velocity components have already been the subject of much interest, in the context of the production and suppression of turbulent energy (see 2.3 and Eq. (2.61)). For a listing of the whole family of variance and covariance equations see for example the article by Donaldson (1973).

As the simplest introduction to the equation consider the vertical flux of material in the one-dimensional time-dependent case, representing the non-steady spatially-uniform release of material over an area, for which case the familiar first-moment equation is

$$\partial\bar{C}/\partial t = -\partial(\overline{w'C'})/\partial z \tag{3.31}$$

and the second-moment equation is

$$\frac{\partial\overline{w'C'}}{\partial t} = \underset{(2)}{-\overline{w'^2}\frac{\partial\bar{C}}{\partial z}} - \underset{(3)}{\frac{\partial\overline{w'^2C'}}{\partial z}} - \underset{(4)}{\frac{\overline{C'\partial p'}}{\rho\partial z}} \tag{3.32}$$

$$+ \underset{(5)}{g\frac{\overline{C'T'}}{T_0}} + \underset{(6)}{\nu\frac{\partial^2\overline{w'C'}}{\partial z^2}} - \underset{(7)}{2\nu\frac{\overline{\partial w'\partial C'}}{\partial x_i\partial x_i}}$$

The terms have the following significance.

(2) the rate of production of correlation between C' and w' by the vertical exchange in the presence of the vertical gradient of C.
(3) the convergence of the vertical transport of this correlation property
(4) the rate at which the correlation is reduced by the action of the pressure fluctuations

(5) the increase or reduction of the $\overline{w'C'}$ correlation by the buoyancy forces
(6) the molecular counterpart of the transport term (3)
(7) presumably a 'decorrelation' term analogous to the molecular dissipation of turbulent fluctuations of velocity

Adoption of the second-moment equation may eliminate the need for the traditional closure at the first-moment stage (i.e. the writing of $\overline{w'C'} = -K\,\mathrm{d}C/\mathrm{d}z$) but only at the expense of introducing extra unknown terms. Further equations could be written for the new unknown terms but this could only have the effect of widening the disparity between the numbers of equations and unknowns. The alternative is to seek independent specification of the new unknown terms at the second-moment stage. Much effort has been devoted to formulating this *second-order closure* by complex mathematical modelling, with the idea that this may be achievable with assumptions and appeals to empiricism less objectionable than the first-order closure.

Application of the technique seems first to have been undertaken in Russian work on the modelling of the dynamics of the boundary layer. In Monin's (1965) analysis the equations for the variances and covariances of the turbulent velocity components are simplified first by making the usual assumptions of stationarity and horizontal homogeneity, neglecting terms containing the Coriolis parameter and also the transport terms (corresponding to term (3) in our equation for $\overline{w'C'}$). The remaining two unknown terms were dimensionally formulated in terms of characteristic velocity and length scales, for which q the square-root of the turbulent energy per unit mass of air and the scale 1 were adopted, with numerical coefficients remaining to be specified empirically. It is especially interesting that this procedure leads to the simple relation for the $u'w'$ covariances already noted in Eq. (3.7) i.e. a simple gradient-transfer expression with $K = \sigma_w{}^2 1/c_2 q$, c_2 being one of the empirical coefficients. This intuitive type of modelling in terms of characteristic velocity and length scales, with empirical coefficients evaluated from the measured properties of various flow conditions, has been adopted and elaborated by various workers, especially in the context of laboratory sheared flows.

An ambitious attempt to apply the second-order closure technique to the dispersion of material from continuous point and line sources has been made by Lewellen and Teske (1976), using length scale and velocity scale dimensional modelling of the transport, pressure correlation and dissipation terms on the lines recommended by Donaldson (1973). In the modelled form of the pressure fluctuation term the scales appear as a quotient q/l multiplying the covariance $\overline{w'C'}$, whereas in the transport term they appear as a product ql multiplying $\mathrm{d}w'C'/\mathrm{d}z$, i.e. in the form of an eddy diffusivity, with numerical coefficients in each case. In a forerunner to their dispersion paper (1973) Lewllen and Teske demonstrate that the Donaldson form of modelling may be used to predict the effects of thermal stratification on several statistical properties of the surface-stress layer, including the forms of the Monin–Obukhov ϕ_M and ϕ_H, so justifying the numerical coefficient values adopted in the modelled terms. The two-dimensional (x, z) forms of the first- and second-moment equations are then solved (1976) numerically, using the same system, but with recognition that the modelling scales must be given effective values q_p and l_p dependent on the dimensions of the cross-section of the plume (σ_z say). It is assumed that $l_p = d\sigma_z$, with d a coefficient to be chosen, subject to the constraint that l_p cannot be greater than the scale of turbulence itself. The numerical solutions for a continuous crosswind infinite line source at ground-level in a neutral boundary layer with specified roughness give $\sigma_z(x)$ curves that are insensitive to the choice of d and in reasonably close agreement with those calculated previously on a simple gradient-transfer/similarity basis, with diffusivity in accordance with the Monin–Obukhov framework for the surface-stress layer and with surface-Rossby-number similarity for the upper part of the boundary layer (see Pasquill, 1978).

Lewellen and Teske's calculations include two aspects of dispersion for which the gradient-transfer hypothesis is especially suspect, and for which therefore an improved representation through the second-order closure technique would be particularly welcome. The first aspect is the dispersion from an elevated continuous source, for which statistical theory indicates that the diffusivity would have to be a function of distance in the early (plume meandering) stages of dispersion. Here the progress does not yet appear to be decisive, in that the calculated dispersion is apparently quite sensitive to the choice of the scale coefficient d. The second aspect is the vertical diffusion in a convectively mixed layer and in this case (for a near-surface release) the calculated concentration distribution is in striking agreement with Deardorff and Willis's (1974) laboratory data, including a downwind rise in the altitude of maximum concentration (which is not predicted by the gradient-transfer approach), though the downwind rise is not as marked as that found in the laboratory.

A much more profound approach to the modelling is advocated by Lumley in several recent papers with co-workers. The crucial difference from the relatively simple velocity-length scaling lies in the recognition that third-order fluxes (such as $\overline{w'^2C'}$) cannot consistently be related to the vertical gradient of $\overline{w'C'}$ alone and must include dependence on the gradients of other properties. Thus, whereas the Lewellen–Teske form of modelling represents the third-order flux effectively in terms of a simple diffusivity ql reflecting only the properties of the field of turbulence, Zeman and Lumley (1976) modify ql by terms containing $\overline{w'\theta'}$ and $d\bar{\theta}/dz$ in the context of vertical heat transfer and, by analogy, terms containing $\overline{w'C'}$ and $d\bar{C}/dz$ in the context of vertical dispersion. The calculations (1976) predict realistic heat transfer properties in a buoyancy-driven-mixed layer, including counter-gradient fluxes of heat and turbulent energy, features that are not predicted by simple transport models. The application to dispersion from a surface area release yields distribution of C and $\overline{w'C'}$ that imply the possibility of very large values of the effective eddy diffusivity in the middle section of the mixed layer. These results are broadly consistent with a conservation analysis by Crane and Panofsky (1976) of the observed build-up of pollution over Los Angeles in convective days, though not to the extent of implying effectively infinite values of the diffusivity as required by the observed upward flux in the presence of apparently zero gradient of $\bar{C}$.

3.3 SIMILARITY THEORY APPLIED TO DIFFUSION

The origins of the similarity approach to diffusion of particles are in the work by Monin (1959) referred to in the previous section, but the whole concept has been crystallized and elaborated by Batchelor (1959, 1964) and others for neutral flow and by Gifford (1962) for stratified flow, and their development of the idea will be followed here.

Essentially the basic hypothesis of *Lagrangian* similarity is that the statistical properties of the velocities of *particles* in the *surface stress layer* of the atmosphere are determined by just those parameters which have been argued to determine the Eulerian properties (see 2.2), i.e. by u_* alone in neutral flow and additionally by the heat flux H in stratified flow, the two being combined in the Monin–Obukhov length scale L.

For passive particles injected singly at a point on $z = 0$ it follows immediately on dimensional grounds that the rate of increase of the *average vertical displacement* ($\bar{Z}$) of an ensemble of particles after each has travelled for a given time must be of the form

$$\frac{\mathrm{d}\bar{Z}}{\mathrm{d}t} = bu_*\Phi\left(\frac{\bar{Z}}{L}\right) \tag{3.33}$$

where b is a universal constant and Φ a universal function, both to be specified. In the special case of neutral flow $\Phi = 1$ and the relation reduces to that originally stated by Batchelor. It is further assumed that the rate of increase of the corresponding mean horizontal displacement $\bar{X}$ is equal to the mean wind speed at a level related to $\bar{Z}$, i.e.

$$\frac{\mathrm{d}\bar{X}}{\mathrm{d}t} = \bar{u}(c\bar{Z}) \tag{3.34}$$

The quantity c is another constant, which was omitted (i.e. implied equal to unity) in earlier formulations and was introduced by Batchelor in his second paper. These two expressions for the mean vertical and horizontal velocities of an ensemble of particles after each has travelled for time t form the entire physical basis of the treatment and the ultimate relations follow from substitution of forms for the wind profile and Φ and from further dimensional considerations.

The relevance to particles released at a finite height h is worth noting here. Obviously the magnitude of h must affect the statistics of particle motion for some time but this effect will be ultimately lost and the statistical properties will then approximate to those for particles released at some point on the surface. Batchelor represents this in a formal extension of the hypothesis which states that the properties after time t from an elevated release will approximate to those $t + t'$ after a ground release, provided $t \gg t'$ and where t' is of order h/u_*. In a practical sense it seems unlikely that the permissible extension of Eq. (3.33) to an elevated release will be very significant. In neutral conditions $h/u_* \cong 10h/\bar{u}$, and the restriction $t \gg 10h/\bar{u}$ is equivalent to a distance restriction $x \gg 10h$. From experience the total vertical spread in neutral conditions is roughly $x/10$ and therefore $\gg h$, and as this must be within the constant stress layer for the treatment to be valid the implication is that h cannot be more than a few metres. Further discussion will therefore be confined to a ground release.

Combination of Eqs. (3.33) and (3.34), with $\bar{u}$ given the Monin–Obukhov form in Eq. (2.48) leads to the following relation between the magnitude of $\bar{Z}$ and $\bar{X}$ coexisting at any particular time after release.

$$\bar{X} = \frac{1}{kb}\int_{z_0}^{\bar{Z}} \frac{[f(c\bar{Z}/L) - f(cz_0/L)]}{\Phi(\bar{Z}/L)}\, \mathrm{d}\bar{Z} \tag{3.35}$$

Thus, given the constants b and c and the universal functions f and Φ the vertical spread from a ground-level source is predictable in principle. For neutral flow, with $\Phi = 1$ and the wind profile function reducing to the logarithmic form [Eq. (2.51)], the integral may be directly evaluated to give

$$\bar{X} = \left[\frac{\bar{Z}}{kb}\ \ln\frac{c\bar{Z}}{z_0} - 1 + \frac{z_0}{\bar{Z}}(1 - \ln c)\right] \tag{3.36}$$

For other conditions some numerical solutions are available (see later).

After some early empirical estimates of 0.1–0.2 for b, it was suggested by Ellison on empirical grounds that $b = k$ (= 0.4). Ellison later gave the proof which follows, on the assumption that the vertical spread of particle is describable by the gradient-transfer relation with a diffusivity equal to the eddy viscosity, i.e. $K = ku_*z$, in the neutral surface-stress layer. For an instantaneous plane source of matter at time $t = 0$ and height $z = 0$ the governing equation is Eq. (3.12) with the x and y terms omitted and $\mathrm{d}\chi/\mathrm{d}t$ reduced to $\partial\chi/\partial t$, i.e.

$$\frac{\partial\chi}{\partial t} = \frac{\partial}{\partial z}\left(K(z)\frac{\partial\chi}{\partial z}\right) \tag{3.37}$$

By definition

$$\bar{Z} = \int_0^\infty z\bar{\chi}\mathrm{d}z \Bigg/ \int_0^\infty \bar{\chi}\,\mathrm{d}z \tag{3.38}$$

in which the denominator is a constant equal to the magnitude Q per unit area of the instantaneous source, and accordingly

$$\frac{\mathrm{d}\bar{Z}}{\mathrm{d}t} = \int_0^\infty z\frac{\partial\bar{\chi}}{\partial t}\,\mathrm{d}z \Bigg/ \int_0^\infty \bar{\chi}\,\mathrm{d}z \tag{3.39}$$

Substituting for $\partial\bar{\chi}/\partial t$ and integrating by parts

$$\frac{\mathrm{d}\bar{Z}}{\mathrm{d}t} = \int_0^\infty \bar{\chi}\frac{\mathrm{d}K}{\mathrm{d}z}\,\mathrm{d}z \Bigg/ \int_0^\infty \bar{\chi}\,\mathrm{d}z \tag{3.40}$$

and since by definition $\mathrm{d}K/\mathrm{d}z$ is independent of z (in neutral conditions) this reduces to the identity

$$\frac{\mathrm{d}\bar{Z}}{\mathrm{d}t} = \frac{\mathrm{d}K}{\mathrm{d}z} \tag{3.41}$$

from which it immediately follows that $b = k$.

Alternatively, it is interesting to note the full solution of Eq. (3.37) which is given by Sutton (1953, p. 216) for the analogous case of heat transfer from a plane source at $z = 0$ with $K(z) = K_1 z^n$. In terms of suspended material the solution is

$$\bar{\chi}(t) = \frac{Qr}{r^{2/r}K_1^{\,1/r}\Gamma(1/r)t^{1/r}} \exp\left(-\frac{z^r}{r^2 K_1 t}\right) \tag{3.42}$$

valid for $0 \leqslant r \leqslant 2$, where $r = 2 - n$. For $K(z) = ku_*z$ this simplifies to

$$\bar{\chi}(t) = \frac{Q}{ku_*t} \exp\left(-\frac{z}{ku_*t}\right) \tag{3.43}$$

Substituting this in the definition for $\bar{Z}$

$$\bar{Z}(t) = ku_*t \tag{3.44}$$

from which again the identity of b and k follows.

The foregoing arguments rely on the assumption that the vertical distributions from the two different sources implied in Eqs. (3.33) and (3.37) can be taken as statistically identical. In the former the particles are released from one position on an ensemble of occasions – in the latter they are released at one instant from a uniform array of different positions. Given that the flow conditions are stationary and horizontally homogeneous, as is assumed in all theoretical treatments of the present type, there is no conceivable reason why the two distributions should be different.

Application to distribution of concentration downwind of a continuous source in neutral conditions

When the foregoing results are used as a starting point from which to deduce the properties of the distribution of material as a function of *position* downwind of a continuous source there are certain difficulties. The first difficulty is that in considering a *point* source there is an implication that the horizontal spread (especially the crosswind spread) is determined by just those physical parameters that determined $\bar{Z}$. This is not immediately obvious and it may be recalled that the Eulerian aspects of the u and v components do not conform to these similarity arguments.

A second source of confusion is the tendency to begin with a statement of the implied form of the concentration distribution from an *instantaneous point* source. Excluding the obvious difficulty of arguing about an instantaneous source which is literally a 'point', and allowing the source to have some finite though small size, there is still the problem that the precise formulation of the subsequent separation of particles involves two-point velocity correlation (see 3.7), about which the foregoing similarity arguments say nothing. However, this misleading involvement of the instantaneous source is actually unnecessary, and indeed does not appear in Batchelor's 1964 paper, which has been taken as the basis for the present discussion.

The first of the foregoing difficulties is also avoided by adapting Batchelor's argument to the two-dimensional case of the infinite crosswind line source, from which particles are released continuously, in neutral flow. The position of any particle at a time t after release will have coordinates $X(t)$ and $Z(t)$ which are random quantities with means $\bar{X}$, $\bar{Z}$. On similarity grounds the statistical departures of X and Z from their mean values are taken to be dependent only on u_* and t, and the probability density of $X-\bar{X}, Z-\bar{Z}$ to have a universal shape scaled with respect to u_*t or $\bar{Z}$. Thus, the probability $P(x, z, t)$ of finding a particle near the position x, z after it has travelled for time t is of the form

$$P(x, z, t) = \frac{1}{\bar{Z}^2} g_1 \left(\frac{x-\bar{X}}{\bar{Z}}, \frac{z-\bar{Z}}{\bar{Z}} \right) \tag{3.46}$$

where g_1 is some universal function.

For a *continuous* line source (cls) the average relative concentration at a position (x, z) is equivalent to the probability of finding a particle near (x, z) *at any time*, and this will have contributions from all particle-travel-times, i.e. is given by $\int_0^\infty P \, dt$. Changing the variable of integration to $\bar{Z}$, using the relation for $d\bar{Z}/dt$, and incorporating the intuitively obvious proportionality of concentration to source strength we write

$$\bar{\chi}_{cls}(x, z) = \frac{Q}{bu_*}\int_0^\infty g_2\left(\frac{x-\bar{X}}{\bar{Z}}, \frac{z-\bar{Z}}{\bar{Z}}\, \frac{d\bar{Z}}{\bar{Z}^2}\right) \tag{3.47}$$

where g_2 is another universal function. Now following Gifford's (1962) development, changing the variable to $(x-\bar{X})/\bar{Z}$ and restricting to $z=0$

$$\bar{\chi}_{cls}(x, 0) = \frac{Q}{bu_*}\int_0^\infty \frac{g_3\left(\dfrac{x-\bar{X}}{\bar{Z}}\right) d\left(\dfrac{x-\bar{X}}{\bar{Z}}\right)}{\bar{Z}\left(\dfrac{x-\bar{X}}{\bar{Z}} + \dfrac{d\bar{X}}{d\bar{Z}}\right)} \tag{3.48}$$

where g_3 is another universal function. It is a reasonable assumption that $\sqrt{\overline{(X-\bar{X})^2}}$ will always be small relative to $\bar{X}$. This means that the function g_3 can be expected to have a fairly sharp maximum near $x-\bar{X}=0$, so that the integral is essentially determined by values of $\bar{Z}$ near $\bar{Z}_x$ at which $\bar{X}=x$. Accordingly Eq. (3.48) reduces to

$$\bar{\chi}_{cls}(x, 0) \cong \frac{\text{const} \times Q}{bu_*\bar{Z}_x(d\bar{X}/d\bar{Z})} \tag{3.49}$$

or, on substituting for $d\bar{X}/d\bar{Z}$ from Eqs. (3.33) and (3.34) with $\Phi = 1$,

$$\bar{\chi}_{cls}(x, 0) \cong \frac{\text{const} \times Q}{\bar{Z}_x\bar{u}(c\bar{Z}_x)} \cong \frac{\text{const} \times Q}{\bar{Z}_x \ln (c\bar{Z}_x/z_0)} \tag{3.50}$$

The reduction of concentration with *distance* from a source follows on substituting $\bar{Z}$ from Eq. (3.36). Taking the approximation for large x (i.e. large $\bar{Z}/z_0$)

$$x = \bar{X} = \frac{\bar{Z}_x}{kb}\ln\left(\frac{c\bar{Z}_x}{z_0}\right) \tag{3.50a}$$

and $$\chi_{cls}(x, 0) \propto 1/x \tag{3.50b}$$

The relevance of this simple approximation is at first in some doubt in that $\bar{Z}$ cannot be taken to be indefinitely large in view of the restriction of the treatment to the surface-stress layer. For a more realistic assessment it is necessary to evaluate (3.50) with the $\bar{Z}$, $\bar{X}$ relation given its full form as in Eq. (3.36). For given c this relation may be put entirely in terms of non-dimensional values, $\bar{Z}'=\bar{Z}/z_0$ and $\bar{X}' = \bar{X}/z_0$ and the graph of the universal form, for $c=0.6$ (see Table 3.II) is given in Fig. 3.3. From Eq. (3.50) it follows that the quantity $\bar{\chi}z_0/c$ is inversely proportional to $c\bar{Z}' \log (c\bar{Z}')$. The reciprocal of the latter quantity, representing the relative magnitude of the surface concentration, is also shown in Fig. 3.3.

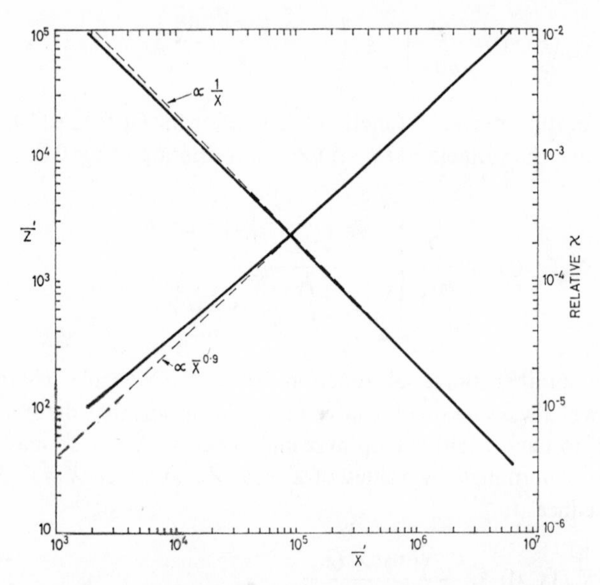

Fig. 3.3 – Variation with distance of vertical spread and relative ground-level concentration, from a ground-level source, according to Lagrangian similarity theory. The graphs represent universal forms

$$kb\bar{X}' = \bar{Z}'[\ln \bar{Z}' - (1 - \ln c)(1 - 1/\bar{Z}')]$$

and $\bar{\chi}$ (relative) $= (c\bar{Z}' \ln c\bar{Z}')^{-1}$ evaluated for $c = 0.6$.

If despite the reservations previously expressed the similarity argument is now followed through for a continuous point source, starting with Eq. (3.46) appropriately modified by adding the variable $Y - \bar{Y}$ in the function g_1 and replacing $\bar{Z}^2$ by $\bar{Z}^3$, the final result for the ground-level concentration on the centre-line of the cloud ($y = 0$) is

$$\bar{\chi}_{\text{cps}}(x, 0, 0) \propto \frac{Q}{\bar{Z}_x^2 \bar{u}(c\bar{Z}_x)} \tag{3.51}$$

and, for large x

$$\bar{\chi}_{\text{cps}}(x, 0, 0) \propto \frac{\ln(c\bar{Z}_x/z_0)}{x^2} \tag{3.51a}$$

Extension to thermally stratified conditions

Gifford's (1962) extension of the analysis of concentration distribution begins with the general forms for $\bar{Z}$ and $\bar{X}$ in Eqs. (3.33)–(3.35). The dimensional arguments for $P(x, y, z, t)$ and $\overline{\chi}_{\text{cps}}$ are otherwise exactly as above for neutral flow, i.e. it is implicitly assumed that the distributions are scaled according to $\overline{Z}$ but that their shape is otherwise independent of $\overline{Z}/L$. On dimensional grounds, to be strictly correct, this dimensionless combination $\overline{Z}/L$ should also be included in the function describing $P(x, y, z, t)$, and the foregoing assumption of independence of $\overline{Z}/L$ is not necessarily acceptable. If it is accepted the development is as given by Gifford and then relation (3.51) holds for thermally stratified conditions also.

There is the further complication that the adoption of the empirical forms so far advocated for Φ and f in Eq. (3.35) necessarily entails numerical solution to give $\overline{Z}$ as a function of $x = \overline{X}$. Monin had previously given values of the integral in Eq. (3.35) for unstable conditions and Gifford extended the numerical evaluation to stable conditions. In both cases the parameter c was taken to be unity. Substitution of the results in Eq. (3.51) gave the variations of concentration with distance, and from these were derived values of the index in the equivalent power law form $\overline{\chi} \propto x^{-m}$. The interesting point is that whereas the treatment gives m slightly less than 2 in neutral conditions the value was then found to be *greater* than 2 in unstable conditions and substantially less than 2 in stable conditions (for distances 50–800 m).

An identification of the universal function $\Phi(\overline{Z}/L)$ has been proposed by Chaudhry and Meroney (1973), on the assumption of gradient transfer and using the solution in Eq. (3.42). The eddy diffusivity was taken to be that describing the vertical transfer of sensible heat in the Monin–Obukhov system, i.e. $K_H = ku_* z/\phi_H$, which was put into the required simple power-law form $K_1 z^n$ by approximating the available empirical form for ϕ_H by the form $A(\overline{Z}/L)^{1-n}$. Substitution of $\chi(z, t)$ from Eq. (3.42) into the definition of $\overline{Z}$ in Eq. (3.38) gives

$$\overline{Z}(t) = (r^2 K_1 t)^{1/r}\,\Gamma(2/r)/\Gamma(1/r) \tag{3.52}$$

from which by differentiation and rearrangement

$$\mathrm{d}\overline{Z}/\mathrm{d}t = K_1^{\frac{1}{r}}\, t^{\frac{1}{r}-1}\, r^{\frac{2}{r}-1}\,\Gamma(2/r)/\Gamma(1/r) \tag{3.52a}$$

$$= (K(\overline{Z})/\overline{Z})\, r\, (\Gamma(2/r)/\Gamma(1/r))^r \tag{3.52b}$$

which reduces to the result in Eq. (3.41) for neutral flow, when $n = 1$ and $r = 2-n = 1$. As shown by Chaudhry and Meroney, the terms containing r in Eq. (3.52b) combine to form a weak function of r and have a resultant numerical value approximating closely to unity for $0.7 \leqslant n \leqslant 1.3$, in which case Eq. (3.52b) simplifies to

$$\mathrm{d}\overline{Z}/\mathrm{d}t = ku_*/\phi_H \tag{3.53}$$

hence $\quad \Phi(\overline{Z}/L) = 1/\phi_H$

In considering the significance of the foregoing similarity relations for vertical spread of passive material it should be noted that basically they do not constitute any advance beyond the gradient-transfer theory, though they may be more convenient for the prediction of vertical spread *per se,* notably in that given the profile of K a solution for $\bar{Z}$ follows with much less effort than say a numerical solution of the parabolic equation of diffusion. It should also be kept in mind that the foregoing assumption of analogy in the vertical transfers of passive material and sensible heat is immediately questionable on the Priestley–Swinbank (1947) argument about the discriminatory transfer of heat by buoyancy forces, though as will be seen in Chapter 4 available experience in field studies is to some extent compatible with that assumption.

A further examination of the similarity approach has been made by Klug (1968) with specific application to the variation of concentration from a continuous point source with the degree of thermal stratification. Klug does include the combination $\bar{Z}/L$ in the probability function corresponding to $P(x, y, z, t)$. Ultimately the ground-level axial concentration is expressed in terms of u_*, Z, the f in Eq. (3.35) and a further unknown function of $\bar{Z}/L$.

Application of similarity arguments to diffusion in convective mixing

An important extension of the foregoing Monin–Obukhov similarity arguments may be made when the vertical heat flux and the associated buoyant production of turbulent energy are the dominant factors. In that case the friction velocity u_* is no longer relevant and, following Yaglom (1972), it may then be argued that the rate of vertical diffusion from a surface release should depend only on height z and the buoyant production term $gH/\rho c_p T$, which on dimensional grounds implies

$$d\bar{Z}/dt \propto (gH\bar{Z}/\rho c_p T)^{\frac{1}{3}} \tag{3.54}$$

In the surface layer one might therefore expect to see some approach to this relation when $\bar{Z}$ reaches a large enough value relative to $-L$. However, a crucial further consideration emerges from the scaling arguments for the whole depth of the boundary layer affected by convective mixing (see 2.2), to the effect that the appropriate vertical scale is not the local height as in Eq. (3.54) but the depth h_c of the mixed layer. From this point Deardorff and Willis make the fundamentally new hypothesis that vertical diffusion should be scaled in terms of h_c and should be related to time of diffusion scaled in terms of the buoyant time-scale h_c/w_*, in a universal similarity form

$$\sigma_z(t)/h_c = f(w_* t/h_c) \tag{3.54a}$$

Deardorff and Willis have discussed in a series of papers the evidence provided for the foregoing hypothesis, and for the form of the function, from calculations of particle dispersion in a numerically simulated mixed layer and from observations of dispersion in a convectively-mixed laboratory tank. For the atmospheric boundary layer the evidence in respect of the field of turbulence has been noted in 2.6 and the rather limited information on dispersion is considered in 4.6. The discussion in 2.6 also includes comments on the conditions of wind speed, surface roughness and heat flux required for mixed-layer scaling to hold.

The mean horizontal speed of particles and the interpretation of dispersion as a function of distance

We consider now the magnitude to be assigned to c in the expression for $\mathrm{d}\bar{X}/\mathrm{d}t$ in Eq. (3.34), which is simply a statement of a hypothesis for the mean horizontal speed of an ensemble of particles after each has travelled for time T. If these diffusing particles are regarded as marked particles of the fluid then by definition

$$\mathrm{d}\bar{X}/\mathrm{d}t = \int_0^\infty u(z)P(z, t)\mathrm{d}z \Big/ \int_0^\infty P(z, t)\mathrm{d}z \tag{3.55}$$

where $P(z, t)$ is the probability that diffusing particles are at height z at time t. If as already assumed in much of the foregoing argument the probability density distribution of the ensemble of particles is identified with $\chi(z, t)$ in the solution of the diffusion equation for an instantaneous plane source of particles at ground level (Eq. (3.37)), with gradient transfer according to $K(z) = K z^n$, then

$$\mathrm{d}\bar{X}/\mathrm{d}t = \int_0^\infty \bar{u}(z)\, \chi(z, t)\mathrm{d}z \Big/ \int_0^\infty \chi(z, t)\mathrm{d}z \tag{3.55a}$$

where $\chi(z, t)$ has the form in Eq. (3.42) or, since $\int_0^\infty \chi(z, t)\mathrm{d}z = Q$ the instantaneous mass of particles per unit area,

$$\mathrm{d}\bar{X}/\mathrm{d}t = \frac{1}{Q}\int_0^\infty \bar{u}(z)\, \chi(z, t)\mathrm{d}z \tag{3.55b}$$

For the neutral surface-stress layer of the atmosphere, with Monin–Obukhov forms of $\bar{u}(z)$ and K_M, diffusivity of material being assumed equal to that for momentum, r in Eq. (3.42) is unity. The solution of Eq. (3.55b) for these conditions was given by Chatwin (1968) and the resulting expression for c_T (the subscript T is included to emphasize the reference to particles that have travelled for time T) is given in Table 3.II. For the details of the integration see the original paper by Chatwin or the later discussion by F. B. Smith (1978).

The corresponding analysis for thermally stratified flow presents further problems in that the Monin–Obukhov expressions for $\bar{u}(z)$ and $K_M(z)$ are not in tractable mathematical form for analytical solution of Eq. (3.37). However, to the extent that these expressions may be approximated by simple power-law forms, as already proposed by Calder (1949) for $\bar{u}$ and by Chaudhry and Meroney (1973) for K in stratified flow, the solution for $\chi(z, t)$ in Eq. (3.42) may continue to be adopted, permitting straightforward integration of the r.h.s. of Eq. (3.55b) and leading to the additional expression for c_T in Table 3.II.

Table 3.II – Summary of theoretical expressions for c and $\bar{Z}$ (c is defined in the hypotheses $d\bar{X}/dt = u(c_T\bar{Z})$ and $dx/dt = u(c_x\bar{Z})$.

For $u(z) = (u_*/k)\ln(z/z_0)$, $K = ku_*z$

$$c_T = e^{-\gamma} = 0.56 \ (\gamma = \text{Euler's constant} = 0.5772)$$

$$d\bar{Z}(\bar{X})/d\bar{X} = ku_*/\bar{u}(c_T\bar{Z})$$

$$k^2\bar{X}' = \bar{Z}'\,[\ln\bar{Z}' - (1-\ln c)(1-1/\bar{Z}')], \quad \bar{X}' = \bar{X}/z_0, \quad \bar{Z}' = \bar{Z}/z_0$$

For $u(z) = u_1z^m$, $K(z) = K_1z^n$

$$s_T = 2-n, \quad s_x = 2+m-n$$

$$p = [s\{\Gamma(2/s)/\Gamma(1/s)\}^s]^{1/(1-s)}, \quad p_T \text{ when } s = s_T \text{ etc.}$$

$$c = [\Gamma((1+m)/s)]^{1/m}\,[\Gamma(1/s)]^{1-1/m}/\Gamma(2/s); \quad c_T \text{ when } s = s_T \text{ etc.}$$

$$d\bar{Z}(\bar{X})/d\bar{X} = K(p_T\bar{Z})/p_T\bar{Z}\,u(c_T\bar{Z})$$

$$\bar{Z}(\bar{X}) = [s_xK_1\bar{X}/u_1c_T{}^m p_T{}^{(1-n)}]^{1/s_x}$$

$$d\bar{Z}(x)/dx = K(p_x\bar{Z})/p_x\bar{Z}\,u(p_x\bar{Z})$$

$$\bar{Z}(x) = [s_xK_1x/u_1p_x{}^{(s_x-1)}]^{1/s_x}$$

Note that the parameter $p = 1$ at $s = 1$, but on either side of that value of s it rises rapidly to 1.52 when $s = 0.99$ or 1.01, and is 1.57 for $s = 2$, and 1.50 for $s = 0.5$. The parameter c rises slowly with increase in s and m, and for example with the combinations $s = 0.5$, $m = 0.05$ (very unstable) or $s = 1.5$, $m = 0.5$ (stable) c changes from 0.41 to 0.83. However, in specifying the mean speed of travel of particles c is effective as c^m, which changes only from 0.96 to 0.92 in the foregoing example. Thus, provided the appropriate shape of wind profile is used, the typical departures of c from the neutral value of 0.56 are of negligible importance, as noted by Chaudhry and Meroney and by van Ulden.

Specification of c_T and hence of $d\bar{X}/dt$ provides for expression of $\bar{Z}(T)$ in terms of $\bar{X}(T)$ for a given time of travel T, and the results are collected in Table 3.II. At this point, however, it is to be emphasized that in the practical context of air pollution the interest is in the distribution of particles at a specified distance x from a source, hence in $\bar{Z}(x)$, which as will be seen below is not identical with $\bar{Z}(\bar{X})$, though the difference is typically small and has tended to be disregarded hitherto. It may be argued of course that if it is necessary to depend on the results of gradient-transfer theory for the ultimate expressions of similarity theory, as appears to be the case at present, then one might as well use the gradient-transfer solutions already available in the x, z format, as provided by Roberts's solution in Eq. (3.20) for a crosswind line source and simple power-law representations of u and K. The merits of this procedure have recently been re-emphasized by van Ulden and Nieuwstadt in their discussion with Horst (Horst *et al.*, 1980). To complete the summary of the developments to date the corrresponding expressions for c_x and $\bar{Z}(x)$ are included in Table 3.II (those for $d\bar{Z}/d\bar{X}$ and $d\bar{Z}/dx$ being in the form given by van Ulden and Nieuwstadt).

Generally the difference between the $\bar{X}(T)$ and x representations are not large in practical terms except for strong wind shear (large m or small $\bar{Z}/z_0$). For neutral flow Hunt and Weber (1979) show that

$$\bar{Z}(x)/\bar{Z}(\bar{X}) = 1 - (2\ln(c\bar{Z}(\bar{X})/z_0))^{-1} \tag{3.56}$$

indicating that $\bar{Z}(\bar{X})$ is an overestimate of the required $\bar{Z}(x)$, but to an extent that decreases with distance and increases with z_0 or m. Broadly similar results for neutral flow have been obtained by F. B. Smith (1978), in analyses that included numerical solutions of the two-dimensional equation for $\chi(z, x)$, and a demonstration of the correctness of the numerical solution by comparing it with the analytic solution in the case of a power-law wind profile. Smith's comparison of $\bar{Z}(\bar{X})$ from analytic solution with $\bar{Z}(x)$ from numerical solution, using the logarithmic velocity profile in both cases (in contrast to Hunt and Weber's comparison using analytic solution and fitted power-law wind profile for $\bar{Z}(x)$) indicates $\bar{Z}(\bar{X})$ greater than $\bar{Z}(x)$ by 15 per cent for $x = 1$ km and $z_0 = 0.1$ m. In this example $cZ/z_0 \cong 200$, for which Hunt and Weber's result in Eq. (3.56) indicates an overestimation of 10 per cent. Smith's comparison using the analytic solution throughout shows the ratio $\bar{Z}(\bar{X})/\bar{Z}(x)$ rising to a maximum of 1.25 when $m = 0.25$ (for arbitrary value of x)

3.4 STATISTICAL THEORY OF DISPERSION FROM A CONTINUOUS SOURCE

Taylor's analysis for homogeneous turbulence

The starting point of the treatment is the essentially mathematical (statistical) analysis of the properties of a continuously varying quantity already considered in 2.1. Taylor's discussion is at first in terms of fluctuations of barometric pressure, though the same argument could be immediately stated for the random fluctuations of *particle* velocity which were the ultimate subject of application, and this discussion includes a demonstration that the usual laws of differentiation may be applied to the mean values of fluctuating variables and their products. Then, if X is the deviation of a typical particle, due to the eddy velocity u', after a time t, and $\overline{X^2}$ the mean square of a large number of values X,

$$\frac{\mathrm{d}\overline{X^2}}{\mathrm{d}t} = 2\overline{X\frac{\mathrm{d}X}{\mathrm{d}t}}$$

$$= 2\overline{Xu'}$$

$$= 2\int_0^t \overline{u'(t)u'(t+\xi)}\,\mathrm{d}\xi \qquad (3.57)$$

If the turbulence is homogeneous and stationary, i.e. if the average properties are uniform in space and steady in time, the velocity product may be replaced by $\overline{u'^2}R(\xi)$, where $R(\xi)=\overline{u'(t)u'(t+\xi)}/\overline{u'^2}$ is a correlation coefficient of Lagrangian type. With $R(\xi)$ and $\overline{u'^2}$ both independent of the time origin the familiar results then follow,

$$\frac{\mathrm{d}\overline{X^2}}{\mathrm{d}t} = 2\overline{u'^2}\int_0^t R(\xi)\,\mathrm{d}\xi \qquad (3.58)$$

and $$\overline{X^2} = 2\overline{u'^2}\int_0^T\int_0^t R(\xi)\,\mathrm{d}\xi\,\mathrm{d}t \qquad (3.59)$$

where X is now the deviation of the particle in time T.

The mean square of the deviations undergone by a single particle in a succession of equal time intervals was thus completely expressed in terms of the mean square eddy velocity of the particle, and the so-called Lagrangian correlation coefficient, $R(\xi)$, between the velocity of the particle at time t and that at time $t+\xi$. Two simple deductions from this result followed immediately, without any assumptions regarding $R(\xi)$, other than the obvious ones that it should be unity when $\xi=0$ and is effectively zero for large ξ. Hence, for small T

$$\overline{X^2} = \overline{u'^2}T^2 \qquad (3.60)$$

and for large T

$$\overline{X}^2 = 2\left(\int_0^\infty R(\xi)\,\mathrm{d}\xi\right)\overline{u'^2}T \qquad (3.61)$$

the integral in the bracket being the Lagrangian *time-scale* t_L.

The meaning of the result in relation to the dispersion of material emitted continuously from a source at a fixed point in the flow may now be considered, with the necessary assumptions that the material particles introduced do not affect the flow, and that they assume the fluid velocity completely. In effect we are now considering a large number of particles as they pass a given *fixed* point in succession. The statistics of the displacements of these separate particles, from the position they would otherwise occupy as a result of the mean flow alone, are clearly identical with those of the displacements of a single particle observed a large number of times. This follows merely by virtue of the original assumption that $\overline{u'^2}$ for the single particle is a constant, independent of time and representative of the whole field of turbulence. Hence, with the usual system of coordinates and eddy velocities, and with the source at (0, 0, 0), the component variances of the particle displacements from the appropriate centre of gravity, i.e. $(x, 0, 0)$, after a time of travel $T = x/\bar{u}$, are (for $\bar{u}$ constant in space and time)

$$\overline{X^2} = 2\overline{u'^2} \int_0^{x/\bar{u}} \int_0^t R_x(\xi) \, d\xi \, dt$$

$$\overline{Y^2} = 2\overline{v'^2} \int_0^{x/\bar{u}} \int_0^t R_y(\xi) \, d\xi \, dt \tag{3.62}$$

$$\overline{Z^2} = 2\overline{w'^2} \int_0^{x/\bar{u}} \int_0^t R_z(\xi) \, d\xi \, dt$$

different values of $R(\xi)$ for the component motions of the particles being recognized. Expressions corresponding to the limiting conditions of Eqs. (3.60) and (3.61) apply with $T = x/\bar{u}$, so that at very short distances the variances are proportional to x^2, and at very long distances to x. Thus the spread of the plume, as represented by the standard deviation in the crosswind and vertical directions, starts off with a linear form (proportional to x) and ultimately tends to a parabolic form proportional to $\sqrt{x}$).

Development and generalization of Taylor's analysis

The results stated above have been the subject of much restatement and elaboration, notably by Kampé de Fériet and Batchelor. Kampé de Fériet (1939) has expressed the equation for $\overline{X^2}$ also in the form

$$\overline{X^2} = 2\overline{u'^2} \int_0^T (T - \xi) R(\xi) \, d\xi \tag{363}$$

and by applying the Fourier-transform relation between $R(\xi)$ and the corresponding Lagrangian spectrum function $F_L(n)$, as in Eq. (2.10), this becomes

$$\overline{X^2} = \overline{u'^2} \int_0^\infty F_L(n) \left(\frac{1 - \cos 2\pi nT}{2(\pi n)^2} \right) dn \tag{3.64}$$

which with further simple transformation may be written

$$\overline{X^2} = \overline{u'^2}T^2 \int_0^\infty F_L(n) \frac{\sin^2(\pi n T)}{(\pi n T)^2} \, dn \tag{3.65}$$

Similar expressions may be written for other components. A statement of this spectral representation, in tensor notation, has been set out by Batchelor (1949) in an extended discussion of diffusion in homogeneous turbulence. The result means that at short time (T), when the integral becomes unity (and $\overline{X^2} = \overline{u'^2}T^2$), the oscillations (of particle velocity) of all frequency contribute to the dispersion exactly as they do to the turbulent energy. At larger value of T the slower oscillations progressively dominate the dispersion; in effect the high-frequency components merely oscillate the position of the particle, whereas the low-frequency components tend to displace it in a more sustained way.

Batchelor's discussion contains an important recognition regarding the applicability of the Kolmogorov theory outlined in 2.5. Since the dispersion is at no time dominated by the high-frequency oscillations, i.e. by the small-scale structure, it follows that it is not possible to apply the theory to the dispersion of material from a *continuous source at a fixed position.* The relation between the properties of dispersion, and those implied by transfer theory, is also discussed. It has been seen that at large T the statistical theory predicts a spread proportional to $\sqrt{T}$, in agreement with Eq. (3.13). Taking this in conjunction with experimental evidence, from studies in effectively homogeneous turbulence in wind tunnels, that the probability distribution of the fluid particle displacement is normal, it follows that the diffusion may be represented by a differential equation of the Fickian type, with the effective K defined by

$$K = \frac{1}{2} \frac{d\overline{X^2}}{dt} = \overline{u'^2} \int_0^{t_1} R(\xi) \, d\xi = \overline{u'^2} t_L \tag{3.66}$$

where t_1 is the value of ξ beyond which $R(\xi)$ remains zero. Formally the same representation may be adopted for all T, but from Eq. (3.58) the effective K is at first zero, increases with time at first linearly and then more slowly, and then finally tends to the constant value of Eq. (3.66). It was recognized at an early stage in the study of the spread of smoke plumes in the atmosphere over short distances that an increase in Fickian K with distance of travel was implied. In terms of the foregoing analysis this result is due to the slower oscillations of the eddy velocities, i.e. the larger eddies. This is not, as is sometimes implied however, a matter of the subsequent action of the larger eddies as dispersion increases, for all frequencies of oscillation are effective from the beginning *when the source is continuous.* The real reason is that slower oscillations of dominating amplitude tend to maintain $R(\xi)$ at its initially high value, whereas, true Fickian diffusion imples $R(\xi)$ falling to zero in a time very small compared with the time of travel involved.

This leads us to consider a simple way of 'correcting' solutions of the diffusion equation to make them consistent with the statistical theories of diffusion. Such a correction would considerably extend the usefulness of these solutions, especially both for elevated sources and where they are in analytical form. The nature of the correction is admittedly a formalism and has no general rigorous basis, it simply extends in application an obvious transformation valid for stationary homogeneous turbulence to general fields. The transformation is to replace the time of travel T_k or downwind distance x_k appearing in the K-theory solutions by transformed variables which represent the true times or distances, using (in the case of time) the relationship:

$$T_k = T - t_L(1 - \exp(-T/t_L)) \tag{3.66a}$$

where t_L is the Lagrangian time-scale at the source height. This relation comes from equating the plume spread, in homogeneous turbulence, derived from K-theory with that from Taylor's statistical theory:

$$\sigma_p{}^2 = \underset{\text{(A)}}{2\,KT_k} = 2\,\sigma_v{}^2\,t_L{}^2\left(\underset{\text{(B)}}{\frac{T}{t_L} - (1 - \exp(-T/t_L))}\right) \tag{3.66b}$$

At large times $T_k = T$ and therefore $K = \sigma_v{}^2\,t_L$ as in Eq. (3.66). At smaller times equality is reached only if T_k is defined as in (3.66a). The transformation is of course extremely simple and reasonably valid provided T_k is close to T by the time the variation of t_L with z becomes important and effectively invalidates the (B) term in (3.66b).

Understanding of the various features of dispersion in homogeneous turbulence is often helped by more detailed reference to the spectrum representation in Eq. (3.65). The spectrum weighting function, which imposes the combined influence of eddy frequency (n) and of time of travel (T), is the same as that already discussed in 2.1 in relation to measurements of turbulent velocities at a fixed point in the flow, and the shape is that given in Fig. 2.2. The effect on the variation of dispersion with time of travel is represented in Fig. 3.4. Here, the outer curve represents the shape of the full spectrum of turbulent velocity (of a particle), the form equivalent to a simple exponential correlogram as in Fig. 2.10 being adopted for convenience. For generality and simplicity the ordinates are $\pi nF(n)$, the maximum value of which is unity, and the abscissae are $2\pi t_L n$ on a logarithmic scale. The integral in Eq. (3.65) is proportional to the area under this curve when it is weighted by the expression $\sin^2(\pi nT)/(\pi nT)^2$, and the inner curves show the effect of this weighting function for various values of T/t_L. When $T/t_L = 0$ the original spectrum curve applies, the integrand is unity, and $\overline{X^2} = \overline{u'^2}T^2$. In general $\overline{X^2}/T^2$ is proportional to the areas under the curves, and thus diminishes systematically with T as the high-frequency side of the spectrum is effectively cut off.

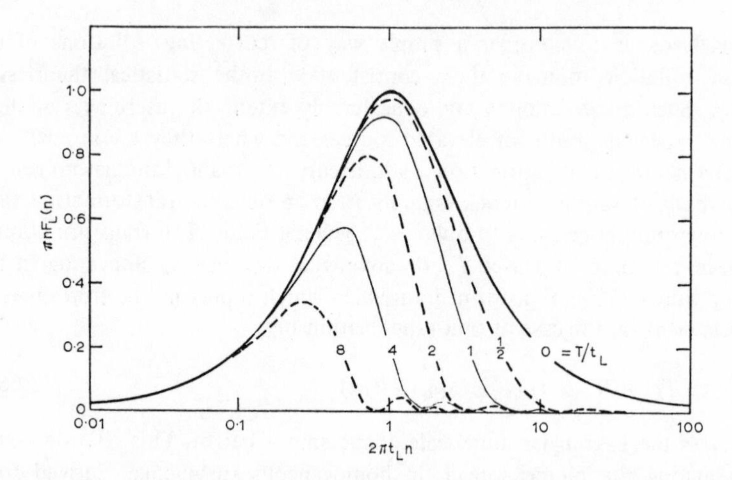

Fig. 3.4 – Spectrum representation of the effect of time of travel on the dispersion of particles from a continuous point source in homogeneous turbulence. T is the time of travel, t_L the Lagrangian time-scale defined by $\int_0^\infty R(\xi)\, d\xi$, and $F_L(n)$ is the corresponding Lagrangian spectrum function, with $R(\xi)$ assumed to be of exponential form.

Referring to the derivation of the appropriate weighting function in 2.1 it will be seen that this function represents the effect of averaging the fluctuating velocity over a specified time interval before determining the variance. Thus, the quantity

$$\overline{u'^2} \int_0^\infty F_L(n) \frac{\sin^2(\pi n T)}{(\pi n T)^2}\, dn$$

is simply the variance of the turbulent velocity when the latter has been averaged over a time T, and following the subscript nomenclature of 2.1, Eq. (3.65) may be written in the simple form

$$\overline{X^2} = \overline{u'^2_{\infty,T}}\, T^2 \tag{3.67}$$

This is merely the definition of X as the product of the time of travel and the velocity of the particle *averaged over the time of travel.* The result is of course implicit in the usual form of Taylor's relation, Eq. (3.59), though this is not obvious immediately. In the form Eq. (3.67) it seems first to have been recognized by Ogura (1952 and later papers), although no discussion was given of the attractively simple interpretation which it provides, in that the dispersion relation in Eq. (3.65) may be obtained directly, by writing the definition of $\overline{X^2}$ as in Eq. (3.67), and inserting the form of the weighting function (for averaging) as simply derived in 2.1 and stated in Eq. (2.21).

So far it has been assumed that the time over which the history of the single particle is observed and, correspondingly, the time over which a continuous source is maintained, are both long enough for all the turbulent velocity fluctuations, however low in frequency, to exert their effect statistically. It is immediately clear from experience, however, that this will often not be so in the case of diffusion in the atmosphere. Very close to the source it is physically obvious that the linear law, say $\overline{Y^2} = \overline{v'^2}T^2$, must apply for release of particles over any arbitrary time τ, with $\overline{v'^2}$ now representing the velocities *at the position of the source* over the time τ. The magnitude of $\overline{v'^2}$ is known to be a function of sampling time, τ, and in many cases of atmospheric turbulence will still be increasing with τ for times which may be of practical interest as regards release of airborne material.

A formal expression in entirely Lagrangian terms follows directly from the relations in 2.1. Using the subscript notation of 2.1 for sampling duration and averaging time, if the spectrum of particle velocity is represented by a total variance $\overline{v'^2_{\infty,0}}$, with spectrum shape $F_L(n)$, then the dispersion $\overline{Y^2}$, observed by measuring particle displacements in a time T from initial positions which are uniformly distributed in a period τ in the trajectory of a typical particle, is given by

$$\overline{Y^2} = \overline{v'^2_{\tau,T}}T^2$$

and writing $\overline{v^2_{\tau,T}}$ from Eq. (2.30)

$$\overline{Y^2} = \overline{v'^2_{\infty,0}}T^2 \int_0^\infty F_L(n) \left(1 - \frac{\sin^2 \pi n\tau}{(\pi n\tau)^2}\right) \frac{\sin^2 \pi nT}{(\pi nT)^2}\, dn \tag{3.68}$$

The practical expression of the dispersion of particles released serially from a fixed point, over a finite time (or the equivalent case of the dispersion from a permanent continuous source when the dispersion is observed or sampled for a finite time), requires that Eq (3.68) should be transformed into Eulerian terms, and this feature is considered later.

Extension of the analysis to specially selected particles

The analysis which leads to Eq. (3.57) *et seq.* refers to a single typical particle, and the subsequent interpretation for a continuous release of particles refers to all particles passing through a specified position. The consequence of restricting attention to only certain particles has been examined by F. B. Smith (1968), applying the following conditions for selection:

(i) specified initial velocity,
(ii) passage through *two* specified positions, i.e. one other position downwind of the source,
(iii) a combination of (i) and (ii).

Smith makes the assumption that the Lagrangian correlation function depends only on the time-lag ξ and not on the particular magnitude of initial velocity that might be selected. In this case there should be a linear regression between $v(0)$ and $v(t)$, and for a demonstration of this use is made of some geostrophic trajectory data. On this basis a whole complex of statistical relations is obtained, including those for mean crosswind displacement and crosswind spread of particles conditioned as above. The results have practical interest in relation to the problem of estimating the track of a cloud of windborne material.

Smith's analysis also suggests a novel basis, which has yet to be exploited, for obtaining the Lagrangian time-scale from measurements of diffusion. It is evident from Eq. (3.60), as will be discussed further in the next section, that when no special selection of particles is made the spread at short distances is independent of the time-scale. The influence of the time-scale appears progressively as the distance is increased and ultimately amounts to ${t_L}^{1/2}$. Generally the quality of diffusion measurements deteriorates with increasing distance and the consequence is that such measurements do not sensitively determine, t_L. Smith's analysis shows that if attention is confined to particles with specified initial velocity, $v(0)$, the crosswind spread of these particles is

$$\overline{Y^2} = 2\overline{v'^2_{\infty,0}} \int_0^T \int_0^r [R(r-s) - R(r)R(s)]\, \mathrm{d}s\, \mathrm{d}r \tag{3.69}$$

For the case of a correlogram of exponential form, and for *small T*,

$$\overline{Y^2} \cong 2\overline{v'^2_{\infty,0}}\, T^3/3t_L \tag{3.69a}$$

i.e. the short-range spread is no longer independent of t_L but is inversely proportional to ${t_L}^{1/2}$. Thus in principle it should be possible to derive t_L fairly accurately from diffusion measurements if provision is made for restricting the source emission effectively to occasions when the wind has a specified v-component. In practice an approach to this might be achieved by a control system which triggers release only when the instantaneous wind direction is near the mean, i.e. $v(0) \cong 0$.

Interest in this conditioned-particle analysis is carried a stage further in Gifford's (1982 a, b) theoretical treatment, based on a form of Langevin's equation. As Smith (1982) has pointed out in subsequent published discussion, his (1968) statistical relations may be used to give exactly the same result as that obtained by Gifford i.e.

$$\sigma_p^2 = 2\, \overline{v'^2_{\infty,0}}\, t_L^2 \left[\frac{T}{t_L} - \left(1 - \mathrm{e}^{-\frac{T}{t_L}}\right) - \frac{1}{2}\left(1 - \frac{\overline{v_0'^2}}{\overline{v'^2_{\infty,0}}}\right)\left(1 - \mathrm{e}^{-\frac{T}{t_L}}\right)^2 \right] \tag{3.69b}$$

with $\overline{v_0'^2} = \overline{v_0^2} - \overline{v}_0^2$ interpreted as the variance for an ensemble of particles with a range of v_0 in place of the single value originally considered. However, the proper interpretation in relation to observational data on dispersion is not a simple matter, as Smith brings out. Suppose an ensemble of observations were available on alongwind sections of a continuous plume, selected as having the same $\overline{v_0}$ and $\overline{v_0^2}$. Complete description of their dispersion requires measurement of the spread σ_{v_0} of the particles in each section and y_{v_0} their mean displacement from the long-term mean centre-line of the plume. In Smith's analysis it is clear that the observational counterpart of Eq. (3.69b) is

$$\sigma_p{}^2 = \lim_{I\to\infty} \frac{1}{I} \sum_{i=1}^{I} (\sigma_{v_0}^2(i) + y_{v_0}^2(i)) - \left(\lim_{I\to\infty} \frac{1}{I} \sum_{i=1}^{I} y_{v_0}(i)\right)^2 \tag{3.69c}$$

in which i defines an individual section and runs from 1 to ∞. In practice, however, observational data refer to the term

$$\lim_{I\to\infty} \frac{1}{I} \sum_{i=1}^{I} \sigma_{v_0}^2(i)$$

The term y_{v_0} has not been specified and its net addition to the summation in Eq. (3.69c) cannot in general be assumed negligible.

In a sense it is perhaps tempting to regard Eq. (3.69b), with $v_0'^2$ small, as representative of the dispersion of a cluster of marked particles, and disregard of the difficulty that the range of v_0 is then open to arbitrary specification may seem justified by two properties, firstly that when $(\overline{v_0'^2})^{\frac{1}{2}}$ is less than say one-tenth of the total velocity fluctuation $(v_{\infty,0}'^2)^{\frac{1}{2}}$ Eq. (3.69b) is a close approximation to Eq. (3.69), and secondly that as already noted the latter equation yields $\sigma_y \propto T^{\frac{3}{2}}$ as $T\to 0$, which is similar to the result deduced (see 3.7) for the relative diffusion of pairs of particles under the action of the inertial subrange of the spectrum. However, apart from the discrepancy noted in the previous paragraph regarding the ensemble averaging, it should be stressed that Eqs. (3.69) and (3.69b) are fundamentally 'single-particle' in character, i.e. every particle can be assumed to diffuse independently of every other particle, whereas relative diffusion is necessarily concerned with the intercorrelation between the motions of pairs of particles, and any assumption that this feature is somehow implicit in Eqs. (3.69) and (3.69b) requires total justification.

3.5 APPLICATIONS AND EXTENSIONS OF TAYLOR'S THEORY

The significance of the shape of $R(\xi)$ or $F_L(n)$

It is of interest to consider first the consequences of different assumptions about the mathematical form of $R(\xi)$ or $F_L(n)$ substituted in Eq. (3.59) or (3.65). In his original discussion Taylor adopted a simple exponential decay of $R(\xi)$

$$R(\xi) = \exp(-\xi/t_L) \tag{3.70a}$$

a form which some have argued to be physically justified. Several alternative forms were later suggested, all of which gave results intermediate between those which follow from Eq. (3.70a) and those which are given by a simple combination of the limiting forms in Eq. (3.60) and (3.61). These were discussed in the previous edition of the book but will not be repeated here, since as will be seen more interest is now attached to the following more recent proposals:

$$F_L(n) = 4t_L(1 + 4nt_L)^{-2} \tag{3.70b}$$

$$F_L(n) = 4t_L(1 + 6nt_L)^{-5/3} \tag{3.70c}$$

$$R(\xi) = (1 + \xi/t_L)^{-2} \tag{3.70d}$$

The spectral form in Eq. (3.70c) has been found to give a reasonable fit to certain fixed-point (Eulerian) measurements of turbulence, while Eq. (3.70b) is a modified form which correctly represents the expected behaviour of the Lagrangian spectrum at high frequency (see Eq. (2.106)). The correlogram form in Eq. (3.70d) was suggested by Phillips and Panofsky (1982) in the course of interpreting observational data on dispersion (see Fig. 4.7).

When the dispersion and the time of travel are expressed in the dimensionless forms $D = (\sigma_y/\sigma_v t_L)^{\frac{1}{2}}$ say and $T' = T/t_L$, originally suggested by Frenkiel (1952a), it is evident from Eq. (3.65) that $D = f(T')$ in which the form of the function f is determined by the shape of the spectrum function $F_L(n)$. Magnitudes of D according to the foregoing forms of $R(\xi)$ or $F_L(n)$ are given in Table 3.III and are shown graphically in Fig. 3.5, no attempt being made to distinguish between those according to Eq. (3.70b) and (3.70d). Although a reduction in D from the values associated with a simple exponential correlogram has been produced the overall variation is still small, the maximum effect being near $T' = 2$.

Table 3.III – Values of the dispersion parameter $D = (\overline{Y^2}/t_L^2\overline{v'^2})^{\frac{1}{2}}$ for a continuous point source in homogeneous turbulence, computed from Eq. (3.59) or (3.65).

$T' = T/t_L$ $R(\xi)$ or $F_L(n)$	0.1	0.2	0.4	1.0	2.0	4.0	10.0
1. $\exp(-\xi/t_L)$	0.098	0.193	0.375	0.86	1.51	2.46	4.24
2. $4t_L(1 + 6nt_L)^{-5/3}$	0.093	0.180	0.338	0.743	1.29	2.12	3.85
3. $4t_L(1 + 4nt_L)^{-2}$	0.096	0.187	0.354	0.780	1.34	2.20	3.93
4. $(1 + \xi/t_L)^{-2}$	0.097	0.188	0.356	0.783	1.34	2.19	3.90

Notes.

For Case 1 $D^2 = 2\exp(-T') + T' - 1$.

For Case 4 $D^2 = 2[T' - \ln(1 + T')]$.

For Cases 2 and 3 D was obtained by numerical integration of Eq. (3.65) rewritten for crosswind dispersion.

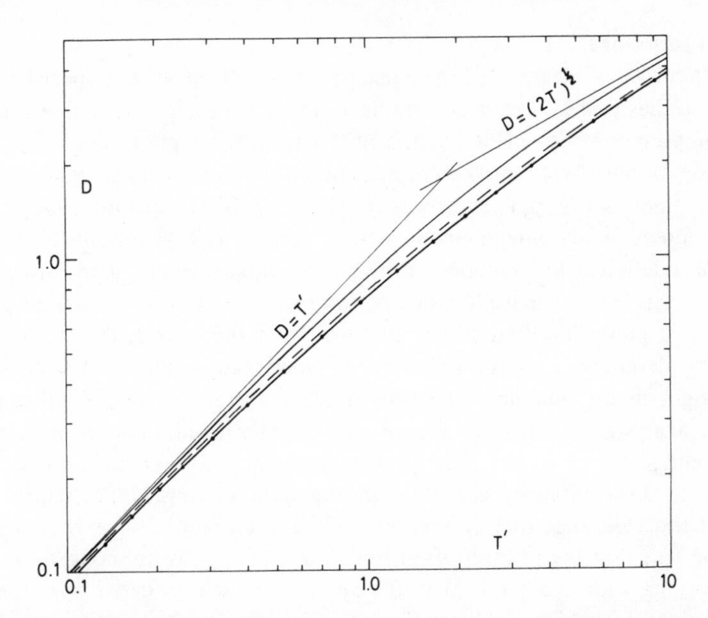

Fig. 3.5 – Examples of the effect of correlogram or spectrum shape on the relation between dispersion and time of travel in homogeneous turbulence. $D = \sigma_y / \sigma_v t_L$ as a function of $T' = T/t_L$, calculated from Eq. (3.59) or (3.65) with various assumed forms of $R(\xi)$ or $F_L(n)$.

———— $R(\xi) = \exp(-\xi/t_L)$

– – – – $R(\xi) = (1 + \xi/t_L)^{-2}$ or $F_L(n) = 4t_L(1 + 4nt_L)^{-2}$

–·–·–·– $F_L(n) = 4t_L(1 + 6nt_L)^{-5/3}$

The simple exponential form of autocorrelation function, as in Eq. (3.70a), has been a popular assumption, though the fundamental justification is far from clear. As pointed out by Tennekes (1978), on dimensional grounds the properties of the high-frequency (inertial subrange) Lagrangian velocity fluctuations should be in accordance with $F_L(n) \propto n^{-2}$ and $R(\xi) = 1 - a\xi$ with a a constant, and support for the spectral form is provided by Hanna's balloon/tetroon results noted in 2.7. The exponential form of correlogram is mathematically consistent with these inertial subrange forms in the $n \to \infty$ and $\xi \to 0$ limits, but so also are the spectral and correlation forms in Eq. (3.70b) and Eq. (3.70d) (see 4.4 for further comment on this point).

Note that in the virtually linear range $\sqrt{(\overline{Y^2})}$ is independent of the scale as well as the shape of the correlogram, and is determined by the intensity of turbulence alone, while in the virtually parabolic range the scale of turbulence is effective only as the square root. This brings out the point that a good first approximation to the dispersion from a continuous point source in a field of homogeneous turbulence will be obtainable from a good estimate of the intensity of turbulence and a rough estimate of the Lagrangian scale of turbulence.

Explicit forms of $R(\xi)$

The difficulties of measuring Lagrangian properties except in very special experimental studies (see 2.7) have already been noted. On the purely theoretical side some progress has been made but the final step in the application of Eq. (3.59) and its developments still rests on some form of indirect empirical argument.

The first useful application was provided by O. G. Sutton (1932, 1934) who proposed easily integrated power-law forms of $R(\xi)$ for substitution in Taylor's relations. In combination with the mixing-length gradient-transfer approach this led to the well-known Sutton formulae (1947a, 1947b) for plume spread as a power-law function of distance from the source, the index being explicitly determined by the index in the power-law variation of wind speed with height in the boundary layer. It should be recalled, however, that from Taylor's analysis, on which the derivation is directly based, the relation between spread and distance is not strictly expressible in power-law form, and could only be so represented by accepting an exponent varying with distance, from unity at short distance to $\frac{1}{2}$ at long distance. The difference is partly concerned with the fact that the integral of Sutton's form of $R(\xi)$ is not convergent, and this together with the point that the analysis is used to derive the stress in sheared flow, for which condition Taylor's theorem does not apply, introduces certain theoretical difficulties.

An alternative approach is contained in the idea of relating the Lagrangian correlogram or spectrum, if only empirically, to the corresponding Eulerian properties defined by the velocity fluctuations at a fixed point. This approach is especially attractive in view of the rapid developments in the techniques of sensing and recording these fluctuations. The general idea was present in latent form for some time, in discussions by Sutton (1953, p. 263) and unpublished analyses by Calder, though in these cases with the special implication that the Lagrangian and Eulerian correlograms were identical. The next step came with attempts to exploit the Taylor expression in reverse, namely to use it with carefully designed diffusion observations in order to deduce the essential properties of R(ξ) for comparison with Eulerian measurements. From wind tunnel experiments on the cross-stream diffusion of a plume of gas Mickelsen (1955) demonstrated the following simple scale relation (in the present nomenclature, different from Mickelsen's)

$$\int_0^{x_2}\int_0^{x_1} R(x)\,\mathrm{d}x\,\mathrm{d}x_1 = \overline{v'^2}\int_0^{\xi_2}\int_0^{\xi_1} R(\xi)\,\mathrm{d}\xi\,\mathrm{d}\xi_1$$

when

$$\xi = \frac{x}{B\sqrt{(\overline{v'^2})}}$$

The right-hand side of this equation [see Eq. (3.59)] is one-half of the measured dispersion of the gas, while $R(x)$ is the Eulerian space-correlation coefficient, actually obtained from measurements at a fixed point by applying the familiar space-time transformation stated in Eq. (2.8). The implied relation in the Eulerian and Lagrangian correlograms is

$$R(x) = \frac{1}{B^2}R(\xi) \quad \text{when } \xi = \frac{x}{B\sqrt{(\overline{v'^2})}} \tag{3.71}$$

With the observed value of B, approximately 0.65, it is clear that the above relation cannot hold near ξ or $x = 0$, but aside from this complication the result means that the Lagrangian correlogram fell off much more slowly than the corresponding Eulerian value.

On the basis of preliminary observations in the atmosphere Hay and Pasquill (1957) concluded that here also the Lagrangian correlation fell off much more slowly than did the autocorrelation of the velocity component measured at a fixed point, a result which is also indicated by balloon measurements (see 2.7). In a later study (1959) Hay and Pasquill adopted the simple hypothesis

$$R_L(\xi) = R_E(t) \quad \text{when } \xi = \beta t \tag{3.72}$$

the subscript L referring to the true Lagrangian autocorrelation, the subscript E to the autocorrelation from measurements at a fixed point. This corresponds to Mickelsen's relation only when $B = 1$, $u/\sqrt{(\overline{v'^2})} = \beta$. Making this time-scale transformation in the Fourier-transform expression for the spectrum function [Eq. (2.10)], it is easily shown that

$$nF_L(n) = \beta n F_E(\beta n)$$

where by definition $\int_0^\infty F(n)\,dn = \int_0^\infty nF(n)\,d\log_e n = 1$. The correlogram and spectrum relations are displayed diagrammatically in Fig. 3.6. Substituting $F_L(n)$ from above in Eq. (3.65), written now in the form appropriate to the y-component of diffusion.

$$\overline{Y^2} = \overline{v'^2}T^2 \int_0^\infty \beta F_E(\beta n) \left[\frac{\sin \pi n T}{\pi n T}\right]^2 dn$$

which reduces to

$$\overline{Y^2} = \overline{v'^2}T^2 \int_0^\infty F_E(n) \left[\frac{\sin \pi n T/\beta}{\pi n T/\beta}\right]^2 dn \tag{3.73}$$

When the duration τ of the release of particles from a fixed point source is long enough to include the effect of the whole spectrum of turbulence, Eq. (3.73) is valid with $\overline{v'^2}$ and $F_E(n)$ prescribed for a sampling duration τ at a fixed

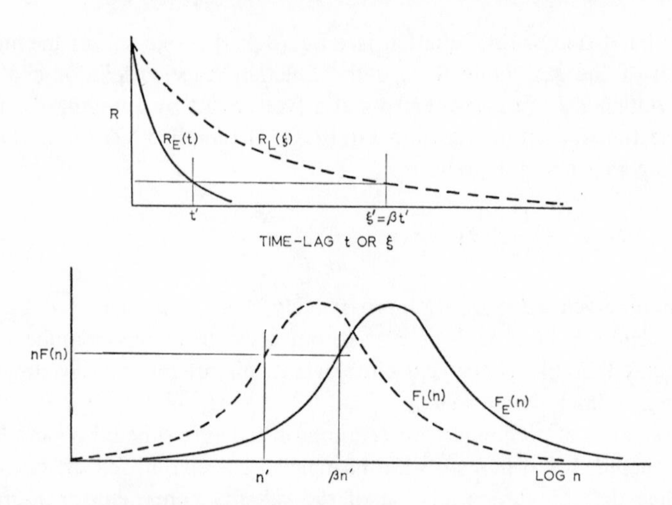

Fig. 3.6 – Hypothetical scale relation between Lagrangian and Eulerian (fixed point) correlograms and spectra. (Hay and Pasquill, 1959).

point. For shorter durations of release, however, the application of the equation requires further qualification. In the treatment developed by Hay and Pasquill it is assumed that $\overline{v'^2}$ [and implicitly $F_E(n)$] can still be identified with a sampling duration equal to τ provided the time of travel T is not substantially larger than τ. This restriction is based simply on the idea that, with the Lagrangian correlation coefficient expected to fall only slowly, the effective variation of v' is dominated by the fluctuations occurring at the point of release and during the period of release.

The influence of a finite duration of sampling on the diffusion of particles has also been discussed by Ogura (1959) in terms of Eq. (3.68). Ogura applies this equation with an assumed specific form of the Lagrangian correlation coefficient, and with $T \leqslant \tau$ (in the present notation, which is different from that used by Ogura). The precise meaning of this restriction is not clear, however, for in Eq. (3.68) τ is implicitly a time in a Lagrangian sense, whereas in the practical case of a fixed point source it is necessary to consider τ in an Eulerian sense. On the other hand Ogura's discussion contains the important physical implication that the restriction is necessary because at greater values of T the diffusion would be occurring relative to a moving origin instead of relative to the fixed position of release.

An approximate argument for the restriction on T in applying Eq. (3.73) is as follows. When τ is the *finite duration of release,* and $T \ll \tau$ it is clear that the dispersion of the particles is determined by their relative velocities as they leave the source, i.e. by the variations of velocity over a duration τ at a fixed point or, approximately, the variations occurring instantaneously over a length $\bar{u}\tau$. When $T = \tau$ the plume of particles is detached from the source, and thereafter travels and grows as an elongated cluster. The spread of the particles will then be determined by relative velocities over a progressively increasing length, and this is a particular aspect of diffusion which is analysed in more detail in the next section. In the present context the important point is that as T increases the effective sampling duration determining $\overline{v'^2}$ is thereby increased beyond the value τ initially set by the duration of release. Similarly, when τ is the finite duration over which an indefinitely maintained plume is *observed*, particles with a separation greater than $\bar{u}\tau$ will be involved, and again the sampling duration determining $\overline{v'^2}$ will be greater than τ. Application of Eq. (3.73) with v' observed over the time τ should therefore be restricted to values of T such that the effective separation of the particles does not exceed $\bar{u}\tau$ by more than a small fraction, say one-tenth, of $\bar{u}\tau$. As a first approximation the additional separation is given by $\sqrt{(\overline{v'^2})}T$, and therefore the required restriction is

$$\sqrt{(\overline{v'^2})}T < \bar{u}\tau/10$$

or

$$T < \tau$$

since $\sqrt{(\overline{v'^2})}/\bar{u}$ in the atmosphere is about 0.1.

The application of Eq. (3.73) does not necessarily require the evaluation of the spectrum function $F_E(n)$, for recalling again the significance of the weighting function $F_E(n)$, the equation may be written shortly as

$$\overline{Y^2}/T^2 = \overline{v'^2_{\tau,T/\beta}} \tag{3.74}$$

where the subscripts refer respectively to the sampling duration and averaging time with which v' is observed. This is simply a statement to the effect that if the magnitude of the Lagrangian variations of particle velocity are identical (statistically) with the variations shown by separate particles as they pass through the point of release, but are on a time-scale longer by a factor β, then the variance of Lagrangian velocity averaged over time-interval T (i.e. $\overline{Y^2}/T^2$) is identical with the variance of velocity at a fixed point averaged over a time-interval T/β. The treatment thus leads to the exceedingly simple result that the dispersion of particles over a useful range of distance downwind of a maintained source is determined completely by appropriately smoothing (averaging) the eddy velocity at the point of release, and then evaluating the variance over the period of release.

From the information in Table 3.III and Fig. 3.5 it appears that if the Lagrangian and Eulerian correlograms are of different shape, the assumption of similar shape as above is unlikely to introduce important error. The only difference which has so far been recognized on both theoretical and observational grounds is in the inertial subranges of the spectra (and correspondingly in the autocorrelation for small lag). As can be seen from Table 3.III the normalized dispersion–time relationship for two spectra which differ only in that respect

(Cases 2 and 3) are negligibly different. On this basis a crucial requirement for the validity of the Hay–Pasquill approach is not that Eq. (3.72) should hold exactly, but that the ratio β of the integral time-scales in the Lagrangian and Eulerian systems should be known. Even in this respect it follows, again from the previous analysis of the dispersion law, that error in the estimation of β (i.e. the error in the estimation of the Lagrangian time-scale) will often not be serious. If the true value is β, and the value assumed is β', it is easily seen that the value of $\sqrt{(\overline{Y^2})}$ derived will be $(\beta'/\beta)^p$ times the true value, where p will vary from zero when T is small to a maximum of 0.5 when T is large.

Application of statistical theory in the atmospheric boundary layer

The foregoing expressions for dispersion are based on the assumptions of stationarity and homogeneity in the field of turbulence, and in respect of the conversion from the original time-dependent form to a distance-dependent form (as in Eq. (3.62)) also on the assumption of uniform mean wind. In the atmospheric boundary layer the flow may often be assumed quasi-stationary for suitably short periods of time (ca. 10 min to 1 hr), but there are variations with height of both the mean wind and the turbulence that cannot always be disregarded.

On the matter of mean wind speed it has already been noted (see the discussion in 3.3) that vertically diffusing particles must be regarded as having an effective speed of travel ($u_e = \overline{X}/T$) which increases as the particles diffuse higher in the boundary layer. Strictly, therefore, in equations like (3.62) the upper limit x/u must be replaced by $\overline{X}/u_e$, and the implications of this requirement will be considered in the discussion of dispersion data in the next chapter.

In respect of the field of turbulence it has for long been held that the characteristic near-linear increase with height of the scale of the w-component in the surface-stress layer makes the Taylor theory inapplicable for vertical spread, except for an elevated source and time of travel small enough for the vertical spread to be small relative to the height of release.

The extension of Taylor's treatment to allow for the increase of scale with height has recently been investigated in an analysis by Hunt and Weber (1979). This starts from the point, brought out in the similarity theory of the previous section, that particles diffusing from the boundary have a non-zero mean vertical speed $\overline{w}$, in contrast to the zero mean value used in describing the statistics of the vertical component of turbulence at a fixed point in the flow. Then in the relation for vertical dispersion, corresponding to Eq. (3.57), for particles released at $T = 0$,

$$\mathrm{d}\,\overline{Z^2}/\mathrm{d}T = 2\int_{-T}^{0} \overline{w(T)w(T+\xi)}\,\mathrm{d}\xi \qquad (3.75)$$

w is replaced by $\bar{w} + \hat{w}$, where $\hat{w}$ is the *Lagrangian* fluctuation about $\bar{w}$ giving

$$\mathrm{d}\,\overline{Z^2}/\mathrm{d}T = 2\bar{w}^2 T + 2\int_{-T}^{0} \overline{\hat{w}(T)\hat{w}(T+\xi)}\,\mathrm{d}\xi \qquad (3\ 75a)$$

Note that the covariance term refers to the variations in velocity of a particle during its travel from the boundary over time T. Beyond this point the develop ment differs from Taylor's treatment in assuming homegeneity in the horizontal and in the vertical only in respect of the variance $\overline{w^2}$, which in spite of the special history of the particles considered is assumed equal to σ_w^2 at a fixed point, the latter quantity being invariant with height in the neutral surface-stress layer. Variation with height of the scale of turbulence is recognized, however and it is here that Hunt and Weber make their crucial assumption, which is in effect that $\frac{1}{\overline{\hat{w}^2}}\int_{-T}^{0} \overline{\hat{w}(T)\hat{w}(T+\xi)}\mathrm{d}\xi$, which defines a special kind of Lagrangian time-scale t_{LT} (the subscript T being included to emphasize that special nature), is related to the Eulerian length scale $l_E(\bar{Z})$ at the mean height of the ensemble of particles, as follows:

$$t_{LT} = \alpha\, l_E(\bar{Z})/\sigma_w \qquad (3.76)$$

Note that this relation is formally identical with that derived theoretically for *homogeneous* turbulence, the coefficient α being equivalent to the ratio it_L/t_E or βi, but in Eq. (3.76) t_L is not a constant for a particular flow and is directly proportional to the time of travel T in the neutral surface-stress layer, through the relations

$$\bar{w} = \mathrm{d}\bar{Z}/\mathrm{d}t = b_1 u_*, \quad \sigma_w = b_2 u_*, \quad l_E(z) = b_5 z$$

which give

$$t_{LT} = (\alpha\, b_1 b_5/b_2)\, T \qquad (3.76a)$$

(Hunt and Weber replace $\alpha\, b_5$ by a single coefficient b_4). Eq. (3.75a) now becomes on integration

$$\overline{Z^2} = \bar{w}^2 T^2 + 2\,\overline{\hat{w}^2}\int_0^T t_{LT}\mathrm{d}t$$

$$= (\bar{w}^2 + \overline{\hat{w}^2}\,\alpha\, b_1 b_5/b_2)\, T^2 \qquad (3.77)$$

Numerical values of the coefficients available in the literature led to the remarkable result that the two separate terms in Eq. (3.77) are exactly equal or that $\overline{Z^2} = 2\,\bar{Z}^2$. This relation corresponds with an exponential vertical distribution as in Eq. (3.43), i.e. as derived by solution of the time-dependent diffusion equation for a plane source at $z = 0$ and a vertical diffusivity, ku_*z, equal to that for transfer of momentum in the neutral surface-stress layer. However, it needs to be emphasized that the values adopted by Hunt and Weber for α (0.5) and b_5 (0.67) are uncertain to an extent that makes the above demonstration of equality in the two terms somewhat fortuitous. The coefficient b_2 is fairly well established at 1.3, and $b_1 = k = 0.4$ is essential for compatibility with Eq. (3.43), so the real point is that Hunt and Weber's statistical representation of vertical spread from $z = 0$ in the neutral surface-stress layer is completely consistent with gradient-transfer in accordance with a diffusivity equal to that for momentum *if* $b_1 = 0.4$ and $\alpha\, b_5 = 0.34$. Note that if we take $b_5\,(= l_E/z)$ to be 0.5 as suggested in Note (i) of Table 3.I then $\alpha\ (= \beta i)$ becomes 0.68, similar to the approximate value already suggested in Table 3.I by a simpler argument, and remarkably consistent with Hanna's (1981a) result from balloon-observations of Lagrangian velocity fluctuations (see 2.7 and Fig. 2.19).

The foregoing statistical treatment of vertical diffusion is applicable only in neutral conditions and contains no allowance for the accelerating or restraining forces imposed on the vertical motion of elements of fluid as a consequence of thermal stratification. For stable flow this aspect has been examined by Csanady (1964, 1973), and his treatment has been developed further by Pearson, Puttock and Hunt (see Hunt, 1982). Csanady starts with the basic equations for the fluctuations of vertical velocity and temperature of a wandering fluid element, and demonstrates that the velocity history of a typical particle (assuming homogeneous turbulence) has an autocorrelation function containing sinusoidal oscillations arising from the restoring forces imposed on displaced fluid elements. Substitution of this form in Eq. (3.62) yields an equation for σ_z which tends asymptotically to

$$\sigma_z^2 = 2\,\overline{w'^2}/N^2, \qquad T \to \infty \tag{3.78}$$

where N is the Brunt–Vaisala frequency. Note that in this model $\int_0^\infty R(\xi)\mathrm{d}\xi = 0$ in accordance with the negative loops in the autocorrelation. The w' variations and the resulting time-mean dispersion of continuously released particles are manifestations of wave motion, the amplitude of which determines the maximum vertical spread.

Pearson *et al*'s development of the treatment leads to the form

$$\sigma_z = (\sigma_w/N)\,[\zeta_z^2 + 2t_L T N^2]^{\frac{1}{2}}, \quad \text{for } NT \gg 1 \tag{3.78a}$$

in which ζ_z is determined by the spectrum of the vertical gradient of the pressure fluctuation and is estimated to be nearly constant with a magnitude 1 ± 0.5. Thus $\sigma_z \approx \sigma_w/N$ may apply over a large range of T but in theory may be succeeded ultimately by the $T^{\frac{1}{2}}$ growth, as a consequence of molecular diffusion of heat and reduction of the buoyancy forces. However, the theory necessarily assumes homogeneity and stationarity in the flow properties but it is not obvious that such conditions can apply in practice to the extent required for the limiting result to be valid.

3.6 STOCHASTIC MODELLING OF DIFFUSION

Markovian random walk methods

Random walk modelling has become increasingly popular in recent years owing to its basic simplicity in concept and application. In its simplest form the particles representing the dispersing substance travel in straight lines until they undergo collision after which the velocity is independent of previous velocities and depends solely on the statistics of the turbulence at the position where the collision occurred. This is the basis of the familiar 'drunkard's walk' and was first used by Einstein (1905) to simulate molecular diffusion. To some degree it was also the basis of the Prandtl mixing-length concepts that led to eddy diffusivity theory. It is also the basis of the integral equation of diffusion discussed in the next section. Thompson (1971) also used this simple form of modelling to simulate three-dimensional eddy diffusion in the atmospheric boundary layer with an Ekman spiral wind variation, and in other situations.

However, more recently there has been a tendency to follow Taylor's (1921) example and treat eddy diffusion as a continuous process. The motion is supposed governed by the Markov assumption, namely by a continuous exchange of 'momentum' with the environment of the particle, so that in the simplest case of one-dimensional homogeneous stationary turbulence:

$$v(t + \delta t) = R(\delta t)\, v(t) + v'(t) \tag{3.79}$$

where R is the Lagrangian correlation function and $v'(t)$ is a random velocity drawn from a Gaussian distribution with zero mean and standard deviation σ_v. This form of the random-walk equation was used by Smith (1968) in a study of conditioned particle motion (see Section 3.4) in homogeneous turbulence. C. D. Hall (1975) applied the method to simulate sea-spray droplet motions and their resulting distribution in the surface layer of the atmosphere. His model included turbulence in the vertical and horizontal downwind directions. The Markov equation (Eq. (3.79)) was applied to the vertical velocities but the horizontal velocity was defined as:

$$u = u(z) \pm 2.2u_*$$

at each time step, where u_* was the friction velocity and the sign was chosen opposite to that of w with probability P, and on all other occasions was positive or negative at random. P was chosen so that the correlation between u and w took its normally observed value of -0.25 in neutral stability. As in all random walk models Eq. (3.79), or its equivalent, is applied to a single particle as it advects away from the source for as long a travel time as is required. The process is then repeated for very many more such particles until reasonably smooth particle distributions are established.

Hall's calculations led to good agreement with measured values of cloud height as a function of distance downwind from a point source out to about

300 metres beyond which they appeared a little too small, as is shown in Fig. (3.7).

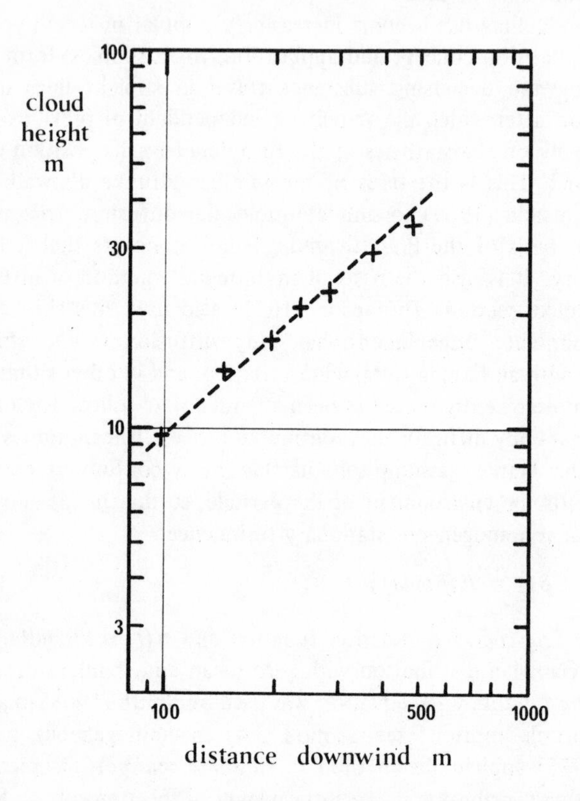

Fig. 3.7 – Cloud heights, in metres, downwind from a ground-level line source in neutral conditions obtained using Hall's (1975) random walk model (the crosses) compared with observed heights over rough downland (the broken line). This comparison was the first atmospheric verification of the random walk technique.

Since 1975 there have been an increasing number of papers reporting work using random walk modelling applied to atmospheric dispersion. Notable amongst these are papers by Hanna (1978), Reid (1979) Durbin and Hunt (1980), Lamb (1982), and Ley (1982). The biggest step forward, however, has been pioneered by J. D. Wilson *et al.* (1981) and carried further by Thomson and Ley (1982) in which dispersion in turbulent fields exhibiting gradients in turbulent energy have been considered. An obvious example is dispersion in a diabatic surface layer.

Durbin and Hunt have explored the behaviour of the average vertical velocity of particles leaving an elevated source in a turbulent shear layer. By expanding the velocity in a Taylor series they concluded the mean height of the particles must increase with time of travel. They also showed that after some downwind distance X, about four times the length scale of turbulence at the source height, the resulting concentration profile using a purely random walk advection process (velocities random at each step) becomes insignificantly different from that obtained using a full Markovian system of the type expressed by Eq. (3.79).

Thomson and Ley (1982) have also investigated the approach of the respective concentration profiles for an elevated source and a ground-level source in a neutral surface-stress layer. Fig. 3.9 shows their results. The concentrations are within indicated percentages of each other at a normalized range x/H (H is the height of the elevated source). Thus for $z_0 = 0.01$ m, $H = 5$ m the concentrations are within about 10 per cent of those for a ground-level source beyond 700 m downstream.

The property that for particles leaving an elevated source within the surface layer $\bar{Z}$ increases with time of travel can also be deduced from Eq. (3.41), since $\mathrm{d}K/\mathrm{d}z > 0$. As Durbin and Hunt point out, the wind shear has the opposite effect of reducing $\bar{Z}$ as a function of x, since particles moving downward advect downstream more slowly than those moving upwards; the net effect of the two processes being normally in favour of upward motion at small x or T.

Lamb (1978, 1982) has made an important contribution to the application of Markovian modelling to diffusion in a field of turbulence defined by the results of Deardorff's (1973) numerical model which simulated the convective atmospheric boundary layer. The basic equation he used is

$$\frac{\mathrm{d}}{\mathrm{d}t} X_i = \hat{u}_i(X_i(t), t) + u_i'(X_i(t), t)$$

where $\hat{u}_i$ represent the velocity distribution given explicitly by Deardorff's model and the u_i' are random, consistent with the statistics of the unresolved sub-scale motions. His results compared favourably with Deardorff's laboratory results. He also considered the effect of adding buoyancy to the stream of particles leaving the source to simulate the behaviour of power station plumes.

Before reporting on further aspects of these models, the validity of the basic equation (3.79) should be considered. Various questions may be asked:

(i) Is the relation between $v(t+\delta t)$ and $v(t)$ linear, as Eq. (3.79) assumes? Hanna (1979) has considered this question in detail by analysing the Lagrangian velocities of tetroons released at Las Vegas and at Idaho Falls during daytime convective conditions. Figure (3.8) is an example of the results. There is no obvious deviation from linearity.

(ii) What are the consequences of assuming an exponential correlogram? Many studies suggest that the exponential form is a useful approximation to reality. If it is then $v(t+\delta t)$ depends solely on $v(t)$ and is dependent on the

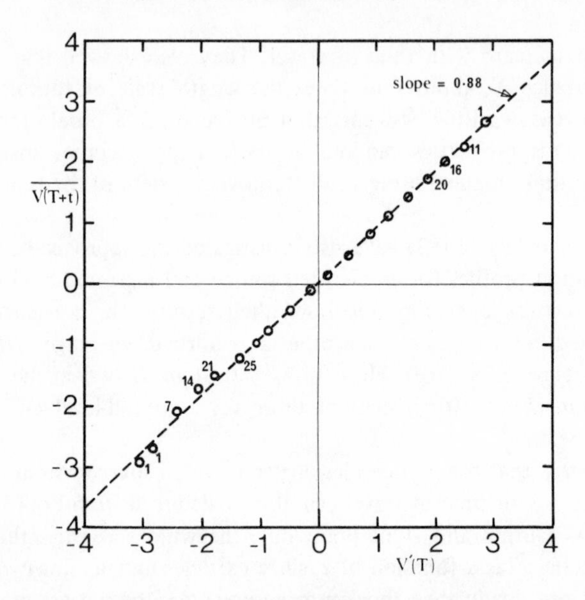

Fig. 3.8 – Hanna's analysis of tetroon runs from Idaho Falls. The Lagrangian lateral wind fluctuations $v'(T)$ and $v'(T+t)$ are normalized by their standard deviation σ_v', i.e. $V'(T) = v'(T)/\sigma_v'$. The points give the average values of $V'(T+t)$ for given different values of $V'(T)$. The numbers attached to some of the points give the number of velocities included in the average when this is rather small (< 27). The close approximity to a straight line supports the linear form of the regression equation (reproduced from the *Journal of Applied Meteorology*, with the permission of the American Meteorological Society).

magnitude of velocities at earlier times only in so far as they act through $v(t)$. In other words $v'(t)$ is totally independent of $v(t), v(t-\delta t), \ldots, v(0)$. The process is then truly Markovian.

However, as Smith (1968) pointed out, the exponential form has a non-zero gradient at zero time and this strictly implies an impulsive process, involving very short-lived infinite accelerations. This may be a fair approximation for molecular collisions and dispersion, but does not seem correct for truly fluid motions. Consequently some deviation from the exponential form may be necessary at very small times, but this may be of no particular consequence here.

(iii) What are the consequences of assuming a non-exponential correlogram? Taken in isolation from the previous history of v, Eq. (3.79) is still valid and $v'(t)$ is independent of $v(t)$. If the previous history is known and needs to be taken into account then Eq. (3.79) is incomplete. For example if the effect of $v(t-\delta t)$ on $v(t+\delta t)$ is to be included explicitly, the appropriate equation is:

$$v(t+\delta t) = R(\delta t)b_0(\delta t)v(t) + b_1(\delta t)v(t-\delta t) + v'(t) \tag{3.80}$$

The term $v'(t)$ will now be closer to a truly random quantity. The coefficients $b_0(\delta t)$ and $b_1(\delta t)$ are

$$b_0(\delta t) = \frac{1-R(2\delta t)}{1-R^2(\delta t)}, \quad b_1(\delta t) = \frac{R(2\delta t)-R^2(\delta t)}{1-R^2(\delta t)}$$

If R becomes exponential: $b_0=1$ and $b_1=1-b_0=0$. The exponential form is one example of a more general class of functions that behave like $R=1-kt$ at small t. Generalizing this still further, functions of the form $R=1-kt^{\alpha}$ may be considered at small enough t. Table 3.IV shows how the coefficients in single, double and treble regression vary with α.

Table 3.IV)– Table of coefficients in multiple regression relationships as a function of α (where the correlogram $R(t)=1-kt^{\alpha}$ for small t). Values have been rounded to two decimal places

$$v(t+\delta t) = R(\delta t)a_0 v(t) + v_0' \qquad \text{single regression}$$

$$= R(\delta t)b_0 v(t) + b_1 v(t-\delta t) + v_1' \qquad \text{double regression}$$

$$= R(\delta t)c_0 v(t) + c_1 v(t-\delta t) + c_2 v(t-2\delta t) + v_2' \qquad \text{treble regression}$$

Note that $a_0 = b_0 + b_1 = c_0 + c_1 + c_2 = 1$

	α	0	0.5	0.8	1	1.2	1.2	2
single	a_0	1	1	1	1	1	1	1
double	b_0	0.5	0.71	0.88	1	1.15	1.41	2
	b_1	0.5	0.29	0.12	0	−.15	−.41	−1
	c_0	0.33	0.66	0.86	0	1.14	1.37	2
treble	c_1	0.33	0.19	0.08	0	−.09	−.25	−.78
	c_2	0.33	0.15	0.06	0	−.05	−.12	−.22

The influence of velocities at times earlier than t on $v(t+\delta t)$ is therefore small, especially when $0.8 \leqslant \alpha \leqslant 1.2$, although the multiple regression coefficients m_i ($m = b, c, \ldots$) only fall off slowly in magnitude as i increases beyond $i=1$. The v' terms approach being truly random the more terms are included in the regression.

It is possible to consider an alternative equation:

$$v(t+\delta t) = \exp(-\delta t/\tau')v(t) + v'(t) \tag{3.81}$$

where the exponential term represents the steady exchange of 'momentum' with the environment. Since this exchange inputs momentum which is correlated with earlier inputs owing to the finite size of eddies, the term $v'(t)$ is not independent of $v(t)$ except for an exponential corellogram. Differentiating (3.81) with respect to time yields the so-called Langevin equation:

$$\tau' a(t) = -v(t) + v''(t) \tag{3.82}$$

where τ' is the timescale of the momentum exchange process and is only equal to the Lagrangian timescale τ when $R(t)$ is exponential. This is the equation considered by Gifford (1982a)

Returning now to the application of random walk modelling to specific situations we may follow Thomson and Ley's 1982 formulation. In two-dimensional turbulence let x be distance downstream and z be distance across-stream (usually in the vertical direction). Let u and w be the corresponding velocities, and i represent the number of time steps of duration Δt accomplished since release.

Then $\quad x_i = x_{i-1} + u_{i-1}\,\Delta t$

and $\quad z_i = z_{i-1} + w_{i-1}\,\Delta t \qquad (3.83)$

If R is the Lagrangian correlogram with a timescale τ then following Reid (1979)

$$\tau = \frac{k_{\text{mass}} u_* z}{\sigma_w^2 \phi_{\text{mass}}} \tag{3.84}$$

in the diabatic surface layer, where $\phi_{\text{mass}}\left(\frac{z}{L}\right)$ is the Monin–Obukhov function for mass corresponding to ϕ_m for momentum and k_m and k_{mass} are the corresponding von Karman's constants. We define

$$\beta = \frac{1}{2}\frac{\phi_m}{k_m} \bigg/ \frac{\phi_{\text{mass}}}{k_{\text{mass}}}$$

Furthermore let $R = R(\Delta t)$

Then $\quad w_i = R w_{i-1} + (1-R^2)^{\frac{1}{2}} \sigma_w \eta_i + \dfrac{\partial}{\partial z}\sigma_w^2 \Delta t \qquad (3.85a)$

$$u_i = R u_{i-1} + (1-R)\bar{u}(z_i) + (1-R^2)^{\frac{1}{2}} \sigma_u \lambda_i \tag{3.85b}$$

where $\quad \lambda_i = r(1-\beta)\eta_i + (1-r^2(1+\beta^2))^{\frac{1}{2}} \epsilon_i \qquad (3.85c)$

and ϵ_i and η_i are independent standard normal random variables. The parameter $r \equiv \overline{u'w'}/\sigma_u\sigma_w$ is the normalized velocity covariance. These equations conserve horizontal momentum $\bar{u}(z)$, the horizontal and vertical turbulence profiles σ_u and $\sigma_w(z)$. They are also consistent with zero mean velocity across any horizontal plane and with the shearing stress, which is assumed constant with height.

In a neutral surface layer σ_w may be considered constant with height but in diabatic conditions this is no longer true and the last term in (3.85a) represents a bias-velocity which counteracts the tendency for there to be a net particle flux down the gradient of σ_w. This bias velocity has been discussed by Wilson *et al.* (1982) and by Thomson and Ley (1982). The nature of the term can be deduced by considering the time-independent turbulent Navier–Stokes equation:

$$0 = -\frac{\partial \bar{p}}{\partial z} - \bar{\rho} g - \frac{\partial}{\partial z}(\bar{\rho}\,\sigma_w^2) \qquad (3.86)$$

(A) (B) (C)

The last term (C) represents the mean Reynolds stress contribution tending to accelerate particles in the z-direction. Terms (A) and (B) are relatively very large and almost balance. The residual balances term (C).

Consider the typical magnitudes of these terms, expressed in mks units:

$$\left.\begin{aligned}(A) &= \frac{\Delta p}{\Delta z} = \frac{10^5}{10^4} = 10 \\ (B) &= \bar{\rho} g = 1 \times 10 = 10\end{aligned}\right\} \text{taken throughout the whole atmosphere}$$

$$(C) = \bar{\rho}\frac{\Delta \sigma_w^2}{\Delta z} = \frac{1 \times 1}{10^3} = 10^{-3} \quad \text{taken across the boundary layer.}$$

(C) is thus four orders of magnitude smaller than the other two terms. The required minute difference between (A) and (B) can be thought of as an additional very small density gradient acting though $\bar{p}$ in (A). It reminds us that the dynamic pressure resulting from normal turbulent fluctuations is really very small compared to the static pressure at ground level resulting from the total weight of air aloft, even if sometimes it is capable of destroying trees and property!

The nature of the final term in Eq. (3.85a) in even more complex fields of turbulence has not yet been resolved in 1982, but the indications are that this present term may not be wholly adequate.

Solutions of Eqs. (3.85) have been obtained by Wilson *et al.* (1982), Ley (1982) and Thomson and Ley (1982). The solutions have been compared with Prairie Grass data (Barad, 1958), with Lagrangian Similarity theory solutions, and with solutions of the parabolic diffusion equation. Details are given in the papers cited and their conclusions may be summarized as follows:

(i) The random-walk equations yield useful predictions efficiently on a modern computer but need around 5000 trajectories to obtain smooth profiles.

(ii) Comparing the results with Lagrangian similarity theory gives assessments of the constants b and c in that theory (see 3.3). Ley advocates $b = 0.39 = k_{\mathrm{mass}}$, and also indicates that in neutral conditions $c \approx 0.5 - 0.53$. Thomson however shows that c varies markedly with stability as is shown in Fig. 3.9(a), changing from about 0.28 in unstable conditions to about 0.9 in stable conditions. The changeover occurs rather rapidly between $-0.2 \leqslant \bar{Z}/L \leqslant 0.2$, and the intersection at $L = \infty$ is rather difficult to determine accurately but does not seem to be inconsistent with the theoretical value $e^{-\gamma} = 0.56$, γ being Euler's constant. This range of c is somewhat wider than that inferred from gradient-transfer theory and noted in Table 3.II, but also as noted there the mean particle speed $d\bar{X}/dT$ is not sensitive to the precise value adopted for c.

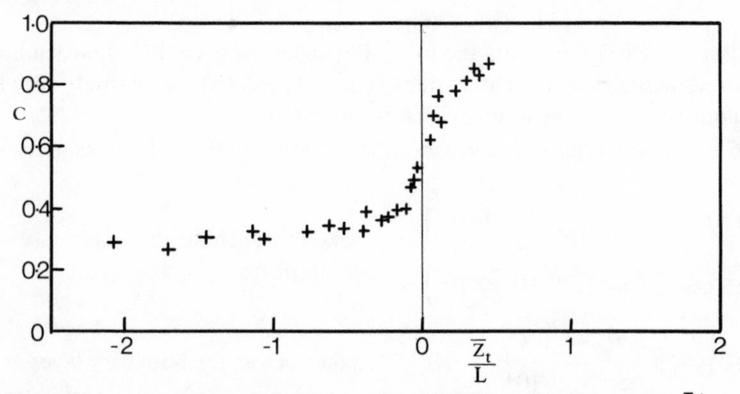

Fig. 3.9(a) – Variation of the Lagrangian similarity parameter c with $\bar{Z}/L$ as implied by solutions of the random walk technique (Thomson, 1982).

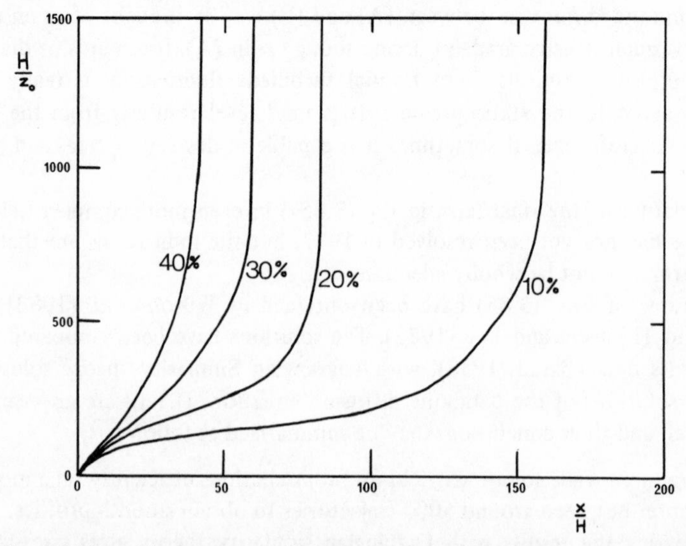

Fig. 3.9(b) – The range at which concentration profiles are within specified percentages of each other for equal elevated and ground-level sources within a neutral surface-stress layer.

(iii) Ley has also tested the model against the analytical solutions (Eq. (3.20)) obtained from the parabolic diffusion equation. In both methods the same wind and K profiles were assumed, namely $u = u_1 z^{1/7}$ and $K = ku_* z$. The agreement out to 500 m is very good.

(iv) The results of comparisons with field data are noted briefly in Table 4.VII.

The integral equation of diffusion

This approach has its origins both in random walk modelling and in the classical basic equation of diffusion:

$$C(\mathrm{r}) = \int S(\mathrm{r}')P(\mathrm{r}-\mathrm{r}')\,\mathrm{d}\mathrm{r}' \tag{3.87}$$

where C is the concentration at point r, S is the source strength at r$'$ and P is the probability transfer-function between r$'$ and r. To give this equation teeth the nature of P has to be defined, and therein lies the problem.

Smith (1982b) has suggested the following solution. Consider a single continuous source in a turbulent stationary airstream. Let the particles leaving the source move in a random walk, a series of straight paths separated by 'collisions' in which all Lagrangian memory is lost. The length of the paths varies from one to another and is determined in each case by a random process equivalent to assuming that during each time-interval δt along the path the probability of collision is $\delta t/\tau$. τ is the local Lagrangian time-scale of the real turbulent field being simulated and may thus vary with position.

In practice, the concentration field may be built up within a grid, whose elemental gridlengths are small compared to the length-scale of turbulence, starting at the source and working progressively downstream. The concentration at any new receptor gridsquare can be calculated from the number of particles being advected through the square which experienced their last collision within upwind squares, intermediate between the source and the receptor square, where the concentration is already determined. Thus for a specified number of particles emitted from the source, the number $\mathcal{N}$ that pass through (X, Z) is given by:

$\mathcal{N}(X, Z) = \Sigma\, \mathcal{N}(x, z)$ × (Probability P' of collision at (x, z))
× (Probability P'' of no further collision between (x, z) and (X, Z))
× (Probability P of the velocity $w = u(Z-z)/(X-x)$ occurring, required to take the particle from (x, z) to (X, Z)).
(3.88a)

For the sake of clarity, it will be assumed first of all that the turbulence is homogeneous, the downstream velocity u is constant and there are no boundaries. The gridlengths are defined as

$$\Delta x = \frac{u\tau}{L}$$

$$\Delta z = \frac{w_m \tau}{L} \tag{3.88b}$$

where L is an arbitrary large constant (10 can be shown to be a good choice, and is used in all calculations), τ is the Lagrangian time-scale and w_m can be identified either with the root-mean-square w-velocity σ_w or, if w has extreme values $\pm w_m$, with this extreme.

Then $$P' = 1/L \text{ and } P'' = \left(1 - \frac{1}{L}\right)^{n-1} \tag{3.89a}$$

where $X - x = n\Delta x, \quad Z - z = -m\Delta z$

and to ensure no loss of particles:

$$P = \Delta\mathcal{T}\left(\frac{m}{n}\right) = \int_{\frac{m-½}{n}}^{\frac{m+½}{n}} p(w)\,\mathrm{d}w \tag{3.89b}$$

where $p(w)$ is the probability density function of the turbulent velocity w. The particles which originate from the source and undergo no collision until reaching (X, Z) make a contribution to $\mathcal{N}$ like all other squares except that P' is put equal to 1 and $\mathcal{N}(0, 0)$ is put equal to $uQ/w_m\Delta x$ (where Q is the source strength, or the total number of particles released).

Expressed in terms of concentrations:

$$C(N, M) = \frac{Q}{w_m \Delta x}\left(1 - \frac{1}{L}\right)^{N-1} \Delta\mathcal{T}\left(\frac{M}{N}\right) + \\ + \sum_{m,n} \frac{1}{L}\left(1 - \frac{1}{L}\right)^{n-1} \Delta\mathcal{T}\left(\frac{m}{n}\right) C(n,m) \tag{3.90a}$$

where $M = Z/\Delta z$, $N = X/\Delta x$.

In the homogeneous field this is exceedingly easy to code for a computer and quick to run because the coefficients:

$$q(n, m) = \left(1 - \frac{1}{L}\right)^{n-1} \Delta\mathcal{T}\left(\frac{m}{n}\right) \tag{3.90b}$$

are independent of N and M and need to be calculated only once. In terms of q:

$$C(N, M) = Aq(N, M) + \sum_{m,n} \frac{1}{L} q(n, m)\, C(n, m) \tag{3.90c}$$

where A is a large constant representing the source strength.

Smith has solved Eq. (3.90c) with $A = 10^4, L = 10$ and with

$$p(w) = \frac{15}{16w_m}\left(1 - \frac{w^2}{w_m^2}\right)^2 \qquad \text{for } -w_m \leqslant w \leqslant w_m$$
$$= 0 \qquad \text{for } |w| > w_m \tag{3.91}$$

This distribution has a standard deviation $\sigma_w = 0.378\ w_m$, a zero mean, and a shape very similar to the Gaussian distribution without the rather unrealistic 'tails' at large w characteristic of the latter distribution. The resulting concentration distribution is very quickly obtained by a computer, and has moments in very close agreement with analytically derived moments. Thus the second moment σ^2 is in complete agreement with Taylor's statistical theory result (Table 3.III, Case 1) with an exponential correlogram with the same Lagrangian time-scale τ.

Higher moments can be obtained analytically from the integral form of Eq. (3.90c):

$$C(X, Z) = \frac{Q}{uT}\exp\left(-\frac{T}{\tau}\right)P(w) +$$
$$\frac{1}{\tau}\int_0^T \exp\left(-\frac{T-t}{\tau}\right)\int_{-\infty}^{\infty} C(x, z)\,P(w)\,dw\,dt \tag{3.92a}$$

where $X = uT$, $Z = wT$ in the first term on the right-hand side, and $x = ut$ and $w = u(Z-z)/(X-x)$ in the second term. Of particular interest is the fourth moment:

$$\mu_4 = \frac{u}{Q}\int_{-\infty}^{\infty} Z^4\,C(X, Z)\,dZ \tag{3.92b}$$

which gives the kurtosis γ (or flatness factor) of the distribution: $\gamma = \dfrac{\mu_4}{\sigma^4} - 3$.

If $\gamma = 0$ the distribution has the same kurtosis as a Gaussian distribution. If, however, $\gamma > 0$ the distribution is more strongly peaked than a Gaussian distribution with the same σ and is called 'leptokurtic', whereas if $\gamma < 0$ it is less strongly peaked and is called 'platykurtic'. For Eq. (3.92a) the concentration distribution can be integrated to give a kurtosis:

$$\gamma = 3(D\,\gamma_w + (K - 1)) \tag{3.93}$$

where γ_w is the kurtosis of the w-distribution,

$$aD = 2v - 6 + (6 + 4v + v^2)\exp(-v)$$

$$aK = v^2 - 6 + (6 + 6v + 2v^2)\exp(-v)$$

$$a = v^2 + 1 - 2v + \exp(-2v) - 2\,(1 - v)\exp(-v)$$

$$v = T/\tau$$

D is very close to 1 until v exceeds about 2, and then falls slowly to zero. K always just exceeds 1, reaching a maximum of 1.224 near $v = 5$. Thus at small T, γ is determined largely by the kurtosis of the w-distribution γ_w but at larger T (near $v = 4$) the particle distribution becomes somewhat leptokurtic owing to the combined effect of the two terms. (Note that the sum of two perfectly Gaussian distributions with different σ's is also leptokurtic.) Eventually the second term in (3.93) becomes dominant and, regardless of γ_w, the value of γ approaches zero. The concentration distribution therefore becomes Gaussian at large T/τ, in accord with the Central Limit theorem of statistics, in spite of the fact that the w-velocity distribution may not itself be Gaussian.

The conditional particle-motion theories developed by Smith (1968) and more recently by Gifford (1982a) as described in Section 3.4, draw interest on to the sub-ensemble of particles that leave the source with a given value of w. Since the width of such sub-ensemble plumes does not depend on the magnitude of w, we may arbitrarily put $w = 0$ for all particles in the first term on the right-hand side of Eq. (3.90c), i.e. $\Delta \mathcal{T} = 0$ unless $M = 0$ when $\Delta \mathcal{T} = 1$. The resulting solutions of (3.90c) enable the width of the sub-ensemble plume to be compared with the width of the full plume (unrestricted w at the source). Table 3.V shows this comparison:

Table 3.V – The normalized width of two plumes are compared at different distances downwind from the source. The first plume contains all particles released whereas the second contains only the sub-ensemble with zero initial w

$x/u\tau$		0.1	0.2	0.5	1	2	5	10
$\sigma/w_m\tau$	whole ensemble	0.040	0.075	0.175	0.324	0.567	1.052	1.541
$\sigma/w_m\tau$	$w(0) = 0$	0.005	0.016	0.065	0.168	0.389	0.918	1.455

Similarly a comparison may be made between the respective values of the kurtosis (see Table 3.VI). For the w-distribution given in Eq. (3.94), the kurtosis $\gamma_w = -\frac{2}{3}$ showing the distribution to be platykurtic (i.e. slightly flattened compared to a Gaussian distribution).

Table 3.VI – A comparison of the kurtosis of two plumes at different downwind distances. The whole-ensemble plume demonstrates decreasing platykurticity reflecting a decreasing influence of the shape of the w-distribution, whereas the sub-ensemble plume, in which the initial w is always zero, demonstrates decreasing peakiness or leptokurticity. The peakiness in the second plume is clearly very strong at small $x/u\tau$

$x/u\tau$		0.1	0.5	1	2.5
γ	whole ensemble	−0.632	−0.500	−0.358	−0.065
γ	sub-ensemble $w(0) = 0$	–	7.292	3.021	0.820

An example of the concentration profile at $x = u\tau$ for particles emitted with zero w is provided in Fig. 3.10. This shows the cumulative concentration as the plume is traversed. Note that 80 per cent of the plume is contained between $z = \pm 2w_m\tau$ compared with 60 per cent between $z = \pm 0.3\ w_m\tau$ showing the highly peaked nature of the distribution. The results for growth of plume width are in complete agreement with Smith's (1968) theory.

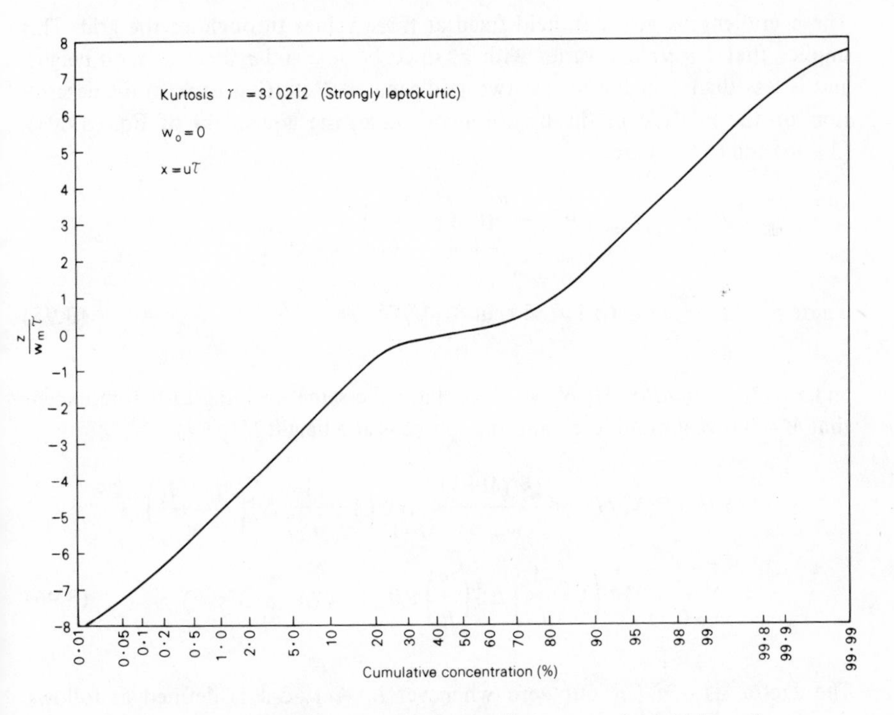

Fig. 3.10 – Cumulative concentration profile at $x = u\tau$ for initial $w = 0$.

The technique can be extended easily to treat diffusion in the neutral surface stress layer in which the wind is varying logarithmically with height:

$$u = \frac{u_*}{k} \ln \frac{z}{z_0} \tag{3.94a}$$

and the Lagrangian time-scale varies linearly

$$\tau = ku_*z/\sigma_w^2 \tag{3.94b}$$

as used by Reid (1979) and others in random walk modelling and discussed earlier in this section. The vertical velocity variance σ_w^2 is assumed constant with height, and the probability distribution $p(w)$ remains as in Eq. (3.91).

A height $z = z^*$ is chosen (typically z^* may be 1000 z_0) within the surface layer and the value of L there is put equal to 10 (as in the homogeneous case). Assuming $\sigma_w = 1.3\, u_*$, Δx and Δz, defined as before, are equal to

$$\left.\begin{aligned} \Delta x &= 408.7\, z_0 \\ \Delta z &= 8.14\, z_0 \end{aligned}\right\} \quad \text{calculated at } z = z^*$$

These gridlengths are then held fixed at these values throughout the grid. This implies that $L = u\tau/\Delta x$ varies with z, since both u and τ increase with height, and is less than 1 in the lowest two grid-squares. Referring back to the description of the method in the homogeneous case, the equivalent of Eq. (3.89a), (3.89b) and (3.90a) are:

$$P' = 1/L_n, \quad P'' = \prod_{i=1}^{n-1} \left(1 - \frac{1}{L_i}\right)$$

where
$$L_i \frac{\Delta x}{u_i \tau_i} = (0.11785\, r_i \ln 81.4 r_i)^{-1} \tag{3.95}$$

and $r_i = M + \frac{1}{2} + mi/n$. M, N, m and n have the same meaning as before except that $M = 0$ is at ground level, and the source is at a height $(M_s + \frac{1}{2})$.

$$u(r_0)\, C(N, M) = \frac{Qu\,(M+\frac{1}{2})}{w_m\, \Delta x} \prod_{i=1}^{N-1} \left(1 - \frac{1}{L_i}\right) \Delta \mathcal{T}\left(\frac{M - M_s}{N}\right) +$$

$$\sum_{m,n} \frac{1}{L_n} \prod_{i=1}^{n-1} \left(1 - \frac{1}{L_i}\right) \Delta \mathcal{T}\left(\frac{m}{n}\right) u(r_n) \times C(N-n, M+m) \tag{3.96}$$

The factor $\left(1 - \dfrac{1}{L_i}\right)$ is put zero whenever $L_i \leqslant 1$. $\Delta \mathcal{T}$ is defined as follows (see Fig. 3.11):

$$\Delta z = \frac{w_m \tau^*}{L^*}, \quad \Delta x = \frac{u^* \tau^*}{L^*}$$

where u^*, τ^* and $L^* = 10$ are defined at $z = z^*$. The particles which collide at A and move to B are assumed to move in straight lines retaining their u and w velocities unchanged and as defined at A.

Thus
$$\frac{w}{u} = -\frac{m\, \Delta z}{n\, \Delta x} = -\frac{m}{n} \frac{w_m \tau^*}{L^*} \frac{L^*}{u^* \tau^*} = -\frac{m}{n} \frac{w_m}{u^*}$$

i.e
$$\frac{w}{w_m} = -\frac{m}{n} \frac{u}{u^*} = -\frac{m}{n} \frac{\ln z/z_0}{\ln z^*/z_0} \tag{3.97}$$

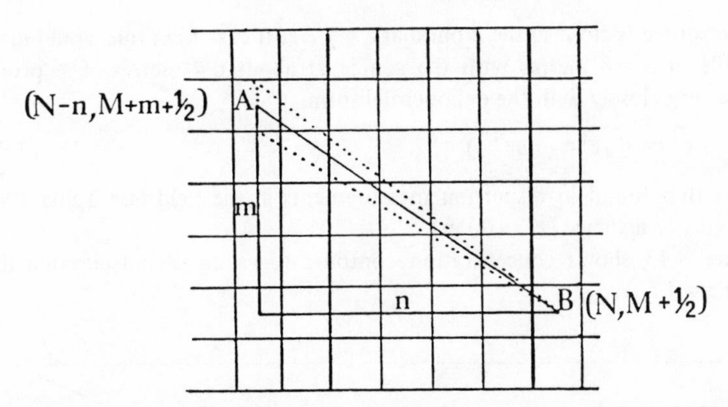

Fig. 3.11 – The 'integration' grid with definition of the notation. B is the receptor square where the concentration is being calculated. A is one of many upwind squares at which the concentration is already known and which is contributing to the concentration of B.

where $z = M + m + \frac{1}{2} = r_n$ and z^*/z_0 is specified. $\Delta \mathcal{T}$ is now defined as before in Eq. (3.89b). Eq. (3.96) can now be solved very easily. Computer time consumed is still very short although clearly longer than for the homogeneous case since the $q(n, m)$ defined in Eq. (3.90c) are now functions of N and M as well as n and m.

Figure 3.12 shows as an example the concentration profile at $x = 10\ \Delta x$

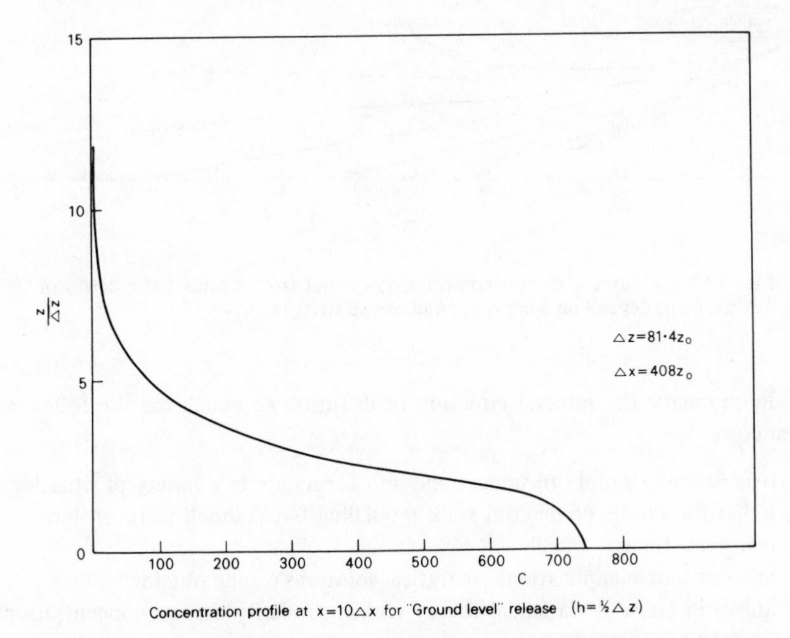

Fig. 3.12 – Concentration profile at $x = 10\ \Delta x$ for a 'ground-level, release ($h = \frac{1}{2}\ \Delta z$).

when the source is close to the ground at $z = \frac{1}{2}\,\Delta z$. If $z_0 = 1$ cm this would mean the profile at $x = 8$ metres with the source at about 0.4 metres. The profile conforms very closely with the exponential form,

$$C = C_0 \exp\left(-az^{3/2}\right) \tag{3.98}$$

similar to that found in dispersion measurements in the field (see Table 4.VI) albeit at greater distance ($x = 100$ m).

Figure 3.13 shows concentration contours for a crosswind elevated line source at $z = 4.5\,\Delta z$.

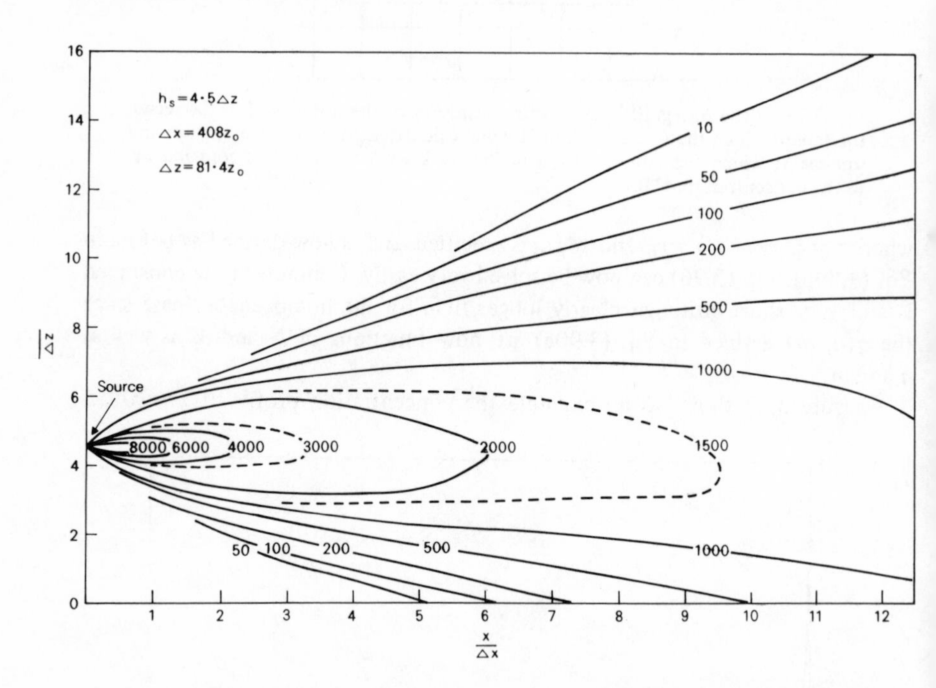

Fig. 3.13 – Contours of concentration downwind from a source at a height of 4.5 Δz. Units depend on wind speed and source strength.

In summary the integral equation of diffusion approach has the following advantages:

(a) it is basically simple to understand and formulate in a variety of situations;
(b) it has the merits of random walk modelling but is much faster in terms of computer time;
(c) in some simple applications, analytical solutions can be obtained;
(d) unlike in classical statistical theory and in similarity theory, concentrations are readily obtained.

However, like all theories of diffusion some knowledge of the Lagrangian parameters, in particular the time-scale, is essential, and these parameters cannot be directly measured and have to be inferred from Eulerian measurements with all the uncertainties this implies.

3.7 THE EXPANDING CLUSTER

We now consider the important differences between the diffusion from a continuous source, in which particles are released in sequence at a fixed position, and that of a single puff or cluster of particles. The particular nature of the latter type of diffusion was recognized by L. F. Richardson at a very early stage (1926). Richardson's analysis was concerned with reconciling the enormous range of Fickian *K*s which was apparently required to explain the whole range of diffusion processes which could be experienced in the atmosphere. In this analysis he introduced the fundamental notion that the rate of separation of a pair of particles at any instant is dependent on the separation itself, and that as separation increases so also does the rate of separation. This meant that the spread of a large cloud of particles could not be built up by superimposing the growths of component elements of the cloud treated separately.

In Fickian diffusion the distribution of particles is described in terms of concentration as a function of distance from a chosen origin and, in the simple case of diffusion of particles along a line, the development of the distribution is determined by the one-dimensional form of Eq. (3.12) with K constant, i.e.

$$\frac{\partial \chi}{\partial t} + \bar{u}\frac{\partial \chi}{\partial x} = \frac{\partial}{\partial x}\left(K\frac{\partial \chi}{\partial x}\right) = K\frac{\partial^2 \chi}{\partial x^2} \tag{3.99}$$

where χ is now a number of particles per unit length of the line. So as to bring in the effect of the separation of the particles in a rational way Richardson described the distribution in terms of a *distance-neighbour* function

$$q(l) = \frac{1}{Q}\int_{-\infty}^{+\infty} \chi(x)\chi(x+l)\,\mathrm{d}x \tag{3.100a}$$

where Q is the total number of particles, and proved that if Eq. (3.99) held q and l were related by the differential equation

$$\frac{\partial q}{\partial t} = 2K\frac{\partial^2 q}{\partial l^2} \tag{3.100b}$$

For non-Fickian conditions Richardson then argued that while Eq. (3.99) could not be generalized, as K could not rationally be regarded a function of position in the atmosphere, Eq. (3.100b) could be generalized by replacing $2K$ by $F(l)$ and regarding $F(l)$ as an increasing function of l, so that

$$\frac{\partial q}{\partial t} = \frac{\partial}{\partial l}\left(F(l)\frac{\partial q}{\partial l}\right) \tag{3.100c}$$

The only observations available to Richardson were in the form of effective values of K, and so the only way to estimate the relation between $F(l)$ and l was to assume that Fickian diffusion applied in these cases, i.e. that $F(l) = 2K$. Richardson then took l equal to the standard deviation of the particles from mean position, or where this was not applicable, in the case of estimates of K from wind profile observations, to a length characteristic of the system of observation. For further details reference should be made to the original paper, the main point to be noted here is that three types of data were involved, giving in all seven values of l as follows:

Molecular diffusion	$l = 5 \times 10^{-2}$
Wind profiles at heights up to 800 m	$l = 1.5 \times 10^3,\ 1.4 \times 10^4,\ 5 \times 10^4$
Scattering of balloons volcanic ash or cyclonic depressions	$l = 2 \times 10^6,\ 5 \times 10^6, 10^8$

The logarithms of K and l were found to lie on a line of slight curvature in the sense that $d(\log K)/d(\log l)$ increased with l, but all except the extreme points could be represented with good approximation by the relation

$$K = 0.2 l^{4/3} \tag{3.101}$$

This simple law obtained by Richardson is especially remarkable in the light of later deductions [seeEq. (3.104d)], though the mixed quality of the observations which provided it should not be forgotten. Indeed it may be noted that if attention had been confined to those observations which were directly representative of scattering of particles on a large scale, so involving only the last of the above groups of l, then the exponent in the power law would have been almost exactly $\frac{5}{3}$. However the really important feature which had been introduced in this and other discussions by Richardson was the idea of a virtually continuous range of eddy sizes, with turbulent energy being handed down from larger to smaller eddies and ultimately dissipated in viscous action. Specific expression of the concept came considerably later in the parallel developments by Kolomogorov (1941), on the basis of his similarity theory, and by Obukhov (1941), on the basis of the equation of energy balance in the spectrum. The relevance to Richardson's relation, as reviewed by G. I. Taylor (1959), is clearly seem on dimensional grounds. If the dimensions length and time are expressed in terms of kinetic viscosity and rate of energy dissipation, as the only two physical quantities relevant to the regulation of energy transfer in the spectrum, then length has dimensions $\nu^{3/4}\epsilon^{-1/4}$ and time has dimensions $\nu^{1/2}\epsilon^{-1/2}$. Richardson's differential equation may be written

$$\frac{\partial q}{\partial t} = A \frac{\partial}{\partial l} \, l^m \, \frac{\partial q}{\partial l}$$

in which A has dimensions $l^{2-m} \; t^{-1}$ or $\nu^{-1+(3/4)m}$, so that if the law of diffusion does not depend on ν it follows that $m = \frac{4}{3}$. This is identical with the empirical result obtained by Richardson, though fortuitously so, partly because of the nature of the data, and partly because in the dimensional argument it is implicitly assumed that the magnitude of l is small compared with the size of the energy-containing eddies, an assumption which is not obviously justified for the largest scale of diffusion considered by Richardson.

The detailed analysis of the separation of particles in homogeneous turbulence

The elaboration and development of the formal mathematical analysis of the relative diffusion of particles was carried further by Batchelor and Brier independently. Batchelor's full discussion, an exposition of the fundamental principles involved in three-dimensional relative diffusion in a field of homogeneous turbulence, is contained in his 1952 paper, but the main achievement as regards application to atmospheric diffusion had been set out earlier (1950). A review of this work on relative diffusion is also contained in the article by Batchelor and Townsend (1956). Brier's (1950) analysis was specifically aimed at a generalization of Taylor's analysis, so as to allow for initial separation of the particles, and as such has stimulated an application by Gifford (1959) to the important problem of the *fluctuation properties* (as distinct from the mean properties) of the plume formed by the continuous release of material at a fixed position. In the discussion here the intricate details of the analyses will be avoided, and only the essential principles and final results will be stated.

Batchelor treats the case of the separation of a pair of marked particles (or volume elements of fluid) as the simplest representation of the diffusion of a group of such particles, and defines a distribution function which is equivalent to Richardson's distance-neighbour function generalized to three dimensions. Attention is then concentrated on the most important parameter specifying this distribution function, namely the variance of the separation of the pair of particles, and its dependence on the velocities of the particles. Batchelor obtains an equation which may be written in the form

$$\overline{y^2} = \overline{y_0^2} + 2\int_0^T \int_0^t \overline{\delta v(t)\,\delta v(t-\xi)}\, d\xi \, dt \qquad (3.102a)$$

where y is the (vector) separation of the pair of particles at time T after an initial separation $\underline{y_0}$ and $\delta v(t)$ the relative velocity of the particles at time t. Apart from the $\overline{y_0^2}$ term Eq. (3.102a) will be recognized as formally identical with Taylor's equation (3.62) when the latter is expressed in terms of a co-variance instead of a correlation coefficient. In the present case, however, the relative velocity is not a random function of time, as is the absolute velocity in Taylor's analysis, since as the two particles separate the range of eddy sizes contributing to δv will increase. Two specific predictions follow immediately for the simple cases of extreme values of T. At very small values of T, when the velocities of the two particles have not had time to change appreciably, the particle trajectories will be approximately straight lines and

$$\overline{y^2} = \overline{y_0^2} + \overline{(\delta v)^2} T^2 \quad (T \text{ small}) \tag{3.102b}$$

where δv is simply the initial relative velocity, at separation y_0. At very large values of T, when the particles have separated so widely that their velocities are uncorrelated

$$\overline{\delta v(t)\delta v(t-\xi)} = \overline{2v'(t)v'(t-\xi)}$$

where v' is now the corresponding absolute velocity of either particle, and

$$\overline{y^2} = \overline{y_0^2} + 4\int_0^T\int_0^{t'} \overline{v'(t)v'(t-\xi)}\,\mathrm{d}\xi\,\mathrm{d}t \quad (T \text{ large}) \tag{3.102c}$$

The second term on the right-hand side of Eq. (3.102c) is identical with twice the quantity in Eq. (3.62), i.e. the dispersion $\overline{Y^2}$ of a single particle, and as $T \to \infty$ this term becomes dominant and

$$\overline{y^2_{T\to\infty}} \to 2\overline{Y^2}$$

Thus after a very long time or distance of travel the mean square separation of a pair of particles tends to a value which is exactly twice the mean square displacement of particles released serially form a fixed position.

For small and intermediate times further predictions are made possible by appeal to dimensional arguments, and to the theory of the small-scale structure of turbulence (see 2.5). When the scalar separation of the two particles is small compared with the size of the energy-containing eddies, i.e. with the scale of turbulence, the development of this separation will depend only on the initial separation and the parameters ν and ϵ. Then, following Batchelor and Townsend's (1956) presentation, from dimensional arguments it is possible to write

$$\frac{\mathrm{d}\overline{y^2}}{\mathrm{d}t} = \epsilon T^2 f(y_0/\epsilon^{1/2}T^{3/2},\ T\epsilon^{1/2}/\nu^{1/2}) \tag{3.103a}$$

If the diffusion is independent of molecular processes ν must disappear from the expression, and Eq. (3.103a) simplifies to

$$\frac{d\overline{y^2}}{dt} = \epsilon T^2 f(y_0/\epsilon^{1/2}T^{3/2}) \tag{3.103b}$$

At small values at T Eq. (3.103b) must be linear in T [conforming to Eq. 3.102b)] and

$$\frac{d\overline{y^2}}{dt} \propto T(\epsilon y_0)^{2/3} \quad (T \text{ small}) \tag{3.103c}$$

Then when T is of intermediate magnitude, i.e. large enough for y to have become independent of its initial value y_0, but not too large in relation to the scale of turbulence, the parameter y_0 must also disappear from Eq. (3.103b) and

$$\frac{d\overline{y^2}}{dt} \propto \epsilon T^2 \quad (T \text{ intermediate}) \tag{3.104a}$$

Integration of Eq. (3.103c) and (3.104a) gives

$$\overline{y^2} \propto [\text{const.} + T^2(\epsilon y_0)^{2/3}] \quad (T \text{ small}) \tag{3.104b}$$

$$\overline{y^2} \propto \epsilon T^3 \quad (T \text{ intermediate}) \tag{3.104c}$$

the constant in Eq. (3.104c) being omitted as negligible in view of the already assumed lack of dependence on y_0. These results bring out the essentially accelerative nature of the relative diffusion, which occurs as long as the separations involved are small compared with the scale of turbulence, and which is in contrast to the case of serial release of particles from a fixed point (see 3.4), when the exponent of T in the expression for the variance of particle spread is initially 2, as in Eq. (3.104b), but thereafter *decreases*. This situation is a direct consequence of the fact that as the separation of the particles increases the variance of their relative velocity also increases, in a manner analogous to the more familiar increase of intensity of turbulence with sampling time.

The internal structure of a diffusing cluster

Batchelor (1952) also re-examines the question of describing relative diffusion by a differential equation, and suggests that it is more reasonable to regard the equivalent diffusivity as a function of $\sqrt{(\overline{y^2})}$, as representing the *statistical* value of the separation of particles, rather than of the value of y (or l in Richardson's analysis), and takes

$$K \propto (\overline{y^2})^{2/3} \tag{3.104d}$$

as consistent with Eq. (3.104c) and (3.13). Solution of the differential equation then gives a probability distribution for the separation, y, of a pair of particles, of the form $\exp(-y^2/2\overline{y^2})$. On the other hand Richardson's analysis leads to a form $\exp(-l^{2/3}/\alpha T)$, where T is the time of travel and α is a constant, i.e one which is much more sharply peaked than the Gaussian form in Batchelor's solution. Yet another form has been suggested by Monin (1955), and observational evidence for the relative adequacy of these distributions has yet to be provided.

The action of a turbulent velocity field alone is to distort a cloud of material into blobs and strands of ever-decreasing thickness. Although the combined volume occupied by the material and intervening clean fluid becomes progressively larger, that occupied by the material (and hence the concentration therein) would remain the same as initially but for molecular diffusion. As Batchelor points out, the effect of molecular diffusion is ultimately to halt the reduction in thickness of the strands at some minimum value.

Obtaining a description of the real concentration distribution within a small cloud or narrow plume of material is very difficult, though the ion-tracer technique recently reported by C. D. Jones (1979) is very promising. So far the indications are that individual distributions are highly irregular, with no sign of the smooth Gaussian shape which (see 4.2) so adequately describes the time-mean crosswind distribution downwind of a continuous point source.

In a recent series of papers Chatwin and Sullivan have reconsidered the theoretical problems and have summarized their ideas in a review (1979). For some admittedly special turbulent flows, and neglecting molecular diffusion, they argue that the probability of finding marked fluid at the centre of mass of an *ensemble* of realizations of a small cluster remains high for a long time, and is of order L_0^{-3} where L_0 is the initial size. Over the ensemble as a whole the probability falls in accordance with the growing volume $L^3(T)$ within which the deformed initial shapes are contained at any time T. The ensemble cluster is thus regarded as having a *core-bulk* structure, in which the initial conditions have an enduring influence on the core, where the ensemble concentration remains high and the main production of fluctuation of concentration occurs. In contrast, in the bulk of the cluster the production of fluctuation is negligible. Their discussion emphasizes, however, that molecular action must ultimately become important, with effects dependent on the precise nature of the turbulent velocity field, and also that in any modelling of the mean or fluctuating properties of the concentration field the representation of the turbulent transfer by an eddy diffusivity is inappropriate.

Explicit formulation of cluster growth

The relative-velocity covariance inside the integral of Eq. (3.102a) involves three kinds of correlation between the absolute velocities of particles, the purely Lagrangian correlation for a single particle, the Eulerian correlation referring to two particles at a given instant, and a mixed correlation referring to two particles at different instants. One of the chief problems is to express this latter type of correlation in more manageable terms. Brier (1950) has attempted this by some speculation about the relation between the various correlations, and expresses the mixed correlation in terms of the Lagrangian and Eulerian forms. Batchelor (1952) has drawn attention to a related and complementary suggestion made earlier in an unpublished note by G. I. Taylor, but suggests that neither of these approximations is likely to be good except at small values of the time interval (of diffusion) involved. Brier carries his analysis further, to the extent of also eliminating the Lagrangian type of correlation, and expressing the relative dispersion entirely in terms of the correlation between particle velocities at the same instant. However, it is not obviously valid to assume that this type of correlation (as a function of particle separation y) is identical with the correlation between velocities measured at *fixed* points y apart, and this is a difficulty which has not yet been satisfactorily resolved in any treatment.

An attempt to make progress in an empirical way, by adopting a simple scale relationship between the Lagrangian and Eulerian variations, on the lines introduced by Hay and Pasquill (1959) for a continuous source,has been presented by Smith [see Smith and Hay (1961)]. Smith starts with the equation for a pair of particles, equivalent to Eq. (3.102a). Taking averages over all pairs of particles in a three-dimensional Gaussian cluster (of standard deviation σ about the centre), and replacing the relative velocity of a *pair* of particles by the velocity (v'') of a particle relative to the mean velocity of all the particles in the cluster, the equation becomes

$$\frac{\mathrm{d}\sigma^2}{\mathrm{d}t} = \frac{2}{3}\int_0^t \overline{v''(t)v''(t-\xi)}\,\mathrm{d}\xi \tag{3.105}$$

Apart from the numerical factor arising from the three-dimensional Gaussian distribution, and that fact that v'' is a (vector) velocity relative to the movement of the clusters as a whole, the equation is identical with Taylor's Eq. (3.58). The assumption is then made that the Lagrangian covariance and the Eulerian covariance, *both appropriate to the finite cluster of particles,* are similar in shape, the ratio of the respective time-scales being β. Development of Eq. (3.105) then leads to

$$\frac{\mathrm{d}\sigma}{\mathrm{d}t} = \frac{2}{3}\frac{\beta}{\bar{u}}\int_0^\infty\int_0^{\bar{u}t/\beta} E(\kappa)\frac{\sin\kappa s}{\kappa s}\cdot\frac{1-e^{-\sigma^2\kappa^2}}{\sigma}\,\mathrm{d}s\,\mathrm{d}\kappa \tag{3.106}$$

where $E(\kappa)$ is the complete Eulerian three-dimensional spectrum function in terms of wavenumber κ. For $x/\beta = \bar{u}t/\beta > \sigma$ this simplifies to

$$\frac{\mathrm{d}\sigma}{\mathrm{d}t} = \frac{\pi}{3}\frac{\beta}{\bar{u}}\int_0^\infty E(\kappa)\frac{1-e^{-\sigma^2\kappa^2}}{\sigma\kappa}\,\mathrm{d}\kappa \tag{3.107}$$

so that $d\sigma/dt$ can now be evaluated, given β, from a knowledge of the Eulerian energy spectrum. For practical use Eq. (3.107) is expressed in terms of the measurable one-dimensional spectrum function. The development is carried further for the particular case when the Eulerian correlogram (actually the so-called trace-correlogram, for an explanation of which see the original paper) is of exponential form with length-scale l. With the function $E(\kappa)$ expressible in terms of l and κ, Eq. (3.107) becomes

$$\frac{d\sigma}{dx} = 2\beta i^2 \int_0^\infty \frac{n^2}{(1+n^2)^2} \cdot \frac{1-e^{-r^2n^2}}{nr} \, dn \tag{3.108}$$

where $i^2 = \overline{u'^2}/\bar{u}^2$ (u being the vector component of eddy velocity), $r = \sigma/l$ and $n = \kappa l$ (the latter not to be confused with the use of n to denote frequency).

The weighting function $(1-e^{-\sigma^2\kappa^2})/\sigma\kappa$ in Eq. (3.107) [and the corresponding term in Eq. (3.108)] is in effect a 'band-pass filter' which modifies the whole three-dimensional spectrum (and the corresponding correlogram) to the form which is effective in determining the rate of growth of a cluster. Finite size of the cluster is taken into account by the cut-off on the low-frequency side, while the cut-off on the high-frequency side represents the diminishing contribution to the dispersion as eddy size decreases to a magnitude small with respect to the cluster.

For clusters which are initially small (say $\sigma_0 < 0.1l$), the first part of the growth, i.e. that which is dependent on σ_0, is a relatively insignificant contribution to the ultimate growth, and this means that numerical integration of Eq. (3.108) yields a virtually universal curve valid for initially small clusters. An example, for $\sigma_0/l = 0.1$, $i = 0.1$ (a common value in the atmosphere) and $\beta = 4$ (a value near this is found to satisfy the observations – see Chapter 4) is shown in Fig. 3.14 with the corresponding curve for a continuous point source. At first, say over a distance up to about ten times the length scale of the turbulence, the rate of spread of the cloud from the continuous source is substantially greater than that of the cluster, because the whole spectrum of turbulence is operative in the former case, while in the latter only those eddies of a size similar to or less than that of the cluster are effective. Thereafter the rate of spread of the continuous cloud falls off noticeably, while that of the cluster is maintained (and even accelerated for a time) by the progressive action of larger and larger eddies. The percentage difference between the two values of σ/l is then reduced, and when $x = 100l$ is less than 20 per cent.

From Eq. (3.108) it may be shown that

$$\left(\frac{d\sigma}{dx}\right)_{max} = \frac{2}{3}\beta i^2 \tag{3.109}$$

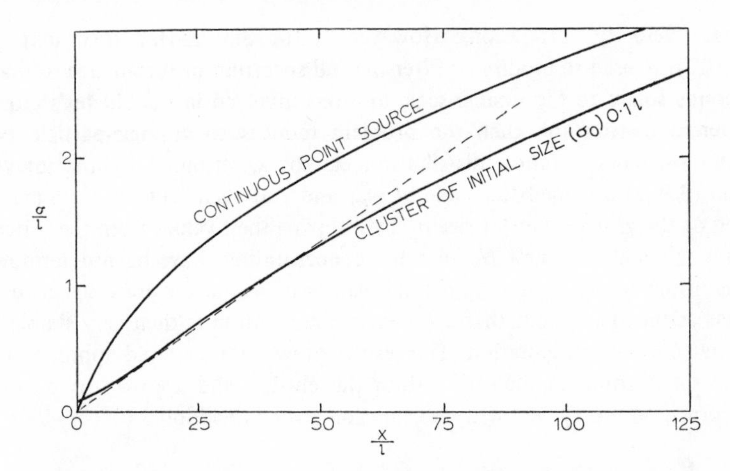

Fig. 3.14 – Dispersion as function of distance of travel for a continuous point source and a small cluster. Correlograms of exponential form are assumed. l is the Eulerian length-scale of turbulence, and the curves are for an intensity of turbulence (i) of 0.1, and a Lagrangian–Eulerian time-scale ratio (β) of 4.

and since, as can be seen from Fig. 3.14, the rate of expansion is almost constant and equal to the maximum value over a wide range of σ/l, this constitutes a simple practical formula for evaluating σ, given β and an estimate or measurement of i. The broken line drawn through the origin in Fig. 3.14 has a slope equal to this maximum value, and is seen to give a good approximation to the cluster curve for values of x/l between say 10 and 80, while at larger values of x/l the values of the continuous source may be adopted as a good approximation for the cluster. It is to be noted that application of the approximation based on Eq. (3.109) does not require any accurate knowledge of the length-scale l, whereas in the application of the true curve (as also of the continuous-source curve), at high values of x/l, the value of σ tends to a dependence on $l^{1/2}$.

The dependence of $(d\sigma/dx)_{max}$ on i^2 is particularly noteworthy in view of the fact that the rate of spread of a cloud from a continuous source is always proportional to i. Qualitatively it can be seen that this extra sensitiveness, to an increase (say) in i, is a consequence of the shape of the energy spectrum. The first effect is that in a given time the cluster attains larger size, but then larger eddies are effective and on that account the rate of growth is larger still. For a general mathematical analysis of the point see the original paper.

It is interesting that the integral equation of diffusion described earlier in the preceding section can be used to explore the nature of expanding clusters and to determine their internal ensemble-mean concentration distribution as a function of cluster size and time of travel. The normal treatment of clusters requires a multi-particle approach involving complex Eulerian–Lagrangian time-space velocity correlations. However, if the Smith–Hay treatment (see Eq. (3.107)) is used to modify or filter the full spectrum of turbulence so that it corresponds solely to the smaller-scale motions involved in the cluster's growth and internal distribution then the problem reduces to a single-particle type problem. The concentration distribution can be determined by the integral-equation (3.90a) provided the value of w_m and τ at each value of N are made a function of the growing cluster size σ_p by inferring their values from the filtered spectrum. Thus at each new N, once the concentrations have been determined, the magnitude of σ_p is determined and values of w_m and τ are assigned to all collisions assumed to occur there. To enact the solution is then very simple for any initial cluster configuration. The results show a rather rapid approach to a Gaussian distribution of material within the cluster, and a growth rate in size which corresponds very well with the solution given in Section 3.6.

3.8 DIFFUSION OF FALLING PARTICLES

All the foregoing discussions are concerned with hypothetical elementary particles of the fluid or alternatively with real foreign particles of the same density as the fluid, whose motions are in effect completely responsive to the whole spectrum of turbulent motion above some wavelength limit of similar size to the particles. When the particles are dense enough and large enough to have terminal velocities v_s which are not obviously negligible, say in relation to the eddy velocities, the distribution of the particles will be affected in various ways. The simplest approximation, first advocated by Schmidt (1925) is to assume that the cloud or section of plume sinks at a rate v_s, that the vertical distribution is otherwise unaffected, and that the ground acts as a permeable surface and retains all material passing through it. Solutions for floating particles may accordingly be adapted by replacing the height z in the vertical distribution by $z + v_s x/\bar{u}$. The model is obviously very crude, but has been used with various elaborations to provide working formulae. In a more formal way the sedimentation of the material may be allowed for in the differential equations of diffusion by introducing a convective term v_s, i.e. the two-dimensional equation of diffusion becomes

$$\bar{u}\frac{\partial \chi}{\partial x} = \frac{\partial}{\partial z}\left(K_z \frac{\partial \chi}{\partial z} + v_s \chi\right) \tag{3.110}$$

Deposition at the boundary (D) must then be correctly represented in the boundary conditions and for this Calder (1961) proposes

$$D = \left[K_z \frac{\partial \chi}{\partial z} + v_s \chi\right]_{z \to 0} = p\chi(x, 0) \tag{3.111}$$

where p is the *deposition velocity*, so-called by Chamberlain (1953). Rounds (1955) gives solutions of Eq. (3.110) for a source at height H, $K_z = ku_* z$,

$\bar{u} \propto z^{\alpha}$ for tractability, and a boundary condition as above but with $p = v_s$. In the form subsequently given by Godson (1958) the concentration at ground level and rate of deposition D are

$$\frac{\chi(x, 0)}{Q} = \frac{D(x)}{v_s Q} = \frac{\gamma}{H\bar{u}_H} \frac{\exp(-A/x)}{\Gamma(1-p)} \left(\frac{x}{A}\right)^{p-1} \tag{3.112}$$

where

$$p = \frac{-v_s}{ku_*\gamma}, \quad \gamma = 1 + \alpha, \quad A = \frac{H^2\bar{u}_H}{\gamma^2 K_H} \tag{3.113}$$

the subscript H referring to values at height H and Q being the rate of emission per unit length of a crosswind line source.

In accordance with the form of K_z the above solution is valid only in neutral conditions. Godson also gives an approximate generalization to other conditions by writing

$$K_z = \epsilon k u_* z \tag{3.114}$$

and assigning to ϵ such a value that the mean K_z is equal to the mean value given by a more realistic form of K_z (Godson actually proposed the use of values corresponding to the K given by Deacon's wind profile, but the procedure could be applied with any form). The result in Eq. (3.112) then applies approximately in general when

$$p = -\frac{v_s}{\epsilon k u_* \gamma}, \quad \gamma = 1 + \alpha, \quad A = \frac{H\bar{u}_H}{\gamma^2 \epsilon k u_*} \tag{3.115}$$

F. B. Smith gives solutions (1962) by Heaviside operational methods for u constant with height and

$$K = \text{constant}, \quad v_s = 0, \quad p > 0, \quad H \geqslant 0 \tag{3.116}$$

$$\begin{aligned} K = k'(H-z), \quad & v_s = 0, \quad p > 0, \quad H = 0 \\ & v_s > 0, \quad p > 0, \quad H = 0 \\ & v_s = p > 0 \qquad H > 0 \end{aligned} \tag{3.117}$$

and gives some examples in graphical form. These show χ increasing with height near the ground when $p > v_s$, and decreasing with height (even as $z \to 0$) when $v_s > p$. The forms of K adopted by Smith for convenience were regarded as reasonable representations for fairly long distances of travel [(Eq. 3.116)] or for vertically bounded diffusion [Eq. (3.117)]. Brock (1962) has provided extensions of these results by analogue solutions, using power and logarithmic forms for the wind profile with the appropriate conjugate forms of K.

Inasmuch as the treatments through the equations of diffusion incorporate the normal eddy diffusivity of the fluid they are obviously suspect, except for very small terminal velocities, because they ignore two other important effects, namely that the particle is continually falling out of the sample of eddies affecting it at any instant, and that its response to eddy motion is reduced by its inertia. Attempts to allow for the former have been made by Yudine (1959) and F. B. Smith (1959) (see also Smith and Hay 1961), in terms of the statistical treatments of dispersion.

Yudine applies Taylor's single-particle analysis for large T, with $R(\xi)$ in effect replaced by an $R(\xi, v_s\xi)$ incorporating the effects of time and vertical displacement $v_s\xi$. Then

$$K = \overline{w'^2}\int_0^\infty R(\xi, v_s\xi)\,d\xi \tag{3.118}$$

According as the effects of time or vertical displacement are dominant the function reduces to either the ordinary Lagrangian or Eulerian correlation coefficients. For these Yudine takes functions in accordance with the theory of the small-scale structure of turbulence, i.e. $1-R \propto \xi$ for the Lagrangian and $1-R \propto (v_s\xi)^{2/3}$ for the Eulerian form, and assumes these relations to extend to values of ξ and $v_s\xi$ at which R becomes zero. Functions approximately representing the upper and lower limits of the function $R(\xi, v_s\xi)$ are then constructed and corresponding limits to the integral in Eq. (3.118) obtained. The results show that K decreases with increase of v_s as would be expected on physical grounds, and at large v_s tends to vary with v_s^{-1}. The general variation is displayed graphically in terms of v_s and the parameters defining the assumed Lagrangian and Eulerian correlation functions. Since the treatment is only concerned with large values of the diffusion time T, at which the process is dominated by the low-frequency components of turbulence, this means that the error arising from neglect of the inertia of the particle will be at a minimum.

An approximate result which is not confined to any particular form of correlation function may be derived by the following simple argument. Let the length-scale of the turbulence be l. The time-scale experienced by a particle will be modified in two ways.

(a) It moves horizontally with the fluid and undergoes Lagrangian variations. If the Lagrangian and Eulerian properties of the fluid are assumed to have the simple scale relation of Eq. (3.72), the variations experienced by the particle will be equivalent to those at a point moving through the fluid at velocity $\bar{u}/\beta$.

(b) The particle also moves through the space spectrum vertically with velocity v_s.

The resultant effect will be equivalent to a movement through the space spectrum with velocity $(v_s^2+\bar{u}^2/\beta^2)^{1/2}$. Assuming that the particle response is perfect its velocity spectrum will be similar to that experienced at a fixed point in the fluid but with frequencies increased in the proportion $(v_s^2+\bar{u}^2/\beta^2)^{1/2}/\bar{u}$, or with an integral time-scale decreased in the same proportion. With K expressed as in Eq. (3.66) it follows immediately that

$$K \propto \left(\frac{v_s^2}{\overline{u}^2} + \frac{1}{\beta^2}\right)^{-1/2} \tag{3.119}$$

which of course tends to $\overline{u}/v_s$ for $v_s > \overline{u}/\beta$. This result is useful only when the diffusion can be treated by the gradient-transfer approach. For a continuous source the statistical treatment of 3.5 is preferable. An expression of the form of Eq. (3.74) may then be used, but with β replaced by

$$\beta' = \left(\frac{v_s^2}{\overline{u}^2} + \frac{1}{\beta^2}\right)^{-1/2} \tag{3.120}$$

Note that this analysis makes the assumption, unlikely to be strictly correct, that the length-scale is independent of the choice of reference line.

Smith's treatment is concerned with the diffusion of a cluster, which is a problem identical with the single particle (i.e. continuous source) case only in the limit of large time of travel. The development falls into two main parts, firstly the determination of the cluster expansion for rapidly falling particles (inertia effects being neglected), and secondly the construction of an interpolation formula for intermediate and small terminal velocities. For the first part (1959) the general nature of the argument is as follows: the loss of correlation in particle velocity due to its vertical fall is the controlling feature when it is substantially more rapid than that associated with the normal Lagrangian decay of correlation for a floating particle. This condition obtains when

$$l/v_s \ll t_L$$

where l is the Eulerian length-scale and t_L the Lagrangian time-scale, i.e.

$$v_s \gg \overline{u}t_E/t_L \tag{3.121}$$

where t_E is the equivalent Eulerian time-scale defined by $l = \overline{u}t_E$ and t_L/t_E is the Lagrangian–Eulerian scale ratio β introduced in Eq. (3.72). In this case the statistics of the velocity fluctuations experienced by a particle will be equivalent to those which would be measured by an instrument moving at speed v_s through the field of turbulence. The correlation coefficient $R(\xi)$ for the particle can accordingly be replaced by $R_E(t')$, the autocorrelation coefficient obtained from measurements at a fixed point, when $ut' = v_s\xi$. Then, for example Eq. (3.58) becomes

$$\frac{\mathrm{d}X^2}{\mathrm{d}t} = 2\overline{u'^2}\,\frac{\overline{u}}{v_s}\int_0^{v_s t/\overline{u}} R_E(t')\,\mathrm{d}t' \tag{3.122}$$

In Smith's (1959) analysis of the spread of a cluster Eq. (3.122) is replaced by

$$\frac{d\sigma^2}{dt} = \frac{2}{3v_s}\int_0^{v_s t} (R_{ii})\, ds \tag{3.123}$$

where σ is the standard deviation of the particles from the centre of the cluster, and (R_{ii}) is the Eulerian vector covariance for distance separation s, appropriately modified for the size σ of the cluster as in the floating cluster problem previously treated. For the floating cluster (see Smith and Hay, 1961), the corresponding equation is

$$\frac{d\sigma^2}{dt} = \frac{2}{3}\frac{\beta}{\bar{u}}\int_0^{\bar{u}t/\beta} (R_{ii})\, ds \tag{3.124}$$

which transforms into Eq. (3.123) when $v_s = \bar{u}/\beta$. Remembering that Eq. (3.123) holds only for $v_s \gg \bar{u}/\beta$ this equivalence in the equations means that it does not matter whether the cluster moves purely horizontally at speed $\bar{u}$, or purely vertically at $v_s = \bar{u}/\beta$, statistically it experiences the same conditions. In other words, if the horizontal scale is compressed by a factor β, then in these transformed coordinates the direction of motion of the cluster is irrelevant. In the general case, therefore, when neither v_s nor $\bar{u}/\beta$ can be neglected, the appropriate result is given by replacing v_s or $\bar{u}/\beta$ by $\sqrt{(v_s^2 + \bar{u}^2/\beta^2)}$, as already argued above, so that

$$\frac{d\sigma^2}{dt} = \frac{2}{3}\frac{1}{\sqrt{(v_s^2 + \bar{u}^2/\beta^2)}}\int_0^{t\sqrt{(v_s^2 + \bar{u}^2/\beta^2)}} (R_{ii})\, ds \tag{3.125}$$

and the solutions for $d\sigma/dx$ are as in Eqs. (3.107), (3.108) and (3.109), with $\bar{u}/\beta$ replaced by $\sqrt{(v_s^2 + \bar{u}^2/\beta^2)}$. Numerical evaluation of σ for typical values of the parameters gives the curves in Fig. 3.15, taken from Smith and Hay's paper. For particles with terminal velocities less than say 0.5 m/sec the reduction in σ due to the fall of the particles is less than 10 per cent at a time of travel of 1 min. On the other hand the effect of terminal velocity is obviously nil at the beginning [since for small t the integral in Eq. (3.125) is proportional to $t\sqrt{(v_s^2 + \bar{u}^2/\beta^2)}$]. Then at large values of t, when this integral is a constant,

$$\frac{d\sigma^2}{dt} \propto \frac{1}{\sqrt{(v_s^2 + \bar{u}^2/\beta^2)}} \quad (t \text{ large}) \tag{3.126}$$

in which case

$$\frac{\sigma}{\sigma_{(v_s=0)}} = \frac{1}{(1 + v_s^2\beta^2/\bar{u}^2)^{1/4}} \quad (t \text{ large}) \tag{3.127}$$

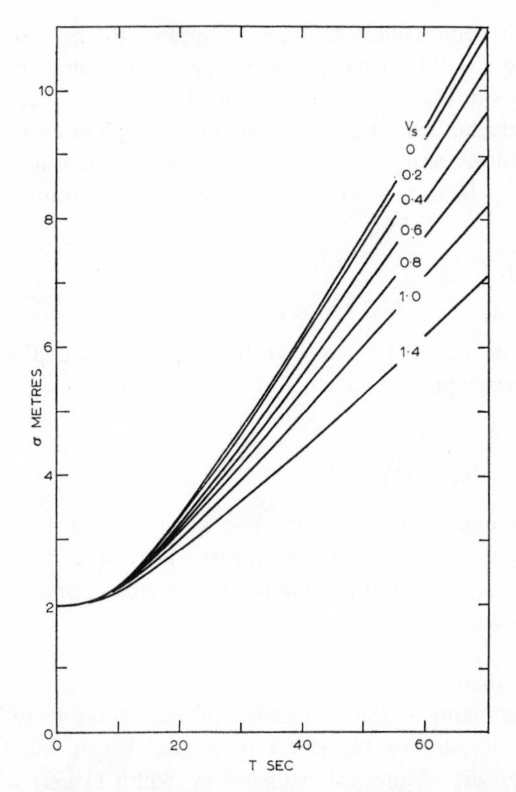

Fig. 3.15 – Dispersion of a falling cluster of particles as a function of time of travel T and terminal velocity v_s. A correlogram of exponential form is assumed. Eulerian length-scale $(l) = 20$ m, $\bar{u} = 5$ m/sec, $i = 0.1$, $\beta = 5$, initial standard deviation $\sigma(0) = 2$ m. (Smith and Hay, 1961).

and for the parameters in Fig. 3.15, and $v_s = 0.5$ m/sec, the magnitude of this ratio is 0.95. Since Eq. (3.125) also applies to a serial release of particles when R_{ii} is the covariance for the whole spectrum of turbulence these generalizations concerning the effect of terminal velocity at limiting times are also applicable to a continuous point source. When $v_s \gg \bar{u}/\beta$ Eq. (3.126) shows that at large times of diffusion σ varies as $v_s^{-1/2}$ and hence K varies as v_s^{-1}, as already deduced.

For the continuous source, with R_{ii} in Eq. (3.125) independent of σ, it is clear that the dependence of the integral on $\sqrt{(v_s^2 + \bar{u}^2/\beta^2)}$ will decrease systematically with increasing t, between linear dependence at small t and independence at large t. Thus it can be said that the dependence of σ on v_s will at all times be between the limits of independence at small t and dependence as in Eq. (3.126) at large t. Hence for $v_s = 0.5$ m/sec and the parameters in Fig. 3.15 the value of 0.95, obtained for $\sigma/\sigma_{(v_s=0)}$ at limitingly large time, may be regarded as a lower limit for all times in practice.

This generalization cannot however be applied to the case of the cluster, because R_{ii} in Eq. (3.125) then depends on σ, and hence on v_s. Further examination of the results in Fig. 3.15 indicates that the ratio $\sigma/\sigma_{(v_s=0)}$ decreases with increase in t throughout the range of t covered, though at $t=60$ sec the rate of change of the ratio is slow, suggesting that the value there is near the minimum. As pointed out by Thomas (1964) the expression for maximum rate of growth becomes

$$\left(\frac{d\sigma}{dt}\right)_{max} = \frac{2}{3}\frac{\overline{u^2}i^2}{(v_s{}^2+\overline{u^2}/\beta^2)^{1/2}} \tag{3.128}$$

This follows from Eq. (3.109) on substituting for β and β' in Eq. (3.120). Integrating and neglecting the constant of integration gives

$$\sigma \cong \frac{2}{3}\frac{\overline{u^2}i^2t}{(v_s{}^2+\overline{u^2}/\beta^2)^{1/2}} \tag{3.129}$$

Thus, at the maximum rate of growth and when $v_s > \bar{u}/\beta$ the result is $\sigma \propto v_s^{-1}$. Also, now taking $v_s = 1$ m/sec for convenience, $\bar{u} = 5$ m/sec and $\beta = 5$ it appears that $\sigma/\sigma_{(v_s=0)}$ is $1/\sqrt{2}$ at the maximum rate of growth and ultimately [from Eq. (3.127)] $1/\sqrt[4]{2}$ at large t.

Effect of particle inertia

The foregoing estimates of the importance of sedimentation still require some qualification on account of the effect of inertia, which will further reduce the turbulent spread. Theoretical estimates by Smith (1959) suggest that for terminal velocities of 2 m/sec the response of the particles will be virtually complete for eddies of wavelength of the order of 1 m and greater.

An analysis of the effect of inertia has also been given by Csanady (1963) and this can be put in more general terms as follows. It is supposed that the particle has a simple exponential response to change in fluid velocity. This means that the spectral response function of the particle is $(1+a^2\omega^2)^{-1}$ where $\omega = 2\pi n$, and the effect of this is roughly equivalent to applying a low-pass cut-off to the turbulent velocity spectrum at $n = 1/2\pi a$ (to be precise this corresponds to a 50 per cent reduction in spectral density – see Table 2.I). So for particle inertia to be unimportant, as regards the overall fluctuation experienced, it is merely necessary that spectral densities at $n > 1/2\pi a$ should make a negligible contribution to the total variance of the turbulent component. From the data on frequency spectra in the lower atmosphere (see 2.6) only a small fraction of the total variance is in equivalent wavenumbers greater than $1/l$, which for a rapidly falling particle will be equivalent to frequencies greater than v_s/l. Hence the approximate condition for neglecting inertia is

$$\frac{v_s}{l} < \frac{1}{2\pi a} \tag{3.130}$$

Assuming Stoke's law, Csanady finds $a=\rho d^2/18\mu$, where ρ is particle density, d diameter and μ the dynamic viscosity of the air. For large particles outside the range of applicability of Stokes' law, this is an overestimate of a. For example, with $d=500$ micron, $\rho/\rho_{\text{air}}=2000$, it is estimated that $2\pi a<1$ and probably about 0.2. The corresponding value of v_s is near 4 m/sec. The condition for negligible effect of inertia is therefore $l>1$ m which should be met in most atmospheric problems.

For a continuous release of particles the condition may be alternatively derived as follows. In terms of the spectral interpretation discussed in 3.4 it follows that only frequencies (of particles velocity fluctuation) less than roughly $1/2T$ contribute significantly to the particle spread at time T. Accordingly, particle inertia will be unimportant if the resulting low-pass cut-off $(1/2\pi a)$ is somewhat greater than $1/2T$, say $5/T$. With the conditions assumed in the previous paragraph this is satisfied when $T \geqslant 1$ sec. In practice therefore it seems that the effect of particle inertia will usually be negligible.

3.9 THE FLUCTUATION OF CONCENTRATION

The discussion has so far been concerned with the *mean* distribution of particles or of the concentration of marked fluid. It is, however, to be expected that the instantaneous value of the concentration will itself be a randomly fluctuating quantity. There are two distinct actions by which this fluctuation may be produced. The first occurs *within* any volume of marked fluid – namely the relatively small-scale intermingling of fluid from different origins. As a result fluctuations are experienced at any position, within the volume, even though this be fixed relative to the centre of mass of the marked fluid. The second action concerns the effect of a continuous release of marked fluid, at a position which is fixed relative to the source. In this case fluctuations arise as a consequence of the varying proximity with which the centres of mass of successive volumes of marked fluid pass this position (concentration tending to diminish with distance from the centre of mass).

Concentration fluctuations in the first of the above categories are completely analogous to the fluctuations of wind velocity, temperature or humidity at a fixed height above ground, where ideally the situation corresponds to an infinite plane source or sink at ground level. In exactly the same way as for turbulent velocity fluctuations a conservation equation may be constructed for the mean square fluctuation of concentration. The equation is of the form of Eq. (2.61), involving a d/dt term, a production term which contains the product of the flux of marked fluid and its mean gradient, a divergence of the flux of mean square fluctuation, and a dissipation term representing the smoothing out of fluctuations by molecular action. For the idealized case of a steady point source with its centre-line along the x-axis the appropriate equation is

$$\bar{u}\frac{\partial\overline{\chi'^2}}{\partial t} = -2\left[\overline{w'\chi'}\frac{\partial\bar{\chi}}{\partial z} + \overline{v'\chi'}\frac{\partial\bar{\chi}}{\partial y}\right] - \left[\frac{\partial\overline{w'\chi'^2}}{\partial z} + \frac{\partial\overline{v'\chi'^2}}{\partial y}\right] - S \quad (3.131)$$

S being the rate of reduction of mean square fluctuation by molecular action. As usual it is assumed that the gradient along the x-direction is negligible compared with those along the y- and z-directions and that molecular diffusivity is negligible. Note the factor of 2 in the production term, which arises from the consideration of *variance* as distinct from half variance (kinetic energy) in Eq. (2.61).

Csanady (1967a) has considered the solution of the above from of equation with various simplifications and assumptions.

(i) A gradient-transfer relation is assumed to describe the flux of both the material (or marked fluid) and the mean square fluctuation, with a common value of the eddy diffusivity K, which is assumed a function of distance x in accordance with the statistical treatment of 3.4.
(ii) The distribution of mean concentration is assumed Gaussian.
(iii) The distribution of $\overline{\chi'^2}$ is assumed self-similar at different values of x, as is that of $\bar{\chi}$, i.e. a universal function of y/L, where L is a length-scale ultimately identified with the standard deviation σ_p of the $\bar{\chi}$ distribution. It is demonstrated that the requirement for self-similarity is that S is equal to $\overline{\chi'^2}/T$ where T is a decay-time scale which is a function of x.
(iv) It is assumed that $\sigma_p{}^2/KT = \alpha$, where α is a number to be determined.

For a given value of α numerical solution is then possible and Csanady gives some examples showing that with a suitable choice of α a reasonable fit may be obtained to observations of fluctuations in an oil fog plume in pipe flow. The same analysis has also been applied to a crosswind line source (Csanady 1967b), in this case to give results which on comparison with laboratory data require α also to be a function of distance. Although the general approach is enlightening this is a feature which obviously needs clarification.

Turning now to the second category of fluctuation, it will be recalled that in the classical mathematical treatments of the continuous source (see Sutton 1953) one of the essential simplifications was to consider the plume from a point source as composed of a succession of overlapping clusters. Physically, it is evident that if this model is to be realistic the 'clusters' or elementary sections of the plume, cannot be regarded as travelling consistently along a single line in the direction of the mean wind. In general, referring to a vertical plane normal to the mean wind at a fixed distance downwind, it is to be expected that the centres of mass of the successive clusters will pass through this plane at different positions, and that the mean distribution of particles at this plane will be determined by a combination of the spread within the cluster and the scattering of the clusters themselves. A relation between the separate and total effects on dispersion is implicit in Brier's (1950) two-particle analysis, and may be expressed as follows: if $\overline{y^2}$ is the mean square separation of *pairs* of particles released simultaneously from the fixed points y_0 apart, and $\overline{D^2}$ is the mean square deviation of the centres of mass of the pairs from their ensemble mean, then at any given time

$$\overline{y^2} + \overline{2D^2} = y_0^2 + \overline{2Y^2} \tag{3.132}$$

where $\overline{Y^2}$ is the mean dispersion of particles serially-released from a single fixed point, as in Eq. (3.62). In principle the generalization of this to successive clusters is applicable either to a finite-size continuous source or to a theoretical point source, where in the latter case the elementary cluster may be regarded as formed by the series of particles which pass through the source-point in some arbitrary small interval of time.

Gifford (1959) has developed a formal treatment of the fluctuating plume, using a model in which, for mathematical convenience, the clusters are represented as discs in the plane normal to the mean wind direction. For further simplification he assumes that the distribution of concentration in the discs is a definite physical function (rather than a probability distribution), so that the mean distribution of concentration is determined by the scattering of the discs. It is also assumed that both this mean distribution and the distribution of the disc centres are of Gaussian form, which implies that the distribution within the disc is also Gaussian. Taking $\overline{Y^2}$ as the variance of the mean distribution, σ^2 the variance with respect to the disc-centre of the distribution in the disc, and $\overline{D^2}$ as the mean square displacement of the disc-centres from a fixed axis through the source,

$$\overline{Y^2} = \sigma^2 + \overline{D^2} \tag{3.133}$$

The instantaneous and mean concentration distribution from a source of strength Q per sec may then be written as

$$\frac{\bar{u}\chi}{Q} = \frac{1}{2\pi\sigma^2} \exp\left[-\frac{(y-D_y)^2 + (z-D_z)^2}{2\sigma^2}\right] \tag{3.134}$$

$$\frac{\bar{u}\bar{\chi}}{Q} = \frac{1}{2\pi(\sigma^2 + \overline{D^2})} \exp\left[-\frac{(y^2 + z^2)}{2(\sigma^2 + \overline{D^2})}\right] \tag{3.135}$$

These equations refer to isotropic diffusion, but they can be generalized for the anisotropic case merely by introducing different values for σ^2 and $\overline{D^2}$ in the two component directions. The variance of the point concentration frequency distribution corresponding to Eq. (3.134), defined as

$$V = \overline{\chi^2} - \bar{\chi}^2$$

is given by

$$V = \frac{Q^2}{(2\pi\bar{u})^2}\left[\frac{\exp\left(-\dfrac{y^2+z^2}{\sigma^2+2\overline{D^2}}\right)}{\sigma^2(\sigma^2+2\overline{D^2})} - \frac{\exp\left(-\dfrac{y^2+z^2}{\sigma^2+\overline{D^2}}\right)}{(\sigma^2+\overline{D^2})^2}\right] \tag{3.136}$$

Relations are also derived for the frequency distribution f of the quantity χ/Q, and the simplest form of this, for positions on the mean plume axis, is the power law relation

$$f\left(\frac{\chi}{Q}\right) = \frac{\sigma^2}{\overline{D^2}}(2\pi\sigma^2\bar{u})^{\sigma^2/\overline{D^2}}\left(\frac{\chi}{Q}\right)^{(\sigma^2/\overline{D^2})-1} \tag{3.137}$$

From Eqs. (3.134) and (3.135) it follows that the ratio of the peak concentration in the instanteneous plume to the average concentration on the mean plume axis is

$$\frac{\text{Peak}}{\text{Average}} = \frac{\sigma^2+\overline{D^2}}{\sigma^2} \tag{3.138}$$

As Gifford points out, since both σ^2 and $\sigma^2+\overline{D^2}$ $(=\overline{Y^2})$ tend to a variation with T (time of travel) for large values of T, the peak/average ratio should tend to a constant value. The argument can be carried further, for as shown by F. B. Smith in an unpublished analysis

$$\frac{\mathrm{d}\overline{D^2}}{\mathrm{d}t} \to 0$$

for large clusters each of many particles. This means that at large T, D^2 tends to a constant value, which eventually implies

$$\sigma^2 \gg D^2$$

and

$$\frac{\text{Peak}}{\text{Average}} \to 1$$

Application of the Eqs. (3.134) and (3.135) is considered further in 5.3 in the discussion of the distribution of effluent emitted from chimneys.

Testing the development of the foregoing theoretical ideas requires high-quality measurements of concentration fluctuations in well-defined experimental conditions. It is only recently that such data have begun to be available, for example from measurements in the field reported by C. D. Jones (1979) and in a wind tunnel boundary layer by Fackrell and Robins (1982a, b). In the first of their papers Fackrell and Robins demonstrate that the Gifford fluctuating plume model provides a reasonably accurate description of the fluctuation properties for an elevated plume. The time-mean and instantaneous dispersion parameters ($\overline{Y^2}$ and σ^2 respectively in Eq. (3.134) *et seq.*) were specified on the basis of the Hay–Pasquill and Smith–Hay treatments respectively (discussed in 3.5 and 3.7), with βi given the value 0.6. It is noteworthy that the latter value, which was deduced from Lagrangian time-scale estimates made from earlier dispersion measurements in the wind tunnel, is close to that most recently indicated by turbulence data in the atmospheric boundary layer (see 2.7 and Table 3.I). The 1982b paper reports the use of the fluctuation measurements in evaluating the terms in the second-moment equations for the variance of the concentration (a more complete form of Eq. (3.131)) and the vertical flux.

3.10 THE EFFECT OF WIND SHEAR ON HORIZONTAL SPREAD

For the case of crosswind diffusion the foregoing discussions have been concerned solely with the direct effect of the lateral component of turbulence, and any generalization therefrom concerning alongwind diffusion would carry a corresponding restriction to the direct action of the longitudinal component. However, a distorting effect of the variation with height of the mean wind, both in speed and direction, is often visibly evident in the development of puffs or plumes of smoke. The effect is most obvious in stably stratified conditions, when the boundaries of the smoke are most definite. When such a condition is followed by relatively vigorous vertical mixing, as with the onset of surface heating after sunrise, the sheared form imparted to the smoke will give rise to enhanced horizontal spread at any level within the layer affected. The significance of this joint effect of shear followed by vertical mixing is obvious in such circumstances, but a more subtle matter is the extent to which the interaction between vertical mixing and velocity shear is continuously effective.

Attention was first drawn to the importance of this continuous interaction by G. I. Taylor in 1953, with reference to the rapidity with which suspended material is stretched out in a flow of liquid along a pipe. It appeared that the interaction between the crosswise diffusion and the velocity profile across the pipe produced a longitudinal extension which was dominant over any direct longitudinal diffusion, not only in laminar flow but also when the liquid was turbulent. Several theoretical analyses have since been made of the effect as it applies in the sheared flow of the atmospheric boundary layer.

The first analysis, by Saffman (1962), is based on the differential equation of diffusion, with the gradient transfer relation assumed as usual to represent the direct action of turbulence. The starting point is the equation which follows from Eq. (3.12), $\bar{w}$ being assumed zero,

$$\frac{\partial \bar{\chi}}{\partial t} + \bar{u}\frac{\partial \bar{\chi}}{\partial x} + \bar{v}\frac{\partial \bar{\chi}}{\partial y} = \frac{\partial}{\partial x}\left(K_x \frac{\partial \bar{\chi}}{\partial x}\right) + \frac{\partial}{\partial y}\left(K_y \frac{\partial \bar{\chi}}{\partial y}\right) + \frac{\partial}{\partial z}\left(K_z \frac{\partial \bar{\chi}}{\partial z}\right) \qquad (3.139)$$

applicable to an instantaneous source of material. To facilitate solution for the magnitude of the horizontal (specifically longitudinal) spread σ_x Saffman uses the method of moments introduced by Aris in the extension of Taylor's work on diffusion in pipe flow. The same method was first applied in the context of atmospheric diffusion independently by F. B. Smith (1957a). The general procedure is to define moments $\boldsymbol{\theta}_{nm}(z, t)$ by the relation.

$$\boldsymbol{\theta}_{nm}(z, t) = \int_{-\infty}^{\infty}\int_{-\infty}^{\infty} x^n y^m \chi \, dx \, dy \tag{3.140}$$

For an instantaneous cloud of material $\boldsymbol{\theta}_{00}$ is the area integral of concentration and on integration with respect to height is equated to the total source. For given height z, $\boldsymbol{\theta}_{10}/\boldsymbol{\theta}_{00}$ is the centroid of the material and $\boldsymbol{\theta}_{20}$ determines the longitudinal spread through the relation $\sigma_x{}^2 = \boldsymbol{\theta}_{20}/\boldsymbol{\theta}_{00} - (\boldsymbol{\theta}_{10}/\boldsymbol{\theta}_{00})^2$.

Integrating Eq. (3.139) over a horizontal plane, first as it stands and then after multiplication by x or x^2, Saffman obtained the following equations for the moments

$$\frac{\partial \boldsymbol{\theta}_{00}}{\partial t} = \frac{\partial}{\partial z}\left(K_z \frac{\partial \boldsymbol{\theta}_{00}}{\partial z}\right) \tag{3.141}$$

$$\frac{\partial \boldsymbol{\theta}_{10}}{\partial t} = \frac{\partial}{\partial z}\left(K_z \frac{\partial \boldsymbol{\theta}_{10}}{\partial z}\right) + \bar{u}\boldsymbol{\theta}_{00} \tag{3.142}$$

$$\frac{\partial \boldsymbol{\theta}_{20}}{\partial t} = \frac{\partial}{\partial z}\left(K_z \frac{\partial \boldsymbol{\theta}_{20}}{\partial z}\right) + 2K_x\boldsymbol{\theta}_{00} + 2\bar{u}\boldsymbol{\theta}_{10} \tag{3.143}$$

Similar equations may be written for $\boldsymbol{\theta}_{11}$ and $\boldsymbol{\theta}_{02}$. The solutions automatically give separate additive terms for the direct action of turbulence (involving K_x) and the interaction effect. For the latter, asymptotic forms are given for the simple case of a linear variation of wind with height ($\bar{u} = \alpha z$) and a constant K_z. According as the vertical diffusion is *fully bounded* (within a layer of height h, $\bar{u}$ ranging from 0 at the ground to αh at the upper bound), *semi-bounded* (source at ground-level) or *unbounded* (source remote from ground), the latter being subsequently derived by F. B. Smith (1965), the results for spread at a given level are:

$$\begin{array}{lll} \text{(a) Fully bounded} & \sigma_x{}^2 = \alpha^2 h^4 T/60K_z \\ \text{(b) Semi-bounded} & \sigma_x{}^2 = 0.036\alpha^2 K_z T^3 \\ \text{(c) Unbounded} & \sigma_x{}^2 = \tfrac{1}{6}\alpha^2 K_z T^3 \end{array} \tag{3.144}$$

(a) is applicable only in the limit $T \to \infty$. On the other hand (b) and (c) are applicable for a height z, measured from the source level, small compared with the vertical spread. Effectively therefore (b) and (c) give the spread at the height of the source at *all* values of T.

The above solutions bring out several important points. First, the dependence on the rate of vertical spread (as represented by K_z) is in opposite senses according as the vertical spread is limited in both directions or unlimited in at least one direction (solution (a) is equivalent to that obtained by Taylor for pipe flow). Second, only in the fully bounded case when $\sigma^2 \propto T$, can the growth be represented by a simple addition to the (assumed) constant diffusivity K_x. Third, in the unbounded cases there is much more rapid growth of σ^2, with T^3, implying that the interaction effect must ultimately dominate the direct effect of turbulence represented in K_x. Finally, the implication that in bounded flow (a) the effective horizontal diffusivity becomes infinite as $K_z \to 0$ is not valid. As Saffman points out the solutions apply for $T \gg h^2/K_z$ and therefore cannot be valid in practice when K falls to a very small value.

The importance of the precise form of wind profile is brought out by Saffman in some additional solutions. For the fully bounded case, with $\bar{u} = (\alpha+1)U(z/h)^\alpha$ (U being the average speed over the whole depth h), and with K_z assumed equal to the conjugate form of eddy viscosity, the asymptotic result for the effect of shear on the spread of the entire cloud (Σ_x^2) in the layer 0-h is

$$\Sigma_x^2 = \frac{2\alpha^3}{(\alpha+2)(3\alpha+2)} \frac{U^3hT}{u_*^2} \tag{3.145}$$

showing that for small α (say in the region 0.05 to 0.15) Σ_x^2 is very sensitive to the shape of the wind profile, being then dependent roughly on α^3.

For unbounded vertical spread Saffman also gives a general dimensional argument for the dominance of the interaction term. Taking

$$\bar{u} = u_0 + \alpha z^a, \quad K_z = k_z z^c, \quad K_x = k_x z^d \tag{3.146}$$

it is found that provided $c<2$, as will normally be the case, the dispersion at a given level is

$$\sigma_x^2 = \alpha^2 T^2 (k_z T)^{2a/(2-c)} f(\rho) + k_x T (k_z T)^{d/(2-c)} f(\rho) \tag{3.147}$$

$f(\rho)$ being some dimensionless function of $\rho = z(k_z T)^{-1/(2-c)}$. The first term (representing the interaction) must ultimately dominate the second (representing the direct action of turbulence) provided

$$2 + 2a > c + d \tag{3.148}$$

With the reasonable assumption that the variation of K_x with height will be similar to that of the integral-scale of the longitudinal component it seems unlikely that d will be as much as 0.5. The exponent c (for K_z) might exceed 1.0 in unstable conditions but probably not substantially so, and with $\alpha>0$ it follows that the condition is likely to hold always.

An alternative and, in some respects. more revealing analysis may be made on the lines of the statistical treatment introduced in 3.4. The problem is then to express the statistical behaviour of a typical particle wandering vertically under the action of turbulence through a mean velocity shear. This approach was first attempted by Högström (1964) with the simplifying assumption that the particle immediately takes up the *mean* horizontal velocity at its level. As will be seen later the error involved in this assumption becomes negligible at large T, but for small T the turbulence in the horizontal and the extent to which the horizontal velocity of the particle retains correlation with its earlier history have to be taken into account. A complete treatment in this respect is given by F. B. Smith (1965).

Referring to Fig. 3.16, the flow is assumed to have a constant streamwise velocity $\bar{u}$ and a lateral velocity $\bar{v} = \psi z$. Smith considers the lateral (y) displacement of these particles released continuously from the source which pass

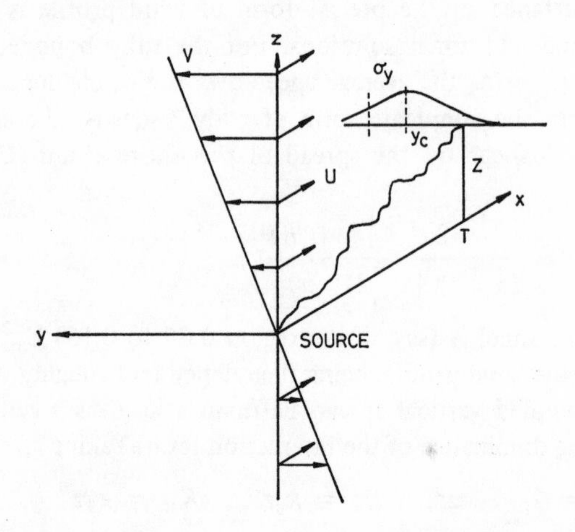

Fig. 3.16 – Schematic representation of flow assumed in Smith's (1965) treatment of the effect of shear on crosswind spread. The irregular line is the trajectory of one of the particles which passes through $z = Z$ at time T and which in ensemble have a centroid at y_c and are spread about this centroid with standard deviation σ_y.

through a crosswind line at distance $\bar{u}T$ and height (above source) Z. Defining $v(t)$ as the velocity in the y-direction of one of the selected particles at a time t when its height is z, for such a particle it follows [exactly as in the analysis leading to Eq. (3.57)] that

$$\frac{\mathrm{d}\overline{y^2}}{\mathrm{d}t} = 2\int_0^t v(t)v(s)\,\mathrm{d}s \tag{3.149}$$

Substituting

$$v(t) = \psi z(t) + v'(t) \tag{3.150}$$

where $\psi z(t)$ is the mean lateral velocity at height $z(t)$ and $v'(t)$ is the deviation therefrom of the velocity of the particle gives

$$\frac{d\overline{y^2}}{dt} = 2\int_0^t [\psi^2\overline{z(t)z(s)} + \overline{v'(t)v'(s)} + \psi\overline{z(t)v'(s)} + \psi\overline{z(s)v'(t)}]\, ds \quad (3.151)$$

The first term is independent of the horizontal turbulence and, as in Högström's analysis, represents the effect of the particle taking up instantaneously the mean velocity at its level. Smith expands these terms using the cosine transform relation between correlation and spectrum functions. For the crosswind spread at a given level after time T the result is finally expressed in three terms:

$$\begin{aligned}\sigma_y^2 &= \sigma_v^2T^2\int_0^\infty G_L(n)\left(\frac{\sin r}{r}\right)^2 dn && \text{(a)}\\ &+ \tfrac{1}{4}\sigma_w^2\psi^2T^4\int_0^\infty F_L(n)\left(\frac{\sin r - r\cos r}{r^2}\right)^2 dn && \text{(b)} \qquad (3.152)\\ &- \rho^2\sigma_v^2T^2\int_0^\infty F_L(n)\left(\frac{\sin r}{r}\right)^2 dn && \text{(c)}\end{aligned}$$

where $r=\pi nT$, σ_w and σ_v have the usual meanings, $F_L(n)$ and $G_L(n)$ are the normalized Lagrangian spectrum functions for the vertical and crosswind components and ρ is the correlation coefficient between the latter components, i.e. $\overline{w'v'}/\sigma_w\sigma_v$.

Term (a) is equivalent to Eq. (3.65) and is the direct effect of the horizontal component of turbulence, reducing to $2\sigma_v^2t_{vL}T$ for large T, where t_{vL} is the integral time-scale for the crosswind component.

Term (b) is independent of the horizontal turbulence and is the effect of the sytematic wind shear, equivalent to the effect of the particle taking up instantaneously the mean velocity at its level. The weighting function applied to $F_L(n)$ is a narrow filter as shown in Fig. 3.17, with a maximum value (0.19) at $\pi nT=2$. For large T this filter transmits only very low frequencies for which $F_L(n)$ tends to the value $F_L(0)=4t_{wL}$ where t_{wL} is the integral time-scale for the vertical component. Accordingly, for large T $F_L(n)$ can be taken outside the integral, which may then be evaluated to give $\frac{1}{6}\psi^2\sigma_w^2t_{wL}T^3$. Writing $K_z=\sigma_w^2t_{wL}$ this becomes $\frac{1}{6}\psi^2K_zT^3$, identical with the solution for unbounded diffusion in Eq. (3.144) except for the replacement of $\alpha(=d\bar{u}/dz)$ by $\psi(=d\bar{v}/dz)$.

Term (c) is the effect of that part of the horizontal component which is correlated with the vertical component. It acts in opposition to (b), being a reflection of the extent to which a particle fails to adjust instantaneously to the mean horizontal velocity at its level. At large T its magnitude is $-2\rho^2\sigma_v^2t_{wL}T$.

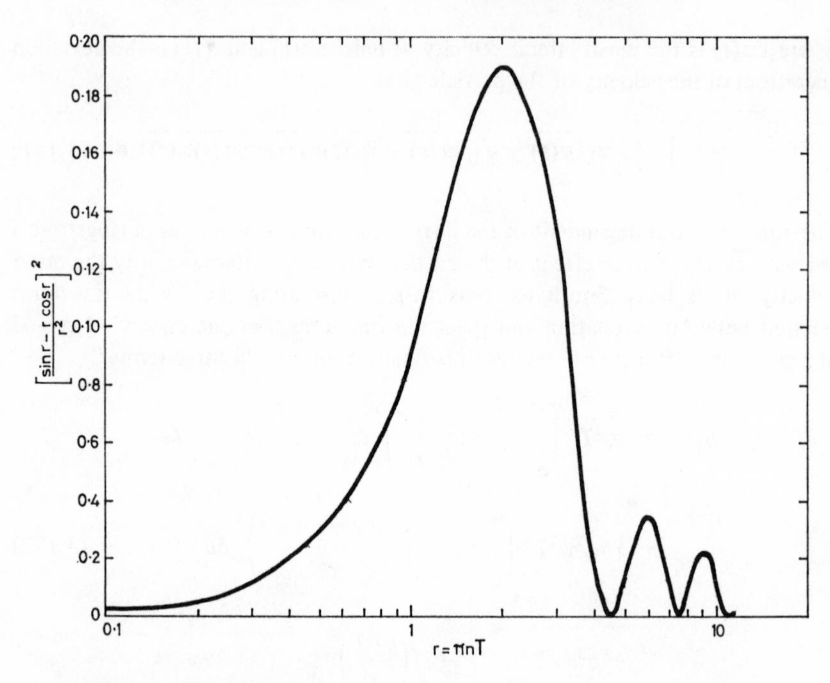

Fig. 3.17 – The weighting function in Smith's (1965) solution for the effect of wind shear on horizontal spread (Eq. 3.152b).

It is obvious that at sufficiently large T term (b) must become dominant. At small enough T this becomes very small and the resultant of terms (a) and (c) becomes dominant (except in the special case of $\rho = 1$). It is especially interesting to note that the direct effect of the crosswind component is apparently partially neutralized by the effect of the correlation between w' and v'. The essential point is that some of the crosswind displacement due to the crosswind component occurs in the form of an initial shearing of the distribution of particles tending to be displaced to one side and downward-moving particles to the other, but in the sense *opposite* to the mean shear which ultimately appears in the particle distribution. The effect can be seen in the expression obtained by Smith for the position of the centroid at any level Z a given time after release, i.e.

$$y_c = \tfrac{1}{2}\psi ZT + \frac{\overline{w'v'}}{\sigma_w{}^2} Z \tag{3.153}$$

in which by definition ψ and $\overline{w'v'}$ must be of opposite sign. Accordingly y_c changes linearly with time from an initial value determined by the correlation between w' and v' to a value of opposite sign, then in the direction of the mean shear.

It is particularly interesting to note the identity of (b) in Eq. (3.152) for large T with (c) of Eq. (3.144) and with Högström's result, together with the fact that neither of the last two results contain anything equivalent to Smith's term (c) in Eq. (3.152). These features are essentially concerned with the validity of neglecting the correlation between the horizontal and vertical components. In the gradient-transfer approach the assumption has been made that

$$\overline{u'\chi'} = -\overline{u'l}\frac{\partial\bar{\chi}}{\partial x} = K_x\frac{\partial\bar{\chi}}{\partial x} \qquad (3.154)$$

whereas strictly the three-dimensional nature of the transfer should be taken into account by writing

$$\overline{u'\chi'} = -\overline{u'l_x}\frac{\partial\bar{\chi}}{\partial x} - \overline{u'l_y}\frac{\partial\bar{\chi}}{\partial y} - \overline{u'l_z}\frac{\partial\bar{\chi}}{\partial z} \qquad (3.155)$$

The additional terms cannot necessarily be neglected if there is a correlation between the horizontal and vertical components of turbulence. For example, if there is correlation between u' and w' there must be correlation between w' and l_z and hence between u' and l_z. Also, if there is correlation between v' and w' there must be correlation between u' and v' and hence between u' and l_y.

Gee and Davies (1963, 1964) have given an extension of Saffman's linear wind shear treatment by including the term $\overline{u'l_z}(\partial\bar{\chi}/\partial z)$, writing $\overline{u'l_z} \cong -l^2(d\bar{u}/dz)$ and $l = kz$. This leads to a *proportionate* reduction in the $\sigma_x{}^2$ of (b) in Eq. (3.144), independent of time, of about 20 per cent. At first sight this result seems inconsistent with Smith's result that the *proportionate* reduction provided by the correlation term (c) in Eq. (3.152) decreases with time and ultimately becomes negligible. The difference must be associated with the point that in Smith's treatment the equivalent l_z is a constant and therefore the growth of vertical spread does not bring as rapid a growth in the correlation as when l increases with height. It may be argued of course that the form $l = kz$ adopted by Gee and Davies is inconsistent with the assumption that K_z = constant.

The solution of Eqs. (3.141)–(3.143) have also been extended by Tyldesley and Wallington (1965) by applying numerical methods, so avoiding the restriction to forms for K_z and $\bar{u}$ which are simple and therefore to some extent unrealistic. Their conclusion is that the significant effect of shear may not be confined to very long times or distances of travel.

The advantage of the statistical treatment followed by Smith is that it provides, at least for idealized flow conditions, a convenient means of estimating the distance of travel at which the interaction effect first becomes important in the crosswind spread of a continuous plume. With this object the first two terms of Eq. (3.152) have been evaluated graphically (Pasquill, 1969) on the assumption that the spectrum functions $F_L(n)$ and $G_L(n)$ both have the form of Eq. (2.96). Although this has been demonstrated to give a good fit to *fixed-point* frequency spectra of the vertical component (see 2.6) its adoption for the

v-component and for the Lagrangian forms of spectrum is questionable. Consequently the analysis provides no more than a rough indication. In this case the disregard of the correlation term is unlikely to be significant. With t_{wL} unlikely to be larger than t_{vL} it follows that the magnitudes of term (c) in Eq. (3.152) will be less than a fraction ρ^2 of term (a). From the data on the relation between surface stress $\overline{w'u'}$ and σ_w and σ_v (see 2.6, and noting that $\overline{w'v'} < \overline{w'u'}_{z=0}$) a reasonable estimate is $\rho^2 \leqslant 0.2$.

Denoting the contributions to the horizontal spread by σ_{yt}^2 for the direct action of turbulence [term (a)] and by σ_{ys}^2 [term (b)] for the interaction with

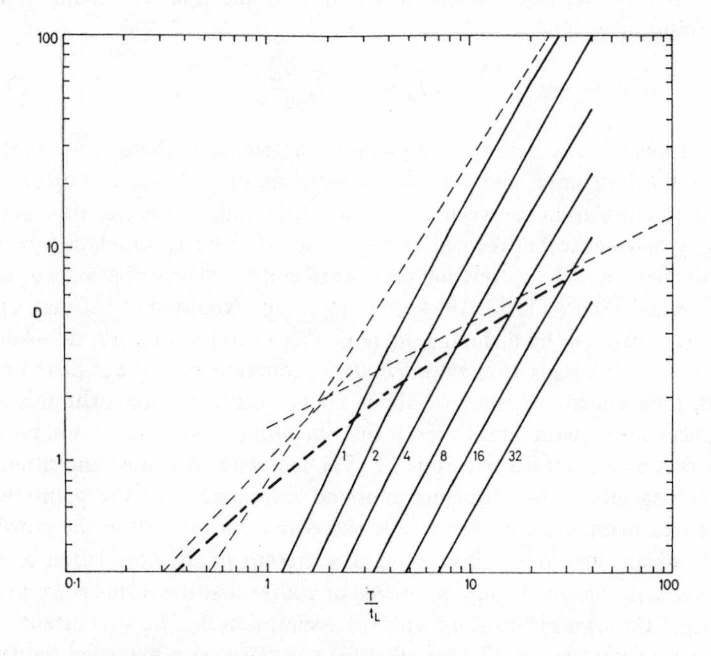

Fig. 3.18 – Effect of turbulence and wind shear on crosswind spread from a continuous point source with unbounded vertical diffusion [flow as in Fig. 3.16, solutions in Eq. (3.152)].

$$– – – – \quad D_t = \sigma_{yt}/\sigma_v t_{vL}$$

$$——— \quad D_s = \sigma_{ys}/\sigma_v t_{vL}$$

where σ_{yt} and σ_{ys} refer to the direct action of turbulence [term (a)] and the interaction with the mean shear [term (b)]. In the abscissae t_L refers to the v component for D_t and to the w component for D_s. The numerals against D_s are values of $1/S$ where

$$s = \tfrac{1}{2}\psi t_{vL} \frac{\sigma_w}{\sigma_v} \frac{t_{wL}^2}{t_{vL}^2} \text{ and } \psi = \bar{v}z.$$

The terms were evaluated graphically assuming a spectrum form as in Eq. (2.96). Asymptotic limits are indicated by faint broken lines – that for D_s referring to $S = 1$.

the mean shear, the dimensionless quantities $D_t = \sigma_{yt}/\sigma_v t_{vL}$ and $D_s = \sigma_{ys}/\sigma_v t_{vL}$ are shown in Fig. 3.18 against the non-dimensional T/t_{vL} and T/t_{wL} respectively. D_s is given for a range of values of the quantity

$$S = \tfrac{1}{2}\psi t_{vL} \frac{\sigma_w}{\sigma_v} \frac{t_{wL}^2}{t_{vL}^2}$$

The asymptotic limits of D_t and D_s for $S = 1$ are also shown, and from these an indication may immediately be obtained of the error which would arise from adoption of the limiting forms at small values of T/t_L.

With the slope of the σ_s, T curves so much steeper than that of the σ_t, T curve it follows that the resultant $(\sigma_s^2 + \sigma_t^2)^{1/2}$ must show a fairly rapid transition from a section approximating closely to σ_t to one approximating closely to σ_s. As a criterion for the significant enhancement of crosswind spread by shear the condition $(\sigma_s^2 + \sigma_t^2)^{1/2}/\sigma_t = 1.2$ has been adopted and the wind shears required to produce this at various distances are given in Table 3.VII for stated values of the time scales and the ratio σ_w/σ_v. Values calculated from the asymptotic forms

$$\sigma_s^2 = \tfrac{1}{6}\psi^2 \sigma_w^2 t_{wL} T^3$$

and

$$\sigma_t^2 = 2\sigma_v^2 t_{vL} T$$

are also given, and for present purposes these are evidently adequate approximations for $T/t_L > 20$ say. The value taken for σ_w/σ_v (0.6) and that estimated for t_{vL} apply to neutral conditions and the assumption that $t_{vL} = t_{wL}$ is probably reasonable for neutral or unstable conditions at heights in the region of 100 m. The result might therefore be regarded as roughly indicative of the effect of an elevated source in such conditions. It is evident that the shear would have to be impossibly large to produce an effect of importance at distances below 1 km. For distances around 10 km however the shear required is not untypical of those observed in practice. The larger shears required for effect at short distances may often be realized in stable conditions, but the quantities σ_w/σ_v and t_{wL}/t_{vL} are then likely to be less than assumed. In both respects the result will be to *reduce* the effect of the shear, in that both bring a reduction in the extent of the vertical spread. It does not follow immediately therefore that stable stratification will necessarily lead to a much more important shear contribution to horizontal spread.

Further discussion of the importance of shear in enlarging horizontal spread will be found in the discussion of the results of long-range diffusion experiments in Chapter 7.

Table 3.VII – Example of calculations from Fig. 3.18 of shear required to produce significant enhancement of crosswind spread at various distances

Assume

$$t_{vL} = t_{wL} = t_L = \beta l_E/\bar{u}$$

$$\beta = 4, \quad \lambda_m/l_E = 4 \quad \text{(see 2.6)}$$

$$\lambda_m \text{ for } v\text{-component} = 300 \text{ m} \quad \text{(see 2.6)}$$

$$\sigma_w/\sigma_v = 0.6 \quad \text{(see 2.6)}$$

The criterion adopted for significant enhancement of the turbulent spread σ_t^2 by the shear contribution σ_s^2 is that

$$(\sigma_t^2 + \sigma_s^2)^{1/2}/\sigma_t = 1.20, \quad \text{i.e. } \sigma_s/\sigma_t = 0.66$$

Adopting the standard values of $S[=\frac{1}{2}\psi t_L(\sigma_w/\sigma_v)$ in the conditions above] values of T/t_L required to satisfy the foregoing criterion are read off Fig. 3.18 and converted to x $(=300T/t_L$ here). The values of S may then be used to derive the quantity $\bar{u}/2\psi$ $(=45/S$ here) which, with $\bar{v}=\psi z$, is the height at which $\bar{v}=\bar{u}/2$, i.e. over which the wind direction turns by approximately 26.5 deg. In the last line are given values deduced from the asymptotic (large T) forms for σ_t and σ_s.

$1/S$	32	16	8	4
T/t_L for $\sigma_s/\sigma_t=0.66$	39	21	11	6
$x=300\ T/t_L$ (m)	11,700	6,300	3,300	1,800
Δz (m) for wind direction turning of 26.5 degrees	1,440	720	360	180
Δz from asymptotic solutions	1,540	830	435	235

4

Experimental studies of the basic features of atmospheric diffusion

The diffusion of material released into the atmosphere is basically described by the dimensions and shapes of the distributions which are developed downwind of the source. In this chapter are collected details of these features, as revealed by experiments which were designed specially for the purpose, and carried out with sources of an idealized nature, usually in selected conditions of terrain and weather. Many important surveys which were concerned with the determination of the maximum or average levels of pollution arising from practical sources are excluded from consideration here, but are discussed later in the more practical context of Chapter 5.

4.1 PRINCIPLES OF TECHNIQUE AND ANALYSIS

Most of the experimental studies of atmospheric diffusion fall into one or other of three main groups in which different techniques are adopted for describing and measuring the effects of diffusion. These are:

(a) *Optical outline* methods, using a suitable form of smoke.
(b) The measurement of trajectories of individual *marked particles.*
(c) Measurement of the concentration of a *tracer element* introduced into the air.

Optical outline methods

The examination of the development of the visible size and shape of a smoke cloud is at first sight the most attractive and economical method of studying diffusion. By taking distant photographs of smoke clouds it is usually possible subsequently to draw some outline around the image of the cloud, and so to specify a shape and size. When the source of smoke is continuously operated at a fixed point, an average or *time-mean* description may be achieved by using prolonged exposures (as with a pin-hole camera), or by superimposing a number of instantaneous photographs. Simple as this technique may be in action, there are however several difficulties in the anlysis and interpretation.

The inherent lack of smoothness in atmospheric diffusion processes results in a frequently ragged appearance of smoke clouds, and it is not always easy to define a usable instantaneous outline without introducing some smoothing of the apparent irregular form. Furthermore, the real meaning of the boundary of the smoke, however unequivocally this may be recognized, is open to some doubt. It obviously depends in a complex way on the optical qualities of the background against which the smoke is observed, the nature of the incident and background illumination, and the absorbing and scattering properties of the smoke material, as well as on the density or concentration of smoke in the plume or puff. In practice it has not been possible so far to take into account the detailed physical features, and progress has depended entirely on simple interpretations of visible outline. The most widely used treatment is that due to Roberts (see 4.8). However, even when a usable simple relation between cloud outline and smoke concentration is adopted, it is still necessary to make some *a priori* assumptions about the distribution of concentration in the smoke cloud. In earlier applications of the method theoretical expressions for this distribution were used to derive expressions for the visible outline, which were then compared with observation. It has, however, been made clear in discussions by Gifford, which are considered in more detail in 4.8, that it is not necessary to anticipate the relation between the *size* of a cloud and the distance or time of travel, but merely to make some assumption about the *shape* of the distribution of concentration in the cloud. This development eliminates a good deal of what has previously seemed to be a rather intractable difficulty in the effective interpretation of observations of puffs and plumes of smoke.

It will be seen in 4.8 that important deductions have been made from observations of smoke puffs, and it seems likely that further developments and use of the techniques might profitably be made (see Gifford, 1957a and Högström, 1964), There is, however, an important limitation in the range over which the process of diffusion can be studied in this way; unless the smoke cloud is diffusing very slowly or unless, as in Högström's study, long puffs produced by maintained releases of smoke are viewed end-on. Although in theory extension of range should merely require an appropriate increase in the amount of smoke released, the improvement in practice may be disappointing. Not only may the observational procedure be complicated by the use of much larger volumes of smoke, but there is also the further difficulty that concentrated smoke clouds might absorb solar radiation in sufficient amount to introduce buoyancy effects and lifting of the cloud.

Trajectories and marked particles

The observation of the trajectories of *markers* floating in the air is theoretically the ideal way of directly observing the basic processes of diffusion, but there are difficulties in making true *Lagrangian* observations, which have already been discussed in principle in 2.7. In practice the problem is to arrange markers which are sufficiently neutral, i.e. which have negligible motion relative to the air surrounding them, and which can then be assumed to display the dispersive action of turbulence components of larger scale than the size of the marker itself. Soap bubbles, thistledown and other light objects have been used for short-range observations, but by far the most widely used marker is the balloon. An advantage of using balloons is that the latest techniques of observing their trajectories, ranging from the traditional visual-reading theodolite systems to elaborate radar-tracking networks, may be available from the operational systems of wind observation on a synoptic scale. The exploitation of the method in the form of the long-range constant-level balloon flights organized in the United States of America (see Angell, 1959) constitutes a major contribution to the study of long-range large-scale horizontal diffusion in the atmosphere. It is also appropriate to mention here the use of computed air trajectories, as initiated by Durst. Although in this case there is no real physical marking of the air, the velocities of a hypothetical air parcel being computed at intervals from successive meteorological charts, the trajectories obtained represent approximations to those of constant-level balloons. Both the observed and computed horizontal trajectories exclude the vertical motion which a real element of air will experience, and thus do not reproduce those contributions to spread which arise from the systematic variation of wind speed and direction with height. Moreover, all trajectory methods suffer from the drawback of requiring a large number of repetitions to provide a statistically satisfactory result, so that apart from the operational effort a large computing effort is also involved.

Measurement of tracer-element concentrations

Undoubtedly the most rewarding experimental studies of diffusion in the atmosphere have been those in which some easily detected *tracer* element is introduced into the atmosphere, and the concentrations occurring at various times and positions downwind are measured by precise chemical or physical techniques. The method possesses the obvious advantage of enabling the distribution within a diffusing cloud to be directly explored, a feature which is not provided at all by the visible-outline technique and is a laborious matter by the trajectory method. It also has the attraction of being most directly relevant to the assessment of the hazards arising from the release of toxic materials. The first known measurements of this type were carried out on Salisbury Plain, England at the War Office Chemical Defence Experimental Establishment, Porton, in 1923. By present standards the method may seem rather crude, but by careful attention to the details of the procedure results which still serve as basic data were obtained. The method comprised the release of a harmless smoke formed by burning 'candles' of a pitch composition, and the estimation of the concentration of smoke in the air by a 'stain-meter' technique, devised by F. J. Scrase on the principle of the Owens Automatic Filter for early measurements of air pollution. In Scrase's method a hand-pump was used to draw the smoke-laden air through a small orifice backed by a filter paper. By comparing the resulting stain on the filter paper with a series of standard stains, obtained by taking samples in an enclosed chamber in which a known quantity of smoke had been released, an estimate of the smoke concentration could be obtained. From samples taken by a team of observers at specified stations downwind of the source, the crosswind distribution of smoke was examined, and in later work the vertical distribution was explored by stationing the observers at various heights on a tower.

Air-sampling methods using gases, smokes and larger airborne particles as tracer elements have since been developed and improved in various ways. One of the most important steps in this process was the application of precise laboratory techniques of chemical estimation to samples obtained by drawing the gas-laden air through a suitable liquid absorbent. However, in practice the use of sensitive gaseous tracers did not easily extend the range over which diffusion could be observed beyond the few hundred metres which were possible with the earlier smoke techniques. None of the gases which were suitable from the point of view of chemical estimation were sufficiently free from irritant or toxic effects at the level of concentration required, and the experiments had to be designed to exclude any possibility of unpleasant or harmful concentrations being set up outside the experimental area. The big step in the extension of the range of diffusion experiments, to the order of 100 km, came with the introduction of the fluorescent-particle tracer, first described in this connection by Perkins, Leighton, Grinnell and Webster (1952) and subsequently reviewed by Leighton *et al.* (1965) with particular reference to the absolute accuracy of the technique. The material which has been used most is a pigment, zinc cadmium sulphide, which can be finely dispersed, one gram providing approximately 10^{10} particles with diameters mainly in the range 1–5 microns. With ultra-violet illumination these particles can be individually recognized and counted under a good laboratory microscope. More recently an alternative to zinc cadmium sulphide for long-range studies has been found in certain gaseous halogenated compounds, notably sulphur hexafluoride (see Niemeyer and McCormick 1968) and Freon-113 (see Jenkins (1983) for an up-to-date review of tracer techniques). Extension of the techniques to even greater distances, with real-time indication of tracer concentration can be achieved using gas chromatography in conjunction with an electron-capture detector.

Air-sampling for particulate tracers is carried out by filtering, impaction, or impingement techniques, for a general authoritative discussion of which see Green and Lane (1957). In the conventional *impactor*, the air is drawn through a nozzle to emerge at very high speed from a narrow slit positioned close to a surface coated with glycerine jelly. Particles impact on the surface, are held there by the jelly, and may subsequently be examined and counted under a microscope. For convenience in obtaining a series of collections at different times or positions, the receiving surface usually takes the form of a cylinder or drum which can be rotated either continuously or in steps. In the *impinger* the jet of air is directed into a liquid in which the particles are retained and from which they are subsequently separated by filtering. Only one sample is obtainable with an impinger, and in the present application the method is preferable to the drum impactor arrangement only when water droplets are likely to be collected with the particles since in this case the particles and jelly coating may be washed off the receiving surface of the impactor. The design and construction of impactors requires considerable care, and as fairly powerful air-pumps are necessary in order to draw air through them at the appropriate rate, sampling in the open at a number of positions simultaneously is a fairly elaborate procedure. For this reason natural impaction techniques, in which the particles are impacted on an object by the action of the wind alone, may sometimes be profitable, especially when fairly large particles are involved. This method was effectively demonstrated in Gregory's (1951) wind-tunnel studies of the collection of *lycopodium* spores on adhesive cylinders, of various diameters, with their axes normal to the air-stream. In this case the number of particles caught on a cylinder is proportional to the number of particles which would otherwise have passed through the cross-section occupied by the cylinder, the proportionality factor, or collection efficiency, being a function of the diameters of the cylinder and the particles, and of the velocity of the air-stream. In field applications the important requirement is that cylinder and particle size should be chosen so that not only is the collection efficiency of a reasonable order (say 50 per cent), but that it does not vary rapidly with wind speed.

Interpretation of tracer measurements

For the direct interpretation of tracer concentrations in specifying the dilution of material by the action of wind and turbulence, it is essential that the tracer be completely inert chemically and that it should not be absorbed or deposited in any way on the ground, i.e. it should be *conserved*. Also it is essential that there should not be any systematic errors arising from the physical or chemical sampling and assay procedures. In practice it would be preferable always to check for satisfactory behaviour in the foregoing respects by obtaining sufficient samples to permit accurate evaluation of the total downwind flux of material. However, this requires very elaborate and inconvenient arrays for collecting samples, and reliable indications have not generally been obtainable in this way at long distances from a source.

An alternative and more convenient procedure which provides a useful indication of the reliability of a tracer is the comparison of the concentrations produced at a given position by releasing different tracers simultaneously from virtually the same position. If the respective source strengths are accurately known then any imperfection in the tracer or technique should be revealed by the observation of concentration ratios different from the corresponding source strengths. If, as may often be the case in practice, the source strengths are not accurately specifiable, an alternative is to make the simultaneous concentration measurements at two or more different distances from the source. Perfect or identically imperfect behaviour in the tracers will then be revealed by constancy with distance of the concentration ratios.

The latter technique was applied by Eggleton and Thompson (1961) to test the reliability of the fluorescent particle technique as applied in tracer studies over several years in the United Kingdom. As the standard for comparison radioactive xenon-133 was selected as being chemically inert and accurately estimable in low concentration. Measurements made at 16 and 60 km from the source in three separate experiments showed that the 'fluorescent-particle/xenon' concentration ratio was consistently about 50 per cent lower at the far distance. Ludwick (1966) has also compared the same two tracers and obtained results which suggested an apparent loss of the fluorescent particle tracer of about 50 per cent at distances up to 1.6 km. Furthermore Niemeyer and McCormick (1968) have made measurements at distances of 0.07–115 km from simultaneous releases of zinc-cadmium sulphide and the gaseous sulphur hexafluoride and give values of the ratios of the concentrations normalized for source strength. At distances 0.07 to 0.2 km the 'gas/particle' ratio was between 0.7 and 1.2 in two experiments in which the releases were at a height of 1.5 m. Higher ratios in a third experiment were regarded as possibly a result of immediate deposition of the fluorescent particles from the lower (0.33 m) release-height. At distances 35, 70, and 115 km, with three releases at a height of 1 m, the ratios were in the range 1.0 to 4.9, though without any systematic trend with distance.

The discrepancies in the foregoing comparative tracer studies have not been satisfactorily explained and their implication as regards the *conservative* nature of zinc cadmium suphide particles in the atmosphere is not entirely clear. However, despite the claims which have been made by Perkins and his co-workers for the absolute reliability of the technique, there are indications which cannot be ignored that in practice concentrations have been measured which must be regarded as suspect in an absolute sense. On the other hand, since it is obviously reasonable to regard any systematic loss in a given experiment as mainly dependent on time or distance of travel, it is also reasonable to regard the measured concentrations in the cross-section of a plume as *relatively* correct. Accordingly, the use of concentrations (in an essentially relative sense) in determining the *dimensions* of the cloud or plume of material across wind or vertically (see below) is much less open to suspicion.

Terminology and diffusion data

Although much of the nomenclature for expressing the effects of diffusion has already been used in the theoretical discussions of Chapter 3, it will be helpful to summarize the various parameters at this stage, in relation to the particular forms of diffusion measurements which are made.

The chemical analysis or physical examination of a *sample*, collected in one or other of the ways briefly outlined above, leads to an estimate of the mass of material, M, or of the number of particles, N, in a given volume of air $S\tau$, where S is the known rate at which air is drawn through the apparatus and τ is the duration of the sampling process. In the case of sampling by natural impaction the effective value of S is given by uAE where u is the air speed relative to the object, A is the area presented normal to the flow, and E is the collection efficiency. Theoretically a correction for the efficiency of collection, retention or absorption, should be made in all the forms of sampling, but in practice the methods in which air is forcibly drawn through a piece of apparatus are deliberately designed to give near 100 per cent efficiency.

If χ (mass per unit volume or number of particles per unit volume) is the instantaneous concentration at the sampling position then the average concentration over the duration of sampling is

$$\bar{\chi} = \frac{1}{\tau}\int_0^\tau \chi \, dt = \frac{M}{S\tau} \text{ or } \frac{N}{S\tau}$$

The quantity $\int_0^\tau \chi \, dt$, which is termed the *dosage, D*, during the time τ, is simply M/S or N/S. When, as in the present chapter, the interest is in the fundamental properties of the size of and distribution within a diffusing cloud of material, it is of course unnecessary to evaluate concentrations and dosages as such, as long as the sampling rate and duration are standard, or corrections for departures therefrom are subsequently made. The quantities M and N, observed as a function of crosswind (Y) or vertical (Z) displacement from an axis along the mean wind direction, may then be used directly in describing the properties of a plume of material extending from a point source. When dealing with a detached cloud or puff these quantities may additionally be observed as a function of distance alongwind (X) from the centre of the cloud.

The crosswind dimension of a cloud is conventionally represented either by the width between positions at which concentration or dosage falls to a given fraction [usually one-tenth] of the central or peak value, or alternatively by the standard deviation of the crosswind displacement of the material, in practice defined by

$$\sigma_y^2 = Y^2 = \frac{\Sigma MY^2}{\Sigma M} - \left[\frac{\Sigma MY}{\Sigma M}\right]^2$$

or by a similar expression involving N. When the distribution is of Gaussian form, and the fraction adopted in specifying cloud-width ($2\,Y_0$) is one-tenth,

$$2\,Y_0 = 4.30\sigma_y$$

Corresponding definitions apply to the alongwind dimension, and also to the vertical dimension of a cloud which is well clear of horizontal boundaries, but when the cloud is generated at an impermeable boundary ($z = 0$), then the cloud-height Z_0 is defined as the height at which concentration or dosage falls to one-tenth (say) of the ground value. The root mean square displacement of material above the boundary, equivalent to the standard deviation in the foregoing cases, is defined by

$$\sigma_z^2 = \overline{Z^2} = \frac{\Sigma M Z^2}{\Sigma M}$$

Alternatively the vertical spread may be defined by the mean height $\bar{Z}$ reached by the particles

$$\bar{Z} = \frac{\Sigma M Z}{\Sigma M}$$

If the fall of concentration with height is according to the Gaussian form

$$Z_0 = 2.15\sigma_z = 2.7\bar{Z}$$

4.2 THE FORM OF THE CROSSWIND DISTRIBUTION AT SHORT RANGE FROM A MAINTAINED POINT SOURCE

The earliest measurement with smoke candles showed that the crosswind variation of smoke concentration at 100 m downwind of the source was basically of 'cocked-hat' shape (see Sutton, 1953, p. 275), with irregularities of some degree or other according to the steadiness of the wind direction. At first sight this basic shape seemed to conform fairly obviously to the *normal error* or *Gaussian* form, i.e. as in the following equation with $r = 2$.

$$\chi(y)/\chi_0 = \exp(-ay^r) \tag{4.1}$$

Fig. 4.1 shows four averaged curves constructed from later dispersion measurements at Porton.

Of the four average distributions two (fluorescent smoke and lycopodium spores at 100 m or yd) contain no obvious deviation from linearity. In the other two (fluorescent smoke at 300 m and sulphur dioxide at 80 m) there are clear deviations from linearity especially at one side of the distribution in each case, but the deviations are in opposite senses in the two cases and correspond to values of r (Eq. 4.1) of approximately 1.5 for the fluorescent smoke distribution and 2.5 for the sulphur dioxide distribution.

A different criterion for the agreement with Gaussian form has been employed by Cramer (1957) in his analysis of the *Prairie Grass* series of diffusion experiments in the U.S.A. These were carried out with sulphur dioxide, and refer to a wide range of stability conditions. For each crosswind distribution, which corresponded to an average over a period of 10 min, Cramer evaluated both the standard deviation of the distribution, and the plume-width as defined

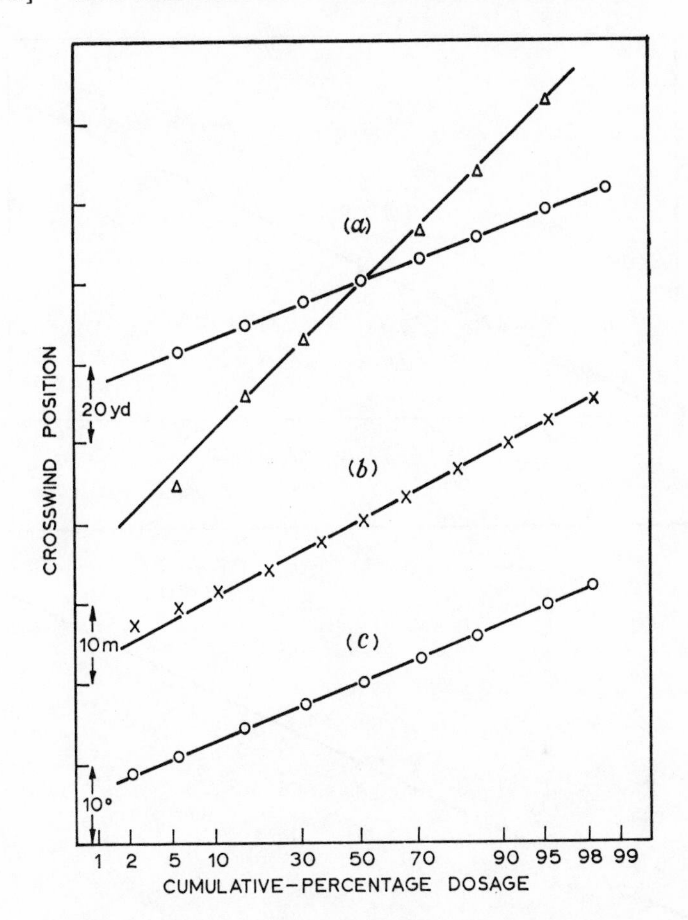

Fig. 4.1 – Average crosswind distributions at short range from a point source at ground level, in near-neutral conditions over downland, Porton, England.

Material	*Duration of release* (min)	*Number of releases*	*Distance of travel* (m)	
(a) Fluorescent smoke	30	7	○ 100 △ 300	Hay and Pasquill, 1959
(b) Sulphur dioxide	4–7	13	80	Unpublished
(c) Lycopodium spores	3	6	100	Hay and Pasquill, 1959 (runs 1,2,3,5,6,7)

by the distance between the points at which the concentration fell to one-tenth of the peak or axial value. For a Gaussian distribution the ratio plume-width/standard deviation is 4.30, and Fig. 4.2 reproduced from Cramer's paper shows that the individual values of plume-width and standard deviation are

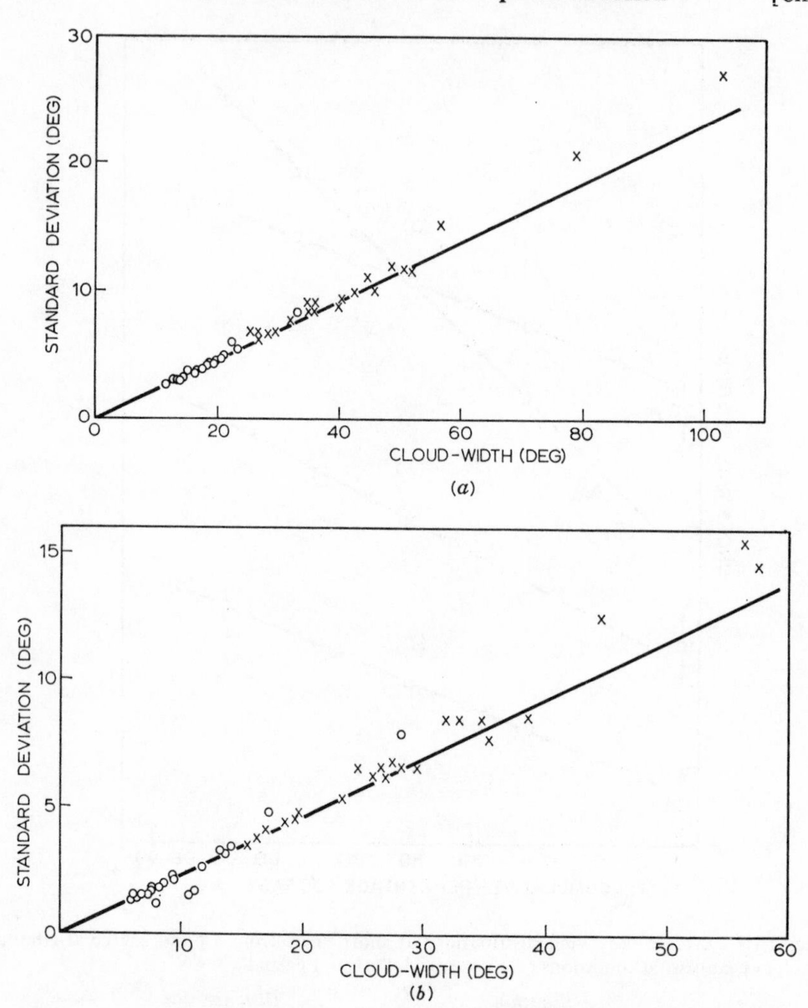

Fig. 4.2 – Relation between cloud-width, appropriate to one-tenth peak concentration, and standard deviation, of crosswind distributions from a continuous point source. From measurements at O'Neill, Nebraska, at distances (a) 100 m; (b) 800 m; × day-time, ○ night. The straight lines correspond to a Gaussian distribution. (Cramer, 1957).

generally quite close to this relation. However, it is necessary to consider the significance of the deviations from the *Gaussian* lines in Fig. 4.2, and to this end the theoretical values of the plume-width/standard deviation ratio for various values of r in Eq. (4.1) are given in Table 4.I. The value of r is evidently very sensitively dependent on the ratio, and it is clear that the departures of the points from the straight lines in Fig. 4.2 represent departures of r of 0.5 and more from the value of 2.0.

Table 4.I – Ratio of width $2\,Y_0$, as defined by one-tenth peak concentration, to standard deviation $(\overline{Y^2})^{1/2}$, for a distribution of material as in Eq. (4.1).

r in Eq. (4.1)	$\dfrac{2\,Y_0}{(\overline{Y^2})^{1/2}}$
2.5	4.34
2.0	4.30
1.5	4.06
1.25	3.77
1.0	3.26

To sum up, the available observational data on the crosswind distribution at short range from a ground-level continuous source in level open country show that individual distributions over periods of the order of 10 min may require values of r in Eq. (4.1) widely different from the Gaussian value of 2.0. Even composite distributions may require values as different as 1.5 and 2.5. However, there is so far no indication of a systematic departure from the Gaussian value, and it is tempting to conclude that the variations in both directions are a reflection partly of the quality of the sampling data and analysis, and partly of the inherent lack of homogeneity in the structure and effects of atmospheric turbulence over the sampling times involved. It is clear at least that on present evidence there is no good reason to adopt an analytical form different from Gaussian, especially as the practical features of the distribution do not vary greatly with quite substantial departures from the Gaussian value of r in Eq. (4.1).

4.3 THE MAGNITUDE OF THE CROSSWIND SPREAD FROM A MAINTAINED POINT SOURCE

Early Porton experiments

The 1923 smoke experiments at Porton provided the much quoted figure of 35 m for the width of cloud, as defined by one-tenth peak concentration, at 100 m downwind of a maintained source at ground level. These experiments were carried out in thoroughly overcast conditions, i.e. in near-neutral conditions of stability, over open, gently rolling downland. The duration of release of smoke in each case was 4 min. It is important to realize, however, that even in the carefully selected near-neutral conditions the individual widths of cloud varied widely, the value of 35 m being the mean of fifty values ranging from 23 to 47 m. Table 4.II shows the distribution of the individual values (this distribution is approximately Gaussian with a standard deviation of 5 m). The individual values are also shown in Fig. 4.3 plotted against the wind speed at a height of 1.8 m, from which it is seen that there is no marked systematic variation with wind speed.

Table 4.II – Frequency distribution of cloud-width at 100 m downwind of a ground-level point source, in neutral conditions of atmospheric stability over downland at Porton, England

Range of cloud-width	20 to 22	23 to 25	26 to 28	29 to 31	32 to 34	35 to 37	38 to 40	41 to 43	44 to 46	47 to 49	50 to 52
No. of cases	0	1	5	9	10	11	8	5	0	1	0
Percentage of cases	0	2	10	18	20	22	16	10	0	2	0

(Each value of cloud-width was derived from an average crosswind distribution measured during a 4-min release of smoke, the boundary of the cloud being identified with a concentration equal to one-tenth of that on the axis of the plume).

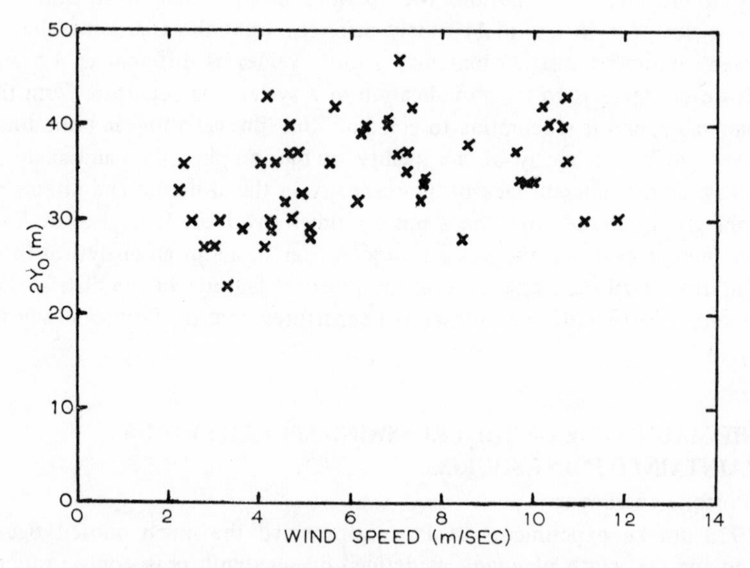

Fig. 4.3 – Observed cloud-width ($2\ Y_0$) at 100 m downwind of a source of smoke, in relation to wind speed at a height of 1.8 m. Over downland, Porton, England (1923). Duration of release 4 min. Near-neutral conditions of stability.

A later series of similar measurements at Cardington in various conditions of atmospheric stability showed a close relation between cloud width and Richardson number, and also a correlation with the lateral dimensions of bi-directional-vane traces (see Fig. 4.4).

The implications of this first demonstration of a relation between the lateral spread of a cloud at short range and a direct though crude measurement

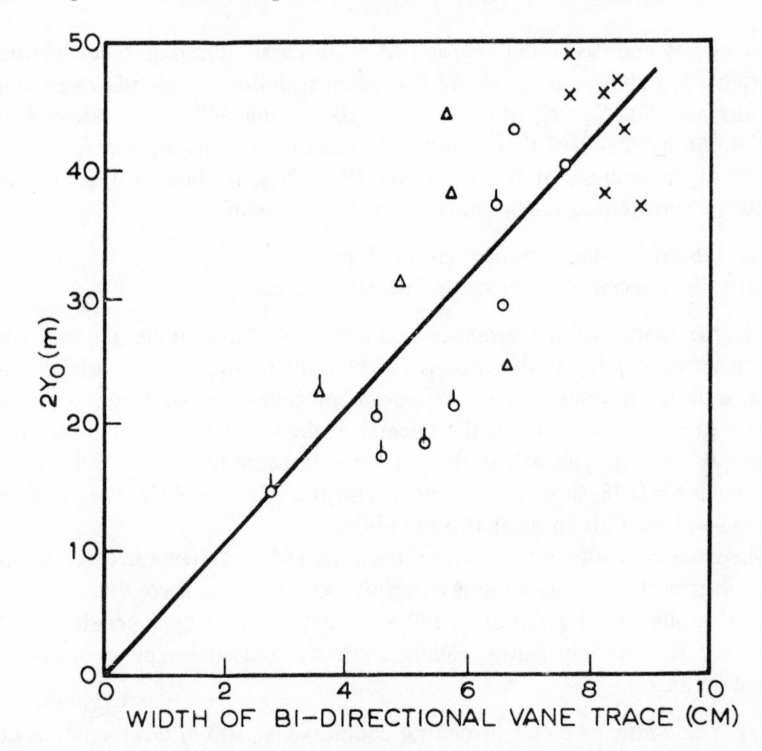

Fig. 4.4 – Relation between cloud-width ($2\ Y_0$) at 100 m from a maintained point source, and turbulence as indicated by the lateral width of a bi-directional-vane trace. The different symbols refer to different dates, the primed symbols to the more stable conditions. (Unpublished data from smoke experiments at Cardington, Emgland, 1934).

of turbulence were not immediately followed up. It is noteworthy that the departures of the individual values of cloud-width from the simple linear relation drawn in by eye are mostly less than 30 per cent – a consistency which now seems quite remarkable in view of the well-known variability in atmospheric turbulence. Furthermore this result can now be seen (see 3.4) to follow directly from the basic statistical treatment of diffusion from a continuous source, according to which the spread of particles at a given distance of travel should be proportional to the intensity of turbulence and hence, to a good approximation, to the corresponding dimension of the bi-directional-vane trace.

Although the early Porton smoke experiments were extended (in 1925) to distances of travel of 300 and 100 m, the main preoccupation was with the variation of the peak or axial concentration, and no figures were given immediately for the crosswind dimensions of the smoke clouds at these distances. The results of eighteen experiments at 300 m were analysed from this point of view some years later, and the mean cloud-width obtained has been quoted by Hay and Pasquill (1959) as 79.1 m, this again referring to near-neutral conditions of stability. If cloud-width is taken to follow a power-law variation with distance, i.e. $Y_0 \propto x^p$, the results for 100 m and 300 m are satisfied by $p=0.74$. An independent though indirect evaluation of this index may also be made from the analysis of the 1923 and 1925 data. In these analyses power-law indices were derived for the variation with distance of

(a) the axial concentration at ground level
(b) the integral of the crosswind distribution curve.

Now if the shapes of the crosswind and vertical distributions are invariant with distance, and if the dimensions of these distributions vary with x^p and x^q respectively, it follows that the power-law index for axial concentration will be $-(p+q)$, while that for the integral of the crosswind distribution curve will simply be $-q$. The values derived were respectively -1.76 and -0.98, from which $p=0.78$, in good agreement with the value obtained directly from the cloud-widths at distances of 100 and 300 m.

The main conclusions to be drawn from the earliest systematic observations of the distribution of smoke concentration downwind of a continuous point source at ground level are thus as follows. They refer to open grassland with either very flat or only gently rolling contours, and to sampling or release times of 4 min.

(a) The width of cloud in neutral conditions at 100 m downwind, as defined by one-tenth axial or peak concentration, has an average value in the region of 40 m, but with individual variation of up to perhaps ± 15 m from this figure.
(b) In the range 100–1000 m, again in neutral conditions, cloud-width increases with distance raised to a power of approximately 0.8.
(c) As the atmospheric stability changes from near-neutral to moderately stable there is a systematic decrease of cloud-width, and a close correlation between cloud-width and the lateral component of the intensity of turbulence.

Statistical data obtained in the U.S.A. in the 1950s on the crosswind distribution from a ground-level point source

Experiments carried out at the Round Hill Field Station of the Massachusetts Institute of Technology (in 1954, 1955, and 1957) and during *Project Prairie Grass*, an extensive co-operative programme carried out in 1956, substantially extended the data on crosswind spread from a ground-level point source. The tracer employed was sulphur dioxide, and an important feature of these investigations was the inclusion in the meteorological measurements of accurate observations of the lateral and vertical wind fluctuations using fast-response instruments. A description of the techniques and instruments is given in the full reports of the investigation by Cramer, Record and Vaughan (1958), Barad (1958) and Haugen (1959). Another important feature is that the two field sites involved are radically different in aerodynamic roughness. The Round Hill site is described by Cramer (1957) as having a roughness parameter [z_0 in Eq. (2.51)] greater than 10 cm, with trees, houses, small buildings and differences in elevation of the order of 30 m within a distance of 0.5 to 1 km immediately upwind of the release area. On the other hand the *Prairie Grass* experiments were carried out on the smooth, level Nebraska plains at O'Neill, on a site for which $z_0 < 1$ cm.

Cramer's analysis of the Round Hill and O'Neill data was largely focused on the empirical relations between the statistics of diffusion and the fluctuations of wind direction near the source. The data on cloud-width at distances of 100 m are displayed in Fig. 4.5, which is taken from Cramer's 1957 paper.

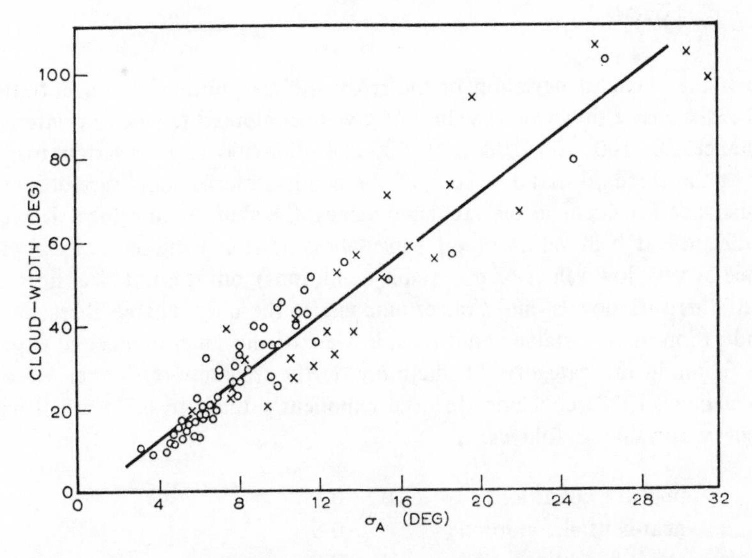

Fig. 4.5 – Relation between angular cloud-width and standard deviation (σ_A) of wind direction. From point source experiments at Round Hill Field Station, Massachusetts (×), and O'Neill, Nebraska (○). Cloud-widths refer to 100 m downwind of source. (Cramer, 1957).

Both the diffusion data and the wind data refer to a sampling time near 10 min, the wind being measured at a height of 2 m near the source. A good correlation exists, the deviation of the points from the regression line being very similar to that found in the analysis of the 1934 Cardington data in terms of cruder measurements of the wind-direction fluctuations. The most striking feature is that the American data from two sites of very different roughness, obtained over a substantial range of atmospheric stability, fit closely to a single relation between plume-width and the wind-direction fluctuation, over a tenfold range of the latter parameter. As Cramer points out, no similar consistency is apparent if the diffusion results are plotted against a stability parameter equivalent to the Richardson number, because in neutral conditions the standard deviation of the wind azimuth at Round Hill is approximately double that at O'Neill. The implication is of course that the simple statistics of wind fluctuation directly reflect the diffusing power of the atmosphere, a feature which, as already noted, is to be expected on any logical statistical treatment of turbulent diffusion.

The O'Neill experiments also included simultaneous measurements of crosswind distribution at 50, 200, 400 and 800 m, and hence provide a major contribution to the knowledge of variation of cloud-width with distance of travel. Cramer (1957) gives values of the index p in the relation

$$(\overline{Y^2})^{1/2} \propto x^p$$

between the standard deviation of the crosswind distribution of concentration and the distance x downwind. Values of p were computed for the four intervals of distance, 50–100, 100–200, 200–400 and 400–800, and for various magnitudes of the standard deviation, σ_A, of the wind azimuth. Some variation of p with distance is evident in the tabulated values, the main trend being a decrease with distance at high values of σ_A (unstable conditions) and an increase with distance at very low values of σ_A (stable conditions), but it is not clear how significant this variation is, and Cramer emphasizes the questionable character of the indication in very stable conditions, in view of the small number of experiments falling in this category. Furthermore, in the final report, Cramer, Record and Vaughan (1958) conclude that the exponent p tends to be invariant with distance, with values as follows,

unstable conditions	0.8–0.9
near-neutral conditions	0.8
stable conditions	0.6

The above analysis of the *Prairie Grass* experiments substantially confirms the early Porton results in neutral conditions, firstly in the sense that the widths of cloud at 100 m are consistent. Cramer's (1957) analysis gives the magnitude of $(\overline{Y^2})^{1/2}$ in neutral conditions as approximately 10 m, and assuming a Gaussian distribution this implies a cloud-width of 43 m, which is to be compared with the values of 35 m and 44 m, obtained at Porton and Cardington respectively, with a shorter sampling time (4 min as compared with 10 min in the *Prairie Grass* measurements). Secondly, the power-law variation of crosswind spread with distance up to 800 m is very similar to that indicated by the Porton experiments at 100, 300 and 1000 m, the index in both cases being near 0.8.

Extension of the study of crosswind spread

The *Prairie Grass* field study provided an important data base for the continued examination of the relation between σ_y and flow conditions over an ideal level surface of small roughness. Further measurements of crosswind dispersion have since been undertaken, using various tracers, and with various more-or-less specific objectives. At the Porton establishment the emphasis was placed on exploring further the dependence of dispersion on the field of turbulence and partly on extending the measurements to much greater range (*ca* 100 km) than hitherto. In the U.S.A. much effort has been put into field studies at various sites with different characteristics, the measurements being made mainly up to a few kilometres downwind. A comprehensive listing of field studies will be found in the United States official document *Meteorology and Atomic Energy* (2nd edn., Slade, 1968; 3rd edn., Randerson, in course of preparation). Another summary, referring especially to studies in airflows possessing a greater complexity, notably in association with topographical features, has been compiled by Draxler (1979).

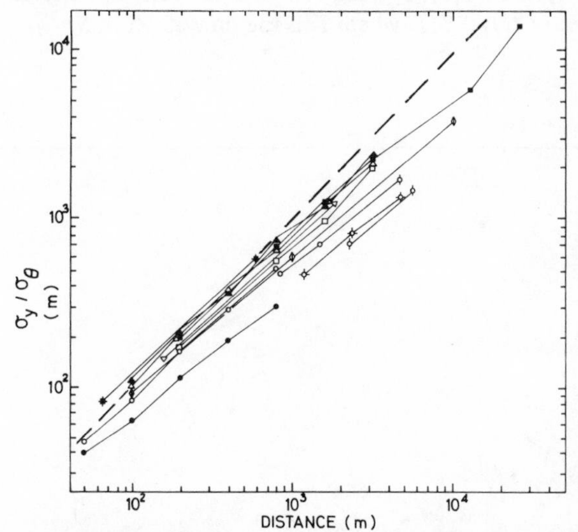

Fig. 4.6 – Crosswind spread (σ_y) normalized according to wind direction fluctuation σ_θ. (Plots are averages for each series of experiments, after Slade 1968.) Sampling durations 10–60 min.

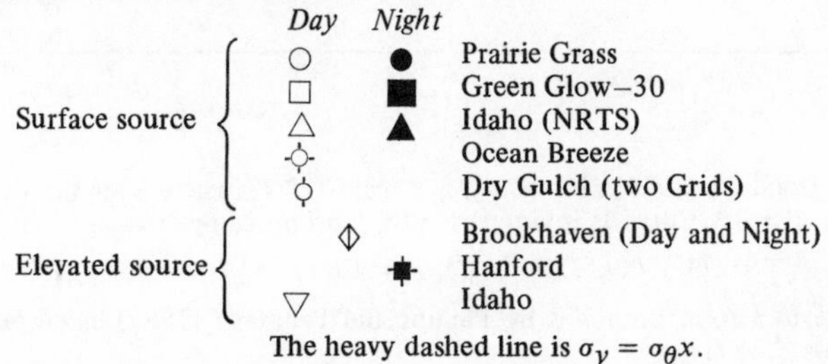

The heavy dashed line is $\sigma_y = \sigma_\theta x$.

An early synthesis of data from several of the studies (Slade, 1968) is reproduced here as Fig. 4.6. In accordance with the correlation with σ_θ already demonstrated by Cramer's analysis of the *Prairie Grass* data, the results are presented as values of σ_y/σ_θ, averaged for each study, as a function of distance downwind. The separation of these average curves is little more than a factor of two in σ_y/σ_θ, which is noteworthy, especially considering that 'day' and 'night' curves are not widely separated and that two of the studies were in far-from-ideal conditions of terrain (*Ocean Breeze* over coastal sand dunes and *Dry Gulch* over rough foothills). In more recent analyses the normalization of σ_y values has been carried further, to the dimensionless form $S = \sigma_y/\sigma_v T$ or its approximate equivalent $\sigma_y/\sigma_\theta x$, and Fig. 4.7 and Table 4.III show examples of such reduced data published by Draxler (1976) and Doran *et al.* (1978). The latter publication is a summary of and selection from much more extensive departmental reports by Nikola (1977) and Horst *et al.* (1979). Draxler's analysis has the special interest that σ_y was considered in relation to *time* of travel, the individual plots of $\sigma_y/\sigma_v T$ being compounded into one diagram by using a normalized time of travel T/T_i where T_i is the time at which $S = \frac{1}{2}$.

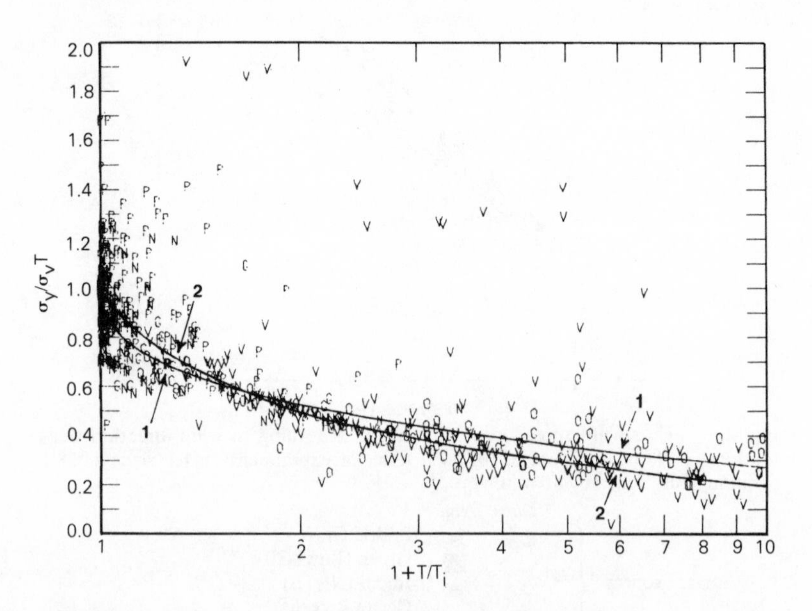

Fig. 4.7 – Draxler's (1976) plot of $\sigma_y/\sigma_v T$ against T/T_i, where T_i is the time at which $\sigma_y/\sigma_v T = 0.5$. Curve 1: interpolation form proposed by Draxler

$$\sigma_y/\sigma_v T = [1 + 0.9\,(T/T_i)^{0.5}]^{-1},\ T_i = 1.6\, t_L$$

Curve 2: revised form proposed by Phillips and Panofsky (1982) based on Eq. (3.70d); $T_i = 5.25\, t_L$.

Table 4.III – $\sigma_y/x\sigma_\theta$ for ground-level releases (Horst *et al.*, 1979)

Test Series		Distance (m)								
		50	100	200	400	800	1200	1600	2200	3200
Hanford 67	No. of Tests			5	4	15	7	11	4	6
	Av. $\sigma_y/x\sigma_\theta$			0.76	0.81	0.63	0.58	0.58	0.50	0.53
	Std. Dev.			0.16	0.14	0.26	0.19	0.20	0.22	0.14
NRTS	No. of Tests		23	32	22	27		18		18
	Av. $\sigma_y/x\sigma_\theta$		1.04	0.98	0.92	0.85		0.77		0.69
	Std. Dev.		0.16	0.15	0.19	0.16		0.17		0.14
Green Glow-30	No. of Tests			35		34		29		32
	Av. $\sigma_y/x\sigma_\theta$			0.94		0.77		0.72		0.67
	Std. Dev.			0.18		0.21		0.24		0.25
Prairie Grass	No. of Tests	44	46	46	46	46				
	Av. $\sigma_y/x\sigma_\theta$	0.88	0.76	0.68	0.58	0.49				
	Std. Dev.	0.14	0.17	0.18	0.20	0.22				

4.4 THE EXPLICIT RELATION BETWEEN CROSSWIND SPREAD AND THE STATISTICS OF THE WIND FLUCTUATIONS

The relation as $T \to 0$

The demonstration of the existence of a good correlation between cloud spread and the intensity of turbulence, irrespective of the condition of surface roughness and stability, was an important step in the general specification of the characteristics of diffusing clouds. The ultimate interest, however, is in the question of the explicit relation between these quantities, particularly in the light of the statistical treatment of diffusion discussed in Chapter 3. From this it will be recalled, that the spread of particles serially released from a fixed point must initially increase linearly with respect to time T or distance x of travel. The absolute magnitude of the spread in this case is given simply by

$$\begin{aligned} \sigma_y^2 &= \overline{v'^2}T^2 \\ &\cong \frac{\overline{v'^2}x^2}{\bar{u}^2} \end{aligned} \tag{4.2}$$

This deduction holds strictly only for homogeneous turbulence which is statistically invariant with time, but it may not be unreasonable to consider it in relation to atmospheric conditions away from the immediate effect of the earth's surface, or even close to the surface in the case of horizontal spread (noting that the variance and scale of the v-component have been found not to have any pronounced variation with height – see 2.6).

A demonstration of the approximate validity of an equation corresponding to Eq. (4.2) was first obtained in a study of the vertical spread from an elevated source (Hay and Pasquill, 1957), which will be discussed later, and from the data already discussed on crosswind spread near the ground it can now be demonstrated that the relation holds to a reasonable approximation in this case also. The statistics of wind direction fluctuation on the Porton site (see Fig. 2.18) give the average standard deviation for a 4-min sampling time and a wind speed of 5 m/sec as 5.25°, in neutral conditions. This means that the intensity of the lateral component of turbulence $\sqrt{(\overline{v'^2})}/\bar{u}$ is approximately 0.09, and that σ_y in Eq. (4.2) is 9 m for $x = 100$ m. Assuming a Gaussian distribution the corresponding cloud-width is 39 m, as compared with the average value of 35 m observed. Again, from the data given by Cramer (reproduced in Fig. 4.5) it is clear that the ratio of angular cloud-width to standard deviation of wind azimuth is about 3.5. Assuming a Gaussian distribution in the cloud, the corresponding ratio of the angular standard deviation of the cloud distribution to that of wind azimuth is 0.8, whereas from Eq. (4.2) the value would be 1.0. The failure to recognize the approximate validity of Eq. (4.2) at a much earlier stage was of course directly attributable to the lack of accurate absolute measurements of the appropriate component of the intensity of turbulence.

A later analysis of the *Prairie Grass* data by Haugen (1966) provides further evidence on the validity and limitations of Eq. (4.2). For each of a selected number of experiments (all in near-neutral or stable conditions) Haugen fitted a regression for the growth of spread with distance and then doubly-differentiated this to derive values of $\overline{v'^2}$ and $R(\xi)$ from the differential form of Taylor's equation [i.e. $d^2(\overline{y^2})/dt^2 = -2\overline{v'^2}R(\xi)$ – equivalent to Eq. (3.58)]. Haugen's declared purpose was partly to test the assumption, implicit in all practical applications of Taylor's treatment, that the Lagrangian and Eulerian standard deviations of eddy velocity (σ_L and σ_E) in his notation) are identical. However, the value of σ_L derived by Haugen is of course equivalent to $(\overline{y^2}/T^2)^{1/2}$ as $T \to 0$ and so we may alternatively regard his comparison of σ_L and σ_E (his Table 4 and Fig. 1) as effectively a test of Eq. (4.2) in which (in contrast to other analyses) an attempt has been made to extrapolate the observed $(\overline{y^2}/T^2)^{1/2}$ to $T = 0$. The comparison is reproduced here in Fig. 4.8. Although Haugen concludes that there is a tendency for σ_L to be greater than σ_E, i.e. in our terms for $(\overline{y^2}/T^2)^{1/2}$ to be underestimated by Eq. (4.2), it is clear from Fig. 4.8 that on average there is little practical significance in the discrepancies.The scatter is of a similar magnitude to that found in the analysis referred to above – otherwise the fairly even distribution of points about the dashed line implies that Eq. (4.2) provides on average a satisfactory representation.

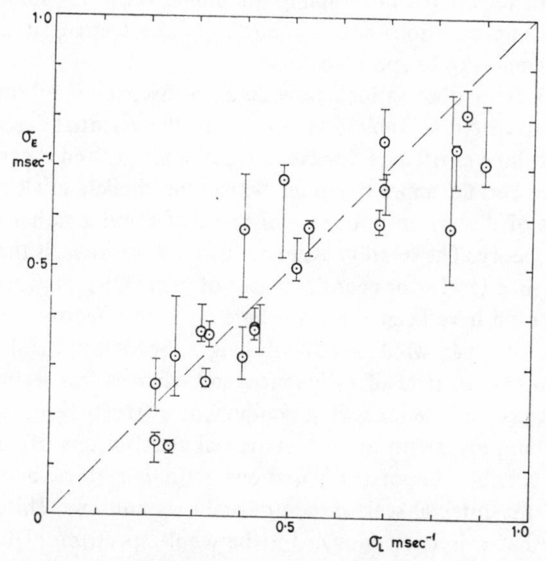

Fig. 4.8 – Standard deviation σ_L of initial crosswind velocity fluctuations of particles (estimated by extrapolation from the observed σ_y, x relation) compared with σ_E measured with vanes and anemometers. The points are medians, usually of 5 separate measurements, but sometimes only of 4 or 3, of σ_E. The vertical lines represent the range of the several values of σ_E. (After Haugen (1966), with permission of the American Meterological Society).

It is also evident from the general summaries of the σ_y data in Figs. 4.6 and 4.7 that near to the source the magnitudes of $\sigma_y/\sigma_\theta x$ or $\sigma_y/\sigma_v T$ do closely approach unity *on average*. However, the scatter in individual estimates of σ_y, already demonstrated in Figs. 4.3–4.5, is brought out again in Draxler's form of analysis in Fig. 4.7, where the magnitude of S at the smallest values of T/t_i range mainly from 0.7 to 1.3. Note that this range is very similar to that in the results of Haugen's special analysis of the *Prairie Grass* data (Fig. 4.8), remembering that Haugen's σ_L/σ_E is $\sigma_y/\sigma_v T$ as $T \to 0$.

Application of the Hay-Pasquill adaptation of Taylor's statistical theory

At larger values of travel-time T the quantity $\sigma_y/\sigma_v T$ depends on the Lagrangian time-scale t_L and also, though probably to a less important extent, on the shape of the Lagrangian correlogram or spectrum. The adaptation of the theory proposed by Hay and Pasquill (1959), set out in 3.5, incorporates this dependence by a combination of turbulence measurements reflecting the Eulerian spectral properties with the assumption of similarity in the shapes of the Lagrangian and Eulerian forms. For practical applications a convenient formulation is

$$\sigma_y = \sigma_v(\tau, T/\beta)T \cong \sigma_\theta(\tau, x/\beta u)x \tag{4.3}$$

in which T/β is the averaging (smoothing) time applied before evaluating σ_v, τ is the duration of release (or of sampling the plume when this duration is much less than the release duration) and β, the ratio of the Lagrangian and Eulerian integral scales, remains to be specified.

Analyses of field observations providing an assessment of the usefulness of the method are listed in Table 4.IV with only the essential descriptions and results. The first three entries are discussed in detail in Hay and Pasquill's original paper. There are certain approximations which are implicit in all the analyses. All observations of dispersion refer to a distance of travel x rather than time of travel T in the theory. The relation assumed, $x = u_r T$ where u_r is the mean wind speed at a reference level at or near the height of release, is not generally correct (see 3.3) and should have been $x = u_e T$ where u_e is an effective speed allowing for the increase of mean wind speed with height. Secondly, the Taylor theory assumes a homogeneous field of turbulence, and whereas this seemed generally acceptable in respect of the crosswind component at the time of these analyses the subsequent improvements in the statistical descriptions of the boundary layer indicate possibly important variations with height of both the r.m.s. turbulence and the integral scale in neutral-stable conditions. Thirdly, Taylor's theory implies that τ is large enough for the whole spectrum of turbulence to be experienced, which is clearly not so, especially for τ in the region of 10 minutes, and so far it has been argued roughly that large error will be avoided by restricting the application to $T \leqslant \tau$.

Resolution of the above difficulties will require further work. In the meantime, remembering that σ_y is only weakly dependent on β, a useful rough generalization supported by Table 4.IV is that σ_y is predicted by Eq. (4.3) with $\beta = 4$. Other evidence (see 2.7 and Table 3.I) points to $\beta i \cong 0.6$.

Table 4.IV – Summary of studies of crosswind spread from a continuous point source, in relation to the Hay–Pasquill adaptation of Taylor's statistical theory, and values of β implied.

Study	β
Values of σ_y (100 m), from measurements of crosswind distribution from 3-min releases of Lycopodium spores, in various conditions of stability, substituted in Eq. (4.3)	$1.1 \leqslant \beta \leqslant 8.5$ $\bar{\beta} \cong 4$ (8 values)
Ratios of σ_y (300 yd)/σ_y (100 yd) from early Porton field studies, 2.3 for a 4-min release and 2.7 for 30-min, demonstrated to be in good agreement with Eq. (4.3) and assumed value of β	$\beta = 4$
Hilst's (1957a, 1957b) photographic measurements of spread of an elevated smoke plume. Values of σ_y after 1-min travel in good agreement with Eq. (4.3) and assumed value of β (but much the same agreement provided by $\beta = 10$)	$\beta = 4$
Analysis by Haugen (1966) of *Prairie Grass* σ_y values at x up to 800 m, 11 cases in near-neutral and stable conditions selected for stationary conditions, to give $R(\xi)$ and hence β by applying Taylor's relation in the form $d^2\sigma_y^2/dt^2 = -2\sigma_y{}^2R(\xi)$	$0.7 \leqslant \beta \leqslant 6.3$ $0.1 \leqslant \beta i \leqslant 0.5$
Measurements in Idaho (Islitzer and Dumbauld (1963) Slade (1968)) of σ_y(100–3200 m) from 1 hr. near-surface releases of uranine dye, best-fitting values of β deduced from Eq. (4.3)	wide scatter $\bar{\beta} = 5$ (unstable conditions)

Note. With the exception of Hilst's study all the above refer to near-surface releases.
The effect of duration of release, implicit in the Hay–Pasquill method, has been brought out again by Doran *et al.* (1978a) in their analysis of the dispersion studies listed in Table 4.III.

Lagrangian time scales implied by the variation of σ_y with distance

For much of the σ_y data summarized in Fig. 4.6 and Table 4.III the specification of the prevailing turbulence is confined to the parameter σ_θ and it is therefore not possible to investigate the connection with the spectral characteristics in the way that has been attempted in the Hay–Pasquill approach. A secondary and more limited approach is to interpret the relation between σ_y and time or distance of travel, which on the Taylor statistical theory originally tends to be simply linear but subsequently falls away from the limiting relation in accordance with the Lagrangian time-scale and the shape of the correlogram or spectrum. Assuming a form for that shape a prescribed value of $\sigma_y/\sigma_v T$ implies a value of T/t_L which may be identified by matching with measured values such as those in Table 3.III. Initial application of this method, summarized in the second edition, showed that on the assumption of a spectrum form as in Eq. (3.70c) the dispersion data for the Green Glow and NRTS field tests were consistent with $ut_L \cong 1$ km. However, it was recognized at the time that this must be an underestimate, on account of the further assumption that $\sigma_y/\sigma_v T = \sigma_y/\sigma_\theta x$, instead of $\sigma_y u_e/\sigma_\theta x u_r$ where u_r is the wind speed at the reference level (at which θ is measured also) and u_e is the effective speed defined by x/T (which increases with T as the mean height of the diffusing particles also increases). Estimating u_e on the basis of similarity theory (from Eq. (3.36)) Hunt and Weber (1979) subsequently estimated that the true value of $u_e t_L$ could be as much as three times the initial estimate.

An up-dated version of the estimation of $u_e t_L$ from averaged σ_y data is summarized in Table 4.V. This brings out in more detail the effect of allowing for the variation of wind speed with height. The increase in the inferred $u_e t_L$ is much greater than u_e/u_r because of the weak variation of $\sigma_y/\sigma_v T$ with T/t_L especially at small T/t_L. Note also the reduction in implied $u_e t_L$ with shortening of released-time.

Table 4.V – Estimates of the Lagrangian time-scale from averaged data (in Table 4.III) on σ_y from surface releases.

	Duration of release	x (km)	Av. $\sigma_y/\sigma_\theta x$	Calculated u_e/u_r (neutral)	$u_e t_L$ (km) assuming $u_e = u_r$	$u_e t_L$ (km) assuming Calc. u_e/u_r
A	10 min	0.8	0.49	1.12	0.15	0.21
B	½–1 hr	0.8	0.81	1.12	1.0	2.4
		3.2	0.68	1.36	1.7	10

Notes:

(1) The values of $\sigma_y/\sigma_\theta x$ are
(A) Overall average of *Prairie Grass* values;
(B) Overall average of NRTS (mainly unstable) and Green Glow (mainly stable) values.

(2) u_e/u_r was calculated from Eq. (3.36) writing $u_e = \bar{X}/T$, $\bar{Z} = ku_*T$ with $k = 0.4$, and $u_* = ku_r/(\ln z_r/z_0)$. The value for (A) refers to *Prairie Grass* conditions $z_0 = 0.006$ m, $z_r = 2$ m, that for (B) is a mean of the values for NRTS, $z_0 = 0.015$ m, $z_r = 4$ m, and Green Glow, $z_0 = 0.03$ m, $z_r = 2.1$ m.

(3) The estimates of $u_e t_L$ $(= xt_L/T)$ are those which on substitution in the crosswind component form of Taylor's Eqs. (3.58) or (3.59), with the Phillips–Panofsky form of $R(\xi)$ (see Eq. (3.70d) and Table 3.III), give the observed average values of $\sigma_y/\sigma_\theta x$. Note that the assumption of $R(\xi) = \exp(-\xi/t_L)$ would give smaller estimates of $u_e t_L$, for example by a factor of roughly 2 in the (B) cases.

Although the analysis by Draxler (see Fig. 4.7) is presented in terms of time of travel it does not include any allowance for variation of wind with height and in that respect is basically no different from the foregoing analyses. It does, however, represent an advance in utilizing the individual values of σ_y (in the form $\sigma_y/x_\theta x \cong \sigma_y/\sigma_v T$) as a function of T/T_i where T_i is the value of T for $\sigma_y/\sigma_\theta x = 0.5$ and is related to t_L. The individual values of T_i have a very wide scatter. For example, for the surface release studies of Fig. 4.6 the range is 400-fold and Draxler proposes the use of an average of 300 sec irrespective of stability. However, it is noteworthy that whereas the *Prairie Grass* unstable values cover a range much the same as the NRTS unstable values and the Green Glow stable values many of the *Prairie Grass* stable values are spread over a distinctly lower range, possibly reflecting the shorter releases of the *Prairie Grass* study.

Values of T_i may be converted to t_L from the form of $F_L(n)$ which gives an acceptable fit to the variation of the normalized dispersion with T/T_i. Draxler's original discussion adopted a fitting to a simple interpolation form for the σ_y, T relation satisfying the limits of Taylor's relation (see Fig. 4.7) but in subsequent consideration Phillips and Panofsky (1982) found that an improved fit could be secured by adopting the $R(\xi)$ form in Eq. (3.70d). This latter form implies $T_i/t_L = 5.25$, which with Draxler's overall average of 300 sec for T_i gives $T_L \cong 1$ min, or ut_L typically 300 m say, which however is to be regarded as a considerable underestimate in view of the neglect of the variation of wind speed with height.

In terms of implied individual values of t_L there is evidently a pronounced disorder, which may of course simply reflect a disorder in the turbulence statistics which one might also expect to appear in the Eulerian scales l_E and t_E. The fact that the Hay–Pasquill approach has yielded estimates of t_L/t_E that are much less scattered than the statistics of T_i is a pointer in this direction. Fortunately for practical purposes the scatter in the inferred t_L values is a considerable amplification of that which is evident in the original σ_y values.

Crosswind spread and mixed-layer similarity

Until recently the fundamental interpretation of crosswind spread has depended entirely on Taylor's statistical theory. Monin–Obukhov similarity theory, which has been so successful in the treatment of vertical diffusion, does not seem to provide a satisfactory interpretation of the horizontal wind fluctuations (see Panofsky *et al.* 1977). However, the introduction of the alternative scaling parameters h_c and w_* as more appropriate for a convectively mixed layer, and the application of these ideas by Deardorff, is now providing a new approach to generalizing about dispersion, in which for example the dimensional prediction for σ_y is

$$\sigma_y/h_c = f(w_* x/uh_c) \tag{4.3a}$$

the form of the function remaining to be specified by observation or numerical simulation of the turbulence and diffusion. Deardorff and Willis (1975) have provided a specification from measurements of dispersion in a laboratory tank and have approximated this in the interpolation form

$$f(X) = [0.26X^2/(1+0.91X)]^{\frac{1}{2}} \tag{4.3b}$$

where $X = w_* x/uh_c$, which satisfies the limiting forms of Taylor's relation as regards variation with x. Results very close to the laboratory data have also been obtained by Lamb (1979) from a numerical simulation of individual particle trajectories in Deardorff's numerically modelled field of turbulence.

A preliminary comparison with data from the full-scale tracer studies was given in a later paper by Willis and Deardorff (1976) and this showed an impressive degree of agreement with Islitzer and Dumbauld's (1963) results (the NRTS data in Fig. 4.6 and Table 4.III), albeit using approximate and indirect estimates of the parameters h_c and w_*. A more detailed and more critical examination has been undertaken by Nieuwstadt (1980a), using 20 *Prairie Grass* cases selected on the criterion $-h_c/L \geqslant 10$, h_c being calculated from upper-level temperature profiles observed during the tracer study, and L and w_* from low-level wind and temperature profiles. A useful degree of agreement with the laboratory form of the function $f(X)$ is confirmed, though with considerable scatter and a tendency for the full-scale dispersion to be on average higher than the laboratory values, by a factor which Nieuwstadt estimates to be 1.2, consistent with Deardorff and Willis's (1974) estimate that modelled σ_v^2 values are smaller than in the atmosphere by a factor in the range 1.2–1.9, as a consequence of exclusion of relatively large-scale fluctuations. Note however that the release time in the *Prairie Grass* study was only 10 min, and from the surface-layer turbulence statistics obtained in the Kansas 1968 experiments (Kaimal *et al.*, 1972) it is clear that a 10-min sampling time there would have missed a substantial fraction (possibly as much as one-half in the lightest winds) of the v-component variance observed in the 1-hr samples. This suggests that for longer release times the underestimation represented by the laboratory model may be even greater than that estimated above.

4.5 VERTICAL DIFFUSION FROM A NEAR-SURFACE RELEASE OF PASSIVE PARTICLES

For obvious practical reasons the direct observation of the vertical distribution presents more difficulty than the crosswind distribution near ground-level, and consequently the major field studies have provided somewhat limited information on the distribution shape and the magnitude ($\bar{Z}$ or σ_z) of the vertical spread. On the other hand, in terms of the crosswind-integrated-concentration from a point source they do provide much more comprehensive information on the main practical significance of vertical spread in the context of air pollution levels. Also, for vertical diffusion, more so than crosswind diffusion, the last decade has seen useful developments in laboratory measurements of turbulent diffusion, especially in respect of the neutral boundary layer and the convectively mixed layer.

The following field studies included measurements of the tracer concentration at various heights above ground, so providing data from which to examine the shape of the vertical distribution or evaluate the magnitude of vertical spread directly.

(a) Field experiments at Porton in 1923–4 (Sutton (1947a)) and at Cardington in 1931 (reported in official records), in which sampling was carried out on a tower downwind of a crosswind line of smoke generators.
(b) The *Prairie Grass* series in U.S.A., which included sampling on six towers set up on an arc 100 m from the point source of sulphur dioxide.
(c) The *Green Glow* study at Hanford, U.S.A., using zinc cadmium sulphide tracer (Barad and Fuquay, 1962).
(d) At Porton (Thompson, 1965) in stable conditions with a source of aniline vapour from liquid sprinkled on the ground on an arc of 100 m or 200 m radius, the sampling being carried out on a tower at the centre of the arc. These experiments probably represent the closest approach to an ideal surface release of passive material.
(e) In the neighbourhood of Cardington, Bedfordshire (Thompson, 1966) in evening stable conditions, with a captive balloon carrying samplers at various heights and a source of zinc cadmium sulphide released from a vehicle travelling along a crosswind line a few kilometres upwind of the samplers.

The shape of the vertical distribution

The shape of the vertical distribution was determined from the early Porton and Cardington series, and the *Prairie Grass* series (the latter in analyses by Elliott (1961) and Nieuwstadt and van Ulden, (1978)) by fitting to the form

$$\chi(z)/\chi(0) = \exp(-bz^s) \tag{4.4}$$

to determine the ***shape exponent*** s. Values of s are given in Table 4.VI. Sakagami (1974) examined the *Green Glow* data and concluded that the distributions are nearer simple exponential form ($s = 1$) than Gaussian. These analyses leave no doubt about the departure from Gaussian shape, and the latest analysis by Nieuwstadt and van Ulden indicates a clear trend with thermal stratification, with $s \cong 1.35$ in neutral flow, a reduction in s on the unstable side, and a close approximation to Gaussian only in the most stable conditions.

Table 4.VI – The shape exponent *s* in the vertical distribution form of Eq. (4.4).

	No. of Expts	Stability	Distance (m)	*s*
Porton 1923–24	7	neutral	100	1.15
Cardington 1931	29	neutral	229	1.5
Prairie Grass 1956				
Elliot (1961)	41	stable–unstable	100	1.49 (± 0.28)
Nieuwstadt and van Ulden (1978)	22			0.95–2.2

(i) Elliott's (1961) result from the *Prairie Grass* data was obtained from two sets of three levels in each of 41 experiments, the figure in parentheses being a standard deviation. Only a slight tendency for an increase in *s* with increasing stability could be discerned against the considerable scatter. Also when vertical distributions were plotted for all 141 cases, and examined for overall fit to *s* 1.4 or 1.5, it was found that 1.4 gave a better fit in 14 unstable cases and 4 stable cases, while 1.5 was better in 15 stable and 4 unstable cases, two stable and two unstable cases being indeterminate.

(ii) Nieuwstadt and van Ulden's analysis refers to a selection (22) of the *Prairie Grass* runs for which the concentration at the highest level was less than one-tenth of that at the lowest level, and for which also the correlation coefficient between the observed concentration and the fitted curve was higher than 0.998. A plot of their value of *s* shows how a clear trend from values near 1.0 for $1/L = -0.1$ to near 2.0 for $1/L = 0.1\ m^{-1}$, with a scatter mostly within 0.15 in *s*. An eye-interpolation of the neutral value is approximately 1.35.

The magnitude of the vertical spread

The *Prairie Grass* data and those of Thompson are shown together in Fig. 4.9 in terms of Richardson Number at a height of 2 m, this being the stability parameter most directly available in the two series of experiments. In the sense that Ri is also expected to be a universal function of Z/L this graph implicitly gives σ_z (at 100 m) as a function of $1/L$. Indeed in unstable conditions the evidence is that Ri and Z/L are of similar magnitude, though the existence of exact equality may still be a matter of debate. The considerable scatter in the σ_z, Ri plot may reflect the inadequacy of representing the stability by Ri instead of $1/L$ but it seems more likely to be a consequence of the inherent patchiness and intermittent quality of turbulence.

The only direct indication which may be obtained at present for the growth of vertical spread with distance in stable conditions is that which follows from combining Thompson's separate experiments at ranges respectively of 100 m and several kilometres. To do this satisfactorily requires that the data should be referred to a common stability parameter and for these data the only parameter available is an approximation to the intensity of the vertical component of turbulence as provided by measurements of the fluctuating inclination of the

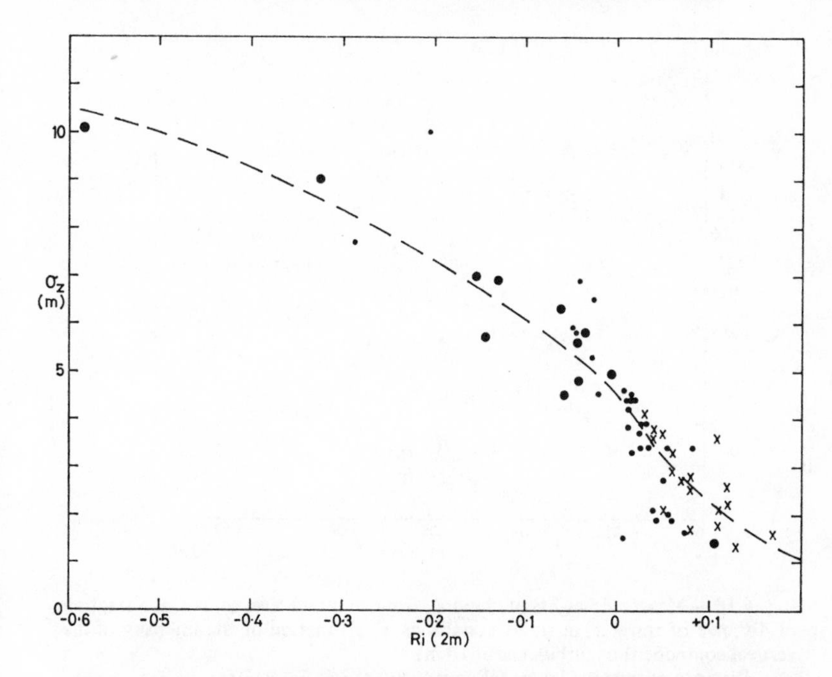

Fig. 4.9 – Data on σ_z at 100 m from a continuous source at ground level. ● average of 3 or more runs • 1 or 2 runs. (From *Prairie Grass* data as analysed by Haugen, Barad and Antanaitis, and tabulated by Pasquill, 1966.)

× Based on Thompson's (1965) measurements of height of cloud Z_0 at Porton and assuming $\sigma_z/Z_0 = 0.475$. The latter ratio is a close enough overall average for vertical distributions with s in Eq. (4.4) ranging from 1.5 to 2.0.

wind. Unfortunately the reference heights for this parameter are different – 2 m in the short-range experiments, whereas 10 m is the lowest height employed in the longer-range experiments. Fortunately the former experiments included mean wind speed measurements at 5 m or 8 m also, and on the reasonable assumption that σ_w did not vary substantially over these heights the value of i_w $(= \sigma_w/\bar{u})$ at 10 m may be estimated from that at 2 m by an extrapolated ratio of $\bar{u}$ at 10 and 2 m based on the assumption of a power-law variation of $\bar{u}$ with z. The resulting variation of the angular height of cloud with i_w at 10 m is shown in Fig. 4.10.

Note that the data at 0.1 km show a correlation with i_w which is remarkably close for data of this type and indeed to a very reasonable approximation the angular height of cloud is equal to i_w. The fewer observations at a few kilometres range are much more scattered but there is an obvious trend for a substantial reduction in Z_0/x, by about two-fold between distances of about 0.1 and 5 km. This comparison is not entirely satisfactory in that whereas all the estimates of i_w refer to much the same roughness (that associated with an open

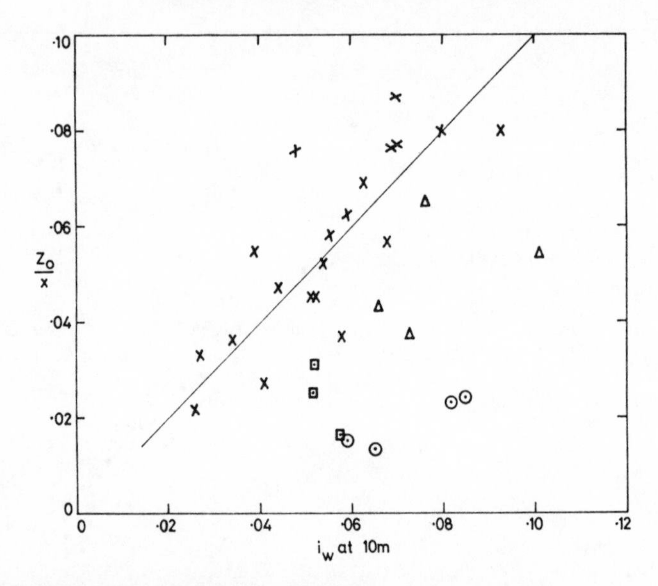

Fig. 4.10 – Measured heights of cloud Z_0 from a ground-level source, as a fraction of distance of travel x, in stable conditions, as a function of the intensity of the vertical component of turbulence at 10 m.

Distance of travel in km as follows: × 0.1, △ 3.9, □ 4.9, ○ 6.3.

For the 0.1 km data it was necessary to estimate the i_w (10 m) values from measurements of i_w at 2 m and wind speed measurements at 2 and 5 (or 8) m. The full line corresponds to $Z_0/x = i_w$ (after Thompson 1965 and 1966).

grassland site) only the short-range diffusion measurements are characteristic of this roughness. For the longer-range diffusion it is certain that the effective roughness would be rather greater, in accordance with the more variable terrain which would inevitably be encountered outside the special site on which the wind and turbulence measurements were made. On this argument the longer-range data have been assigned values of i_w which must be too low, and accordingly the reduction of Z_0/x with distance (for given roughness) is probably even greater than that noted above.

4.6 GENERALIZATIONS IN TERMS OF THEORIES OF VERTICAL DIFFUSION

In retrospect it is clear that Calder's (1949) combination of gradient-transfer theory (in the form of Robert's solution in Eq. (3.20)) with the wind profile laws in neutral boundary-layer flow still stands as the first successful rational generalization for vertical diffusion of material released at the ground. Moreover, in this treatment the form of the probability distribution of concentration is prescribed analytically. In this respect the data summarized above on the 'shape' exponent s (Table 4.VI) are to be compared with the theoretical value, on

Calder's treatment, $s = 1 + 2\alpha$, where α is the exponent in the best fitting power law for the variation of wind speed with height over the height-range of the diffusion. The value of α (0.187) specified by Calder for the neutral wind profile over the Porton terrain, and presumably reasonable also for the not greatly dissimilar Cardington site, leads to $s = 1.37$. There is also a simplification of the theoretical result which has some merit, i.e. to take n in Eq. (3.15) as unity, which is physically more acceptable than the $1-\alpha$ of Eq. (3.24) adopted for formal consistency with the 'conjugate-power-law' argument in the considerations of momentum transfer in relation to the wind profile. This leads to $s=1+\alpha= 1.19$ in the above example. The data support the larger of the two theoretical values of s, but the main point is that the theory is supported to the extent that the distribution is not Gaussian but intermediate between that and a simple exponential form ($s=1$).

Note that in general Eq. (3.20) implies that, since m in Eq. (3.15) increases with increasing stability and n decreases, the index r in Eq. (3.20), s in Table 4.VI, should increase in stable flow and decrease in unstable flow. Elliott's and Nieuwstadt and van Ulden's analyses of *Prairie Grass* data do show a tendency in the predicted direction. It is also a consequence of the theory in neutral flow, emphasized subsequently by Hunt and Weber (1979), that the index s should be a decreasing function of distance from the point of release, since with the progressive increase of the vertical spread the exponent α providing best overall fit with the logarithmic wind profile must decrease. However, there has not so far been any demonstrations of this variation of s.

Turning to the magnitude of the vertical spread, either in its direct representation as $\bar{Z}$ or σ_z, or as implied in the crosswind-integrated-concentration from a point source, we now consider these properties also in terms of the Monin–Batchelor Lagrangian similarity theory (Eqs. (3.35), (3.36) and (3.50)), which has the special merit of including the correct form of wind profile and of leading to the relatively simple universal relation between $\bar{Z}/z_0$ and $\bar{X}/z_0$ (Eq. (3.36)). However, in the consideration of *absolute* magnitudes of $\bar{Z}$ and crosswind-integrated-concentration from a point source (the latter being equivalent to the concentration from a continuous line source of infinite extent across wind – $\bar{\chi}_{cls}$ in Eq. (3.50)) the similarity approach is still tied to the gradient-transfer theory for the theoretical specification of the constant b and the function $\Phi(\bar{Z}/L)$ in Eq. (3.33). Also it needs to be kept in mind that dispersion measurements refer to tracer 'particles' that have travelled a fixed distance x from the point of release, in contrast to a fixed time T (and average distance $\bar{X}$) in the Lagrangian similarity representation, and it has been seen (3.3) that $\bar{Z}(\bar{X})$ is somewhat greater than the $\bar{Z}(x)$ calculated from gradient-transfer solutions. Another aspect affecting both theoretical approaches is the continuing dispute about the precise value assigned to von Karman's constant k.

Testing of the Lagrangian similarity/gradient-transfer theories for neutral flow is summarized in Figs. 4.11 and 4.12. The first of these shows a universal

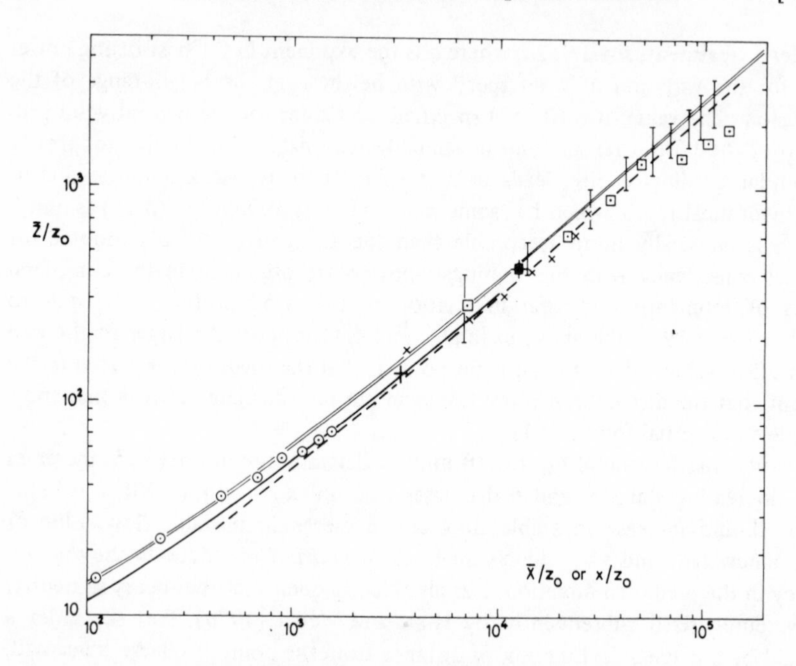

Fig. 4.11 – Tests of Lagrangian similarity/gradient transfer theory for vertical spread $\bar{Z}$ from a surface release in the neutral surface-stress layer.

═════ Lagrangian similarity relation (see Eq. (3.38))

$$kb\bar{X} = \bar{Z}\,[\ln(c\bar{Z}/z_0) - 1 + (1 - \ln c)z_0/\bar{Z}]$$

with $c = 0.561$, and with $b = k = 0.4$ *(lower line) or* $k = 0.35$, $b = 0.35/\phi_H$, $\phi_H = 0.74$ as used by van Ulden (1978) and Horst (1979) (upper line).

———— F. B. Smith's (1978) numerical solution of the two-dimensional equation of diffusion, with $K = ku_*z$ and $u(z) = (u_*k)\ln(z/z_0)$, and $k = 0.4$, giving values of $\bar{Z}$ as a function of distance x downwind of the point of release.

– – – – – van Ulden's adaptation of Roberts's solution (Eq. (3.20)) to give $x = (0.74/k^2)\,\bar{Z}(\ln 0.6\,\bar{Z}/z_0)$ with $k = 0.35$.

Atmospheric boundary layer average values of $\bar{Z}/z_0$ from observed vertical distributions, + Porton, England, $z_0 = 3 \times 10^{-2}$ m, ✚ *Prairie Grass* study, U.S.A., $z_0 = 8 \times 10^{-3}$ m (see Pasquill (1966) for the origins of these estimates of $\bar{Z}$. Note that the *Prairie Grass* point represents two independently derived estimates of $\bar{Z}$, 3.25 and 3.55 m. The more recent analysis by Nieuwstadt and van Ulden (1978) indicates a value close to 3.4 m).

Wind tunnel observations of vertical distribution (Robins and Fackerel (1979), with additional data provided in private communication)

	○	×	I	□
z_0(m)	4.4×10^{-3}	2.8×10^{-4}	7.2×10^{-5}	7.2×10^{-5}
h/z_0 (h = boundary layer depth)	450	4300	8300	5000

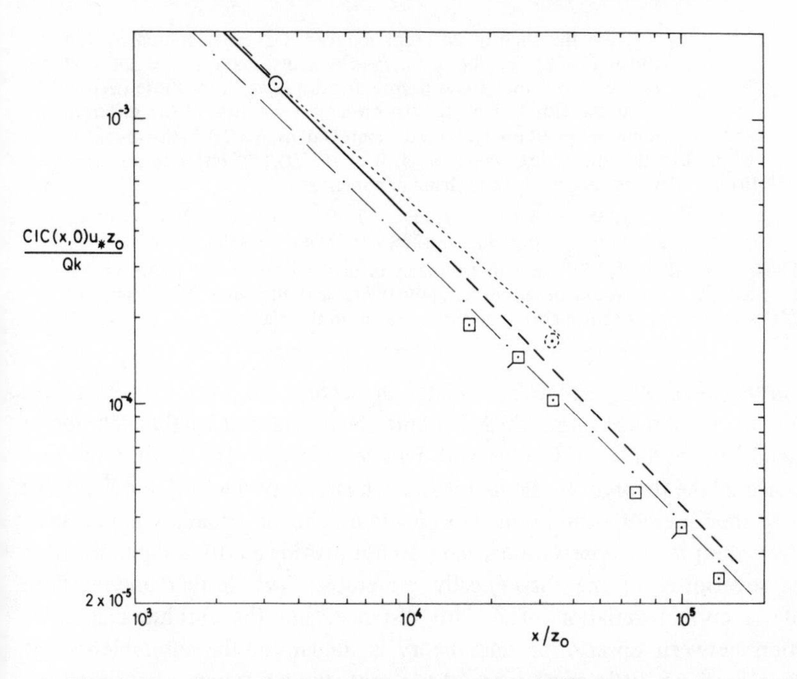

Fig. 4.12 – Crosswind-integrated-concentration (CIC) from the Porton and Prairie Grass field studies, in comparison with Lagrangian similarity/gradient theory, for neutral flow.

- - - - - - - - - - Calder's (1949) adaptation of Roberts's solution of the two-dimensional equation of diffusion (see Eq. (3.20) *et seq.*).

The remaining theoretical lines are all representations of the relation

$$CIC(x,\,0)/Q \;=\; A/(\bar{Z}\,u_{px})$$

which follows from van Ulden's adaptation of Roberts's solution, with A taking the value 0.7304 appropriate to a vertical distribution of the form $\exp\,[-(Bz/\bar{Z})^{1.5}]$ $u_{px}=u(cZ)$ with $c=0.561$ (i.e. the Lagrangian similarity form for $\mathrm{d}X/\mathrm{d}t$, here used as $\mathrm{d}x/\mathrm{d}t$; see 3.3 for discussion of the distinction between the $\bar{X}$ and x references to distance), and with other details as follows:

———————— using F. B. Smith's (1978) numerical solution values of $\bar{Z}(x)$ as in Fig. 4.00;

– – – – – – using van Ulden's (1978) form of $\bar{Z}(x)$ (as in Fig. 4.11) which leads to the relation

$$CIC(x,0)\,u_{*}z_0/Qk \;=\; 4.41\;z_0/x$$

– · – · – using the Lagrangian similarity form for $\bar{Z}(\bar{X})$ as $\bar{Z}(x)$, with k = 0.35, b = 0.35/0.74. This differs from the form used by Horst (1979) only slightly, as a consequence of a slightly different value of c.

○ Porton observations at x = 100 m as reported by Calder (1949). The dotted symbol is the value at $x = 1000$ m in accordance with the reported average variation of CIC with $x^{-0.9}$.

□ *Prairie Grass* observations at $x = 100$, 200, 400 and 800 m, using 14 of the cases tabulated by Horst *et al.* (1979), with L^{-1} ranging from -0.0145 to $0.0153\ \mathrm{m}^{-1}$, to interpolate a value for $L^{-1}=0$.

Continued overleaf

from the slightly different reduced data as tabulated by van Ulden (1978) for the same *Prairie Grass* cases at $x = 200$ and 800 m. Note that these points are displaced from those derived from the Horst *et al.* tabulation simply because of the different values adopted for z_0, 8×10^{-3} (van Ulden), 6×10^{-3} (Horst *et al.*).

Note that the theoretical value of A (i.e. $s\Gamma(2/s)/[\Gamma(1/s)]^2$) is not very sensitive to the value adopted for the shape exponent s:

s	1.00	1.10	1.25	1.50	1.75	2.00
A	1.00	0.9196	0.8239	0.7304	0.6739	0.6367

Thus, if a value of 1.35 were to be taken instead of the round figure of 1.5 adopted above A would be approximately 0.775 and the theoretical values of *CIC* would be 6 per cent higher than those drawn on the diagram.

---

relation between $\bar{Z}/z_0$ and x/z_0 broadly in accordance with the theoretical predictions, for a range of z_0 which is impressively widened by the inclusion of the wind tunnel data of Robins and Fackrell (1979). The most important difference in the theoretical calculations shown is as between $\bar{Z}(x)$ and $\bar{Z}(\bar{X})$, the effect of the different assumption about b and k being of secondary importance. However, the present observational data do not provide a critical demonstration of the superiority of the theoretically preferable $\bar{Z}(x)$. Both representations provide a correct variation of $\bar{Z}$ with distance, and the absolute-magnitude deviation between observation and theory is similar for the ensemble-average full-scale data, i.e. little more than 10 per cent, though in opposite directions. It is interesting to note that the wind tunnel measurements extend to substantial values of $\bar{Z}$ in relation to boundary layer depth h, there being no obvious sign of departure from the theoretical form for $\bar{Z}/h$ as large as 0.2, though the theoretical treatment is restricted to the surface-stress layer.

The comparisons of the crosswind-integrated-concentration at ground level, show more important discrepancies. Firstly, in contrast to the close consistency of the Porton and *Prairie Grass* observed values of $\bar{Z}$, the corresponding normalized observed values of CIC differ substantially (by about 60 per cent at $x/z_0 = 3.3 \times 10^4$). Secondly, although the *Prairie Grass* observations show a close approximation to a variation with x^{-1}, as predicted (closely by Smith's numerical solution and exactly by van Ulden's adaptation of the analytic solution), they fall below all the theoretical lines, the closest fit being provided by the Lagrangian similarity specification of $\bar{Z}(\bar{X})$ (as used by Horst (1979) in tests of the whole range of thermal stratification covered in the *Prairie Grass* study). Calder's theoretical variation, $CIC \propto x^{-0.86}$, is presumably a reflection of his adoption of $K \propto z^{1-m}$ instead of the more acceptable $K \propto z$. In comparing the observed levels of concentration with theory it may be significant that different tracers were used (a chemical smoke at Porton and sulphur dioxide gas in *Prairie Grass*). Loss by deposition was probably greater for the gas than for small particles (see the discussion on deposition in 5.2).

Further light is shed on the foregoing point by the latest numerical solutions published by Gryning *et al.* (1983), in which gaseous deposition is included through the boundary condition $K\partial\chi/\partial z = (v_d\chi)_{z=z_0}$. Using either Businger's (1973) ϕ_H (with $k=0.35$) or Dyer's (1974) ϕ_H (with $k = 0.41$), adoption of $v_d = 0.05\ u_*$ (implying an acceptable magnitude of v_d typically around 1 cm sec^{-1}) reduces the theoretical neutral-flow magnitude of *CIC* at the 800 m distance and 1.5 m height by about 40 per cent and provides a close fit to the *Prairie Grass* data. If it is reasonably assumed that the deposition velocity of smoke particles is substantially less than that of sulphur dioxide gas this would largely remove the foregoing apparent inconsistency in the Porton and *Prairie Grass* observations.

For thermally stratified flow the theoretical background is provided not only by the gradient-transfer and Lagrangian similarity theories, but also by the original Monin–Obukhov similarity framework (which identifies boundary-layer parameters likely to be significant), by free-convection scaling, by Deardorff mixed-layer similarity in strong convection and, in the latest studies, by Markovian random-walk methods. The main steps in the progressive use of these ideas are summarized in Table 4.VII.

Table 4.VII – A chronological digest of comparisons of field experimental data on vertical diffusion from a surface with theoretical treatments for stratified flow

---

Tyldesley (1967a)
Numerical solution of the two-dimensional diffusion equation (3.14) with $K = ku_*z/\phi_M$, $k=0.4$ and $\phi_M = 1+\alpha Ri$, providing estimates of vertical spread in good agreement with the trend in Thompson's (1965) data, a decrease by a factor of three as Ri(2m) increased from 0 to 0.2, with $\alpha = 5.5$.

Chaudhry and Meroney (1973)
Integration of Lagrangian similarity equation (3.35) for $\bar{Z}$, with $\Phi(\bar{Z}/L) = \phi_H^{-1}$, and Businger *et al's* (1971), Businger's (1973) forms of ϕ_H and wind profile function f, providing $\bar{Z}$(100 m) in reasonable agreement with trend of *Prairie Grass* values with $1/L$, but over-estimating systematically in stable flow.

Nieuwstadt and van Ulden (1978)
Numerical solution of Eq. (3.14) with K either K_H or 1.35 K_M (the factor 1.35 being K_H/K_M in neutral flow) and with Businger's (1973) forms of the Ks, both providing agreement with the trend of $\bar{Z}$(100 m) with $1/L$ from *Prairie Grass* data, greatest mean deviation (8%) for $K = K_H$, r.m.s. deviations near 10 per cent for both Ks. Also agreement (mostly within ±25%) with Thompson's (1965) measurements.

---

van Ulden (1978)
Use of Robert's solution (3.20) of the two-dimensional diffusion equation to express $d\bar{Z}/dx$ in terms of $\bar{Z}$, $K(\bar{Z})$ and $u(\bar{Z})$ and the vertical distribution shape

exponent s, integration of this with Businger's *et al.*'s forms of K_H and u, with empirical value of s (1.5), to give analytic form of $\bar{Z}$ which is then substituted in van Ulden's re-expression of Robert's solution for $\chi(x, z)$, for comparison with *Prairie Grass* data on $CIC(x, 0)$ at 50, 200 and 800 m, showing agreement within ±30% for most cases.

Briggs and McDonald (1978)
Examination of the vertical diffusion scale $Q/uCIC$, derived from *Prairie Grass* data, in relation to boundary-layer parameters, demonstrating best ordering in terms of u_* and L. Also ordered by mixed layer scales w_* and h, not quite so well but improving with distance. Useful ordering by $\Delta\theta/u^2$, especially in stable flow, but not by $\Delta\theta$ alone.

Horst *et al.* (1979) and Horst (1979)
Adoption of Chaudhry and Meroney's approach, with Businger's ϕ_H and f, to give $\bar{Z}$ as a function of $\bar{X}$, for substitution in a $\chi(x, z)$ equation identical with that used by van Ulden (1978), to derive $CIC(x, 0)$ showing good agreement on average with *Prairie Grass* data at 50, 100, 200, 400 and 800 m (overall mean fractional error of prediction +0.08) but tendency for substantial over-estimation in unstable conditions at the longer ranges. More scatter in comparisons also made with NRTS and Hanford data.

Nieuwstadt (1980)
Examination of *Prairie Grass* data on $uhCIC/Q$ in strongly convective conditions ($-h/L > 10$) in terms of convection scaling, showing good ordering with $t_* = w_*x/uh$ for $0.01 < t_* < 2.0$, but falling below the Deardorff–Willis (1975) laboratory curve over $0.3 < t_* < 1$. On a log-log plot the relation is slightly curved but is fitted well by $t_*^{-3/2}$, implying no dependence on h, over $0.03 < t_* < 0.2$.

Wilson *et al.* (1981), Ley (1982), Thomson and Ley (1983)
Comparison of random-walk modelling with *Prairie Grass* data on vertical distribution, indicating that k (mass) is near 0.4 and that $\phi_{\text{mass}} \neq \phi_M$. Wilson *et al.* claim that $\phi_{\text{mass}} \neq \phi_H$ also, but Thomson and Ley find that equality of ϕ_{mass} and ϕ_H is adequately supported.

Gryning *et al.* (1983)
Extension of Nieuwstadt and van Ulden's (1978) numerical solution of Eq. (3.14), with gaseous deposition included by the condition $K\partial\chi/dz = (v_d\chi)_{z=z_0}$. Use of $v_d = 0.05\,u_*$ improves agreement with observed values of shape factor s at $x = 100$ m, and with CIC especially at 800 m. $k = 0.41$ and Dyer's (1974) ϕ_H (which differs only marginally from Dyer and Bradley's (1982) form) provides agreement at least as good as $k = 0.35$ and Businger's (1973) ϕ_H.

The position reached in the interpretation of data on vertical spread from surface release may now be summed up briefly as follows:

(a) The assumptions $\Phi = 1/\phi_H$ and $K = K_H$ provide alternative partly successful generalizations in stable and slightly unstable conditions, especially when reasonable allowance is made for surface deposition, but deviations remain to be resolved in strongly unstable conditions.
(b) In unstable conditions there appears to be an intermediate range of distance over which a useful representation is provided by free-convection similarity, with heat flux H the dominant parameter.
(c) At longer range in unstable conditions there are clear signs that the foregoing systems fail, and that mixed-layer similarity is the more appropriate framework, though more crucial testing would be desirable.
(d) In general it seems that a useful ordering of the dispersion data may be achieved in terms of Monin–Obukhov L and also, more simply, in terms of $\Delta\theta/u^2$.

Physical understanding of the foregoing results is still not entirely clear. Aside from the basic objections to the concept of gradient-transfer *per se*, it does not seem obvious why, for *passive* material, the eddy transfer action should have a closer relation with that for heat than with that for momentum. It is readily understandable that K_H and K_M may differ in stratified flow in the atmosphere, and that K_H and K_W (for natural water vapour) are similar, but not that the transfer of strictly passive material is necessarily analogous to that of natural water vapour, the concentration of which is not generally independent of temperature. Nieuwstadt and van Ulden (1978) attempt to circumvent this difficulty by using K_M multiplied by the ratio K_H/K_M in neutral conditions, irrespective of the actual stratification, on the basis that in such conditions the relatively small temperature fluctuations cause negligible fluctuations in buoyancy and the heat is accordingly transferred as a passive property. However, they agree (see the discussion by Horst, van Ulden and Nieuwstadt (1980)) that in strongly unstable conditions K_H does provide a better representation, while pointing out that the gradient-transfer concept is then most suspect. In any case Hörst's own results do not show good support for $K = K_H$ in such conditions, and the crucial question yet to be answered is whether or not the agreement in other conditions is more than fortuitous support for the gradient-transfer theory.

4.7 DISPERSION FROM AN ELEVATED SOURCE

The behaviour of an elevated plume of diffusing material may be considered in three regimes:

(a) at short range such that σ_z is a small fraction of the height of release z_s, there then being no material concentration at ground-level. In this case the vertical variation of the turbulence properties affecting the plume should usually be minimal and it is a reasonable expectation that the vertical spread, as well as the crosswind spread, should be in accordance with the Taylor statistical theory. Relations corresponding to Eqs. (4.2) and (4.3) should then hold, i.e.

$$\sigma_z = \sigma_w(\tau, s)T = \sigma_\phi(\tau, s)x, \quad s = T/\beta \;, \tag{4.5}$$

which tends to $\sigma_\phi(\tau, 0)\,x$ in the short-time limit.

(b) At long range such that the overall depth of the plume is much greater than z_s, the ground-level concentration is decreasing with x and is approaching asymptotically the magnitude for a surface release, for which the generalizations of the foregoing section are applicable.

(c) An intermediate range encompassing the appearance of material concentration at ground-level and the build-up to the characteristic maximum immediately preceding region (b). Regime (c) is the most complicated of the three, in the sense that it is not describable in terms of either the simple statistical theory (because of systematic changes in the scale of turbulence with height) or the gradient-transfer theory (because of the fundamental objections discussed in 3.1).

Experimental examination of the vertical distribution presents even greater difficulty than for a surface release, but some useful studies (a digest of which is given in Table 4.VIII) have been accomplished, from which two general conclusions may be drawn.

(d) At short range, for σ_z/z_s up to the order of 0.1, Eq. (4.5) evidently provides a useful approximation to the observed σ_z. In the data of Hay and Pasquill (1957) the values of σ_z/z_s at $x = 500$ m were 0.17 and σ_z was in good agreement with the full value of σ_w as measured by the sensitive vane instrument then employed.

(e) On average the shape of the vertical distribution approximates fairly closely to the Gaussian form, similar to the crosswind as reasonably to be expected, but in contrast to the vertical distribution from a surface release. It is somewhat surprising, however, that, according to the Horst *et al.* (1979) results at $x = 200$ m, this convenient conclusion also applies even when the lower edge of the plume has reached the ground (σ_z/z_s then being roughly 0.5 and the ground-level concentration roughly one-tenth of the maximum in the vertical profile).

Table 4.VIII – Some experimental studies of vertical dispersion from an elevated source. (The data below all refer to a continuous source, where necessary by combining the puff size with the displacements of the puff centres).

Reference	Conditions†	Technique	Principal results
Crabtree and Pasquill (unpublished)	(a) 100 (b) 124 (c) u	Series of smoke puffs released from tethered-balloon-borne generator, tracked by theodolite and camera obscura	Negligible departure from linear variation of σ_z with x so confirming good approximation provided by Eq. (4.5).
Hay and Pasquill (1957)	(a) 150 (b) 500 (c) n to u	Lycopodium spores from tethered-balloon-borne dispenser, collected on adhesive cylinders spaced on cable of another balloon	Vertical distribution approximating closely to Gaussian on average. $\sigma_z/\sigma_\phi x$ ranging from 0.55 to 1.50 and averaging 1.07 (10 cases) so confirming Eq. (4.5) as useful approximation
Hilst and Simpson (1958)	(a) 56 (b) 1500 (c) s	Fluorescent pigment released from tower, collected on adhesive cylinders spaced on tethered balloon cable	Demonstration of σ_z increasing with distance more slowly than linearly and ultimately even more slowly than $\sqrt{x}$.
Högström (1964)	(a) 50 and 87 (b) 5000 (c) n to s	Smoke puffs of 30-sec duration released at intervals. Photographed from source position to provide area and displacement of centre in plane normal to wind	Demonstration of near-linear growth at short range followed by close approach to $\sigma_z \propto \sqrt{x}$, The group of data in the most stable conditions, shown in Fig. 4.13, implies $ut_L \cong 100$ m.
Horst *et al.* (1979)	(a) 26 (b) 800 (c) n to s	Various tracers released from a tower and sampled on arrays of towers on crosswind arcs.	Observed vertical distributions fitted reasonably well by Gaussian-plume model of elevated source (Eq. (6.17)) σ_z (200 m) computed from this model and the observed CIC ($z = 0$) shown to be a useful approximation to σ_z evaluated from the vertical distribution.

†(a) Height of release, *m*. (b) Maximum downwind distance, m. (c) Atmospheric stability: u, unstable, n neutral, s, stable.

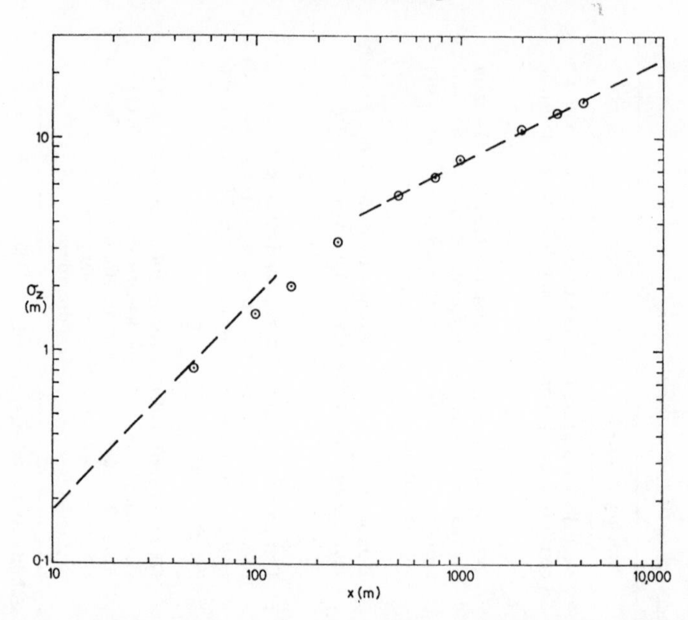

Fig 4.13 – Vertical spread (σ_z) measured by Högström (1964) from a source at a height of 50 m at Agesta, Sweden. These values were obtained by combining the displacements and growths of a series of smoke puffs and are regarded as representative of a continuous source. The data are those in the stability category $\lambda = 2.40$ in Högström's Table 3. The broken lines have slope unity and one-half corresponding to the short time and long time limits of statistical theory. (Reproduced with permission of the Swedish Geophysical Society.)

Interpretation of ground-level concentration from an elevated source

In so far as a convenient form may be assumed for the vertical distribution (and provided there is no significant loss of material after release, e.g. by deposition at the ground) the magnitude of σ_z may be inferred from suitable measurements of concentration at ground level, given also the strength of the source and the wind profile, simply on the principle of conservation. This is a familiar exercise, undertaken in many studies of the ground-level distribution of concentration from an elevated release of a tracer, notably by Smith and Singer (1966), Islitzer (1961), Vogt *et al.* (1973), Vogt (1977a, 1977b), Thomas *et al.* (1976a, 1976b) and Horst *et al.* (1979) who also provide a useful discussion of the earlier studies. In all these cases the classical *image source* model, with Gaussian distribution is assumed (Eq. (6.17)). In providing apparent values of σ_z analyses of this type serve a useful comparative purpose, but they have not so far added greatly to the basic understanding of elevated-source dispersion, primarily because of the lack hitherto of an acceptable theoretical framework applicable to the most complicated stage (the intermediate range of distance defined at

(c) above). There are, however, some interesting considerations to be noted at this point.

Horst *et al.*'s (1979) study includes some simultaneous measurements with two different tracers released at different heights, 26 m and 56 m. From 11 cases, in near-neutral to stable conditions, the apparent values of σ_z at $x = 400$ m were consistently greater for the higher release, the average value of their ratios being 1.5. At short range Vogt *et al.*'s (1978) results, for release heights of 100 m and 50 m, also show an increase of σ_z with release height in neutral-stable conditions, but a decrease in unstable conditions. However, at longer range (up to 10 km) the effect tends to be smaller in magnitude and much less systematic in relation to stability, which is consistent with the expectation that as vertical spread becomes large relative to release height (or alternatively as the vertical distribution progresses toward uniformity) the influence of release height is eliminated.

Despite the support apparently provided for the indirect estimates of σ_z, by the agreement shown with estimates made from observed vertical distributions at shorter range (200 m) in the Horst *et al.* study, the foregoing results are at first sight inconsistent, at the shorter ranges, with the observed decrease of σ_w with height in stable conditions and the increase with height in unstable conditions, coupled with the increase of wind speed with height. However, these strange results may be associated with the assumption that the mean height $\bar{Z}$ of the particles remains constant (equal to the release height) until the effects of 'reflection' from the ground come into play. However, in stable flow, it is arguable that there is a reduction of $\bar{Z}$ with distance from the source, as a consequence of a reduction of σ_w with height, or alternatively as a consequence of the wind shear as suggested by Hunt (1982), who also points to possible evidence for such a decrease in some of the observations of the vertical profile of concentration reported by Horst *et al.* (see also Doran *et al.* (1978b)).

Although the basis now exists for carrying out more detailed and searching examination of the point, in the numerical-experimental (random-walk) techniques described in 3.6, the application so far has been mostly for convectively mixed conditions (notably by Lamb (1979) and Hanna (1981b), and calculations bearing on the above results in neutral-stable conditions have yet to be provided. Hanna's calculations are particularly relevant to a consideration of Vogt *et al.*'s data in that three heights in the surface layer are used (z_s/h_c = 0.01, 0.025, 0.05), as well as the greater heights also used by Lamb. At the shortest range ($x/h_c \leqslant 0.1$) there is no pronounced systematic effect on σ_z, but with increasing range (up to $x/h_c = 1$) an increase of σ_z with z_s develops, the ratio for $z_s = 0.05$ and 0.1 reaching about 1.8 (for $h_c/z_0 = 10^4$, $h_c/L = -10$). Qualitatively these results are not generally consistent with Vogt *et al.*'s indirect estimates, which show a marked decrease of σ_z with z_s at very short range but this is progressively reduced with increasing range and ultimately converted into a slight increase. Obviously much remains to be clarified and the numerical-experimental (random-walk) technique may be expected to play a decisive role in the process.

As regards crosswind spread Horst *et al.* (1979) (see also Doran *et al.* (1978)) also examined the relation between ground-level σ_y and σ_θ for releases at different heights. For near-surface releases it has been seen that σ_y is well correlated with σ_θ at low level. Plots of average $\sigma_y/\sigma_\theta x$ for the three release heights 2, 26 and 56 m were found to be more closely coincident when σ_θ at low level (1.5 m) was used than when σ_θ near the release height was used. It is emphasized that these observations, all in neutral to stable conditions, were not generally carried out with simultaneous releases at the three heights. Also, the fact that use of σ_θ at z_s does not separate the plots with respect to z_s in a consistent sense (i.e. the effect is greater for z_s 2 m and 26 m than for z_s 2 m and 56 m) means that on average the decrease of σ_θ with height must have been considerably steeper when z_s was 26 m.

4.8 THE GROWTH OF CLUSTERS OF PARTICLES

Reference has already been made to the existence of two distinct forms of dispersion, manifested in the meandering of a plume of smoke and in the widening of the plume itself, and the distinction between these two forms has been emphasized in the theoretical discussions of Chapter 3. The preceding sections of the present chapter have been concerned with the combined effect of the two processes, and it is now appropriate to consider the observational data relating to the latter form of diffusion alone, i.e. the relative separation of the particles or elements of a diffusing cloud. In its elementary form this relative diffusion is represented in the growth of an individual cluster of particles, but it also determines the *instantaneous* distribution in the cross section of a continuous plume of material.

The visible growth of smoke-puffs

Because of the difficulty of measuring the instantaneous values of concentration simultaneously at a sufficient number of positions the observation of relative diffusion has tended to rely mainly on the visual methods, and especially on the observation of the growth and dissipation of puffs of smoke. Earliest examples are contained in discussions by Roberts (1923) and Sutton (1932), in which photographic observations of the growth of anti-aircraft shell bursts were used to test theoretical treatments. Later and more extensive series of observations of this type have been reported by Kellogg (1956), Frenkiel and Katz (1956), and Högström (1964). In Kellogg's experiments in New Mexico, vials of titanium tetrachloride and water, attached to trains of balloons, were exploded at predetermined altitudes by small charges of cordite set off by baroswitches. The altitudes of release ranged from 23,600 to 63,000 ft and the resulting puffs of smoke were observed by photo-theodolites, from the records of which both the positions and visible sizes of puffs were obtained at intervals. Eighteen usable sets of observations were reported in the form of graphs of visible diameter against time, up to a maximum time between 3 and 11 min. The smoke puffs used by Frenkiel and Katz were released within the first 100 m above the surface, over water off Maryland, U.S.A., in unstable conditions. The puffs were generated by exploding small pill-boxes of gunpowder carried by a tethered balloon. Positions and sizes of the smoke puffs at 1-sec intervals were obtained from cine-photographs, and the data on smoke-puff radius were tabulated by Frenkiel and Katz for nineteen cases, with total durations of observation ranging from 7 to 20 sec. In experiments of this type the puffs obtained are usually not of perfectly spherical shape, and some system has to be adopted for obtaining a consistent measurement of puff size. The system adopted by Frenkiel and Katz was to measure the visible *area*, and from this obtain the radius of the equivalent circular area. In Kellogg's experiments the puffs were elongated horizontally by wind shear, and it was argued that the minimum diameter was the relevant dimension for the diffusive spread. However, it was found to be more satisfactory to determine this dimension in an indirect way, by measuring the visible area and the *maximum* diameter, and calculating the minimum diameter for an assumed elliptical shape.

Högström's technique was to release smoke for periods of 30 sec at a time and to photograph the cross-section of the elongated puffs from the elevated point of release. In these cases the behaviour was initially as for a continuous source but tended subsequently to that of an instantaneously generated puff.

In general the smoke puff results show a rate of visible growth which is fairly constant to start with, and is then progressively reduced. This is well illustrated in the average curve taken from Kellogg's paper and reproduced here as Fig. 4.14. Some cases actually show the visible size reaching a maximum value and then diminishing with the further lapse of time. However, as pointed out by Gifford (1957a) in a reappraisal of Kellogg's data, many of the observations in the stratosphere show a distinct tendency for the graph of diameter against time to be slightly convex to the time axis in the early stages (see Fig. 3 of Kellogg's paper). The same effect is indeed slightly evident even in the average curve, for times between 0.5 and 3 min. This property immediately points to an accelerated growth of the cloud, as predicted in 3.7.

For the proper interpretation of observations on the growth of smoke puffs, it is essential to make some assumptions about the way in which the visible outline is related to the smoke concentration in the cloud. The method which is most commonly employed is that introduced by Roberts (1923). In this treatment of *opacity* it is assumed that the smoke particles produce a loss of intensity of the light from some further object or background by interception, and that the background or object is obscured (i.e. the smoke puff is outlined) when this intensity is reduced to a certain fraction of its magnitude in the absence of intervening smoke. Thus, for obscuration to occur the integrated

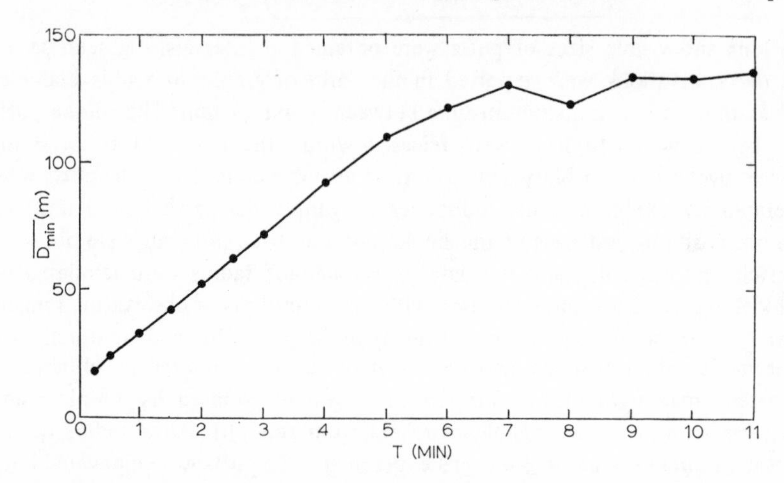

Fig. 4.14 – Growth of the visible diameter of a smoke-puff as a function of time of travel T. Average curve from eighteen releases at heights ranging from 23,600 to 63,000 ft over New Mexico (Kellogg, 1956).

smoke concentration along the line of sight must have a certain limiting value. This limiting value must depend on such factors as particle size distribution and the nature of the background, but for given circumstances it is assumed to be constant. This is obviously a considerable simplification of the complex of optical effects which applies in reality and some discussion of the way these contribute to the contrast between a cloud of particles and its background has been given by Jarman and de Turville (1966).

The procedure followed by Roberts, Sutton, Kellogg, and Frenkiel and Katz, was to use a theoretical relation for the distribution of concentration in the puff as a function of time, in order to derive a relation for the variation of visible size with time. Comparison with the observations then enabled the correctness of the original diffusion treatment to be assessed, and parameters such as diffusion coefficients and intensities of turbulence to be evaluated. Roberts's treatment of diffusion used the concept of a constant eddy diffusivity which is not generally acceptable while the remaining treatments were all based ultimately on Taylor's statistical treatment which, as discussed in Chapter 3, is applicable to the diffusion from a fixed continuous source, but not to the relative diffusion now under discussion. However, it has since been recognized and emphasized by Gifford (1957a), that in order usefully to interpret smoke puff data it is not necessary to adopt any particular solution for the variation of concentration with time, though it is necessary to make some assumption about the *shape* of the distribution of smoke concentration within the puff.

Following custom Gifford assumes that the statistical distribution of concentration in the puff follows the Gaussian form, with variance σ^2 at time t

after release. For a given quantity of smoke Q the concentration at x, y, z relative to the puff centre is then given by

$$\chi(x, y, z, t) = \frac{Q}{(2\pi\sigma^2)^{3/2}} \exp\left[-\frac{x^2+y^2+z^2}{2\sigma^2}\right] \tag{4.6}$$

If the puff is viewed from some distance along the y axis, the visible outline will be determined by the condition

$$N_0 = 2\int_0^\infty \chi \, dy = \frac{Q}{2\pi\sigma^2} \exp\left[-\frac{r^2}{2\sigma^2}\right] \tag{4.7}$$

where $r = (x^2+z^2)^{\frac{1}{2}}$ is the visible radius of the puff, and N_0 is a constant independent of t. Differentiation of (4.7) with respect to time leads to

$$\frac{dr^2}{dt} = \frac{d\sigma^2}{dt}\left[\frac{r^2 - 2\sigma^2}{\sigma^2}\right] \tag{4.7a}$$

so at the time of maximum visible radius r_m it follows that

$$r_m^{\ 2} = 2\sigma^2. \tag{4.8}$$

Substitution in (4.7) gives

$$N_0 = \frac{Q}{e\pi r_m^{\ 2}} \tag{4.9}$$

and equating 4.7 and 4.9, and rearranging

$$\sigma^2 = \frac{r^2}{2}\left[\ln\left(\frac{r_m^{\ 2}e}{2}\right) - \ln\sigma^2\right]^{-1} \tag{4.10}$$

Thus the absolute values of σ may be derived as a function of time from observations of the growth of a smoke puff, provided these observations are maintained for a sufficient period to enable the maximum radius to be specified. This treatment has assumed isotropic growth of the puff, but as shown by Högström (1964) the method can be extended to anisotropic growth at the expense of a little more complication in the procedure.

In the re-examination of the smoke puff data reported by Kellogg, and by Frenkiel and Katz, Gifford found that in both cases a number of observations were characterized by the existence of a definable maximum radius, so that Eq. (4.10) could be applied. The values of σ^2 obtained by Gifford ($\overline{Y^2}$ in his notation) are shown in Fig. 4.15. The curves correspond to the theoretical relations for the separation of a pair of particles, which follow from Batchelor's (1952) application of the Kolmogorov similarity theory, i.e.

$$\overline{y^2}(T) - \overline{y^2}(0) \propto T^2 \qquad \text{small } T \tag{4.11}$$

$$\overline{y^2}(T) \propto T^3 \qquad \text{intermediate } T \tag{4.12}$$

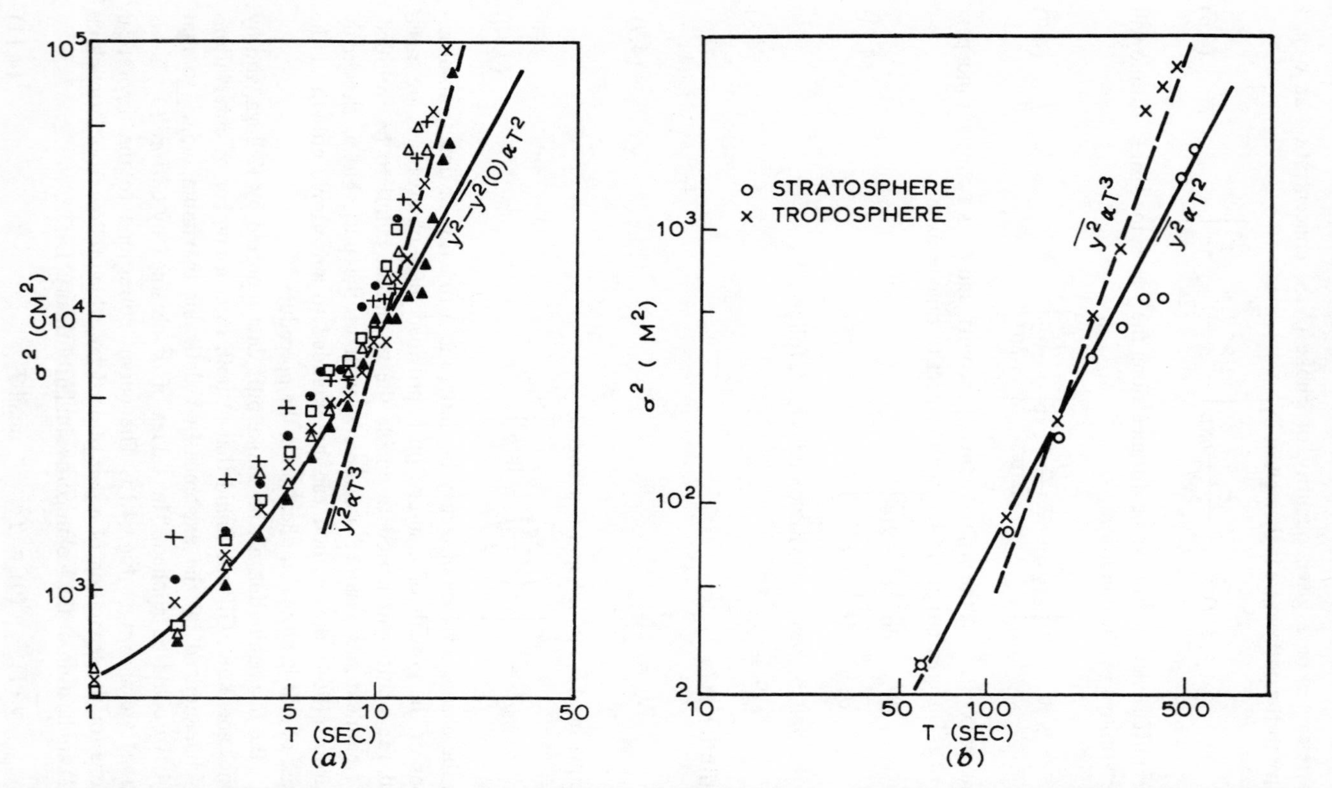

Fig. 4.15 – Gifford's (1957a) re-analysis of data on the growth of smoke-puffs, for comparisons with predictions from the Kolmogorov similarity theory. σ^2 is the variance of the distribution in a smoke-puff. (a) Frenkiel and Katz (1956), individual puffs; (b) Kellogg (1956) averaged data.

and which have been discussed in 3.7. It was assumed that $\overline{y^2}(0) \ll \overline{y^2}$ for all applicable T values in the case of (4.12), and even in the case of (4.11) for Kellogg's data. In the fitting of (4.11) to the results of Frenkiel and Katz a reasonable value of σ_0^2 (=400 cm^2) was adopted, but it was pointed out that any other value between zero and the obvious upper limit of 550 cm^2 would give an equally acceptable fit. It can be seen that Gifford's representation of the data strongly supports the existence of the predicted t^2 and t^3 regimes. Apart from the agreement with these specific laws Gifford's analysis is particularly notable in demonstrating an increase of dispersion (σ) more rapid than linear with time, whereas all previous analyses of diffusion data have been considered to show a growth slower than linear. Another good example of a t^3 regime has been reported by Nappo (1979) from a similar analysis of photographs of a large oil fog plume.

The average values of instantaneous vertical spread deduced by Högström from his photographic measurements show an extensive range of near-linear growth (see Fig. 4.16) followed by a reduction in rate of growth. Absence of accelerated growth may be merely a consequence of the large (elongated) initial form of the puffs.

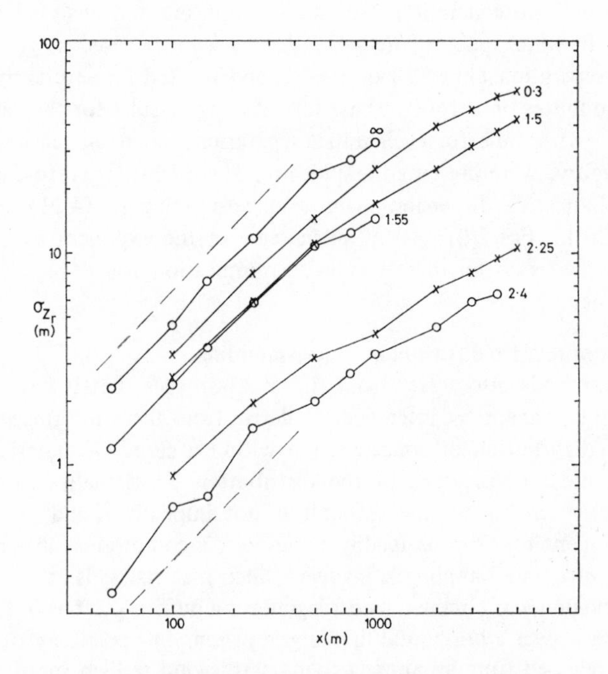

Fig. 4.16 – Vertical spread of elongated smoke puffs (from Högström, 1964). × Studswick, source height 87 m; ○ Agesta, source height 50 m. Numerals are values of Högström's λ. The broken lines represent $\sigma_z \propto x$. (Reproduced with permission of the Swedish Geophysical Society.)

Trajectory studies

The more direct but analytically more laborious technique of evaluating the variation with time of the relative separation of pairs of *particles* has been followed by Charnock (1951) and Wilkins (1958). Charnock used data obtained by Durst (1948) from camera obscura observations of the trajectories of pairs of smoke puffs fired from an aircraft flying at heights in the range 2000–5000 ft. The results tabulated by Charnock include values of

$$l_0 = \overline{l(0)}$$

and $$F(t, l_0) = \overline{[l(t)-\overline{l(0)}]^2}/t$$

for $t = 10$, 20, 30, 40 and 50 sec, where $l(0)$ is the initial separation between two puffs at an arbitrary time zero (not necessarily the time of release), and $l(t)$ the projection on a horizontal plane of the separation t seconds later. The bar implies averaging over a number of observations with similar initial separation. The quantity $F[t, l(0)]$ was examined as a function of time for comparison with the similarity-theory prediction corresponding to Eq. (4.11), i.e. $F(t, l(0)) \propto t$. Charnock found that there was some support for the relation in winds less than 5 m/sec, but that the results obtained with stronger winds were too scattered to provide any definite conclusion. Wilkins has reported and analysed analogous experiments in which pairs of neutral balloons were released at ground level, with initial separation either 10 or 100 m, and tracked for several minutes by pairs of theodolites on a 1600-m base line. Average results for the variation of $[l(t)-l(0)]^2$ with time for each initial separation, based on ten experiments in each category, were presented graphically. These confirm Gifford's previous analysis in indicating dispersion régimes corresponding to (4.11) and (4.12) for $l(0) = 100$ m. For $l(0) = 10$ m an increase in the exponent of T was evident, but owing to scatter there was no clear indication of a T^2 régime followed by a T^3 régime.

Observations of relative diffusion by tracer-sampling

The visual methods also suffer from the disadvantages that it has not so far proved possible, except by inference, to derive from them information on the form of the distribution of concentration within a cloud of material. On the other hand direct observation of the distribution by virtually instantaneous sampling measurements is very difficult if not impossible, and probably the best measurement that can be readily achieved is a compromise in which some element of *finite-time* sampling is involved. Such material as is at present available for discussion was obtained by arranging for a puff of gaseous or particulate material to pass over a horizontal line or grid of sampling positions, or for a line of material released from an aircraft flying acrosswind at high speed to cross a vertical array of sampling positions. In either case the results are expressed in the form of dosages, or numbers of particles collected, during some interval, and error will arise unless the motion of the cloud is exactly normal to the sampling line. If the standard deviation of the particles from the line of motion is σ_{true}, the apparent value observed on the sampling line will be approximately σ_{true} sec θ, where θ is the angle between the direction of motion and the normal to the sampling line. This discrepancy in angle may arise from a misjudgment of the mean wind direction when setting out the crosswind sampling line, but even a magnitude of 20° will introduce an error of only about six per cent in σ. Additionally, the turbulent fluctuations in the cross wind component will result in fluctuations of θ for successive clouds. Then for clouds which on average cross the sampling line normally the observed standard deviation will average approximately $\bar{\sigma}_{\text{true}} \overline{\sec \theta'}$, i.e. $\bar{\sigma}_{\text{true}} (1 + (\overline{\theta'^2}/2))$ for small values of θ'. The quantity $(\overline{\theta'^2})^{1/2}$, i.e. the intensity of the lateral component of turbulence, will normally be in the region of 0.1 radians, and the error in $\bar{\sigma}$ will then be negligible. This means that provided the experiment is arranged so that the cloud travels in a general direction fairly close to the normal to the sampling line, and that cases of excessive intensity of turbulence are avoided, the error involved in determining the spread within the cloud from an integrated sample is likely to be unimportant.

Sampling data of the above type, from field tests conducted at the Dugway Proving Ground, U.S.A., have been reported by Tank (1957). Clouds of gas released explosively were arranged to pass over arrays of gas samplers on a flat site. Diagrams of the isopleths of total dosage during the passage of the cloud are given for each of seven tests, for positions within about 300 m downwind of the point of release. Tank's analysis of these data was carried out in terms of Roberts's (Fickian) solution of the instantaneous point source (see Sutton, 1953, p. 134), modified so as to allow for the initial finite size of the cloud. The general assumption of Fickian-type diffusion for this case is of course entirely unacceptable, and the results as presented by Tank are for this reason of somewhat limited empirical value. An analysis in the light of Batchelor's applications of the Kolmogorov similarity theory has been given by Gifford (1957b), using the total dosage on the alongwind axis of the ground-level dosage pattern. This total dosage is taken to be inversely proportional to the mean-square particle dispersion [$\overline{y^2}$ in the notation of Eq. (4.12)]. From a plot of axial dosage against time T after burst Gifford concludes that with the exception of two tests in stable conditions all cases showed a T^{-3} régime, which corresponds to a T^3 régime in $\overline{y^2}$. It is noteworthy that in two stable cases the fall of dosage with time after burst was slower than as T^{-1}, implying that the increase of $\overline{y^2}$ was slower than as T, i.e. slower than would be expected from Fickian-type diffusion, but this was probably owing to the fact that the finite initial size of the clouds was not negligible in these cases.

Smith and Hay (1961) have given the results of an experimental study in which clusters of *Lycopodium* spores were formed near the ground by catapulting the material from a small container. Within a few metres of travel the spores were dispersed into a cloud approximately 1–2 m in height and 2–4 m in downwind length, and these were taken to be the initial dimensions of the cluster, subsequent growth being ascribed to turbulent diffusion. Samples of the cloud were collected on adhesive cylinders set out on crosswind lines at distances of 100, 200 and 300 m downwind of the point of release. The experiments are of particular interest in that in conjunction with these observations of lateral diffusion fairly detailed records were taken of the fine-structure variation of wind direction θ at a height of 2 m, at four positions with crosswind separation ranging from 5 to 35 m. From the numbers of spores collected on a specified length of cylinder, which were taken to be measures of the total dosage during the passage of the cloud, the standard deviation of the crosswind distribution was computed in each case. The results are summarized in Table 4.IX with other relevant data. To a close approximation the values of $\sigma_\theta[=(\overline{\theta'^2})^{1/2}]$ represent the intensity of turbulence $i=(\overline{v'^2})^{1/2}/\bar{u}$. The values of scale of turbulence, which showed no systematic connection with the other meteorological factors observed at the time, were obtained from the correlograms of wind direction against crosswind separation, using an equation corresponding to (2.6). Both the intensities and scales of turbulence are appropriate to averaging and sampling times (see 2.1) of 1 sec and 3 min respectively.

It is noteworthy that at each distance the sizes of the cloud as represented by the standard deviations show roughly a twofold variation, without any obvious relation with the intensity of turbulence. This is perhaps not surprising in view of the relatively small range of intensity, and the fact that these esti-

Table 4.IX – Experimental data on the expansion of particles in near-neutral conditions. (Smith and Hay, 1961)

Expt. No.	σ_p in m at 100 m	200 m	300 m	σ_θ (rad)	$\bar{u}$ (m/sec)
1	4.76	10.80	17.70	0.136	6.1
2	3.16	7.64	–	0.153	5.6
3	3.68	6.80	13.98	0.151	6.1
4	5.04	10.28	14.40	0.126	5.0
5	5.38	11.92	–	0.147	4.3
6	–	5.80	9.24	0.140	5.9
7	3.48	–	–	0.113	9.5
8	4.70	–	–	0.085	9.1
9	2.20	–	–	0.091	9.7
10	5.37	–	–	0.095	9.0

(σ_p is the standard deviation of the crosswind distribution, σ_θ is the standard deviation of wind direction for an averaging time of 1 sec and a sampling time of 3 min. Horizontal scale of turbulence 5–25 m. All wind measurements at a height of 2 m.)

mates of intensity, by virtue of the sampling time of 3 min, include the effect of *eddy sizes* much larger than those which could possibly have contributed to the spread of the clouds. The final dimensions (i.e. about six times the standard deviation) of the clouds were about 100 m, which would be equivalent to a sampling time of 10–20 sec. In any case it is known from the observations of turbulence discussed in 2.7 that the intensity of turbulence for a sampling time even as long as three minutes fluctuates widely over short intervals of time or space, so that the diffusion and turbulence data of the type given in Table 4.IX can be compared only on a statistical basis.

On the assumption of a Gaussian distribution within the cloud the average value of the standard deviations of the particle distribution at 100 m, i.e. 4.2 m, would correspond to a cloud-width (as defined by one-tenth peak concentration) of 18 m. This may be compared with the average 'instantaneous' cloud width of 25 m measured in the experiments with continuous sources at Cardington in 1931. A further point is that the data of Table 4.IX contain evidence of an accelerated increase in cloud dimension with distance of travel, in the sense that the exponent in the power-law variation of standard deviation with distance tends to be greater than unity. There is, however, no clear indication of an exponent as large as 1.5, corresponding to a T^3 régime as in Eq. (4.12). This feature is not surprising when it is remembered that the T^3 régime is predicted for diffusion which is dominated by eddies in the inertial sub-range. In the conditions of Table 4.IX it is clear that *eddy sizes* of 10 m or more would be effective, and at the heights of a few metres involved it seems unlikely that these eddies would have the properties of the inertial sub-range.

The search for further evidence for a $\sigma \propto T^{3/2}$ régime has been continued by Gifford (1977, 1982b) in an examination of several sets of dispersion data summarized by Hage (1964) and Crawford (1966) and other data heretofore overlooked or obtained subsequent to the Hage–Crawford reviews. The 1977 overall plots, with time of travel ranging from 2 sec to about 100 hr, contain large scatter with σ growing slightly more rapidly than linearly with T over the whole range on average (in his analysis Crawford suggested $\sigma \propto T^{1.2}$). From an inspection of Gifford's replot of Crawford's figures, with stratospheric measurements omitted, it seems that only two of the ten sets of dispersion data can be said to show individually a good approximation to $T^{3/2}$. These are the Frenkiel and Katz smoke puff data already noted, in which the time of travel is up to about 20 sec, and Crawford's nuclear cloud data for T in the range 10–25 hours roughly. The Smith–Hay data support an exponent greater than unity, but not as large as 1.5, and the remaining seven sets do not support 1.5 as well as 1.0. Thus while there is undoubtedly some evidence for limited ranges of the $T^{3/2}$ régime, at widely different magnitudes of T, it seems questionable to interpret the overall plot as a universal curve with an *extensive* intermediate range approximating to $T^{3/2}$, especially when the data have not been discriminated in terms of intensity and scale of turbulence.

Gifford's 1982b paper considers observations on the spread of a smelter plume in Australia for travel times up to 55 hr and demonstrates a fit, though with large scatter, with Eq. (3.69b). However, although that equation contains a small range for which $\sigma_y \propto T^{3/2}$ the fit is over a range of T/t_L for which the equation indicates a near-linear variation of σ_y with T at first, followed by a slowing-up of the growth and a tendency ultimately to a growth with $T^{\frac{1}{2}}$.

Comparison of observed rates of growth with those estimated from F. B. Smith's treatment

As seen in 3.7 the theoretical considerations of the growth of a cluster of particles in isotropic turbulence provide an estimate of the absolute magnitude of the cluster spread in terms of the intensity and scale of the turbulence. The maximum rate of growth [Eq. (3.109)] has been suggested as a simple practical approximation valid over a considerable intermediate range, beyond the early stages of growth, but before the subsequent reduction rate of growth becomes important (see Fig. 3.14). In the second edition version of Table 3.I it was suggested that for the quantity β the generalization $\beta i = 0.44$ be adopted, this β (say β_2) referring to the *transverse* Eulerian (fixed point) correlogram. The β in Eq. (3.109) (say β_3) refers to the *trace-correlogram*, as defined by Smith, involving both the transverse and longitudinal scales, and for isotropic turbulence $\beta_3 = \frac{3}{4}\beta_2$ (see p. 87 of Smith and Hay's 1961 paper). With these substitutions Eq. (3.109) becomes

$$\left(\frac{\mathrm{d}\sigma}{\mathrm{d}x}\right)_{\max} \cong 0.22i \tag{4.13}$$

The values of $\mathrm{d}\sigma/\mathrm{d}x$ according to Eq. (4.13) may be compared with the finite-difference ratios $\Delta\sigma/\Delta x$ from Table 4.IX, or, in rougher terms with σ/x when value of σ are given at only one distance. Grouping Expts. Nos. 1–6 gives a mean i of 0.14, hence a $(\mathrm{d}\sigma/\mathrm{d}x)_{\max}$ of 0.031 to be compared with an observed $\Delta\sigma/\Delta x$ averaging 0.052. The remaining experiments (7–10) have a mean i of 0.096, giving $(\mathrm{d}\sigma/\mathrm{d}x)_{\max}$ 0.021, to be compared with an average σ/x of 0.040. Thus Eq. (4.13) appears to underestimate cluster spread by a factor near two in these cases. However, if the higher estimate of $\beta i = 0.6$ now suggested in Table 3.I is adopted the numerical coefficient 0.22 on Eq. (4.13) becomes 0.3 and provides somewhat better agreement.

The examination of the Smith–Hay and Högström observational data with Smith's theoretical treatment has been followed up in more detail by Sawford (1982), in the course of a reconsideration of the theory, with special reference to the approximation adopted by Smith for the two-particle Lagrangian covariance, which is different from that suggested earlier by Taylor (1935) in an unpublished note referred to by Batchelor (1952). Sawford's comparisons, reproduced here in Figs. 4.17(a) and 4.17(b), bring out several interesting points:

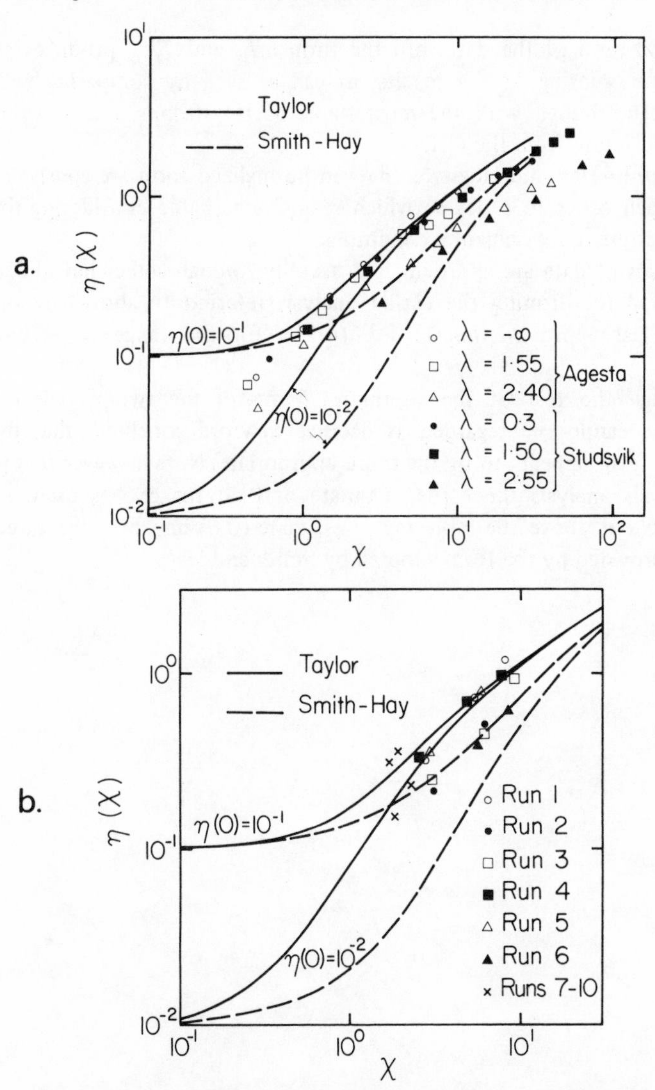

Fig. 4.17 – Comparisons of the theoretical growth of a small cluster of passive particles with dispersion observations in the atmospheric boundary layer (after Sawford, 1982).

$$\eta = \sigma_p/l_E, \quad \chi = x/\beta l_E$$

The theoretical curves, for initial values of η as indicated, were derived following Smith's theoretical analysis (see Smith and Hay, 1961), with an assumed form of the two-particle Lagrangian covariance either as adopted by Smith and Hay or as proposed earlier by Taylor. (a) Comparison with Högström's (1964) σ_p values from photographs of the y–z cross-sections of x-elongated smoke puffs released from an elevated source. (b) Comparison with Hay's σ_p values from crosswind sampling of small clouds of *Lycopodium* spores generated suddenly near the ground.

(i) Normalization of the data into the form σ/l_E and T/t_L produces an effective ordering of the results, as can be seen by comparing with Fig. 4.16, though with the most stable of Högström's data showing some deviation from the rest.

(ii) The Smith–Hay and Högström data in normalized form are consistent with each other to a degree which seems remarkable considering the quite different experimental conditions.

(iii) Both sets of data are approximated usefully, though somewhat underestimated (confirming the earlier analysis referred to above) by the theoretical approximation in Eq. (4.13), for the range $0.1 \leqslant ix/l_E \leqslant 3.0$.

(iv) Although the test of the alternative forms of the two-particle covariance cannot be regarded as decisive Sawford concludes that the Taylor form appears to be the more appropriate. Note however that in Sawford's analysis the earlier estimate of 0.44 for βi was used. As pointed out above the more recent estimate (0.6) improves the agreement provided by the form adopted by Smith and Hay.

5

The distribution of windborne material from real sources

Release of material into the atmosphere may be acompanied by significant ascent, as in the case of hot gases discharged from industrial chimneys. On the other hand the material may be in the form of solid particles large enough for sedimentation and fall-out to occur, or in the form of smaller particles, gases and vapours which may be retained or absorbed by vegetation and other surfaces. These complications are now considered with emphasis on practical aspects and as a preliminary to discussion of various features of the incidence of air pollution.

5.1 THE RISE OF HOT EFFLUENT

The greater the height of the axis of a plume of effluent the more the dilution by vertical (and crosswind) spreading which must occur before the effect is noticed at ground level. As will be seen later the ground-level concentration from a source at height H is inversely proportional to H^2, according to the simple theoretical model usually adopted. This means that considerable importance is attached to the magnitude of H and especially to the increase above chimney height which may result from the forced or buoyant ascent of the effluent. In practice the buoyancy effect is the more important for the maintenance of the ascent and much effort has been devoted to the theoretical and observational study of this aspect. This process is very complex and it is not surprising that no complete and exact theory has emerged. The combination of several semi-empirical theories with a wide range of observational studies of variable quality has led to a great number of 'plume rise' formulae, the relative merits of which have been the subject of much argument and confusion.

A full review of these arguments would serve little purpose in the present context and the aim here is rather to underline the basic principles and difficulties and to summarize some significant observations and analyses.

The physics of the problem

The essence of the problem, both as regards the features which are reasonably clear and those which remain confused, is easily seen from a quite simple thermodynamical approach based on Scorer's discussions (1958, 1959a and b). Consider the plume to be an expanding tube, bent over nearly horizontally in the wind, with the heat content uniformly distributed in the cross-section of radius r, producing a uniform temperature excess ΔT relative to the surrounding air, which is assumed to have uniform temperature T (strictly the immediately following analysis assumes *potential* temperature constant with height). Equating the rate of change of upward momentum of unit length of the plume (along the wind) to the buoyancy force

$$\frac{\mathrm{d}}{\mathrm{d}t}(\rho w \pi r^2) = w\frac{\mathrm{d}}{\mathrm{d}z}(\rho w \pi r^2) = \frac{\pi r^2 g \rho\, \Delta T}{T} \tag{5.1}$$

Assuming no loss of heat from the plume

$$\frac{\mathrm{d}}{\mathrm{d}z}(\pi r^2 \rho c_p\, \Delta T) = 0 \tag{5.2}$$

Now assuming the plume radius to grow linearly with height (a relation for which there is some support from both laboratory and full-scale observations), i.e.

$$r = az \tag{5.3}$$

Eqs. (5.1) and (5.2) are satisfied by

$$w = bz^{-1/2}, \quad \Delta T = cz^{-2} \tag{5.4}$$

where

$$c = \frac{Q_H}{\pi \bar{u} a^2}, \quad b = \left(\frac{2gQ_H}{3\pi a^2 \bar{u} T}\right)^{1/2}$$

and Q_H the rate of output of heat, is simply $\pi r^2 \bar{u}\, \Delta T$, it being assumed that except close to the chimney exit the sections of plume move with the mean speed of the wind. From these results it follows that

$$\frac{\mathrm{d}z}{\mathrm{d}t} = w \propto \left(\frac{Q_H}{\bar{u}z}\right)^{1/2}$$

and therefore (writing $x = \bar{u}t$) that

$$z(x) \propto \frac{{Q_H}^{1/3} x^{2/3}}{\bar{u}} \tag{5.5}$$

The foregoing analysis does not imply any termination of the rise of the plume. It is clear that there must be a limit when the atmosphere is stably stratified, but for a neutral atmosphere (potential temperature constant with height), as is implied in the previous analysis, the existence of a limit is not immediately obvious (there is further discussion of this point below). A practical approach which has been suggested for overcoming this difficulty is to regard the rise as effectively complete when the upward velocity is some small fraction of the vertical eddy velocities in the wind, so that the downward transfer of the effluent is effectively determined by diffusive spread. Note also that the analysis has ignored the possible effect of atmospheric turbulence in diluting the plumes.

The functional form in Eq. (5.5) is of course supportable on dimensional grounds alone and the dimensional approach has been exploited in some detail by Scorer (1959 a and b) and extended more recently by Briggs (1965, 1969). Following Briggs (1965) notation the determining physical parameters are

the 'stack parameter' $F = \dfrac{gQ_H}{\pi \rho c_p T_0}$

the stability parameter $s = \dfrac{g}{T}\dfrac{d\theta}{dz}$

the wind speed $\bar{u}$

some parameter representing the effect of atmospheric turbulence.

Note that $\pi\rho F$ is the initial vertical flux of buoyancy, i.e. $(\rho w \pi r_0^2 g \,\Delta T/T_0$ at the stack exit. The wind speed enters as a representation of the along-axis dilution which occurs when the plume is bent over in the wind, and accordingly the parameters F and $\bar{u}$ are combined in the form $F/\bar{u}$. Still neglecting the effect of atmospheric turbulence dimensional analysis of the remaining variables gives the results for plume rise ΔH as follows:

Incomplete rise in neutral windy atmosphere

$$\Delta H \propto F^{1/3}\bar{u}^{-1}x^{2/3} \tag{5.6}$$

Final rise in stable windy atmosphere

$$\Delta H \propto F^{1/3}\bar{u}^{-1/3}s^{-1/3} \tag{5.7}$$

Final rise in stable calm atmosphere

$$\Delta H \propto F^{1/4}s^{-3/8} \tag{5.8}$$

For the problematical case of final rise in a neutral atmosphere omission of t (or x) from the analysis leads automatically to a variation with $F/\bar{u}^3$, and the same result has been argued by Briggs (1965) in a further dimensional analysis in which the friction velocity u_* is introduced as the parameter representing the role of atmospheric turbulence in diluting the plume. As will be seen later there is no obvious evidence for this much more sensitive dependence on both F and u.

Progress in the specification of the constants in the forms above depends either on fitting observations or on more sophisticated treatment of the equations expressing conservation of momentum and heat. In the latter respect the crucial problem lies in the correct representation of the entrainment process by which the plume section is expanded. When this process is predominantly a consequence of the turbulence induced by the vertical motion of the plume relative to its environment, as in calm or non-turbulent air or even in turbulent air for some initial stage, a simple assumption that the average rate at which air enters the plume surface is proportional to the vertical velocity gives satisfactory results. This hypothesis leads to the linear relation between plume radius and height of rise, as assumed in the simplified analysis outlined at the beginning. The problem is more difficult when atmospheric turbulence is effective, as must be expected ultimately except in calm or stable conditions. An attempt to overcome this difficulty in a realistic way was first made by Priestley (1956) in a two-stage working theory in which the effect of atmospheric turbulence (second stage) was represented in terms of the eddy viscosity. The second stage solution was matched by trial and error to the first stage, which incorporated the assumption of self-induced entrainment noted above. Priestley's first stage solution has since been put into the asymptotic form $F^{1/4}u^{-1}x^{3/4}$ (see Briggs, 1969) and the graphical solution originally required for the second stage was subsequently shown by Lucas, Moore and Spurr (1963) to be analytically representable in the form ${Q_H}^{1/4}/\bar{u}$ multiplied by a complex function of x. These forms contain the same inverse dependence on $\bar{u}$ as Eq. (5.6) and a not very different dependence on Q_H.

A detailed reconsideration and redevelopment of the basic theory has since been set out in a U.S. Atomic Energy Commission Critical Review by Briggs (1969). His latest treatment incorporates the idea already mentioned of relating the entrainment by atmospheric turbulence to the rate of dissipation of turbulent kinetic energy, as representative of the relevant small-scale turbulence. Approximate solutions in the form of Eqs. (5.6)–(5.8) are obtained, with numerical constants 1.8, 2.4 and 5.0 respectively, and with an additional distance function in Eq. (5.6) representing the stage of rise controlled by atmospheric turbulence. Briggs's review also gives expressions for the penetration of sharply defined inversion layers and includes a useful summary of other plume-rise theories.

An informative discussion of the basic equations [equivalent to Eqs. (5.1) and (5.2)] and of the physical structure of plumes has been given by Moore (1966). This brings out the importance of the breaking up of the plume into more-or-less discrete lengths, a process which Moore argues to have significant influence on the rate of dilution of plume buoyancy and thereby on the existence of bounded solutions for the height of rise. As already noted for a continuous plume a dilution in accordance with $r \propto z$ leads to an unbounded solution unless the environment is stably stratified. For a bounded solution the vertical velocity of the plume must ultimately decrease more rapidly than t^{-1} or x^{-1}.

Moore's arrangement of the basic equations brings out the point that this requires the volume of a plume element containing a given amount of heat to increase more rapidly than x^2. In the simple analysis of a continuous plume the assumption $r \propto z$ leads to $r \propto x^{2/3}$ and hence to a volume proportional to $x^{4/3}$. There is no doubt that the growth (of r) may exceed $x^{2/3}$ for some distance, but theoretical considerations (and Högström's data – see 5.3) insist that the limiting rate is $x^{1/2}$. This suggests that in practice the termination of rise must be due to penetration into a layer which is at least slightly stable.

Observations of the ascent of plumes

The first full-scale observations to be reported were made by Bosanquet, Carey and Halton (1950) and gave a total of only seven estimates of plume height. Since then much more extensive measurements have been made by various workers. A summary of the usable observations has been given by Briggs (1969) and this is reproduced in Table 5.I.

Probably the most outstanding feature of the data obtained is the absence of unequivocal estimates of *final* rise, except for some cases in stable air. This state of affairs has persisted despite the considerable increase of the downwind travel involved, from the 600 ft of Bosanquet *et al.*'s observations to the nearly 6000 ft of the Central Electricity Generating Board's survey at Northfleet. It is therefore not surprising that the numerous attempts made to justify formulae for *final* rise in terms of such heterogeneous data have often resulted in more confusion than clarification.

The interpretation and generalization of the accumulated data by Briggs (1969) has been used as the main basis for the following summary. The first step was to consider the dependence on wind speed, since establishment of an acceptable relation would then facilitate the examination of the separate effects of heat outputs and distance from stack. In Briggs's monograph twenty plots of plume rise (at a specified distance in near-neutral conditions) are given, referring to fourteen of the sources in Table 5.I. As examples the plots for the Bosanquet and Central Electricity Generating Board data are reproduced in Fig. 5.1. Some of the plots contain more scatter than these examples. However, Briggs claims that a slower reduction than u^{-1} is supported only by the Duisburg data and a substantially more rapid reduction only by the Shawnee and Widows Creek data, and concludes that on the whole $\Delta H \propto u^{-1}$ is the best elementary relation.

Accepting the foregoing reciprocal relation with wind speed, the next stage in the analysis was to derive values of the product $\bar{u}\ \Delta H$ (a procedure first followed by Holland (1953) in his analysis of the Oak Ridge data leading to the well-known formula of that name). These values (for near-neutral conditions) were then plotted against distance downwind for several of the sources, to give the diagram reproduced here in Fig. 5.2. This diagram leaves no doubt about the continued increase of plume height over the whole range of distances

Table 5.I – List of plume rise studies in near-neutral conditions, including values of $u\ \Delta H$ at maximum distance x (after Briggs, 1969), (notation as in Table 5.II)

Code	Source	Number of stacks	h_s (ft)	D (ft)	w_0 (ft/sec)	Range of u, (ft/sec)	Q_H stack, (10^6 cal/sec)	x (ft)	$u\ \Delta h$ (ft-ft/sec)	Reference
B	Ball†					2–14	0.0096	60	112	Ball, 1958
HA	Harwell A†	1	200	11.3	32.6	14–30	1.10	2950	4,430	Stewart, Gale and Crooks, 1958
HB	Harwell B	1	200	11.3	32.6	17–38	1.10	1900	3,980	
BO	Bosanquet†	1		6.5	31.9	14–33	1.54	600	2,450	Bosanquet, Carey and Halton, 1950
DS	Darmstadt†	1	246	7.5	15.7	16–25	0.855	820	2,150	Rauch, 1964
DB	Duisburg	1	410	11.5	28.0	15–29	1.88	1150	3,400	
T	Tallawarra†	1	288	20.5	12.0	20–23	2.93	1000	5,500	Csanady, 1961
L	Lakeview†	1	493	19.5	65.0	25–49	11.6	3250	22,100	Slawson, 1966
	CEGB plants									
E	Earley	2	250	12.0	18.3	14–35	1.54	4800	5,580	Lucas, Moore and Spurr, 1963
E	Earley	2	250	12.0	56.0	14–35	4.72	4800	8,150	
CD	Castle Donington	2	425	23.0	40.9	10–26	11.95	4800	14,800	
CD	Castle Donington	2	425	23.0	54.7	10–35	16.0	4800	18,600	
N	Northfleet†	2	492	19.7	46.3	13–52	7.9	5900	10,900	Hamilton, 1967
N	Northfleet†	2	492	19.7	70.0	13–52	11.95	5900	11,150	
	TVA plants									
S	Shawnee†	8	250	14.0	48.7	8–29	5.45	2500	6,210	Carpenter *et al.*, 1968, see also Thomas *et al.*, 1969
C	Colbert†	3	300	16.5	42.9	10–17	6.74	1000	7,200	
J	Johnsonville	2	400	14.0	94.8	6–22	10.8	2500	10,100	
WC	Widows Creek†	1	500	20.8	71.5	8–21	16.8	2500	8,000	
G	Gallatin	1	500	25.0	52.4	7–34	16.9	3000	14,250	
G	Gallatin	2	500	25.0	23.7	5–39	8.55	2000	7,850	
P	Paradise	1	600	26.0	51.3	6–55	20.2	4500	21,200	
P	Paradise	2	600	26.0	57.2	12–34	21.9	4500	20,000	

Observational techniques: B, Visual. BO, DS, DB, T, L., Photography. T.V.A. plants, Infra-red photography. HA, HB., Vertical survey of radioactivity. E, CD., Neutral balloons in plume. N, Lidar observed, and despite the variability in the ΔH, x slopes the overall impression is in keeping with the predicted $x^{2/3}$ variation.

Briggs then considers the agreement between the observed plume heights and several of the formulae which have been proposed for neutral conditions, restricting attention however to those formulae which include the $\Delta H \propto \bar{u}^{-1}$ relation. Ratios of the calculated to observed values of $\bar{u}\ \Delta H$ were derived for each source. Table 5.II lists the median values and mean percentage deviations for each formula considered. Of the eight formulae included (1), (2) and (3) are entirely empirical and are implicitly for final rise (though in view of previous remarks unrealistically so). Those at (4), (5) and (6) in the table all have theoretical support from the Priestley (1956) 'working theory', while (7) and (8) are from Briggs's theoretical treatments. In accordance with these results Briggs concludes that his $x^{2/3}$ formulae provide the best overall representation of plume behaviour in neutral conditions. However, it is evident that Lucas's (1967) adaptation of Priestley's theoretical formulation is not greatly inferior.

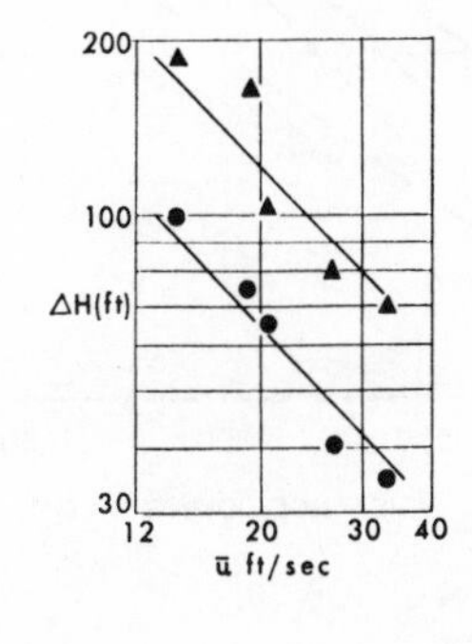

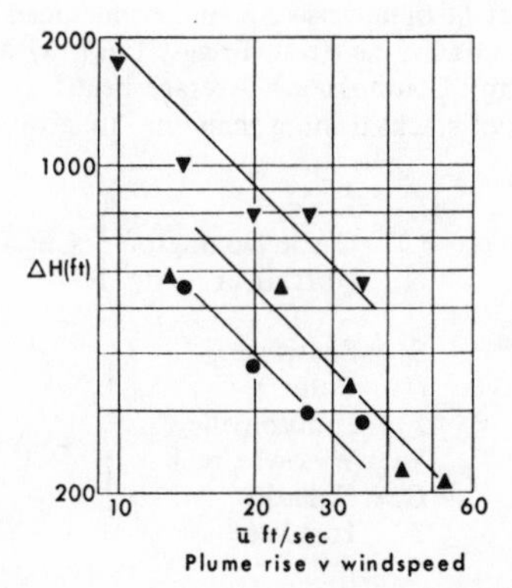

Fig. 5.1 – Examples of data on variation of plume-rise with wind speed in near-neutral conditions (Briggs, 1969). Upper graph – Bosanquet's data. Lower graph – data for Earley, Castle Donington and Northfleet power stations.

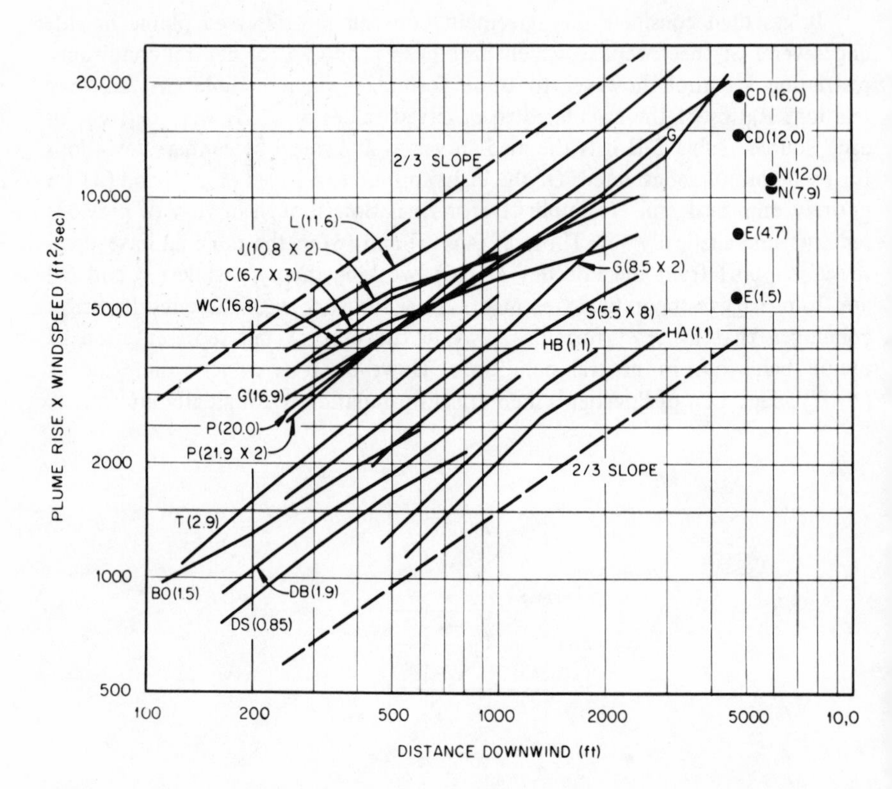

Fig. 5.2 – Data on product of plume-rise ΔH and wind speed u against distance from stack in near-neutral conditions (from Briggs, 1969, with permission of the United States Atomic Energy Commission). Average heat flux per stack, in units of 10^6 cal/sec, and number of stacks if more than one, are given in parentheses.

Key to sources:

HA	Harwell A	E	Earley	CEGB plants
HB	Harwell B	CD	Castle Donington	
BO	Bosanquet	N	Northfleet	
DS	Darmstadt			
DB	Duisberg	S	Shawnee	TVA plants
T	Tallawarra	C	Colbert	
L	Lakeview	J	Johnsonville	
		WC	Widows Creek	
		G	Gallatin	
		P	Paradise	

Table 5.II – Comparison of calculated and observed values of $u\,\Delta H$ (after Briggs, 1969)

Formula for $u\,\Delta H$ and reference	Median and percentage deviation of calc/obs ratio of $u\,\Delta H$: All sources in Table 5.I	Median and percentage deviation of calc/obs ratio of $u\,\Delta H$: Selected sources (see Table 5.I)
(1) $1.81\left[\dfrac{\text{ft}^2/\text{sec}}{(\text{cal/sec})^{1/2}}\right] Q_H^{\,1/2}$ Moses and Carson, 1967	0.54 ± 34	0.48 ± 19
(2) $1.5w_0D+118\left[\dfrac{\text{m}^{1/2}}{\text{sec}}\right]D^{3/2}\left(1+\dfrac{\Delta T}{T_0}\right)^{1/4}$ Stümke, 1963	0.79 ± 27	0.72 ± 24
(3) $1.5w_0D+4.4\times10^{-4}\left[\dfrac{\text{ft}^2/\text{sec}}{\text{cal/sec}}\right]Q_H$ Holland, 1953	0.44 ± 37	0.47 ± 26
(4) $2.7(\text{ft/sec})^{1/4}\,F^{1/4}x^{3/4}$ Priestley 1956	1.44 ± 26	1.41 ± 18
(5) $258\left[\dfrac{\text{ft}^2/\text{sec}}{(\text{cal/sec})^{1/4}}\right]Q_H^{\,1/4}f(x)$ Lucas, Moore and Spurr, 1963	1.36 ± 21	1.24 ± 22
(6) As (5) multiplied by $0.52+0.00116h_s$. Lucas, 1967	1.18 ± 20	1.16 ± 14
(7) $1.8\,F^{1/3}x^{2/3}$ Briggs, 1969	1.17 ± 23	1.17 ± 12
(8) As (7) for $x\leqslant x^*$ then $1.8F^{1/3}x^{*2/3}f(x/x^*)$ for $x>x^*$ Briggs, 1969	1.09 ± 19	1.09 ± 7

Notes:

(4) Asymptotic form of Priestley's first stage solution, see Briggs, 1969.

(5) $f(x)=1-\left(0.6+0.2\dfrac{x-x_1}{1000\text{ ft}}\right)\exp\left(-\dfrac{x-x_1}{1000\text{ ft}}\right)$ where x_1 is distance of transition to Priestley's second stage, estimated to be 600 ft.

(8) $f\left(\dfrac{x}{x^*}\right)=\left[\dfrac{2}{5}+\dfrac{16}{25}\dfrac{x}{x^*}+\dfrac{11}{5}\left(\dfrac{x}{x^*}\right)^2\right]\left(1+\dfrac{4}{5}\dfrac{x}{x^*}\right)^{-2}$

$x^* = 0.43\,F^{2/5}(\beta^{-1}u\epsilon^{-1})^{3/5}$, $\beta \simeq 1.0$

$F = gQ_H/\pi\rho c_p T_0$

Q_H = Rate of emission of heat with stack gases

D = internal stack diameter

w_0 = efflux velocity of stack gases

h_s = stack height

ϵ = rate of dissipation of turbulent kinetic energy per unit mass of air

T_0 = emission temperature of stack gases.

For practical purposes Briggs recommends use of his formulae with the coefficients of 1.8 reduced to 1.6 to give optimum fit. Also, as a working approximation for large plants (Q_H 20 Mw or more) he recommends the use of (7) in Table 5.II up to a distance of 10 stack heights and the assumption of no further rise thereafter. In a later review of the empirical estimates which he and others have derived for the constants in Eqs. (5.6) and (5.7) Briggs (1972) confirms that values of 1.6 and 2.4 are good average values, being slightly on the conservative (low) side.

Much further discussion will be found in the literature as regards the relative merits of various plume-rise formulae. Most of this fails to provide any significant additional insight into the physical aspects of plume behaviour, being instead preoccupied with mainly statistical appraisals of the fit provided by particular formulae. It is important to note on the one hand the discrepancies between observation and any current theory on individual occasions and on the other hand the discrepancies which appear to be associated with non-ideal surroundings of the plant. An indication of the first type of discrepancy is provided by the deviations from a $\Delta H \propto \bar{u}^{-1}$ law in Briggs's (1969) plots. At best these are ± 10 per cent but may be as much as 50 per cent. Also, in Thomas *et al.*'s (1969) fitting of the T.V.A. data to various formulae the individual ratios of observed to calculated value ranged between 0.5 and 1.8. In the second respect Briggs estimates that the existence of a terrain discontinuity may be responsible for a ±40 per cent deviation from his formula.

Plume rise in thermally stratified air

There is not much definite information about plume rise in unstable air, presumably partly because of the much more irregular behaviour of plume elements in the more-or-less vigorous natural convection then prevailing. The extreme effect is very evident in the typical *looping* condition (see Fig. 6.2). As regards the effect on the buoyancy of a hot plume it should however be noted that the existence of a lapse rate which is markedly superadiabatic is usually confined to the first ten metres or so above the ground. Thus at the heights typical of effluent stacks the extra maintenance of the buoyancy force may not be very significant and may be more or less compensated by more rapid expansion as a consequence of the higher intensity of small-scale turbulence generated in the convective conditions. However, the direct evidence from the plume rise studies is inconclusive. According to Briggs the Lakeview data indicate a slightly higher average rise than in natural conditions but no significant effect is obvious in the T.V.A. data. In Thomas *et al.*'s (1969) analysis of the T.V.A. data in terms of the form (8) in Table 5.II, the average values of the numerical constants for the 'stable' and 'neutral and unstable' groups used by these authors differed by only 3 per cent. Pending further evidence, Briggs concludes that the neutral formula is the most acceptable approximation for unstable conditions also but emphasizes that larger individual fluctuations must be expected.

For buoyant rise in a calm stable fluid with fairly uniform temperature gradient there are measurements ranging from the scale of a brine tank (about 1 ft) to that of a large oil fire (about 10,000 ft rise). According to Briggs's summary (see his Fig. 5.7) these are well represented by the form in Eq. (5.8) with the constant 5.0 as originally estimated theoretically. In the present context the bent-over buoyant plume is of more interest and for this data are available in stable air from some rocket ignitions as well as from six of the T.V.A. observations with large single stacks. Briggs finds the final rise to be well represented by Eq. (5.7), with the original constant of 2.4 increased to 2.9, as indicated in Fig. 5.3, taken from Briggs's Fig. 5.8. It is also noteworthy that the rise to the final level is closely in accordance with the relation for neutral conditions [Eq. (5.6)].

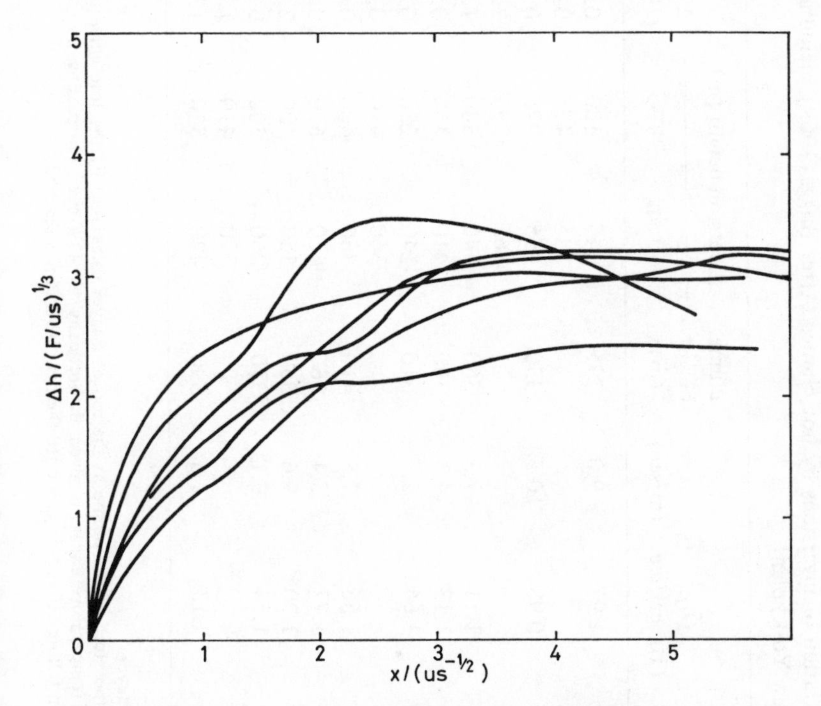

Fig. 5.3 – Rise of buoyant plumes in stable air in a crosswind at the TVA Paradise and Gallatin plants (from Briggs 1969, with permission of the United States Atomic Energy Commission). Curves refer to plume centre line and are in non-dimensional form in accordance with Eq. (5.7).

Information on the penetration of inversion *layers* by bent-over buoyant plumes has been provided by observations at a plant in New York City. The data as summarized by Briggs (1969) are reproduced here in Table 5.III. The occurrence of penetration appears to be reasonably consistent with a prediction procedure suggested by Briggs and summarized in the notes below Table 5.III.

Table 5.III – Data on penetration of inversions by hot plumes. After Briggs (1969), including summary of Simon and Proudfit data for the Ravenswood (New York) plant

Date	Time	$\bar{Q}_H$, (10^7 cal/sec)	$\bar{u}$ (m/sec)	Plume height (m)	Inversion height (m) Bottom	Top	ΔT_i (°C)	Calculated θ' (°C)	Penetration
May 25	1825	1.97	9.0	295	145	180	0.2	15	Yes
					325	475	0.7	−0.5	No
July 20	0552–0559	0.98	10.5	350	255	275	0.3	0.05	Yes
					365	395	2.0	−2.0	No
	0617–0820	1.11	7.3	360	540	580	1.9	−1.9	No
July 21	0600–0724	1.13	4.3	360	410	450	0.6	−0.45	No
	0828	1.64	2.7	510	240	280	0.6	1.7	Yes
					360	410	0.4	0.0	Yes
September 8	0648–0930	1.66	7.5	410	360	400	0.8	−0.6	?
	1000–1020	1.77	5.4	560	620	650	0.4	−0.3	No
September 9	0640–0705	1.20	9.6	350	360	400	2.1	−2.0	No
	0747–0850	1.54	9.1	370	260	300	0.7	−0.2	Yes
					370	410	1.6	−1.6	No
	0930–1000	2.13	9.6	390	420	530	1.8	−1.7	No

Notes:

$\Delta T_i = \Delta T$ through inversion layers.

θ' is the residual excess temperature of the plume at the top of a layer calculated by the following procedure suggested by Briggs. For rise initially through neutral air the temperature excess is given theoretically by $4TF/guz^2$. On passing through an inversion layer θ' is reduced by ΔT_i but must remain positive then falling off as z^{-2} again until the next inversion is reached.

A useful updated brief summary on plume rise will be found in Hanna, Briggs and Hosker's (1982) *Handbook on Atmospheric Diffusion.*

Moore's formula for plume rise

Another formula for predicting the rise of buoyant plumes has been used quite extensively in the U.K. and was developed within the Central Electricity Research Laboratories by D. J. Moore (1980). The distinctive aspect of this formula is that it recognizes that plumes often break up during the plume-rise phase into blobs that are only rather loosely interconnected. This implies three-dimensional mixing with the environment rather than the two-dimensional mixing implied implicitly by Briggs. Consequently the rise is proportional in Moore's formula to $Q^{\frac{1}{4}}$ rather than $Q^{\frac{1}{3}}$. Which of the two alternatives is best supported by actual data is not totally clear, and maybe the small difference in the exponents makes such a distinction of more academic than practical interest. Overall Moore's formula is rather more empirical than Briggs's, and attempts to interpolate between rise conditions in various meteorological situations. Some definitions are required and these are now listed, without explanation of their origins:

(i) Q_H is the heat flux in megawatts in the stack gases. Very roughly Q_H is 1/7th of the electrical output of the station. Thus a 1000 MW station has a heat output of about 140 MW. Q_H may be calculated directly from the efflux velocity V_s of the gases at stack top, the stack diameter d, and the temperature difference $(T_s - T_a)$ between the stack gases and the environment:

$$Q_H = \frac{\pi}{2} \rho c_p\, d^2\, V_s\, (T_s - T_a) \times 10^{-6}$$

Usually $(T_s - T_a)$ is about 100° K for modern power stations.

(ii) U is the wind speed (m s^{-1}) at stack height, $g = 9.81$ is the acceleration due to gravity, h_s is the stack height in metres, and $\Delta\theta$ is the increase in potential temperature of the environmental air over 100 m above the stack top. Then let

$$U^* = \max.(0.2, U)$$

$$\Delta\theta^* = \max.(\Delta\theta, 0.08)$$

$$h^* = \min.(h_s, 120)$$

$$f = \begin{cases} 1 \text{ if } \Delta\theta/U^2 > 0.0025 \\ 0.16 + 0.007h^* \text{ otherwise} \end{cases}$$

$$T' = \max.(12, T_s - T_a)$$

$$x_N = \min.(4224, 1920 + 19.2h_s)$$

$$x_m = \frac{U\, V_s\, T_a}{g(T_s - T_a)}, \qquad x_s \fallingdotseq \frac{120\, U^*}{\sqrt{\Delta\theta^*}}$$

$$x_r = \frac{x_s\, x_N}{(x_s^2 + x_N^2)^{\frac{1}{2}}}, \qquad x^* = \frac{x x_T}{(x^2 + x_T^2)^{\frac{1}{2}}}$$

The plume rise Δh as a function of distance downwind x is then given by

$$\Delta h = 2.25\, Q_{H'}^{\frac{1}{4}} x^{*\frac{3}{4}} \frac{f}{U^*} \left(\frac{T'}{110}\right)^{\frac{1}{8}} \left[1 + \frac{27d + 1.5\, x_m}{x^*} + \frac{54\, d\, x_m}{x^{*2}}\right]^{\frac{1}{4}}$$

The formula, although apparently rather involved, is of course readily usable given the basic parameters.

5.2 DEPOSITION OF AIRBORNE MATERIAL

Deposition of material during airborne transport may occur in three main ways – general sedimentation of particles or droplets, retention at the ground by processes such as impaction or absorption, with consequent downward turbulent transport to the sink thereby provided, and washout in association with rain or other forms of precipitation. In progressively depleting the material content all three processes ultimately hasten the reduction of airborne concentration otherwise caused by diffusion, but the associated contamination of the ground and, in the case of sedimentation, the possibility of increasing the concentration close to the source, present additional features of practical importance. As the rate of sedimentation is obviously represented, at least to a first approximation, by the terminal velocities of the particles it is clear that this mechanism is important only for relatively large or dense particles. The theoretical terminal velocity (v_s) in air of a spherical particle of diameter (d) 1 μ and density (ρ) 5 g/cm^3 is 0.016 cm/sec. For a Reynolds number ($v_s d/\nu$) less than about 0.5 v_s is proportional to ρd^2. Thus for very finely divided particulate material, as well as for gases and vapours, effective deposition must depend on the other processes mentioned above.

The complete theoretical treatment of the effect of rapid sedimentation in proper association with the diffusion process, i.e. in particular with correct allowance for the effect of the motion of the particles relative to the air and of particle inertia, has yet to be achieved. Also, the complex physics of impaction and absorption is not easily represented in the equations of transfer. The consequence is that many practical applications ultimately rely on relatively crude adjustments of the original solutions in which sedimentation and deposition were neglected.

Deposition of large or heavy particles

We first consider the case of particles which have free-fall velocities similar to or greater than the commonly occurring vertical eddy velocities, i.e. tens of centimeters per second or more, for which direct gravitational deposition is the ultimate controlling factor. Experimental data on which to test ideas about this process are not very plentiful but there are three studies in Canada which have been reported by Hage (1961a and b), Walker (1965) and Stewart and Csanady (1967). In all cases glass microspheres were released from an elevated position on a tower (at heights 15 m, 15 or 7.4 m and 18.6 m respectively) at a steady rate (over various periods in the range 10 to 60 min). The distribution on the ground was measured by collecting sample deposits on flat adhesive plates set out on arcs concentric about the emission point, at radii up to 800 m in the first two studies and up to about 240 m in the third study.

The simplest aspect to be considered is the crosswind spread of the deposited particles and in the analyses by Walker and Stewart and Csanady this is examined in relation to measurements of wind direction fluctuation near the emitter.

Stewart and Csanady give graphs of the standard deviation of the crosswind distributions, which on the whole conform to the simple rectilinear spread relation in Eq. (4.2). This applies to the two sizes of particles used (mean diameters 99 and 217 μ, mean free-fall velocities 0.5 and 1.4 m/sec) and to ranges of about 100 m. The authors point out that this result implies absence of the reduction in turbulent spread which might have been expected as a result of the motion of the particles relative to the air. It is however worth noting an estimate of the magnitude of this effect which may be made as follows from the approximate theoretical treatment in 3.8. As argued there the effective Lagrangian/Eulerian time-scale ratio β' is $(1/\beta^2 + v_s^2/\bar{u}^2)^{-1/2}$. In accordance with Eq. (4.3) the growth of particle spread σ_p with time T is given by

$$\sigma_p{}^2 = [\sigma_v{}^2]_{T/\beta'} T^2 \tag{5.9}$$

in which the connotation is that the lateral velocity fluctuations have been first averaged over T/β' before evaluating the variance. Adopting the empirical form $\sigma_p \propto T^\alpha$, and noting that changing T at constant β' is formally equivalent to changing β' in the same ratio at constant T, it follows that for different particles sizes

$$\sigma_p/T \propto \beta'^{(1-\alpha)} \tag{5.10}$$

Adopting the above form of β', and using subscripts 1 and 2 to denote the different particles, it immediately follows that

$$\left(\frac{\sigma_1}{\sigma_2}\right)_{\text{given } T} < \left(\frac{v_{s_2}}{v_{s_1}}\right)^{1-\alpha}, \quad v_{s_2} > v_{s_1} \tag{5.11}$$

In Stewart and Csanady's work the ratio of the free-fall velocities is 2.7 which means that for α 0.8, 0.9 or 1.0 the respective ratios of particle spread would be expected to be <1.22, <1.10 and 1.0. As the observations support a value of α near unity it is not surprising that σ_1 and σ_2 are indistinguishable against the magnitude of scatter in the dispersion plots.

Walker examined his results (for mass-mean terminal velocities) ranging from 0.14–0.58 m/sec in terms of Eq. (4.3). A constant value of 4 for β gave good agreement on average, but as can be seen from Fig. 5.4 individual values

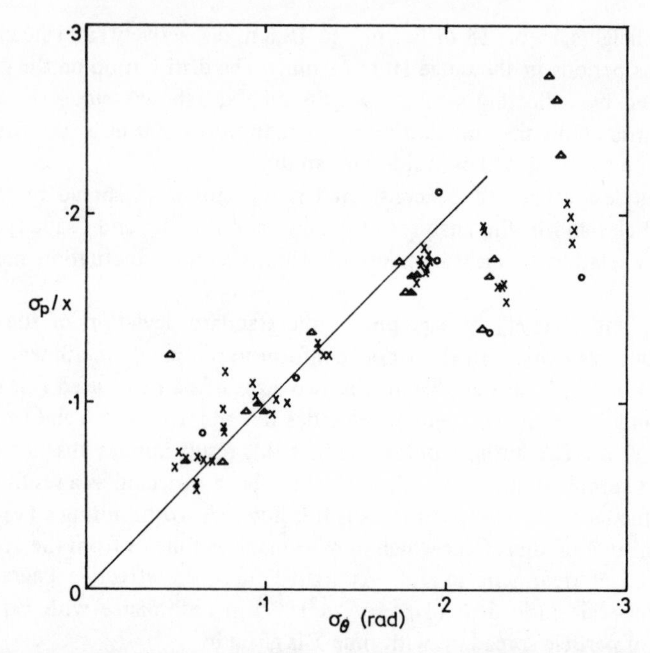

Fig. 5.4 – Data on crosswind spread σ_p of glass microspheres, released at a steady rate from a height of 7.4 or 15 m, in relation to wind direction fluctuation σ_θ. Mass-mean terminal velocities 0.14–0.58 cm/sec. Distance x downwind (m) ○ 27 and 46, × 73–146, △ 201–805. σ_θ refers to an averaging-time $x/\bar{u}\beta$ with $\beta = 4$, and a sampling time equal to the duration of release of particles (from Walker, 1965).

show appreciable disagreement and a tendency for Eq. (4.3) to underestimate at low values and overestimate at high values. Also, as in the study of non-depositing tracers at short range the comparison is not sensitive to the value adopted for β.

The *density* of deposition of particulate material in the crosswind spread pattern is otherwise determined by the extent of spreading in the general direction of the wind. In a non-turbulent atmosphere application of the simplest ballistic principle, with particle inertia neglected, places the ground impact of particles at a distance $U_H H/v_s$ downwind, where U_H here represents the mean speed between the ground and the height of release H. Observations immediately show that the ground deposit pattern is at roughly this distance but with an alongwind spread which may be several times H. Also it is important to note that Hage's earlier study confirmed that in turbulent (near-neutral) conditions this alongwind spread is much greater than could be explained by the variations in particle size in the nominally monodisperse sample of particles used.

The theoretical derivation of the distribution of deposited material, using the gradient-transfer method, has been outlined in 3.8. Alternative approaches which are formally simpler, and probably no less realistic physically, are represented schematically in Fig. 5.5. At (a) is the conventional 'tilted plume' method, in which the distribution of concentration is assumed identical with that for non-sedimenting material, but the effective height H of the plume is replaced by $H - v_s x/\bar{u}$. Adopting the Gaussian distribution discussed in more detail in 6.3,

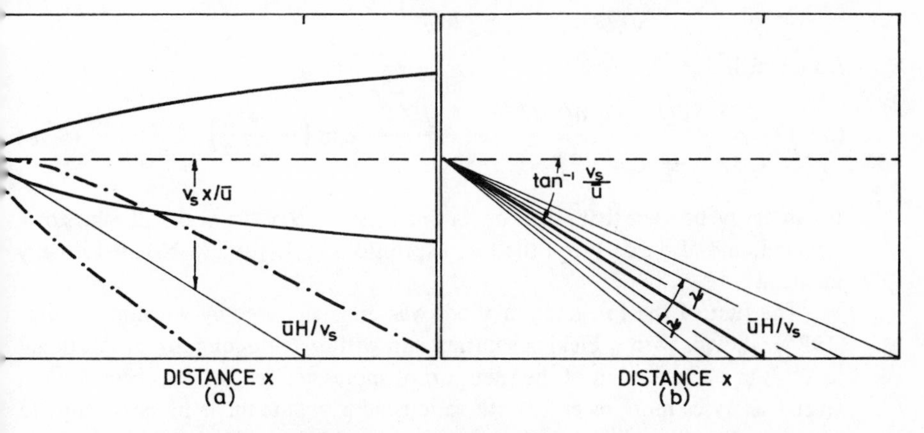

Fig. 5.5 – Schematic representation of treatments of deposition of heavy particles. (a) Tilted plume method, (b) ballistic method.

the expression for the concentration $\chi(x, y, 0)$ in the air at ground level is otherwise given by Eq. (6.17) with the image term omitted, since material reaching the ground is supposed to be retained there, and the rate of deposition $D(x, y)$ is $v_s\chi(x, y, 0)$. Thus

$$D(x,y) = v_s\chi(x,y,0) = \frac{v_s Q}{2\pi\bar{u}\sigma_y\sigma_z} \exp\left(-\frac{y^2}{2{\sigma_y}^2}\right) \exp\left(-\frac{(H - v_s x/\bar{u})^2}{2{\sigma_z}^2}\right) \qquad (5.12)$$

Since the derivation assumes $\bar{u}$ constant with height the interpretation of $\bar{u}$ in practice is somewhat arbitrary. The same expression applies to the total deposit density if Q is the total amount of material emitted [rather than rate of release as in Eq. (5.12)]. Also the corresponding expression for the crosswind integrated deposit density $\int_{-\infty}^{+\infty} D(x, y)\,dy$ is

$$\text{C.W.I.D.} = \frac{v_s Q}{\sqrt{(2\pi)}\bar{u}\sigma_z} \exp\left(-\frac{(H - v_s x/\bar{u})^2}{2{\sigma_z}^2}\right). \qquad (5.13)$$

At (b) in Fig. 5.5 is represented a method which may be described as a simple *ballistic* model, in which particles are assumed to have rectilinear trajectories with inclinations distributed normally about the value $\tan^{-1}(v_s/\bar{u})$.

Thus if the deviations from this value are ψ and the total mass of particles emitted is Q we may write

$$\psi = \tan^{-1}\frac{H}{x} - \tan^{-1}\frac{v_s}{\bar{u}} \cong \frac{H}{x} - \frac{v_s}{\bar{u}} \tag{5.14}$$

$$\mathrm{d}Q = \frac{Q}{\sqrt{(2\pi)}\sigma_\psi} \exp\left(-\frac{\psi^2}{2\sigma_\psi{}^2}\right) \mathrm{d}\psi \tag{5.15}$$

from which

$$\text{C.W.I.D.} = -\frac{\mathrm{d}Q}{\mathrm{d}x} = -\frac{\mathrm{d}Q}{\mathrm{d}\psi}\frac{\mathrm{d}\psi}{\mathrm{d}x} = \frac{HQ}{x^3\sqrt{(2\pi)}\sigma_\psi} \exp\left(-\frac{\psi^2}{2\sigma_\psi{}^2}\right) \tag{5.16}$$

It can easily be seen that when σ_ψ is small (hence $H/x \cong v_s/\bar{u}$ for all values of x affected; and also $x\sigma_\psi = \sigma_z$) the two expressions (5.13) and (5.16) are formally identical.

The first of the foregoing methods was originally used by Wilhelm Schmidt (1925) (though with a Fickian solution and with σ_z consequently proportional to $x^{1/2}$) in his treatment of the transport of spores and seeds. It has been used in several analyses more recently with various adjustments made in an attempt to allow more realistically, though without precise justification, for the progressive depletion of the plume as deposition proceeds. The latest attempt in this direction (by Csanady) is summarized in Stewart and Csanady's paper, and in this case the amendments to Eq. (5.13) amount to multiplication of the right-hand side by a factor $(1+a)/B$, and the replacement of the $2\sigma_z{}^2$ in the denominator of the exponential term by $2B^2\sigma_z{}^2$, where

$$a = \frac{1}{2}\left(\frac{u}{v_s}\frac{H}{x} - 1\right), \quad B^2 = 1 + \frac{v_s{}^2\sigma_x{}^2}{\bar{u}^2\sigma_z{}^2} \tag{5.17}$$

Hage (1961b) and Walker (1965) give tabular summaries of the observed crosswind integrated deposits (C.W.I.D.) at the various distances, with theoretical values calculated in the former case from Rounds's and Godson's gradient-transfer solutions [Eqs. (3.112)–(3.115)] and in the latter case from the simple ballistic model [Eq. (5.16)]. Stewart and Csanady (1967) give graphs of the C.W.I.D. as a function of distance as observed and as calculated using the 'tilted plume' solution in Eq. (5.13) with the adjustments referred to above [Eq. (5.17)]. To provide a common basis for an overall assessment the data in Hage's and Walker's tables have also been plotted. From all the graphs estimates have been made of the peak value of the crosswind integrated deposition, the distance x_m of this peak from the point of emission, and the alongwind spread of the cloud as prescribed by one-tenth of the peak C.W.I.D. The ratios of the calculated and observed values of these parameters are summarized in Table 5.IV with other experimental details.

Table 5.IV – Experimental tests of solutions for the deposition of large particles (glass microspheres) from an elevated source.

Series	Hage, 1961b	Walker, 1965			Stewart & Csanady 1967
No. of comparisons	9	12			7
Height of release (m)	15	15		7.4	18.6
Particle size (micrometres)	107 (S.D. 6.8)	107	49	56	99 and 217†
Range of $\bar{U}_L$ m/sec	4.4–9.1	2.4–8.2			2.5–6.4
Range of i_w	0.01–0.17	0.03–0.12			0.04–0.06
Theoretical solution	Eqs. (3.112)–(3.115)	Eq. (5.16)‡			Eq. (5.13)§
Mean and (range) of calc/obs values for					
(a) peak crosswind integrated deposit	0.6 (0.5–0.9)	1.3 (0.8–2.0)			0.8 (0.5–1.2)
(b) distance of peak C.W.I.D. from source	0.9 (0.5–1.0)	0.8 (0.5–1.0)			0.9 (0.8–1.2)
(c) alongwind spread¶	1.6 (1.1–2.2)	0.8 (0.3–1.1)			0.9 (0.8–1.1)

† Simultaneously released.
‡ Taking $\sigma_\psi = [\sigma_\phi]_{T/\beta}$, i.e. corresponding to Eq. (4.3) with $\beta = 4$.
§ With Csanady's modification as indicated above Eq. (5.17), using Eq. (4.2) for σ_z and σ_y, and assuming $\sigma_x = \sigma_y$.
¶ Defined by C.W.I.D. one-tenth of peak value.

From Table 5.IV it appears that the simplest of the three treatments (i.e. that used by Walker) is not significantly inferior to the other two. This result, together with that already discussed concerning the crosswind spread, suggest that a useful estimate of dust deposition from an elevated source may be made in a simple way from measurements or estimates of wind fluctuation.

Retention of small particles and gases at the ground

Turning now to the case of particles with settling velocities *small* compared with typical eddy velocities, the first fundamentally important step was the recognition that the rate of deposition might be greater than that provided by settling. In the context of atmospheric transport this seems first to have been considered by Gregory (1945) in an examination of data on the travel of spores. Gregory used observations by Stepanov of the deposition of spores artificially released from a point source over a lawn and collected on glass slides, coated with glycerine jelly, distributed on the ground around the source at distances up to 40 m. It was found that the observed variation of the deposit with distance was consistent with a modified form of Sutton's formula for the distribution downwind of a point source at ground level. Gregory's modification amounts to assuming that the shape of the vertical distribution is unaltered, but that material is deposited at a rate proportional to the local ground-level concentration. As noted later in this section the effect is equivalent to replacing the original source strength by an effective strength decaying exponentially with distance of travel downwind.

In his analysis Gregory introduces a *deposition coefficient p* which corresponds to $D/\bar{u}\chi$ in the present notation. The ratio D/χ, which has the dimensions of velocity, and can be regarded as the rate of settling of the cloud equivalent to the actual rate of deposition, was later termed *deposition velocity* by Chamberlain (1953). Gregory's estimate of the value of p was 0.05, for wind speeds which were not clearly specified, but which were unlikely to be less than 1 m/sec, implying that the deposition velocity v_d was about 5 cm/sec, whereas for the spores used the settling velocity v_s was known to be about 1 cm/sec.

In view of the uncertainty of Gregory's estimate, depending as it did on a crude theoretical estimate of the concentration of spores in the air, the implication that in turbulent air light particles may be deposited more rapidly than in accordance with their settling velocity was not decisive. Moreover, subsequent attempts to establish the result by measuring both deposition and concentration of *Lycopodium* spores in the field (Chamberlain, 1953) at first proved unsuccessful. However, the matter seems now to have been settled beyond any reasonable doubt in a later and much more searching study by Chamberlain (1966b) of the transport of small particles to rough surfaces. An essential part of this study was the use of radioactive 'tagging' to facilitate the measurement of the amount of material deposited on a surface.

Several important features of Chamberlain's results are brought out in Figs. 5.6 and 5.7. Fig. 5.6, reproduced from Chamberlain's paper, shows how the deposition velocity of *Lycopodium* spores varies with the friction velocity of a real grass surface, both in the field and in a wind tunnel. For small u_* the deposition velocity is evidently close to the settling valocity, i.e. about 2 cm/sec, but at large u_* there is a significant addition to the settling velocity. The importance of particle size and settling velocity is brought out in Fig. 5.7. Here it is seen that the excess of v_d over v_s becomes more marked as particle size and settling velocity are reduced, and indeed there is ultimately a tendency for v_d to reach a lower limit and even increase with further decrease of v_s.

These deposition measurements using small particles and also others using vapour (which will be discussed shortly) indicate the importance of the retaining quality of the surface and of the precise mechanism by which the material is finally conveyed to the surface. Recalling the definition of deposition velocity – the ratio of the net *vertical flux* to the surface and the *concentration* of the property at a specified height we may, following Chamberlain, define another

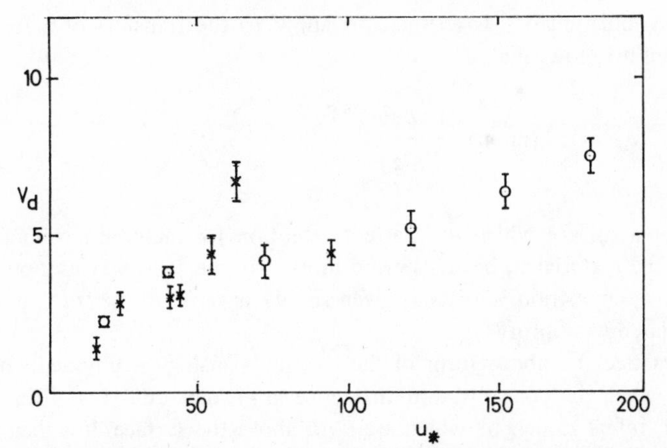

Fig. 5.6 – Deposition of spores to grass (from Chamberlain, 1966b, with permission of the Royal Society). ○ Wind tunnel results corrected to 60 cm reference height; × Field experiments.

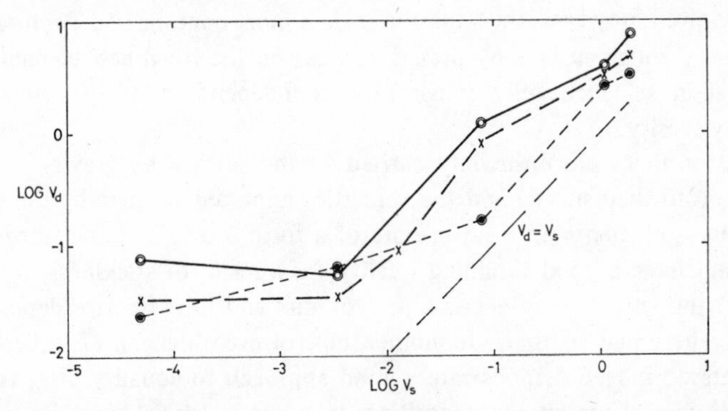

Fig. 5.7 – Deposition velocity v_d of particles as a function of settling velocity v_s, for a natural grass surface with friction velocity u_* (cm/sec) as follows: ○ 140, × 70, ● 35. (Chamberlain, 1966b).

quantity, the *transport velocity*, using instead the difference in concentration between the surface and a specified height, i.e.

$$v\,(\text{trans}) = \frac{D}{\chi(z) - \chi(0)} \tag{5.18}$$

If the surface acts as a perfect sink, a condition represented by setting $\chi(0)$ to zero, the deposition and transport velocities are of course identical. For momentum the 'deposition' or vertical flux at the surface is $\tau(0)$ (the surface shearing stress), the 'concentration' of momentum is $\rho\bar{u}(z)$, and accordingly

$$v_d\,(\text{mom}) = \frac{\tau_0}{\rho[\bar{u}(z) - \bar{u}(0)]} = \frac{\tau_0}{\rho\bar{u}(z)} = \frac{u_*^{\,2}}{\bar{u}(z)} \tag{5.19}$$

If the Reynolds analogy is assumed to apply to the transfers of material and momentum it follows that

$$v_d = v_d\,(\text{mom}) = \frac{u_*^2}{\bar{u}(z)} \tag{5.20}$$

Hence, for a surface which has perfect retention for material reaching it, and assuming the material to be transferred in exactly the same way as momentum, the effective deposition velocity is given simply in terms of the friction velocity and associated wind profile.

In practice the above form of the Reynolds analogy is physically satisfactory only when the concentration-difference in the denominators of Eqs. (5.18) and (5.19) refers to heights which are *both* above the surface. It is then reasonable to adopt the notion that the ratio of the rate of transfer to the vertical gradient is independent of the nature of the property and is determined entirely by the turbulent mixing action of the air. When the ultimate transfer to the surface is involved, however, the situation is then more complicated. Momentum is *absorbed* at a rough surface by pressure forces on the roughness elements, at a rate which in so-called fully rough flow is independent of the *molecular* property of viscosity.

Material particles are ultimately carried to the surface by gravity and by impaction due to their inertia and the velocities imparted to them by the wind. If the roughness elements on a surface are of a form providing good impaction efficiency and have a good retaining quality (as a result of stickiness, wetness or hairiness) the surface may act as a perfect sink and the effective deposition velocity may be equal to that for momentum (or even higher). Chamberlain's measurements do indeed demonstrate a good approach to equality of v_d and v_d (mom) for *Lycopodium* spores depositing on sticky artificial grass in a wind tunnel.

For gaseous material however there is no *absorbing* process analogous to dynamic pressure on a bluff body or to impaction by inertia forces, and the retention at the surface is entirely dependent on molecular penetration or cohesion. Thus, for a rough surface there is accordingly less resistance to the deposition of momentum, than to the ultimate deposition of gaseous material (or of sensible heat, which is similarly controlled by molecular conductivity). This extra resistance arising from molecular interaction at a surface is now conventionally expressed in terms of the dimensionless reciprocal Stanton number B^{-1} introduced by Owen and Thompson (1963), which may be defined by the relation

$$v_d = \frac{u_*^2}{\bar{u}(z)}\left(1 + \frac{u_*}{\bar{u}(z)}B^{-1}\right)^{-1} \tag{5.21}$$

or alternatively

$$\frac{u_*}{v_d} - \frac{u_*}{v_d\,(\text{mom})} = B^{-1} \tag{5.22}$$

The alternative form is the more expressive in that the reciprocals of deposition velocity may be regarded as equivalent to resistances (if gradients instead of finite differences of χ and u were used in the definition of v_d, then v_d would be effectively a diffusivity).

The measurements made by Chamberlain (1966a), of 'perfect sink' vapour deposition on grass and other surfaces in a wind tunnel used thorium-B vapour. From measurements of the concentration of thorium-B in the air close to the surface, and estimates of u_* derived from velocity profile measurements and assuming the neutral logarithmic wind profile law, the terms on the left-hand side of Eq. (5.22) and their difference B^{-1} were obtained. From regression analyses of the values of B^{-1} at various friction velocities the values appropriate to u_* 25, 50 and 100 cm/sec were obtained and these are summarized in Table 5.V. This shows B^{-1} increasing with u_* for short grass (both real and artificial) and rough glass, but essentially independent of u_* over the range studied for long grass and towelling. Also for the surfaces with fibrous roughness elements (i.e. excluding the rough glass) there does not appear to be any significant difference at given u_*, despite the 20-fold variation in roughness length z_0. On the other hand B^{-1} is evidently much greater for the less fibrous form of roughness element represented by rough glass and the value obtained in this case confirmed results obtained earlier by Owen and Thompson using camphor vapour.

Table 5.V – Laboratory data on the resistance to the transport of thorium-B to various surfaces (from Chamberlain, 1966a). The figures are mean values and standard errors of the quantity B^{-1} defined in Eq. (5.22).

Surface	z_0 (cm)	B^{-1} for u_* (cm/sec)			S.e. of
		25	50	100	B^{-1}
Grass (*ca.* 7 cm high)	0.7	–	7.7	–	0.4
Short grass	0.2	7.9	8.7	10.3	0.6
Artificial grass	1.0	7.0	8.0	10.1	0.3
Towelling	0.045	–	8.0	–	0.6
Rough glass	0.02	28.5	30.7	35.3	1.4

Extension of these investigations to field conditions is not straightforward and for the moment only rather scanty data is available on the values of B^{-1}.

Chamberlain and Chadwick (1965) have reported experiments using radioactive iodine vapour over the grass surface of an airfield. For measurements at distances of 50 and 100 m from the source they obtain a mean value of 15, which however they regard as an overestimate because of evidence to the effect that the 'perfect sink' condition, $\chi(0) = 0$, was not fully satisfied.

When deposition is controlled by downward turbulent transport it follows that the concentration in the air must ultimately decrease as the ground is approached. In principle this offers a possibility of indirectly estimating deposition from measurements of concentration at two or more levels in the air, now assuming analogy with the transfer of momentum without any complication from surface interaction. At a height sufficiently small in relation to the distance from the source the rate of deposition $v_d\chi(z)$ may be equated to the flux $K(\partial\chi/\partial z)$, and with a K (2m) of say 5×10^3 cm^2/sec (typical of long grass) and a deposition velocity of say 5 cm/sec the quantity $(\partial\chi/\partial z)/\chi$ would be 10^{-3} cm^{-1}. Thus even with such a substantial rate of deposition the increase in concentration between 150 and 250 cm would only be about 10 per cent. This means that the measurements of the vertical profile of concentration need to be made with the same sort of precision as is required in wind profile measurements. It is noteworthy that in Chamberlain's experiments with *Lycopodium* spores the appropriate increase of concentration with height was measured in the *wind tunnel* (at heights 5 and 10 cm) but not in the field, where on the whole the concentration *decreased* slightly over the heights 30, 60 and 100 cm. The reasons for this latter discrepancy are not clear and the result serves as a warning of the difficulties likely to be encountered in applying the profile methods in the field. Nevertheless the technique has apparently been used with some success in terms of carbon dioxide measurements with a sensitive infra-red gas analyser (Monteith and Szeicz (1960), and its use in terms of sulphur dioxide has been reported by Garland *et al.* (1973). These measurements provided profiles of sulphur dioxide over grass, for the height range 0.2 to 2 m, clearly showing an increase of concentration with height (the fractional change in concentration being roughly $\frac{1}{10}$ over the whole height range). From wind and temperature profiles observed at the same time estimates were obtained of the eddy viscosity K according to Eq. (2.47) and used with gradients of sulphur dioxide concentration to give the downward flux. The corresponding deposition velocities (referring to $z = 1$ m) lie in the range 0.1–2 cm/sec. In relation to Eq. (5.22), Garland *et al.*'s data imply total resistance $(1/v_d)$ generally greater than the sum of the 'aerodynamic resistance' $[1/v_d$ (mom)] and the 'surface resistance' for a perfect sink $(1/u_*B)$.

Washout in rain

The collection of suspended particles by falling drops is an aspect of the problem of impaction which has received detailed attention, both with regard to the physics of aerosols [see Green and Lane (1957)] and the coalescence and growth of raindrops [see Mason (1957)]. In general only a fraction of the particles in the path of a drop will collide with it and the remainder will be swept clear in the streamlines around the drop. A theory of the process has been developed mainly by Langmuir, and this specifies the fraction collected as a function of the terminal velocities of the raindrop and particle. The fraction is alternatively referred to as *collection* efficiency or *collision* efficiency; the latter term is preferred since collision does not necessarily ensure retention of the particle.

Disregarding for the moment the distinction between collection and collision the rate of removal of particles from the air by raindrops may be expressed in an approximate way as follows. Considering drops of radius between s and $s + ds$, let the number per unit volume of air be N_s and the terminal velocity V_s. Since the number of drops which fall through a cube of unit volume in 1 sec is $N_s V_s$, then the amount of the unit volume which is actually swept is $\pi s^2 N_s V_s$, and if the collection efficiency for particles of radius r is $E_{(r,s)}$, the fraction of particles removed in 1 sec by the whole spectrum of raindrops will be

$$\Lambda = \int_{s=0}^{s=\max} \pi s^2 N_s V_s E_{(r,s)} \, ds \tag{5.23}$$

Using Langmuir's theoretical values of $E_{(r,s)}$ and Best's table of the size distribution of raindrops for various rates of rainfall, Chamberlain has evaluated Eq. (5.23) to give the *washout coefficient* Λ as a function of rate of rainfall and terminal velocity of particle v_s, and has constructed the series of curves in Fig. 5.8.

A small number of field measurements of the washout by natural rain of *Lycopodium* spores released from an artificial source have been described by May (1958) and the results used to determine the effective Λ for comparison with the theoretical values calculated by Chamberlain. Spores marked with a radioactive tracer were released from a height of 3.3 m, and the rain was collected in small basins set out around the source on a circle of 10 m radius. After several experiments had been carried out it was realized that the assumption of negligible deposition by processes other than washout was unjustified and thereafter *dry* deposition was also measured in basins shielded from rain. The results of the subsequent five experiments are summarized in Table 5.VI, with the theoretical values of Λ.

May discusses the errors which might have arisen from the electrostatic collection of the spores by raindrops, and also from the fact that the smallest raindrops affecting the plume of *Lycopodium* spores probably did not reach the ground within the sampling distance. Estimates of the possible total error from these two sources range from 0 to + 20 per cent. Furthermore, Langmuir's values of collection efficiency, as used in calculating Λ, neglect the finite size of the particles. A correction for this on lines suggested by Mason (1957, Appendix A) increases the theoretical value of Λ by an amount which depends on the rate of rainfall, but which is always less than 10 per cent for the cases in Table 5.VI.

Table 5.VI – Measurements of washout coefficient Λ for *Lycopodium* spores. (After May, 1958).

Date	$\bar{u}$ (cm/sec)	Rate of rainfall (mm/hr)	Type of rain	Λ 10^{-4}/sec	
				Observed	Theoretical
16.8.56	320	3.91	frontal	10.2	9.7
27.9.56	543	1.12	frontal	4.2	3.6
25.10.56	845	14.1	heavy frontal	30.8	26.8
11.12.56	334	1.01	frontal	3.2	3.2
31.12.56	332	3.64	continuous rain of showery type	8.9	9.2

From the data in Table 5.VI the *washout coefficient* appears to have been correctly calculated but in general the position is not entirely clear. According to Engelmann (see *Meteorology and Atomic Energy*, 2nd edn., 1968) washout coefficients calculated using *measured* collection efficiencies are lower than those calculated using Mason's theoretical efficiencies, by a factor ranging from about 4/3 in very light rain to about 2 in rain at a rate of 4 mm/hr. Evidently then a precise confirmation of the theoretical analysis has not yet been achieved, which is perhaps not surprising in view of the difficulty of the measurements required. Nevertheless, it has been demonstrated that at least useful rough estimates of deposition in rain may now be made, and the relatively powerful action of rain in relation to dry deposition of particles is immediately obvious from the following simple example. If particulate material is distributed uniformly at concentration χ over a height h' then rate of deposition by rain which has fallen through the whole depth will be $\Lambda h'\chi$, and the equivalent deposition velocity as previously defined will be $\Lambda h'$. For particles of terminal velocity 0.1 cm/sec, and rainfall at the modest rate of say 1 mm/hr, Fig. 5.8 gives $\Lambda = 10^{-4}$ and with h' 100 m this would correspond to a deposition velocity of 10 cm/sec.

It should however be borne in mind that the foregoing theoretical estimation of washout may be an overestimate, since collision may not be followed by retention of the particle on the raindrop. This situation may arise if there is sufficiently high interfacial tension between water and the particulate material, that is, if the particle is not easily wettable. A discussion of this feature, including the results of some field experiments with wettable and non-wettable fluorescent powders, has been reported by McCully *et al.* (1956). Particles were

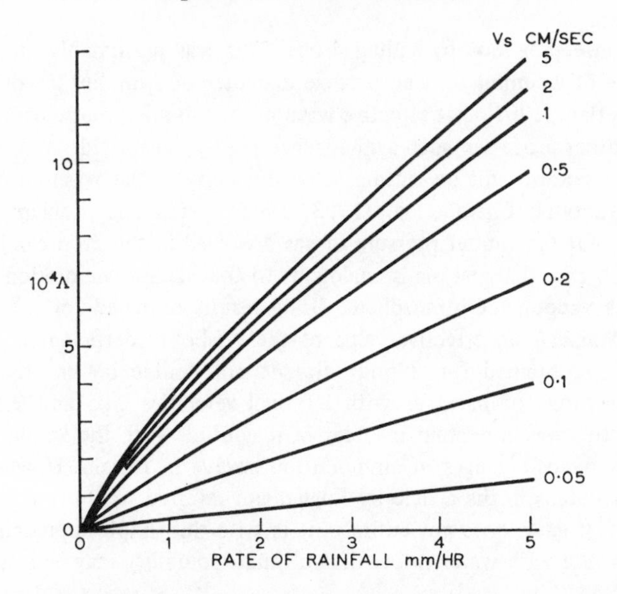

Fig. 5.8 – Washout coefficient, Eq. (5.23), as a function of rate of rainfall, for particles of terminal velocity v_s as indicated (Chamberlain, 1953). From AERE HP/R 1261, United Kingdom Atomic Energy Authority, Copyright reserved.)

released in rain, from an aircraft flying acrosswind at a height of 500 ft, and the rain was collected on a parallel line 4000 ft downwind. The results indicated collection efficiencies of the non-wettable dust well below the theoretical collision efficiencies. A theoretical examination by Pemberton (1961) suggests that for particles of median diameter 2 μ, and a rainfall intensity of 2.5 mm/hr, the fraction of non-wettable material removed in a few hours may be only about 25 per cent, as compared with about 90 per cent of wettable material. This theoretical work has been extended by McDonald (1963) to include partially wettable material (angle of contact $< 180°$). The work required for complete penetration (which would have to be provided by the particle kinetic energy relative to the raindrop) is derived as a function of surface tension, angle of contact and depth of penetration. Examples are worked out for angles of contact 90–180°. Further work is apparently necessary for angles 0–90°, though McDonald argues that collection efficiency is unlikely to vary appreciably over this range.

The above considerations have referred only to the collection of particles by raindrops *falling through* the cloud of material. An additional mechanism, discussed by Greenfield (1957) in relation to the washout of atomic debris, is the capture of very small particles by *cloud droplets* and the subsequent deposition of these droplets as rain. Greenfield's calculations for assumed typical conditions indicate that the bulk of deposition was nevertheless provided by the process of sweeping out by falling drops. This was presumably an inevitable consequence of assuming a mean particle diameter of $2\mu m$, but it would appear (see Chamberlain, 1961) that effective washout of sub-micrometre particles must depend on other processes, such as that suggested by Greenfield.

Another feature still presenting some difficulty is the washout of soluble gases and vapours; Chamberlain (1953, 1961) treats this problem with the assumption that the vapour pressure of gas dissolved in the drop can be neglected, in which case the analysis is analogous to that of the evaporation of falling drops into a vapour-free atmosphere. Using results obtained for the latter by Ranz and Marshall an effective value of the washout coefficient is obtained. The values so obtained for sulphur dioxide and iodine lay mostly between those for washout of *particles* with terminal velocities 0.05 and 0.1 cm/sec. Apparently this was consistent in order of magnitude with the amounts of SO_2 collected in deposit gauges in air pollution suveys in Britain. However, from more recent details in the article by Englemann referred to above it seems that the washout of gases does not entirely fit in with this simple approach and that the reaction rate with water rather than ultimate solubility may be a controlling factor.

Estimation of the effect of deposition on airborne concentration

Although further improvement may yet be sought in the details of effective deposition velocities it is clear that for practical purposes useful progress has been made, and the question now to be considered is how such information may be utilized in estimating the effect on the concentration of material remaining airborne. The application is greatly simplified by adopting the assumption that the vertical distribution of the material is independent of lateral transport processes (which is reasonable for sources of large crosswind extent and also for the plume from a point source until shear effects become important). With this qualification the analysis may then proceed entirely in terms of the two-dimensional case, i.e. the concentration from a crosswind line source of infinite extent (or the equivalent crosswind-integral of the concentration from a point source). For this case the continuity condition is

$$\int_0^{x_1} \chi(x, z_1) v_d \, dx = Q - \int_0^{\infty} \chi(x_1, z) \bar{u} \, dz \tag{5.24}$$

in which Q is the original source strength and strictly $\bar{u}$ should be regarded as a function of height also. The last term on the right-hand side may be regarded as an effective (reduced) source strength $Q(x)$, and on differentiating

$$\chi(x, z_1) v_d = -\frac{dQ(x)}{dx} \tag{5.25}$$

Also we may write

$$\chi(x, z_1) = Q(x)f(x) \tag{5.26}$$

where

$$\frac{1}{f(x)} = \int_0^\infty \frac{\bar{u}(z)\chi(x, z)}{\chi(x, z_1)} \, dz$$

and is thus determined by the shapes of the vertical profiles of χ and u. Accordingly

$$\frac{dQ(x)}{Q(x)} = -\,v_d f(x)\,dx \tag{5.27}$$

and

$$Q(x_1) = Q(0) \exp\left(-\int_0^{x_1} v_d f(x)\,dx\right) \tag{5.28}$$

Eq. (5.28) may be integrated by making various assumptions.

In the simplest idealized case, when mixing to the stage of uniformity throughout depth h' is assumed and the wind is also taken to be uniform with height, it follows from the definition that $f(x) = 1/\bar{u}h'$ and then (5.28) gives

$$Q(x_1) = Q(0) \exp\left(-\frac{v_d x_1}{\bar{u}h'}\right) \tag{5.29}$$

with $\chi(x_1) = Q(x_1)/\bar{u}h'$.

If alternatively the horizontal flux $\bar{u}(z)\chi(z)$ is assumed invariant with height up to $z = h'$ and thereafter zero, it follows that $f(x) = 1/\bar{u}(z_1)h'$, which leads to Eq. (5.29) with $\bar{u}$ replaced by $\bar{u}(z_1)$ and the need to assume wind constant with height has been eliminated. There is, however, no *a priori* justification for the assumption concerning the horizontal flux and this must be justified empirically.

Another working assumption, referred to earlier in this section, is that the vertical distribution of concentration is unchanged by the deposition process. Then, for example, using the simple Gaussian forms for the distribution from sources, the required expression for $f(x)$ follows from the crosswind-integrated form of Eq. (6.17), i.e

$$f(x) = \frac{2^{1/2}}{\pi^{1/2}\bar{u}\sigma_z} \exp\left(-\frac{H^2}{2\sigma_z^2}\right) \tag{5.30}$$

Taking a simple power-law growth of σ_z with distance, i.e. $\sigma_z \propto x^q$ Eq. (5.28) may be integrated to give:

Ground level source ($H=0$)

$$Q(x) = Q(0)\exp\left[-\frac{2^{1/2}v_d x}{\pi^{1/2}(1-q)\bar{u}\sigma_z}\right] \tag{5.31a}$$

Elevated source

$$\ln\frac{Q(x)}{Q(0)} = \frac{b}{m}\left(-\frac{e^{-\xi}}{\xi^m} + \Gamma_\infty(-m+1) - \Gamma_\xi(-m+1)\right) \tag{5.31b}$$

where

$$m = \frac{1-q}{2q}, \quad b = \frac{1}{2^{1/2q}\pi^{1/2}}\left(\frac{H}{\sigma_z}\right)^{(1-q)/q}\frac{v_d x}{\bar{u}\sigma_z}, \quad \xi = \frac{H^2}{2\sigma_z^{\ 2}}$$

The equation may alternatively be integrated numerically for arbitrary forms of σ_z and this procedure has been followed (see *Meteorology and Atomic Energy*, 1968, p. 204) using the semi-empirical forms suggested by Pasquill.

The assumption that the form of the vertical distribution is unaffected by the deposition process is an approximation which may be expected to be acceptable when v_d is small compared with $d\sigma_z/dt$. Thus, in neutral conditions in the lower atmosphere, when for a ground-level source $d\sigma_z/dt \cong u_*/2$, it should be permissible to use the foregoing formulae for deposition velocities less than say $u_*/20$. From Fig. 5.7 this would include particles with settling velocities about 1 cm/sec or less. On the other hand, recalling the definition of B^{-1} in Eq. (5.22), and noting that at low heights over grassland in neutral conditions $u(z)/u_*$ is typically 10, the foregoing condition is equivalent to $B^{-1}>10$. In this case it appears from Table 5.V that the method becomes questionable for 'perfect sink' vapour absorption over grass.

As a check on the validity of Eqs. (5.26), (5.30) and (5.31a) the resulting ground-level concentrations may be compared with the exact analytical solution given by Smith (1962) under the following conditions: $\bar{u}$ is constant and equal to 5 ms^{-1}, K is constant and equal to 10 m^2 s^{-1} and the height of the source H is zero. Smith's solution is for zero z:

$$\chi(x,0) = \frac{Q}{\sqrt{(\pi uKx)}} - \frac{v_d Q}{4Kx}\exp\left(\frac{v_d^{\,2}x}{Ku}\right) erfc\left(v_d\sqrt{\frac{x}{Ku}}\right)$$

A comparison of the solutions yields:

x (m)	10	100	1000	10000	100000
χ/Q (Smith)	2.51×10^{-2}	7.81×10^{-3}	2.51×10^{-3}	5.7×10^{-4}	2.66×10^{-5}
Ratio of the 2 solns*.	1.004	1.005	1.015	1.227	6.333

*former solution to Smith's solution for $v_d = 0.008$ ms^{-1}.

The ratios are very close to unity except at very long range, supporting the reasonableness of the above approach in more general conditions for which no analytical solutions exist.

Various modification of the foregoing simple approach have been proposed in attempts to achieve greater realism. Overcamp (1976) extending earlier work by Csanady, adopts the same device of depleting the plume over its whole depth, but only in respect of that part associated with the image source. Horst (1977, 1978) modifies the equations in a more rigorous fashion by introducing a depletion-plume, so representing the effect of extracting the material over a deep layer to represent deposition far upwind of the receptor and over shallower layers for the deposition at closer positions. Whereas the equations above may be written in the form

$$\chi(x, z) = f(x, z, H) Q(0) - f(x, z, H) \int_0^x v_d \chi(x, z_1)\, \mathrm{d}x$$

Horst's *surface-depletion* form of equation is

$$\chi(x, z) = f(x, z, H) Q(0) - \int_0^x f(x-x', z, 0)\, v_d \chi(x', z_1)\, \mathrm{d}x' \qquad (5.31c)$$

in which the deposition is effectively represented as occurring at all positions between the receptor and the source, instead of implicitly all at the source. Unfortunately the integral is more troublesome than in the preceding (*source-depletion*) equation, and although it may be evaluated numerically, Horst suggests a modified form of the source-depletion equation in which the diffusion function f is multiplied by a second function $P(x, z)$ to allow for the change in vertical distribution caused by the deposition. An expression for P is derived through certain approximations, firstly that the deposition flux is from a vertical distribution otherwise quasi-constant with height (so that $f(x) = 1/\bar{u}h'$), secondly that the flux is in accordance with gradient transfer, and thirdly that the corresponding $K(z)$ is relatable to the $\sigma_z(x)$ of the Gaussian plume equation (which is strictly reasonable only for a surface release). The result is

$$1/P(x, z_1) = 1 + (v_d/h') \int_0^{h'} R(z, z_1)\, \mathrm{d}z\ ,$$

$$\bar{u} R(z, z_1) = (2/\pi)^{\frac{1}{2}} \int_{x(z_1)}^{x(z)} (1/\sigma_z(x'))\, \mathrm{d}x \qquad (5.31d)$$

In his review paper (1979a) Horst gives examples of calculations showing that the closest approach to his surface-depletion equation is provided by his modified source-depletion equation (5.31d), emphasizing however that no test against observation has yet been possible.

The effects of washout on deposition and diminution of airborne concentration may also be expressed is a simple way. In this case the assumption that the *shape* of the vertical distribution is unaffected is more reasonable, provided the rain is falling through the whole depth of the suspended material. As before the analysis may be carried out for a crosswind line source of infinite extent, and the result applied to a point source. At any distance x downwind the rate of deposition per unit area, by rain which has fallen through the whole depth of the particulate cloud, will be

$$\Lambda \int_0^\infty \chi(x, z)\, \mathrm{d}z = \frac{\Lambda Q(x)}{\bar{u}} = -\frac{\mathrm{d}Q(x)}{\mathrm{d}x} \tag{5.32}$$

where $Q(x)$ is by definition the reduced source strength effective at distance x, and wind speed is assumed constant with height. Accordingly

$$Q(x) = Q(0) \exp\left(-\frac{\Lambda x}{\bar{u}}\right) \tag{5.33}$$

where $Q(0)$ is the original source strength. Substitution of this $Q(x)$ in the appropriate diffusion formula then gives the airborne concentration as modified by washout. Some examples of calculation of washout using this principle are given in *Meteorology and Atomic Energy* (1968, Ch. 5).

5.3 THE DISTRIBUTION OF GASEOUS EFFLUENT FROM INDUSTRIAL STACKS

The dispersion of the gaseous effluent issuing from the stack of a modern power station or industrial plant is a complex process dominated first by the rise and induced growth of the plume, as a consequence of its upward momentum and buoyancy, and then ultimately by the natural dispersive action of the atmosphere. Before considering the interpretation of data on the concentration of effluent near the ground, which is usually of most concern practically, it will be useful to summarize some of the known features of plume behaviour and the information available on the induced and natural spread.

The dilution of the momentum, heat and pollutant concentration within the plume is of course inversely proportional to the cross-sectional area. Most data on the dimensions of this cross-section have been obtained from visual or photographic observations, necessarily over a somewhat limited range of travel from the stack. Of the early surveys the only one to give a more direct measure of the cross-section was that reported by Stewart, Gale and Crooks (1954, 1958), in which the vertical distribution in the radioactive plume of the Harwell BEPO reactor was surveyed with instruments carried on the cable of a captive balloon. Recently the development of *lidar* has shown considerable promise in the surveying of plume size and position. This is strikingly evident from the example reported by Hamilton (1969) and reproduced here in Fig. 5.9. It is particularly interesting to note that in this example the mean *instantaneous* size of the plume and the area over which the successive positions of its centre are scattered are similar, and that in both respects the crosswind dimension exceeds the vertical. It is also noteworthy that the frequency of incidence of the plume at various distances from the mean position of its centre, which may be

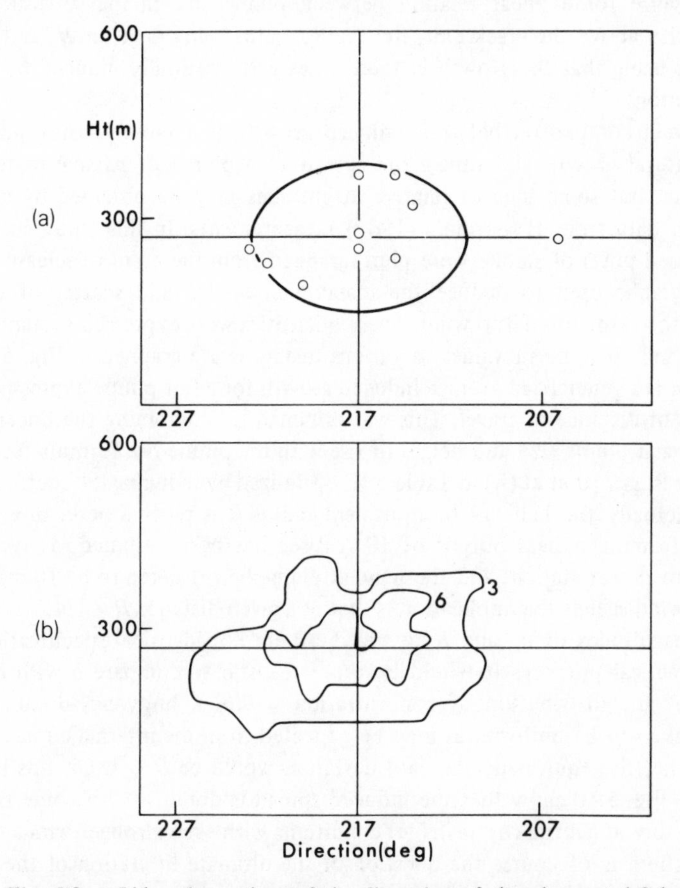

Fig. 5.9 – Lidar observations of the dispersion of the plume at 1.8 km from Northfleet power station (from Hamilton, 1969). The diagrams are a summary of ten lidar scans, 1530–1630 BST, 6 May 1966. (a) Points ○ denote successive positions of the centre of the cross-section of the plume and the contour represents the average instantaneous cross-section. (b) Numbers of occasions that a position lay within the plume cross-section. Outer and inner contours define regions in which plume was observed for more than 3 or more than 6 occasions. The abscissae represent the direction from which the plume was blown but are also on the same scale as the height.

regarded as an approximation to the relative time-mean concentration of pollutant, has a distribution in the horizontal close to Gaussian. In the vertical the overall distribution is skewed, presumably in keeping with more effective dispersion above than below the mean position, but the distribution *below* this mean position is also fairly close to Gaussian.

More comprehensive data on plume size is available from an extensive series of photographic observations of plumes from the Tennessee Valley Authority plants (Carpenter *et al.*, 1968). This gives convincing support in a statistical sense for a linear relation between plume size in the vertical and height of rise above the stack exit, the average ratio being near unity, and the implication being that the growth in these cases was essentially induced by the upward motion.

No direct comparison between induced growth of a rapidly rising plume and the natural growth and time-mean spread of a plume of *passive* material are available, but some idea of relative magnitudes may be obtained by comparing with data from Högström's (1964) measurements. In this study successively released puffs of smoke were photographed from the point of release and the photographs used to deduce the instantaneous size and scatter of puff centres, in the plane normal to wind. These quantities were expressed as standard deviations and their mean values at various distances are graphed in Fig. 5.10. Also shown is a generalized average induced growth for a hot plume expressed as a function of distance of travel. This was obtained by combining the linear relation between plume size and height of rise with the plume-rise formula recommended by Briggs [that at (8) in Table 5.II, optimized by reducing the coefficient to 1.6]. Actually the half-size or equivalent radius R is plotted here. In evaluating the formula a heat output of 10^7 cal/sec has been assumed, as typical of a modern power station, and the wind at plume height taken to be 10 m/sec. For other wind speeds the implication is that at a given distance $R \propto 1/\bar{u}$.

The magnitudes of σ_z and R in Fig. 5.10 are not identical specifications, and for practical purposes it would be more realistic to compare σ with $R/2$. Actually, if the distribution of concentration *within* a buoyancy-dominated plume is taken to be uniform, as may be expected from the internal circulation in the plume, the equivalent standard deviation would be $R/\sqrt{3}$. On this basis the data in Fig. 5.10 show that the induced spread is dominant for some time, except possibly in neutral (or unstable) conditions with even stronger winds than assumed. There is of course the question of the ultimate limitation of the rise (and hence of the induced growth) of the hot plume, and here it may be recalled that the plume-rise studies are inconclusive on this point, though clearly the rate of rise is much reduced at distances of 1 or 2 km. Beyond these distances the natural dispersion of the plume may be expected to overtake the induced growth and ultimately become the dominant process.

There is also the possibility that in practice the time-mean spread in the vertical also includes a contribution from variations in height of rise associated

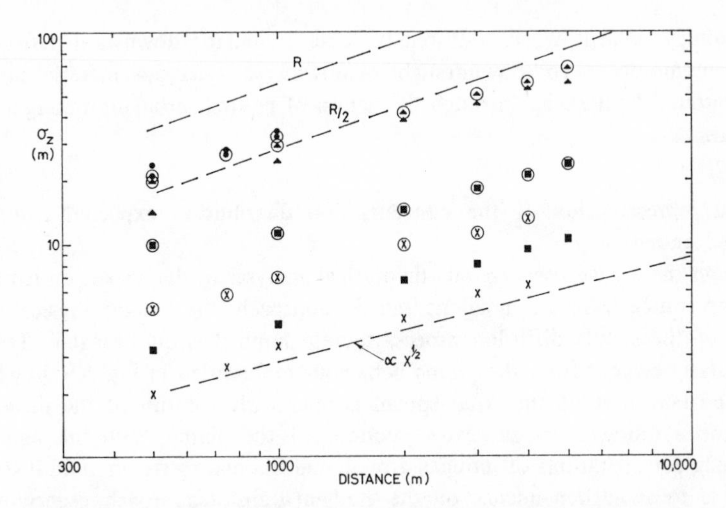

Fig. 5.10 – Vertical growth of passive smoke puffs (points without circles) and vertical scatter of centres (points encircled) expressed as standard deviations (Högström, 1964).

	Stability	*Site*	*Ht of release* (m)
●	neutral	Agesta	50
X	stable		
▲	neutral	Studsvick	87
■	stable		

R is half vertical size of a hot plume.

with relatively low-frequency variations of wind speed. If the root-mean-square of these variations is $\sigma_u(\text{LF})$ say, and the height of rise in a steady wind is $a/\bar{u}$ it follows that the root-mean-square variation in the height of rise will be roughly $\sigma_u(\text{LF})a/\bar{u}^2$. With a value of $\sigma_u(\text{LF})/\bar{u}$ of 0.1–0.2, which is not unlikely, this is comparable with the standard deviation ($\cong a/4\bar{u}$) equivalent to the induced growth of the plume. Moreover, this fluctuation in height of rise is likely to be accentuated by the usual (negative) correlation between the longitudinal and vertical components of the wind in the atmospheric boundary layer. It should also be emphasized that whereas induced *growth* affects the resultant value of both σ_z and σ_y the fluctuation in height of rise affects only σ_z.

From all the above it is evident that the resultant growth functions of σ_z and σ_y will generally be rather complex. Also, it is evident from Högström's study that the growth functions for natural dispersion are not generally the same for the vertical and crosswind components. This is contrary to the widely used assumption that σ_z/σ_y is independent of distance and, as will be seen later, has important consequences.

Finally, it is probably worth underlining a fairly obvious point concerning the significance of the induced growth of a plume. It has been seen that the correspondingly equivalent radius is on average only one-half of the plume-rise.

Accordingly, except when additionally some systematic downward deflection of the plume occurs (by downdraught or downwash – see 5.4) material cannot reach ground level except through the action of natural turbulent mixing in the vertical.

Formal representation of the concentration distribution expected from an elevated source

Although there have been various theoretical analyses of the dispersion from an elevated source using the gradient-transfer approach, the general physical relevance of the simple diffusion process therein implied is questionable. This is particularly evident from the plume behaviour represented in Fig. 5.9, in which an important part of the total spread is manifestly a result of the different trajectories followed by successive sections of the plume. With this and the foregoing considerations of induced growth and fluctuating rise in mind it seems desirable to avoid dependence on the gradient-transfer approach, especially in the important early stages when the plume is progressively nearing the ground. The alternative which on the whole seems preferable is to retain the more conventional approach – i.e. the assumption of a simple plume geometry, with formally convenient crosswind and vertical distributions and perfect 'reflection' from the ground surface (the 'image' source method as introduced by O. G. Sutton, 1947b). On this basis, and adopting a Gaussian distribution as suitable for both vertical and horizontal distribution from an elevated source, the solution for the crosswind peak in the concentration at ground level ($y = z = 0$) is

$$\chi(x, 0, 0) = \frac{Q}{\bar{u}\pi\sigma_y\sigma_z} \exp\left(-\frac{H^2}{2\sigma_z{}^2}\right) \qquad (5.34)$$

Substituting simple power-law forms for the crosswind and vertical spread

$$\sigma_y = \sigma_y(1)\left(\frac{x}{x_1}\right)^p, \quad \sigma_z = \sigma_z(1)\left(\frac{x}{x_1}\right)^q \qquad (5.35)$$

Eq. (5.34) may be differentiated w.r.t x to give the distance x_m at which $\chi(x, 0, 0)$ is a maximum as

$$x = x_m \text{ when } \frac{H^2}{\sigma_z{}^2} = 1 + \frac{p}{q} \qquad (5.36)$$

and the magnitude of this maximum as

$$\chi_m = j^{j/2} \exp\left(-\frac{j}{2}\right) \frac{Q}{\pi\bar{u}} \frac{\sigma_z^{j-1}}{\sigma_y} \frac{1}{H^j} \qquad (5.37)$$

where $j = 1 + p/q$. Note that σ_z^{j-1}/σ_y is independent of distance and also that Eq. (5.37) may be rewritten, on substituting for $\sigma_z(x_m)$ from Eq. (5.36)

$$\chi_m = j \exp\left(-\frac{j}{2}\right)\frac{Q}{\pi\bar{u}H^2}\left(\frac{\sigma_z}{\sigma_y}\right)_{x_m} \tag{5.38}$$

$$= j^{1/2} \exp\left(-\frac{j}{2}\right)\frac{Q}{\pi\bar{u}H(\sigma_y)_{x_m}} \tag{5.38a}$$

When $p = q$ the foregoing expressions for x_m and χ_m reduce to the simpler and more familiar forms which have for long been used as the basis on which to interpret data on concentrations from elevated sources:

$$x = x_m \quad \text{when} \quad \frac{H}{\sigma_z} = \sqrt{2} \tag{5.39}$$

$$\chi_m = \frac{2Q}{e\pi\bar{u}H^2}\frac{\sigma_z}{\sigma_y} \tag{5.40}$$

With $p/q > 1$, as may usually be expected, the more correct form in Eq. (5.37) implies a dependence on H more sensitive than the familiar inverse square law.

A change in p/q has opposing effects on the first two factors on the right-hand side of (5.38) or (5.38a) and their product changes only slowly [e.g. it falls from 0.74 to 0.67 as p/q changes from 1 to 2 in Eq. (5.38), and from 0.52 to 0.39 in Eq. (5.38a)]. Accordingly, for $1 < p/q < 2$, a most likely condition except in very stable conditions, it follows that to a good approximation χ_m is proportional to σ_z/σ_y at the position of maximum concentration and, furthermore, that the magnitude of χ_m is dependent on the properties of σ_z primarily to the extent that these properties determine x_m. However, since x_m and therefore $(\sigma_z/\sigma_y)_{xm}$ in principle depend on H, the simple inverse proportionality to H^2 does not apply irrespective of p/q, as already noted.

Whatever value is adopted for p/q in the foregoing mathematical representation there is an important condition to be observed regarding the specification of the effective height of source H for a buoyant plume. In so far as this is implicitly a time–mean value (recalling the fluctuations evident in the heights of different plume sections) so also is the concentration distribution necessarily a time–mean property. Clearly the sampling period required to give steady values is that which has admitted virtually all the low-frequency contributions to the spectra of both the vertical and the longitudinal components of turbulence. For plumes tending to rise on average to some hundred metres or more the time scale of turbulence is some tens of seconds for both components in well-mixed conditions. Considering the role of sampling time in relation to the time-scale it is easily ascertained that the required sampling time is therefore in the region of tens of minutes at least. As sampling time is reduced the average ground-level concentration may be expected to display fluctuation of increasing magnitude, especially near the source and in conditions of vigorous vertical mixing, when extreme peaks of short duration may be caused by the occasional descent of isolated plume sections more or less directly to ground level.

Another factor to be kept in mind is that whereas the formal treatments assume a given value of H, for buoyant plumes H continues to increase for appreciable distances. Strictly therefore, especially when the time-mean vertical spread is large, the progressive change in H should be taken into account in estimating the pattern of concentration at ground level.

Analyses of early studies of distribution from stack-sources

The first major analyses of the concentration distributions from elevated sources were carried out on the basis of Sutton's (1947b) formula, which is effectively provided by Eqs. (5.34) and (5.35) with $p=q$. In its detailed format σ_y and σ_z were given specific forms in which $p=q=1-n/2$, with n determined by the vertical profile of wind speed. These specific formulations have physical relevance only in so far as n is basically related to the vertical profile of wind speed and there no longer appears to be any foundation for this in the context of horizontal spread or even vertical spread if the source is elevated. Accordingly there would appear to be no advantage over the simplest and most direct formulation as contained in Eqs. (5.34) and (5.35) with $p=q$, and these have been adopted in the immediately following analysis.

The main practical interest lies in the magnitude χ_m and distance x_m of the maximum ground-level concentration. From Eq. (5.40) it follows that

$$\frac{\chi_m \bar{u} H^2}{Q} = 0.24 \frac{\sigma_z}{\sigma_y} \tag{5.41}$$

Chamberlain (1961a) has collected the values of the non-dimensional normalized concentration represented by the left-hand side of (5.41) as provided by the well-known studies of Thomas, Hill and Abersold (1949), Holland (1953), Stewart, Gale and Crooks (1954, 1958) and the Brookhaven National Laboratory studies reported by M. E. Smith. The values range from 0.02 to 0.28, but the low values (from Thomas *et al.*'s data) are attributed to the use of release-height instead of actual plume height for H, and to the possibility that the measured concentrations (at fixed positions) would not necessarily represent the maxima with respect to distance. Bearing in mind that σ_z/σ_y at x_m may be expected to be less than 1.0, though not greatly so except in stable conditions, the observations of χ_m appear to be roughly in accordance with the simple theoretical treatment.

The values of σ_z/x which have been found to fit the data from several early studies are summarized in Table 5.VII. These estimates of σ_z/x are consistent in a broad way, apart from the large value from Stewart *et al.*'s data at the

Table 5.VII – Estimates of σ_z/x provided by early studies of distribution of concentration from stack emissions†.

Source of data	Assumed plume height (m)	Order of distance (km)	Conditions	$100\ \sigma_z/x$
Thomas *et al.* (1949)	100–200‡	3	Unspecified	3
Gosline (1952)	25‡	0.2	Unstable	10
			Neutral	7
Holland (1953)	100	2	Neutral	3
		5	Mod. inversion	0.5
Stewart *et al.* (1954)	120	1	Lapse	9§
			Neutral	7§
			Inversion	4§
		8	Lapse and neutral	4
			Inversion	<1
Stratmann (1956)	150‡	2	Neutral	5
Meade and Pasquill (1958)	150	1.5	Summer average	11
			Winter average	6
Leonard (1959)		1		4
		10	Stable	0.7
		100		0.1

†The pollutant was sulphur dioxide in all cases except Gosline (nitrogen oxides), Holland and Stewart *et al.* (Argon 41).

‡Figures are stack heights, hence for these in particular the values of σ_z/x may be underestimates.

§These are measured values, the remainder have all been estimated indirectly, essentially by fitting the ground-level concentration data to Eqs. (5.34) and (5.35) with $p=q$.

short range in inversion conditions, which as suggested by Stewart *et al.* is in keeping with a contribution to σ_z from fluctuations in the plume rise. The overall indication is that at distances of a few thousand metres, in the absence of vigorous convection or marked stabilization of the atmosphere, the values of σ_z/x for plumes at a height of 100–200 m are near 0.05, with a tendency for a slow decrease with distance. This is somewhat larger than the total σ_z values observed by Högström using a non-buoyant smoke. According to Fig. 5.10, combining σ_{zc} and σ_{zr} by adding squares, the total values of σ_z/x range from about 0.045 at 1 km to about 0.02 at 5 km. The discrepancy is at least qualitatively in keeping with a contribution from induced growth and fluctuations in plume rise. Returning to Table 5.VII and the effect of thermal stratification, in convective conditions there is evidently at least a two-fold increase in σ_z/x and in conditions of moderate stabilization a reduction by 5-fold or more, the latter factor being reasonably consistent with Högström's measurements. Also the reduction of σ_z/x with distance is much more obvious in stable conditions, and in this respect it may be noted that the data reported by Leonard extends to 100 km, at which distance apparently σ_z/x had fallen to 0.001.

The results summarized in the foregoing paragraph are useful as a first step in the sense that they indicate a rough consistency between the vertical spreads apparent or implied from studies of real effluent plumes and those which have been measured or estimated as a result of atmospheric turbulence alone. However, there are more recent studies in which the effects of meteorological conditions have been brought out in more detail and these will now be considered.

The effects of wind speed and turbulence

Recalling the dependence of plume rise on wind speed it is immediately evident that variations in mean wind speed will have two opposing effects – on the one hand an increase of wind speed produces directly an increase in the dilution of the effluent [the term $\bar{u}$ in the denominator of Eq. (5.40)], while on the other hand the effective height of the source is reduced and the concentration at ground level tends thereby to be increased.

If there are no other effects of wind speed Eq. (5.40) reduces to the simple form

$$\chi_m = \frac{A}{\bar{u}H^2} = \frac{A}{\bar{u}(H_s + \Delta H)^2} \tag{5.42}$$

where H_s and ΔH are respectively the stack height and height of rise, and A is independent of wind speed. If generally ΔH is supposed to decrease with wind speed according to a simple power law

$$\Delta H = \frac{B}{\bar{u}^r} \tag{5.43}$$

(the preferred value of r according to 5.1 actually being 1.0) it follows that the maximum concentration at ground level will not decrease monotonically with wind speed, but will initially increase to a maximum when

$$\bar{u}^r = \frac{B(2r-1)}{H_s} \tag{5.44}$$

$$\Delta H = \frac{H_s}{(2r-1)} \tag{5.45}$$

The magnitude of the maximum is then

$$\chi_{\text{abs. max.}} = \frac{A}{B^{1/r}} \frac{1}{H_s^{(2-1/r)}} \frac{(2r-1)^{(2-1/r)}}{4r^2} \tag{5.46}$$

The foregoing expressions reduce to very simple forms when the preferred value of unity for r is inserted.

There are however other influences of wind speed which have not so far been taken into account. Observations at heights of hundreds of metres (see 2.6) have shown that the intensity of the vertical component (and hence implicitly the behaviour of σ_z/x) is not essentially independent of wind speed as it is at heights much closer to the ground. A dependence on wind speed also appears in the lateral component according to some of the observations reported by Moore (1967) from measurements on a 187 m tower. The precise effects of these variations have yet to be fully evaluated but the possibility exists that σ_z/σ_y and the exponents p and q in Eq. (5.35) are significantly affected, in which case the immediately foregoing relations will no longer hold.

Some observational evidence for the resultant effect of wind speed on χ_m has been summarized by Moore (1969). These results, based on systematic surveys of the ground-level distribution of sulphur dioxide downwind of the Central Electricity Generating Board generating stations at Northfleet and Tilbury, are reproduced in Fig. 5.11. The outstanding features of these data are

(a) the relatively small difference between daytime and night-time values of χ_m, except in light winds,
(b) the tendency for all except daytime values in very light winds to *increase* in a near-linear fashion with wind speed,
(c) the agreement with the simple theoretical value according to Eq. (5.40), with $\sigma_z/\sigma_y = \frac{1}{2}$ independent of distance, *only* at the highest wind speeds,
(d) the existence of observed values well below this simple theoretical value in the wind speed range 3–10 m/sec.

Moore argues that the disagreements with the simple theoretical value imply not merely a departure of σ_z/σ_y from the value of $\frac{1}{2}$ but also a difference in the growth functions of σ_z and σ_y with distance x, contrary to the earlier assumption and to Smith and Singer's (1966) conclusion based on dispersion studies at the Brookhaven National Laboratory in U.S.A.

There is no difficulty in justifying an assumption of quite different growth functions for σ_y and σ_z. Reference has already been made (4.7) to the indications, from Högström's (1964) measurements, of an asymptotic approach to $x^{1/2}$ variation for σ_z, and the closeness to such a variation is also evident in the summary of Högström's data given in Fig. 5.10 earlier in this section. On the other hand data on the growth of crosswind spread have in general displayed a tendency to a power-law growth with an exponent much greater than one-half. In particular, Högström's Studsvick data in neutral conditions may be fitted over the range $\frac{1}{2}$–5 km by power laws with exponents 0.85 and 0.55 respectively for the y and z components.

However, as clearly demonstrated in the laboratory and numerical modelling studies of Willis and Deardorff (1976) and Lamb (1982), a small exponent for

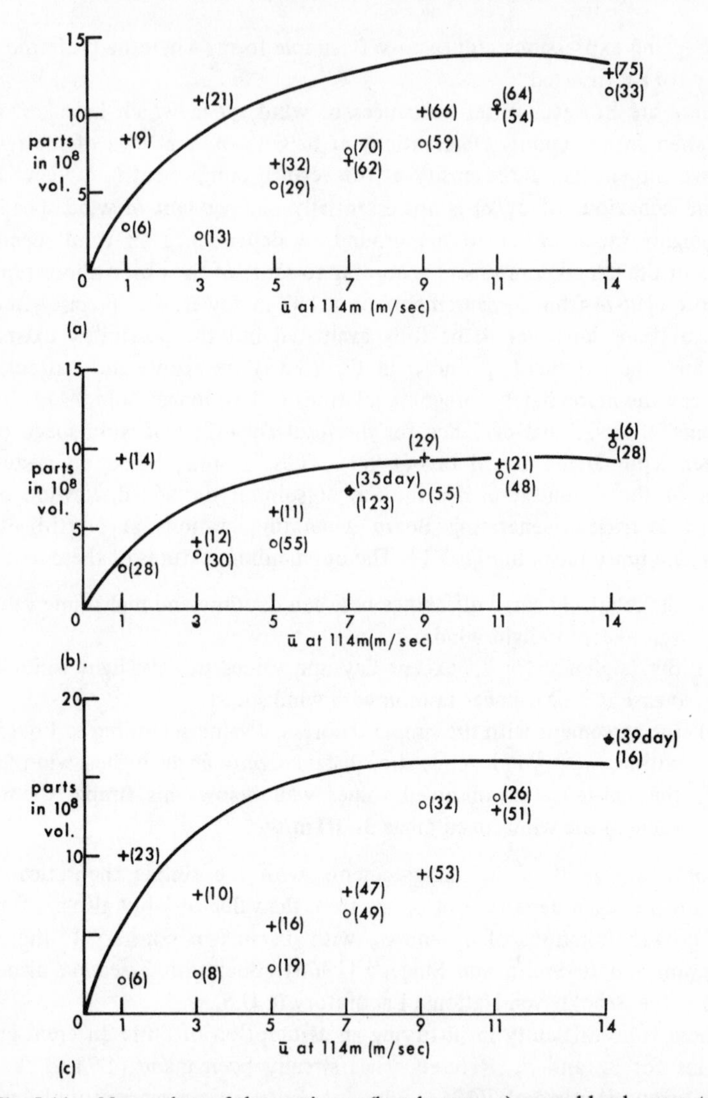

Fig. 5.11 – Mean values of the maximum (hourly average) ground-level concentration, from two C.E.G.B. power stations, as a function of wind speed (from Moore, 1969). (a) Northfleet – mean load 552 MW; (b) Northfleet – mean load 332 MW; (c) Tilbury – mean load 291 M.W. + between 0800 and 1700 local time. O between 1800 and 0700 local time. Number of observations in parentheses. The curves correspond to the simple form in Eq. (5.40) with $\sigma_z/\sigma_y = \frac{1}{2}$.

the σ_z growth in a convectively mixed layer will not apply except ultimately as a consequence of the restraining action of the overhead stable layer.

The Tilbury and Northfleet data, reprocessed and reported in further publications by Moore (1974, 1975 and 1976) provide a basis for the development and testing of realistic models for the prediction of ground-level maximum concentrations (χ_m) from power-station plumes. In Moore's latest approaches the Gaussian-plume formulation (Eq. (5.34) *et seq.*) is retained, but with the exponents in Eq. (5.35) given values $p = 1$ and $q = \frac{1}{2}$. In principle the crosswind spread is determined by the wind-direction fluctuation, with dependence on sampling duration, whereas the vertical spread is expressed in terms of a vertical diffusivity. The latter assumption (with which $q = \frac{1}{2}$) is unjustified at short range and in convective conditions at all ranges. In practice, however, Moore uses the theoretical results which follow from this assumption with certain arbitrary coefficients selected to provide the best overall fit to the data. The fit contains wide discrepancies in terms of the individual one-hour samples but (see Moore, 1976) the ensemble mean values for various wind speed and stability groups (ranging from 4 to 20 p.p.h.m.) are fitted with a mean error of about 1.5 p.p.h.m.

Interpretation in terms of the field of turbulence

In an earlier analysis Moore (1967) considered the accuracy with which the diffusion properties obtained in the Tilbury survey could be calculated from the turbulence measurements using the Hay–Pasquill approach in the form of Eq. (4.3). It was found that crosswind spread was given reasonably well on average with $\beta = 4$, the agreement being best of all in strong winds. Also it was found that combination with the computed values of σ_z in strong winds (using $\beta = 6$ as appropriate to the lower intensity of turbulence) gave σ_z/σ_y about 0.5 for 1-hr samples, agreeing with that required to give satisfactory correspondence between observed χ_m and the value calculated from Eq. (5.40).

The importance of the varied trajectories followed in the vertical motion of sections of plume in response to low-frequency fluctuations of the w-component was brought out clearly in studies by Barad and Shorr (1954) and Davidson and Halitsky (1958), in which the incidence of sections of an experimental smoke plume was observed at ground level at various distances downwind. A particularly interesting feature of the second of these studies is the demonstration that the duration of smoke at ground level could be predicted from measurements of wind inclination near the site of release.

In Davidson and Halitsky's study the plume is assumed to expand linearly with an angle α subtended on the release point. Successive elements are assumed to have linear trajectories determined by the wind vector averaged over 10-sec intervals. The fraction of time that plume segments released at a height H will cross a given arc (radius R) on the ground is taken to be twice the probability that the inclination of this wind vector is downward and numerically greater than $(H/R) - \alpha$. The factor of two is based on observations that the speed of plume segments on the ground was about one-half of the mean wind speed at the release point, which was at a height of about 100 m. Values calculated in this way, taking $\alpha = 0.055$ on the basis of visual estimates of plume size, are shown compared with observations of the duration of appearance of smoke in Fig. 5.12. Some over-simplification may have been introduced here, but the exercise is significant in confirming the essential mechanism of the intermittency of effluent concentration at short range from an elevated source and in demonstrating the possibility of estimating this intermittency from a knowledge of the wind fluctuation distribution.

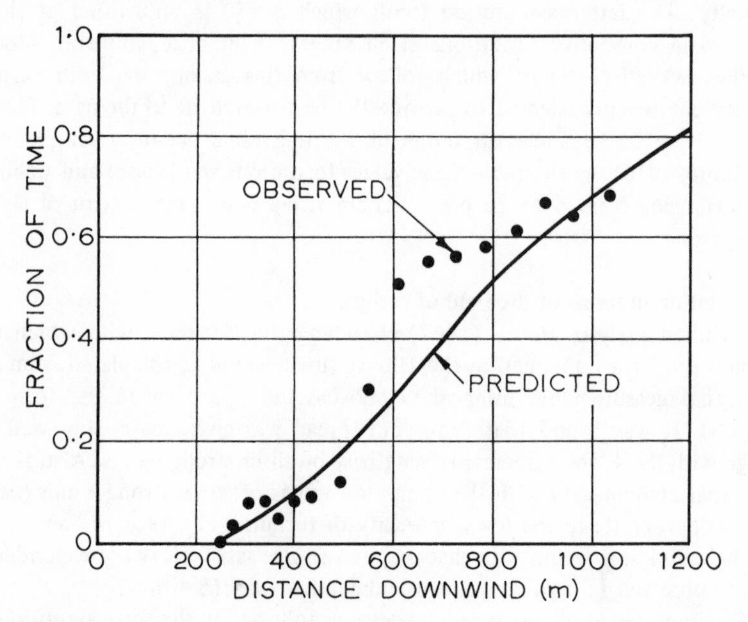

Fig. 5.12 – Observed and computed duration of smoke, at ground level, from an elevated source. (Davidson and Halitsky, 1958).

More sophisticated developments of the foregoing simple approach have been stimulated by the similarity generalizations about turbulent mixing in a convectively stirred boundary layer and by the emergence of numerical-statistical (random-walk) techniques for calculating the spread of ensembles of particles. Some significant steps are noted below.

(1) Venkatram (1980) and Venkatram and Vet (1981) assume an average downdraft velocity equal to half the convective-velocity scale w_*, and from this and the Briggs form for the ascent of a buoyant plume calculate an average 'touch-down' distance of plume elements. The alongwind distribution of 'touch-down' is assumed to be log-normal, with a standard deviation in accordance with Deardorff's numerically modelled statistics of turbulence in a convectively mixed layer. Combination of this distribution with plume spreads

$$\sigma_z = h_c(1 - \exp(-1.5\,X)), \quad \sigma_y = 0.45\,Xh_c \tag{5.46a}$$

where $X = w_* x / u h_c$, provides a specification of the concentration along the ground-level centre-line of the plume.

(2) Lamb (1979, 1982) has used a random-walk technique to calculate crosswind and vertical distributions of concentration as a function of X and for a range of release heights. Initially these calculations were restricted to *passive* material but they nevertheless provide several important and innovative features, especially the virtually rectilinear descent of the plume centre-line to ground level, a behaviour which has been confirmed in laboratory studies by Willis and Deardorff (1978). In accordance with this behaviour the ground level concentration has a maximum which occurs at a distance $2z_s u/w_*$ (i.e. the mean downward velocity of the passive plume is $w_*/2$) and there has a value inversely proportional to $z_s h_c$ (in contrast to the z_s^2 of the classical image-source treatment).

(3) The statistical approach by Misra (1982) is essentially the same as Venkatram's, differing only in treating the effects of updrafts separately (assuming reasonably that they contribute to the ground-level distribution an amount equivalent to a vertically uniform distribution) and in using Lenschow's (1970) statistics on the vertical velocities in updraughts and downdraughts. A reasonable agreement with Lamb's numerical and Willis and Deardorff's laboratory data is shown.

Venkatram has tested his models against surveys of the ground-level concentrations of sulphur dioxide from two large power plants in the U.S.A. and a

Table 5.VIII – Testing of Venkatram and Vet's (1981) model against observations in large sulphur dioxide plumes in convectively mixed conditions.

Source	No. of Observations	Percentage with C(obs.)/C(pred.) between 0.5 & 2	Mean (S.D.) of C(obs)/C(pred)
Dickerson power plant, Md., U.S.A.	26	85	0.98 (0.34)
Morgantown power plant, Md., U.S.A.	22	86	1.20 (0.37)
Smelter, Sudbury, Ontario	25	80	1.02 (0.27)

The figures in the last column refer to the observations fulfilling the factor of two agreement. Overall range of distance from source: 2–40 km.

smelter in Canada, at distances up to about 30 km, and has demonstrated the agreement summarized in Table 5.VIII. Apparently almost as good agreement was also demonstrated with a much simpler version of the model in which uniformity in the vertical is assumed, and which would be expected to provide a closer approximation with increasing distance. It is interesting to note that the agreement does not obviously deteriorate with decreasing distance even to the smallest distance (corresponding to X values near 0.3). This result is broadly consistent with Lamb's calculations for passive plumes, which give cross-wind integrated ground-level concentrations more than twice the vertically uniform value only when X is below about 0.7. It should be kept in mind however that the agreement may be somewhat fortuitous as no allowance was attempted for deposition of the sulphur dioxide.

The fluctuation of concentration

The extent to which the *instantaneous* concentration downwind of a point source is greater than the *time-mean* value is essentially determined by the relative magnitudes of the instantaneous and time-mean spreads, and by the position of the receptor in relation to the centre of the time-mean distribution in the plane normal to wind. On Gifford's simple model [see 3.9 and Eqs. (3.133)–(3.137) for the formal representation, also Gifford (1960) for further discussion thereof] the ratio of instantaneous to time-mean concentration at positions x, y, z relative to a continuous point source is

$$\frac{(\chi)_i}{(\chi)_\tau} = \frac{(\sigma_y\ \sigma_z)_\tau}{(\sigma_y\ \sigma_z)_i} \exp\left[\frac{1}{2}\left(\frac{y^2}{(\sigma_y)_\tau{}^2} + \frac{z^2}{(\sigma_z)_\tau{}^2}\right)\right] \tag{5.47}$$

where the subscripts i and τ denote instantaneous values and values averaged over a sampling time τ. This relation can easily be seen to follow by writing Eq. (6.2) for both the instantaneous and time-mean case and taking $y = z = 0$ in the former case to represent coincidence of receptor and puff centre.

The expected magnitudes of $(\chi)_i/(\chi)_\tau$ may easily be estimated in certain circumstances from the data available on σ_y and σ_z. Both theory (see Fig. 3.14) and experience (see Höström's 1964 data) indicate that close in to a point source in ideal neutral flow the instantaneous spread is roughly half the time-mean spread as typically observed over tens of minutes, except in the case of vertical spread from a *ground-level source,* for which there is no indication of a difference either from theory or observation. As a consequence [putting $y = z = 0$ in Eq. (5.47)] we may expect to have values near 4 downwind of (and at the same level as) an elevated source but only 2 downwind of a ground-level source. In practice, apart from departure from the ideal flow conditions the short-period sampling of concentration is not instantaneous but typically occupies a finite sampling time of some 10 sec or more. Data quoted by Gifford (1960) for *peak/average* ratios at the same level as the source, and for 'average' and 'peak' sampling times in the ratio 20 or more, are between 2 and 3, and therefore not obviously inconsistent with expectation.

It is evident from Eq. (5.47) that the more striking variations in $(\chi)_i$ will occur when the vertical (z) or crosswind (y) displacement of the receptor from the centre of the time-mean distribution is large. For ground positions in the mean wind direction from an elevated source, and assuming for simplicity that both $(\sigma)_i$ and $(\sigma)_\tau$ grow linearly over a short range of distances, it follows that $(\chi)_i/(\chi)_\tau$ will be greater than at the level of the source by the factor exp $(z^2/2(\sigma_z)_\tau^2)$ where z is now the height of the source H. Thus, when σ_z is $H/\sqrt{2}$, which corresponds to the position of the ground-level maximum (on a time-mean basis) for σ_z/σ_y independent of distance, the factor is e. When $\sigma_z = H/2.15$, which corresponds to the conventional 'height of cloud' as defined by a concentration one-tenth of the axial value, and therefore may be regarded as roughly defining the position close in to the source at which its effect is first significantly experienced on the ground, the factor is $e^{2.3}$. This means that the ratio $(\chi)_i/(\chi)_\tau$ may be expected to increase from about 10 near the point of maximum $(\chi)_\tau$ to about 40 closer to the source.

The data available do not provide for any precise test of the foregoing estimates, but the values summarized by Gifford show peak/average ratios falling generally from 30–50 at short distances (about 200 m) to about 2 at distances of 2–5 km downwind. It will be recalled that the theoretical limit at long distance is unity (see 3.9), when the fluctuations are supposed to arise from the different trajectories of sections of a plume, but of course this neglects the contribution from the patchiness *within* the plume section itself. Not much is known about this aspect, which may have considerable importance in the odour properties of a plume of gas.

The magnitude of $(\chi)_{\tau_1}/(\chi)_{\tau_2}$ in general, where τ_2/τ_1 may take on any value, is of considerable practical importance – for example, in facilitating the comparison of data obtained with different sampling times. As in the case of the peak/average ratios discussed above there will be an important dependence on the distance downwind and on the vertical or crosswind displacement from the source position. For positions very close to a source the appropriate magnitudes of σ_y and σ_z as a function of sampling time τ may be estimated from Eq. (4.2) and the corresponding dependence of σ_v and σ_w (or σ_θ and σ_ϕ, the corresponding angular fluctuations of wind). This dependence also involves the characteristic time-scale of the turbulence, as discussed in 2.6. In general the relation between σ_v etc. and sampling time τ is determined by the shape of the spectrum. As pointed out in 2.6 the simple power law $\sigma_\theta \propto \tau^{0.2}$ provides a good fit over the range $20 < \bar{u}\tau < 400$ m. However, outside this range, and for the rather more irregular individual spectral shapes which are often encountered (especially at lower frequencies), this simple law cannot be expected to apply. Furthermore, when use is made of such data in conjunction with an equation such as (5.47), to estimate the effect on *concentration* ratios, as already seen the result must depend to an important degree on the relative positions of source and receptor.

The long-term distribution around an isolated source

The long-term average distribution of effluent concentration around an isolated source reflects the slow changes in wind direction which occur, over periods greater than about one hour, as a result of changes in the general weather situation. For the simple hypothetical case when the wind direction over a long period is uniformly distributed around 360 degrees, the long-term average concentration $(\chi)_\infty$ at radius x from a ground-level source, or at long distance from an elevated source is related to the peak value for a short duration τ, $(\chi)_\tau$, as follows. With the usual Gaussian distribution for the crosswind distribution over time τ and assuming $(\sigma_y)_\tau$ to be small compared with x,

$$2\pi x(\chi)_\infty = 2(\chi)_\tau \int_0^\infty \exp\left[-y^2/2(\sigma_y)_\tau{}^2\right] dy \tag{5.48}$$

from which

$$\frac{(\chi)_\infty}{(\chi)_\tau} = \frac{(\sigma_y)_\tau}{(2\pi)^{1/2}x} \tag{5.49}$$

For example, with $\sigma_y/x = 0.1$, which is appropriate for $\tau \cong 4$ min and ranges near 1 km in near-neutral conditions over flat terrain

$$\frac{(\chi)_\infty}{(\chi)_{4\ \mathrm{min}}} = 0.04$$

In reality the long-term distribution of wind direction will normally not be uniform, and the long-term average distribution of χ around a source will depend on the actual wind direction frequencies in combination with speed and diffusive conditions. An analysis of this feature was carried out by Meade and Pasquill (1958) in connection with attempts to detect the effect of variations in coal consumption on the long-term averages of sulphur dioxide given by routine surveys around a power station. The analysis starts with the assumption as above that the short-term crosswind spread is small relative to the distance x, so that the average around the circle is given by

$$2\pi x\bar{\chi}\,(\text{circle}) = \int_{-\infty}^{+\infty} \chi(x, y, 0)\, dy \tag{5.50}$$

with $\chi(x, y, 0)$ for an elevated source given by Eq. (6.17). Accordingly

$$\bar{\chi}\,(\text{circle}) = \frac{1}{2\pi x}\, g(H, x)\frac{Q}{\bar{u}} \tag{5.51}$$

with

$$g(H, x) = \frac{\sqrt{2}}{\sigma_z\sqrt{\pi}} \exp\left(-\frac{H^2}{2\sigma_z^2}\right) \tag{5.52}$$

Then if $f(\theta)$ is the percentage frequency of wind direction in any octant the long-term average concentration over that octant will be

$$\bar{\chi}\,(\text{octant}) = \frac{f(\theta)}{12.5} \frac{1}{2\pi x} g(H, x) \frac{Q}{\bar{u}} \tag{5.53}$$

If there is a background of pollution of magnitude a from other more distant sources, the total pollution p averaged over any octant around a power station discharging sulphur dioxide at a rate S should therefore be related to wind speed, wind direction and rate of discharge, by the form

$$p = a + b\frac{f(\theta)S}{\bar{u}} \tag{5.54}$$

When S has practical units of tons per month, $\bar{u}$ is in knots, x in yards, p and a in mg SO_3 per day per 100 cm^2 (as derived from the lead peroxide candle measurement), and using an empirical relationship between concentration of SO_2 and deposition on the lead peroxide candle, it follows that

$$b = \frac{8.5 \times 10^3}{x} g(H, x) \tag{5.55}$$

with $g(H, x)$ still in c.g.s. units.

From routine wind records, obtained at the meteorological station nearest to the power station at Staythorpe, values of $f(\theta)/\bar{u}$ were obtained for octants 0–45°, 45–90° etc., for each of six 5-month seasons covering the summer and winter periods from 1951 to 1954. Pollution diagrams were constructed to show average values over each season at fourteen observation sites. Smooth isopleths of pollution were drawn in, and used to estimate the average pollution on each 45° arc at a radius x of 1500 m. With these data, and the estimated rates of emission of sulphur dioxide, the plots of p against $f(\theta)S/\bar{u}$ were as in Fig. 5.13. An approximation to a linear relation, as in Eq. (5.54), is supported and the regression equations (shown with the diagrams) were found to be statistically significant. The values of b were used with a graphical solution of Eq. (5.52) to derive the average values of σ_z/x discussed earlier in this chapter (see Table 5.VII). The values of the constant a were in good agreement with the seasonal background values of pollution observed at Staythorpe before the power station began to function. It is particularly interesting that the analysis on these lines revealed a systematic effect of the variations in S from year to year which had apparently been unrecognizable from an examination of the overall average pollution around the power station.

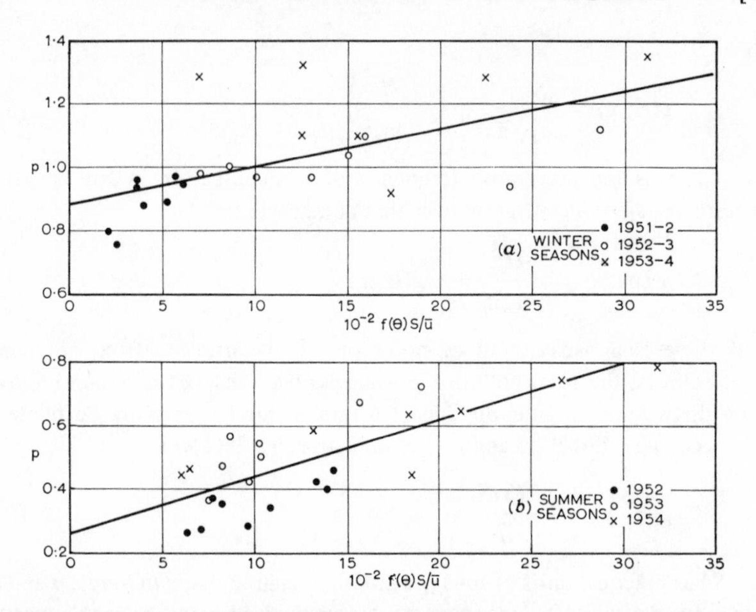

Fig. 5.13 – Average sulphur pollution, p, at 1500 yd from a power station, as a function of wind direction frequency, $f(\theta)$, wind speed $\bar{u}$, and rate of emission, S. (a) Winter season; (b) Summer season. S is in tons SO_2 per month, p in mg SO_3 per month, $\bar{u}$ in knots, $f(\theta)$ is the percentage frequency in a prescribed octant. The lines represent the equations

$$\text{(a) } p = 0.89 + 1.16 \times 10^{-4} f(\theta)\, S/\bar{u}$$
$$\text{(b) } p = 0.26 + 1.76 \times 10^{-4} f(\theta)\, S/\bar{u}$$

Meade and Pasquill, 1958).

Fumigation from power station plumes

The term *fumigation* refers to the well-known phenomenon of the rapid transfer to ground level of effluent material which has remained aloft in stable conditions, in association with the convective breakdown of the stable situation as a result of ground heating after dawn. Assuming that the material is thereby mixed uniformly between the ground and the height of the top of the plume H' at the onset of fumigation, the magnitude of the concentration $\bar{\chi}_F$ is given by integrating Eq. (6.2) with respect to z from 0 to ∞ and dividing the result by H'. The result is

$$\bar{\chi}_F = \frac{Q}{(2\pi)^{1/2}\bar{u}\sigma_y H'} \exp\left(-\frac{y^2}{2\sigma_y{}^2}\right) \tag{5.56}$$

On the centre-line of the plume ($y = 0$) this temporary concentration is easily seen by comparison with Eqs. (5.40) and (5.39) to be $e\pi^{1/2}/2 (\cong 2.5)$ times the maximum ground-level concentration associated with vertically unbounded dispersion from the same source with effective height H' and similar conditions of crosswind spread.

The foregoing considerations assume a definite downwind travel of the plume. In *calm* stable air the plume does not travel significantly in any preferred direction but may be expected to rise near-vertically and then spread out horizontally when the limiting height is reached. Briggs (1965) has used a simple model of this process in an attempt to interpret some observations during fumigations at the Tennessee Valley Authority plants in very light winds. It is assumed that the plume rises by a height ΔH in accordance with Eq. (5.8). It is further assumed that the material spreads out evenly over an area A which is determined by the same parameters, F and s, as determine the height of rise. Thus, if the plume rise and outward spread continue for a time τ the concentration produced on fumigation will be

$$\bar{\chi}_F = \frac{Q\tau}{A(H_s + \Delta H)} \tag{5.57}$$

where, as in Eq. (5.8), with the magnitude of constant recommended by Briggs,

$$\Delta H = 5.0\, F^{1/4} s^{-3/8}$$

and from further dimensional analysis

$$A = cF^{1/2} s^{-1/4} \tau \tag{5.58}$$

A few observations at three T.V.A. plants give values of $c = 20$, 16 and 25, suggesting an overall value near 20. Clearly some further consideration of this interpretation would be desirable but the present limited test is encouraging.

5.4 THE EFFECTS OF SURFACE FEATURES

The properties of the underlying surface may be important either in deflecting the airstream or in modifying the rate of mixing and consequent dilution of the material carried with it. The various effects may be considered under four main headings:

1. The pattern of flow and dispersion near individual buildings.
2. The overall effect of the aerodynamic roughness of an area.
3. Modification of the flow by urban heating.
4. Topographical influences on flow, circulation and dispersion.

The effect of individual buildings

Deflection and disturbance of the wind field by a building will obviously have a controlling influence on the distribution of pollutant from a source on or close to the building. The broad characteristics of the resultant flow pattern around an isolated building are now well-known from aerodynamic studies: an upwind *displacement zone* in which the approaching airstream is deflected around the obstacle, a zone immediately to the leeward which is relatively isolated from the main flow (termed the *cavity* region by Halitsky (1968)), and further downstream the highly distürbed *wake*. A simplified and idealized representation for a sharp-edged building is shown in Fig. 5.14.

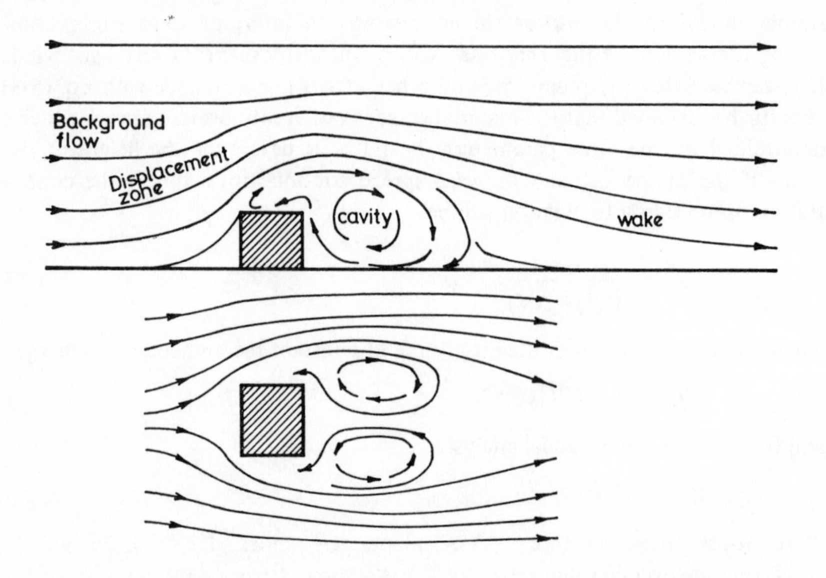

Fig. 5.14 – Simplified schematic flow patterns around a sharp-edged building (following Halitsky, 1968).

One of the most important practical considerations is the extent to which the effluent from a chimney stack on or adjacent to the building is entrained in the characteristic flow pattern. For this aspect the potential of wind tunnel model studies has been widely accepted and exploited in studies such as that of Davies and Moore (1964). Their investigation, which was concerned with the effects on effluent plumes from two particular nuclear power stations, was notable in demonstrating that avoidance of entrainment (and consequent appearance of effluent at ground level at some specified distance) was dependent on exceeding a critical value for the *ratio* R of the vertical efflux velocity w_E to the wind velocity. For a stack projecting very little above general roof level the critical value was in the region of 3–5. From a purely functional standpoint a much higher chimney in relation to building height is desirable, and there has for long been a 'rule-of-thumb' that the chimney should extend to a height about two and a half times the height of adjacent buildings. In practice this tends to ensure that the chimney exit is clear of the flow which turns downward on the lee-side (the *downdraught*). There is, however, also the risk that effluent may be drawn down the lee face of the chimney in the vortex motion generated there (*downwash*) and so reach the region affected by downdraught. The practical means of avoiding this is to keep the ratio R sufficiently high. Efflux velocities quoted by Hawkins and Nonhebel (1955) correspond to $R \cong 1.3$, much lower than those determined by Davies and Moore simply because they refer to chimneys which are much taller in relation to building height.

When the pollutant is entrained in the disturbed flow pattern one might expect the concentration to be basically *scaled* according to the rate of emission, the wind speed and an area A (or length L squared) characteristic of the cross-stream section of the building. In other words the normalizing of concentration in terms of the quantity $Q/\bar{u}L^2$ may be expected to reveal a field of concentration which is universal for a given geometrical shape and disposition of building relative to wind, providing certain hydrodynamic scaling criteria are met. Thus, Halitsky (1968) defines a non-dimensional concentration coefficient K_c by the relation

$$\chi = \frac{K_c Q}{\bar{u}L^2} \tag{5.59}$$

Halitsky's article contains a detailed discussion of the flow fields near buildings, of the scaling requirements in terms of Reynolds and Froude numbers and of extensive wind tunnel surveys of the distribution of K_c around buildings of various shapes and proportions.

The details of the concentration distributions are somewhat complex and do not lend themselves to any brief generalizations. The main point is that the aerodynamic principles and wind tunnel techniques provide a useful basis for assessing special effects in the immediate vicinity of buildings, for which the detailed structure of the ambient airflow is of secondary importance. With increasing distance downwind of a building the intensity of induced turbulence in the wake decreases and the dilution of the effluent by spreading becomes increasingly dependent on the general level of turbulence outside the zone of influence.

For practical purposes in estimating likely *downwind* concentration it may sometimes be adequate to calculate an upper limit by neglecting the elevation of the source and the initially rapid cross-stream spread which may result from the aerodynamic effects discussed above, in other words to treat the emission as a *ground-level point source.* Such a procedure has been considered by Culkowski (1967), who gives examples showing measured concentrations lower than those estimated for a ground-level point source but approaching them with increasing distance downwind.

Further experience is provided by the experimental study conducted by Munn and Cole (1967) on the Canadian Research Council Site in Ottawa. Measurements were made of ground-level concentrations of fluorescent tracer injected into one of the 10 ft-high stacks on the 60 ft-high flat roof of the central heating plant. The samplers were operated for 20-min periods at ground-level along a line to which the normal distance from the source was about 500 ft. Bi-directional vane records were taken with instruments mounted at a height of 80 ft about 300 ft upwind of the source on relatively open ground, and on a 33-ft tower about 400 ft from the source in the wake area generally downwind of the heating plant and other buildings. These records were used to derive values of the standard deviation of wind direction (σ_θ) and inclination (σ_ϕ) for various averaging times (smoothing times) ranging from 5 to 100 sec. The values of σ_θ and σ_ϕ for an averaging time of 5 sec were greater in the downwind disturbed area than in the upwind area by factors of about 1.2 and 1.5 respectively.

In four of Munn and Cole's experiments the sampling line was at right-angles to the wind direction and it was possible to estimate the width of cloud defined by concentrations one-tenth of the peak. The results are shown in Table 5.IX, with values of $\theta/4.3$ ($\cong 57.3\ \sigma_y/x$ assuming a Gaussian distribution) and corresponding values of σ_θ at the downwind site. The values of $\theta/4.3$ and σ_θ are remarkably similar indicating that the crosswind diffusion was represented tolerably well by the gustiness measurements in the building wakes.

Table 5.IX – Data on angular cloud-width θ and the standard deviation of the wind direction fluctuation σ_θ over a built-up site (from Munn and Cole, 1967).

Trial	Distance downwind (ft)	θ (deg)	$\theta/4.3$ (deg)	σ_θ (deg)
2	500	30	7.0	7.9
3	535	64	14.8	14.0
5	550	58	13.5	16.7
6	525	50	11.6	13.1

Munn and Cole also compared the magnitudes interpolated for the peak in the crosswind distribution of concentration with values calculated from various formulae. Using Eq. (4.3) and the corresponding form for the vertical spread, with $\beta = 4$, and Eq. (6.17), the calculated concentrations were always substantially higher than those observed. The ratios of the observed and calculated values ranged from 0.025 to 0.40 with a median value of 0.10. Note that this overestimation of the concentration by calculation occurred despite the allowance attempted for effective height of source. Since the magnitude of σ_y was evidently correctly calculated by this procedure it would be reasonable to suppose that so was the vertical spread σ_z, but for the effects of downdraught and downwash and the variation with *height* of the wake intensity in relation to the effective height of the plume. Furthermore, where absolute concentrations are concerned there is always the possibility of overestimation of the *effective* source strength because of loss or decay of the material in some way, though in the present case Munn and Cole state that there was no reason to expect this. Much more detailed observations than are available from the report would be required to clarify all these aspects, but it is worth noting that the build-up of ground-level concentration with distance immediately downwind of an elevated source is sensitively dependent on σ_z/H. For example, using Eq. (6.17), an *overestimation* of σ_z/H by factor of 2 could cause an *over-estimation* of concentration by a factor of 4 or more at distances less than that (x_m) of the ground-level maximum. The fact that the peak concentrations observed by Munn and Cole show a sharp increase from virtually zero with increase of wind speed over the range 5–20 ft/sec is strongly suggestive that the sampling distance was less than x_m on many occasions.

To sum up, Munn and Cole's study emphasizes the great complexity which must usually exist in the dispersion of effluent immediately downwind of the source-building, but it does not conflict with the reasonable policy of using ground-level point source calculations to provide a 'safe' estimate of concentration.

The effects of roughness over an area

In principle the effects of the general degree of aerodynamic roughness over an area on the vertically diffusive action of the flow are represented in terms of transfer-theory (with K a function of roughness) or Lagrangian similarity theory as in 3.3. There are however important limitations which reflect on the distance from a source for which these representations are valid.

(a) For the present form of similarity theory, and also for transfer-theory in which the K is realistically related to the wind profile, the treatments are valid only for diffusion *within* the surface-stress layer.

(b) The representation of the diffusive conditions is not valid *within* the rougness elements themselves.

(c) There is an implicit assumption of horizontal uniformity in the theories, whereas in practice more or less severe changes of roughness commonly arise over terrains of most interest.

Limitation (a) means that consideration should be restricted to vertical spread up to some tens of metres at most, while this in combination with (b) means that rigorous consideration of vertical diffusion over a city area is virtually ruled out.

Regarding (c) the important point is that following a sudden change of roughness there is a redevelopment of the boundary layer in which two zones can be recognized. There is an *internal boundary layer* of growing depth in which the turbulent and diffusive properties are affected by the new roughness. Over a shallow layer adjacent to the ground there is a new *equilibrium* layer within which the relation between the stress, wind profile and other properties of the flow are effectively those for an unlimited area of the new roughness. The rates of growth of these two layers have attracted much theoretical investigation. A treatment by Peterson (1969) seems to avoid some of the difficulties of earlier treatments of natural boundary layer development and provides some useful estimates. In neutral conditions, and within the surface-stress layer for the flow upstream of the discontinuity, the depth of the internal boundary layer is very close to one-tenth of the distance of travel. Taking 90 per cent adjustment (of the *change* in stress) as defining the new equilibrium layer, then for a 'smooth to rough' change this layer is between 0.1 and 0.2 of the depth of the whole internal boundary layer and thus, to be conservative, roughly one-hundredth of the distance of travel. On the other hand 50 per cent adjustment (again for 'smooth to rough' change) holds up to a height of about one-thirtieth of the distance. Corresponding heights for equilibrium after a 'rough to smooth' change are considerably less, especially using the more severe criterion.

Practical experience indicates that the slope of the top of a ground-based plume is about one-tenth in neutral conditions. This means that the vertical distribution of the plume will be within the depth of new equilibrium flow only for distances of travel a small fraction ($\frac{1}{3}$ or $\frac{1}{10}$ according to the adjustment criterion adopted) of the upwind fetch over the new roughness. Proportionality of stress and turbulent *energy* (i.e. mean square velocity fluctuation) is assumed in Peterson's treatment. It follows therefore that the predicted adjustment of the root-mean-square velocity fluctuation over the heights and distances enumerated above will be substantially more complete than the 90 per cent or 50 per cent.

Good experimental determinations of vertical spread over natural rough surfaces have been confined to grassed areas showing a difference in vertical spread of only about 20 per cent. This difference is broadly predicted by similarity theory but the test cannot be regarded as a critical one. For the much greater roughness of an urban area there are as yet no useful direct measurements of vertical spread. The only experimental values at present available for this case are those inferred from the measurement of the ground-level concentration of a tracer, assuming a shape for the vertical distribution and invoking conservation of tracer material. McElroy and Pooler (1968) have given estimates of σ_z obtained in this way from measurements of a fluorescent particle tracer at distances up to about 7 km from a ground-level point source in St. Louis, U.S.A. Their results for neutral conditions have been converted to values of $\bar{Z}$ for direct comparison with similarity theory predictions in accordance with Eq. (3.36), and are reproduced in this form in Fig. 5.15. On the same diagram are lines representing an empirical estimate for open downland and similarity theory values for $z_0 = 3$ cm and 3 m.

The open-country empirical curve corresponds to the *D*-category curve in tentative estimates of vertical spread (Pasquill, 1961, Fig. 2). It is based, with

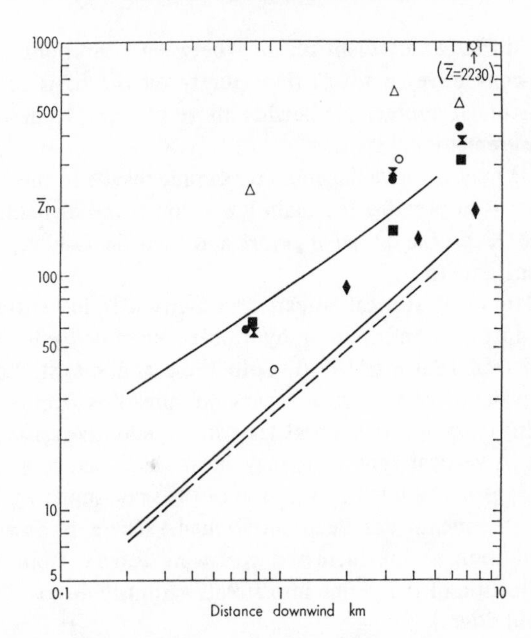

Fig. 5.15 – Vertical spread $\bar{Z}$ as a function of distance from a source near ground level (Pasquill, 1970).

——— From Lagrangian similarity theory, $z_0 = 3$ m (upper curve), 0.03 m (lower curve) $c = 0.6$.

– – – – – Earlier semi-empirical curve for open country (Pasquill, 1961).

Points are from St Louis dispersion study, McElroy and Pooler (1968) experiments No. 16 (●), 24 (○), 35 (△), 37 (■), 42 (◆), 43(⧗) open points being for daytime experiments, blocked-in points for evening. (Reprinted from Air Pollution Control Office Publication No. AP-86, U.S. Environmental Protection Agency, 1970).

some extrapolation, on the Porton field data for which the z_0 value was assessed to be 3 cm (see Calder, 1949). There is good agreement with the similarity-theory curve for this z_0. The urban values from the St Louis night-time measurements are systematically higher by a factor of about two and show a trend more in accordance with, though slightly lower than, the curve for $z_0 = 3$ m. Examination in more detail of the theoretical dependence on z_0 suggests that the best fit would occur with a z_0 of about 1 m, which is quite consistent with estimates made from wind profile and turbulence measurements over an urban area. It has to be admitted however that as the comparison stands at present the agreement is quite fortuitous, since it is obvious that the limitations discussed above must have been exceeded for all data except possibly those at the shortest range. Another complication which awaits explanation is that the 'daytime' values which were also obtained in neutral conditions are substantially larger than the 'evening' values just discussed. It has been suggested tentatively that this may reflect the difficulty of distinguishing between conditions of convective mixing and non-convective (neutral) flow purely on the basis of temperature profile measurements at appreciable heights above ground (40 m and 150 m in the St Louis measurements).

In principle it may be advantageous to examine results of this type in terms of transfer theory – in practice the main limitation is the difficulty at present of specifying the K profile in an *a priori* and realistic fashion, avoiding the assumption of constant stress.

The general trend of vertical spread over a city area in neutral conditions has been given further confirmation by tracer experiments in Bedford and Cambridge in England (Barrett, 1970). Both these and the St. Louis, U.S.A., results contain evidence of increased crosswind spread as compared with flat open country. This is to be expected at least in a qualitative sense for the same reasons as apply to vertical spread (it may be recalled that σ_v is roughly proportional to u_*, as is σ_w, though application of the same similarity principles to the horizontal components has been questioned.) There is however a more immediate contribution to the increased crosswind spread, from wake effects and from the deflection of the plume into streets variously disposed with respect to the general wind direction.

The statistical effect on crosswind spread of the enhanced turbulence from trees, small buildings and variations in contour has been well demonstrated in Cramer's observations of turbulence and diffusion at Round Hill Observatory, Massachusetts. These observations have already been summarized in 4.3, and the point to be recalled here is that the plot of crosswind spread and σ_θ falls into line with that of the generally smaller magnitudes of turbulence and diffusion observed over the much smoother site at O'Neill, Nebraska. Barrett's experiments also included measurements of σ_θ at the emission site, but the relation between σ_y and σ_θ shows wide scatter and, according to Barrett, a significant dependence cannot be claimed. There are, however, two features in relation to other data which should be noted. Firstly, if Barrett's data are 'normalized' in terms of the wind fluctuation measurements the relation with distance is entirely consistent with the overall representation for a number of other (non-urban) sites in U.S.A. Fig. 5.16 shows the Bedford and Cambridge data plotted on a graph which also bears the boundary lines from the series of graphs of average σ_y/σ_θ versus distance already given in Fig. 4.6. Secondly, if a plot of angular crosswind spread against wind fluctuation is compared with that of *Prairie Grass* data at the comparable distance of 800 m [see Record and Cramer (1958)] it will be seen that the scatter is not essentially different. Taking the two sets of data together the values of σ_y/x are mostly within ± 40 per cent of a linear relation with σ_θ, over a range of about 10:1 in σ_θ.

In summary, for near-neutral conditions over flat terrain a fairly consistent description of the effect of roughness on crosswind and vertical spread is now available. A rough but useful statistical prescription of the magnitude of spread

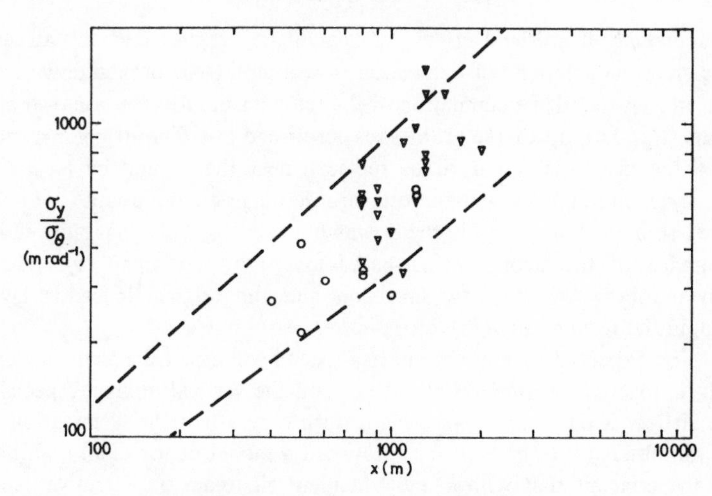

Fig. 5.16 – Variation with distance of crosswind spread σ_y normalized according to wind direction fluctuation σ_θ.

Observations in towns (Barrett 1970)
○ Bedford, England
∇ Cambridge, England

[Crown copyright, reproduced by permission, from W.S.L. Report LR 117 (AP).] Broken lines encompass range of σ_y/σ_θ, x plots in Fig. 4.6 for various relatively smooth sites in the U.S.A.

may be given from measurements or estimates of turbulence. This generalization also applies to the variation of crosswind spread with thermal stratification (excepting extreme cases): the position is not so well established for vertical spread.

The urban heat-island

There is abundant evidence that the temperature regime within an urban area is measurably different from that in open country. Mid-urban air temperatures are known to reach values higher than those in surrounding rural districts, by amounts ranging from a few tenths of a degree C to several degrees, the maximum effect (in U.K. cities) being found on calm clear nights in summer. There are probably two important contributions in the production of this *heat island*. The first is concerned with the natural heat transfers, in that the balancing of the net supply or loss by radiation must be provided predominantly by ground conduction and turbulent transfer in the air, whereas in vegetated areas the loss of heat by evaporation is an important and at times controlling contribution. As a consequence *surface* temperatures may reach appreciably higher values in towns and in cities than in country districts. There is also an additional 'man-made' supply of heat in cities which will contribute further to the enhancement of urban temperatures.

The detectable heat-island effect on clear nights implies that the surface-based inversion characteristic of open country will tend to be broken down over the urban area. A detailed examination of the temperature distribution over and near Cincinatti, Ohio, by Clarke (1969) has confirmed this. The strong inversion upwind of the city was found to be replaced near the ground by an *urban boundary layer* containing a temperature profile ranging from a lapse over the central area to a weak inversion in the downwind suburbs. This was followed by a reappearance of the strong surface-based inversion, the urban layer there being elevated between this new inversion and the original inversion layer higher still, in the form of an *urban heat plume*.

It is to be expected that the urban heat island will affect the airflow over a city, both as regards the horizontal pattern and the vertical mixing, especially when the airflow is naturally weak. This in turn may affect the distribution of pollution. Summers (1967) has in fact proposed a model of urban air pollution based on the concept that with a stable incident airstream the effective depth of mixing over a city is essentially determined by the heat released in the city from domestic heating and industrial operations.

In the first instance it is useful to note the magnitudes of heat transfer which are likely to be involved. Summers estimates that over the densely built-up area of Montreal the heat output from space-heating and industrial activities is on average about 160 Wm^{-2}, and points out that in winter this amounts to an appreciable fraction (about $\frac{1}{4}$) of the heat received from the sun. In open country even with strong insolation the rate of transfer of heat from the surface by turbulent and convective transfer is probably not much larger than the urban heating rate quoted by Summers – for example, measurements of turbulent heat flux range up to about 330 Wm^{-2}. On the other hand on clear nights the transfer of heat to the surface by turbulent transfer is probably only about 40 Wm^{-2}.

It is probably a reasonable assumption that in the absence of unnatural heat supplies the turbulent transfer of heat over a city would be rather larger than over open country by day, and of similar or smaller magnitude by night. The net effect of including also a fraction (say half) of the unnatural output of heat might therefore range from a modest increase in the (upward) turbulent heat flux H during the day and a marked decrease or even reversal of the (downward) heat flux at night. A first-order estimate may then be attempted by regarding the effect on the rate of vertical spread as characterized by the Monin–Obukhov scale L, which is proportional to u_*^3/H. From the trend of the *Prairie Grass* data on short-range vertical spread (see Fig. 4.9) it seems that a substantial reduction or reversal of the downward H (hence of $1/L$) at night is likely to be more effective in proportionately increasing vertical spread than a corresponding increase of the upward H during the day.

Pasquill (1970) has examined the estimates of vertical spread derived by McElroy and Pooler from the surface concentrations in the St. Louis experiments in various stabilities. Although there is no doubt about the apparent enhancement of vertical spread in *all stabilities* encountered, the data are clearly not adequate to test the foregoing speculation that the effect should be more marked in stable conditions.

In Summers's model referred to above the urban temperature profile is assumed to have a dry adiabatic lapse rate. Its depth at any position is determined by equating the increase in horizontal flux of heat (over that in the up-wind area) to the total amount of heat injected from urban domestic and industrial heating. Pollution is then assumed to be uniformly distributed with height throughout this layer. The model may be expected to be a considerable over-simplification in several respects and the degree of approximation which it provides in the estimation of pollutant concentration has yet to be demonstrated.

A useful general review of urban climatology has been given by Oke (1974).

Topographical influences

For the case of generally rugged terrain a survey of the air-flow in relation to dispersion processes was carried out at the United States Atomic Energy Commission site at Oak Ridge, Tennessee (Holland, 1953). This site is located in a large valley running roughly NE–SW in the Southern Appalachians. The observations display typical régimes of valley and drainage winds, and include some interesting data on the vertical motions, as derived from double-theodolite observations on neutral balloons. A special analysis was made of flights with trajectories either down-valley or across-valley, in the latter case over a ridge rising 100 m above the release station to a parallel valley some 2 km to the north-west. During the day, and especially in very unstable conditions, the ridge appeared not to have any mechanical effect comparable with the thermal eddies of large vertical extent, which alone would quickly disperse any material released in the valley. At the other extreme, on clear nights with light winds, vertical flow within the valley was obviously dominated by the circulations generated by cooling of the valley slopes, and was virtually isolated from the general airflow above the ridge.

Some quantitative indications of diffusion on a valley are provided by Hewson's (1945) well-known systematic study of the effluent of the Trail smelter in the Columbia River valley, about seven miles north of the boundary between Canada and U.S.A. In this area the valley is about 2500 ft deep and 1–3 miles wide, and the 400-ft stacks of the plant are located on a site about 100 ft above the valley floor. Effluent from these stacks had apparently caused damage to crops and vegetation at distant points downstream. Continuous records of the concentration of sulphur dioxide in the air at places up to over thirty miles downstream had shown a striking diurnal pattern. In the spring and summer especially, pronounced maxima tended to occur at about 8 a.m. at all places simultaneously. Moreover, the examples quoted by Hewson show maximum concentrations of similar magnitude at the different places, despite the fact that these places were at distances in the ratio of 4:1 and more.

The special investigations carried out by Hewson included an exploration of the three-dimensional distribution of effluent in the valley, using sampling instruments based on the ground and also some carried aloft by aeroplane and kite-balloon, and a detailed study of the air motion in the valley by numerous pilot balloon observations. From these Hewson was able to build up the following general picture of the travel and dispersion of the effluent plume. In the early morning hours, with light winds, the plume rose to about 500 ft above the tops of the stacks, and thereafter moved downstream within the valley without any appreciable spreading. Apparently any air movements and circulations generated by the drainage of cold air down the valley side had no important effect, and the plume remained undisturbed until it became entrained in the cross-valley circulation set up by the heating of the valley sides. As a result of this, effluent appeared relatively suddenly and at much the same time at all positions along a considerable length of the valley floor. It also seemed possible that these 'fumigations' could result merely from the spread upwards of the usual convective régime, which occurs irrespective of the ground configuration. Such fumigations have been recognized as a characteristic diurnal effect over level country (see 5.1 and 6.1). In any case it seemed that the result was to disperse the plume rapidly throughout the cross-section of the valley, giving similar concentrations at various places, the approximately exponential reduction thereafter being a consequence of the progressive ventilation of the valley by the air-stream above.

Hewson's study establishes the essential pattern of behaviour for a well-elevated source in a valley. For low-level sources the dispersion pattern may be expected to be different and some aspects of this have been brought out in meteorological and air quality surveys in the long narrow industrialized valley of Johnstown, Pennsylvania.

On clear autumn and winter nights well-established down-valley drainage flows were consistently observed, but with a progressively deepening mixed layer with temperature roughly constant with height, above which was a temperature inversion. The feature seems to be primarily a consequence of urban heating influence, as it is stated to be only slightly in evidence or entirely absent in the summer season. This was indicated by the typical trend of pollutant levels – in particular the absence of pollutant concentrations from the elevated (industrial) sources remaining in summer (see Ball and Muschett, 1967).

The diffusive behaviour in the urban-influenced drainage flow has been demonstrated in a tracer study with a low-level source of zinc cadmium sulphide particles (D. B. Smith, 1968). Exploration of the vertical distribution at a distance of 2 km, with samplers carried on the tethering cable of a balloon, showed an essentially uniform profile of concentration up to about 100 m with a rapid decrease thereafter in association with the encountering of the inversion layer. Note that this depth of uniform pollutant concentration is similar to the 'height of cloud' from a ground-based source in *neutral* conditions over flat relatively smooth country (as represented by the z_0=3 cm curve in Fig. 5.15, height of cloud being roughly 2½ times the mean vertical displacement). It is also quite similar to the vertical spread observed in the St Louis study in stable conditions.

The crosswind spread observed in Smith's study was also similar to that in neutral conditions over open country for the first 2 km or so, but by 6 km the tracer was uniformly distributed across the valley. On the basis of a limitation in crosswind spread Panofsky and Prasad (1967) suggest a simple model for the Johnstown pollution, in which pollution level is determined primarily by the reciprocal of the product of the wind speed (as reflecting the streamwise dilution) and the vertical gustiness (as representing the extent of vertical mixing). Statistics of the 12-hr average smoke pollution are broadly in accordance with this idea. However, since Smith's study showed that several kilometres were required for a single plume to be uniformly spread across the valley it seems likely that any uniformity from multiple sources would be more a consequence of the cross-valley distribution of these sources.

Studies such as those considered above leave no doubt about the lack of ventilation of deep narrow valleys in the generally stable conditions encountered on clear nights with light winds. However, even in the absence of nocturnal cooling it seems that valley air tends to be fairly effectively isolated from the main airflow above unless the wind is strong. This was clearly demonstrated by observations on the persistence of smoke released at the Shippingport reactor site in Pennsylvania (Hosler, Pack and Harris, 1959). The same effect is presumably responsible for the point source distribution observed by Hinds (1970) in mountainous terrain in Southern California. Apparently the most commonly observed pattern was one with the plume crossing ridge–canyon systems, showing canyon-floor concentrations roughly one-half of the ridge concentrations. Valley sheltering from higher-level transport of air pollution is also claimed by Lawrence (1962) to be evident in the distribution of ground-level concentration of SO_2 with height of ground, in a Lancashire hilly rural area surrounded by industrial sources.

Egan's (1975) review provides a summary of dispersion experiments designed to indicate the general effect of mountainous terrain. Analysis of the data is complicated by a lack of unambiguous comparison with the behaviour over smooth level terrain and of suitable reference measurements of the turbulence. For example, the useful tracer study in mountainous terrain by Start, Ricks and Dickson (1976) provides measurements of crosswind spread at distances up to nearly 5 km that are almost twice as great as the Pasquill 1961 tentative general estimates for level terrain, but it is not clear how much of this difference is a consequence of the different sampling durations represented in the measurements and the estimates. The general conclusion that topographical complexity imposes greater dilution of pollution is qualitatively reasonable, but full quantitative assessment needs further consideration, including an improved appreciation of the effects of individual hills on individual plumes (see later).

Plume behaviour near individual hills

Much practical interest has grown in the possibility that sites in hilly country may be subject to particularly intense local pollution as a consequence of the early impact of elevated plumes originating from plants at lower levels. The problem divides naturally into two general aspects. First there is the matter of the deflected trajectory of the plume as the high ground is approached and the extent to which the plume makes contact with local ground level at much shorter range than if the terrain had remained level. Then there is the effect of the flow disturbance on the intensity of turbulence, which affects the rate of spread of the plume and the distance at which the contact with ground-level is achieved. The early engagements with the problem have been mostly with the first of these aspects, in consideration of simple extensions of the traditional ideas of dispersion in undeflected flow, without any serious attempt at allowance for possible effects on the intensity of turbulence. Such an approach is clearly most relevant to the practically important case of a power-plant plume in strongly stable flow, when the ambient turbulence is relatively unimportant and the dispersion of the plume is primarily determined by buoyancy-induced entrainment in the course of plume rise. Flow distortion in the lee of a hill may also be of crucial importance. If the lee slope is steep enough (say greater than 1/5) there is a tendency for separation into a turbulent recirculating zone. An elevated plume within this zone is likely to be dispersed very rapidly to ground level. Even if the slope is too gentle for separation to occur the increase in turbulence may still be enough to bring plume sections to ground much closer to the source than over level ground. (See Chapter 7 for further discussion of some of these issues.)

The mathematical formalities have been developed especially by J. C. R. Hunt and co-workers in a series of papers discussing

(a) the reformulation of the two main working theories (the statistical theory and the gradient-transfer theory) in the coordinates of a converging–diverging curved flow,

(b) the application, especially of the gradient transfer approach, in potential flow around cylindrical and spherical shapes representative of ridges and hills, providing thereby some insight into the behaviour in the absence of separation of the flow, i.e. on the upwind slope and top of a hill.

Detailed reviews of these approaches have also been given by Hunt, Britter and Puttock (1979) in respect of the theoretical basis and Egan (1975) in the context of the overall practical aspects of dispersion over complex terrain.

The principal effects indicated by the potential flow considerations may be summarized as follows.

(i) Convergence of the streamlines reduces the local rate of spread of a plume (divergence having the converse effect) and the resultant rate of spread is a combination of streamline convergence (or divergence), which *per se* does not alter the concentration in the plume, and diffusion across the streamlines.

(ii) The reduction in the effective height of a plume from an elevated source on upwind level ground is more marked in the case of an isolated hill, than in the case of a ridge across the flow, because of the combined effect of convergence ('vertically') and divergence (acrosswind) in the case of the hill.

(iii) For strongly stable conditions (low Froude number), when the flow may plausibly be treated as horizontal flow around a vertical cylinder, the concentrations produced on the hill surface are critically dependent on the crosswind offset of the source from the 'stagnation' streamline.

(iv) In (iii) the maximum concentration C_m on the hill surface, from a source on the stagnation line, is theoretically greater than the plume centre-line concentration C_0 at the same distance in the absence of the hill and with the same diffusivity (Hunt, Puttock and Snyder, 1979). This is a consequence of *increased* cross-section of plume (due to the pronounced divergence near the stagnation line) which *per se* does not alter the concentration but leads to decreased cross-plume gradients of concentration and hence *decreased* cross-plume flux and rate of dispersion. The calculated C_m/C_0 is not generally much greater than unity, but may be 2 or more when the source is very close to the hill, though this may be an exaggeration introduced by the neglect of eddy distortion. In any case it is emphasized that this factor of two does not arise as a consequence of 'reflection' of the plume in the sense formally introduced in the conventional treatment of an elevated source.

(v) For an elevated source upwind of a 'half-cylindrical' ridge, in near-neutral flow, when the ground-level maximum concentration C_m does occur on the ridge surface its magnitude is found to be similar (slightly less) to that which would occur over level ground. In this case lowering of the plume relative to the surface appears to be approximately compensated by reduction in the plume dimension normal to the surface, so that, roughly speaking, at a given distance the ratio of σ_z to the local plume-height is unaffected.

(iv) For a hemispherical hill in near-neutral flow C_m/C_0 may be quite large, depending on the height of release in relation to the height of the hill. The contrast with a ridge is ascribed to the much closer approach of the plume to a hill surface in combination with increased rate of dispersion normal to the surface, the latter effect being a consequence of greater concentration gradients in the converged plume (diffusivity again being assumed unchanged).

Quantitative confirmation of these indications, from dispersion measurements either in the wind tunnel or at full scale in the atmospheric boundary layer are still awaited. In the meantime the possible relevance of the physical factors neglected in the potential flow treatments, especially the occurrence of flow separation and the topographical modification of the field of turbulence, are already receiving attention. On the former point numerical solutions of the Navier–Stokes equations and of plume dispersion as a result of constant diffusivity (Mason and Sykes, 1981) take the theoretical modelling a stage further and bring out the importance of flow separation on the upwind face of a hill. The probable importance of induced changes in the field of turbulence has been given emphasis by full-scale measurements recently reported by Bradley (1980) and Jenkins *et al.* (1981). In the first of these studies approximate doubling of all three components (σ_u, σ_v and σ_w) was observed near the surface ($z-d=9$ m, where d is an estimated zero-plane-displacement of 7 m) at the crest of a 170-m high hill in neutral conditions. Jenkins *et al.*'s measurements over a steep-sided 330-m high island also showed increased turbulence (relative to levels previously observed over the open sea) in the flow around the side of the island, in neutral conditions, most noticeably in σ_v. Further useful indications are provided by recent tracer studies (Jenkins and Whitlock, 1982) in the Sirhowy valley area of S. Wales, with the source at ridge-crest. These include increased lateral plume widths, lower centre-line concentrations and, by implication, increased vertical spread, relative to estimated values for level ground.

Clearly the detailed appreciation of plume dispersion in the vicinity of a hill is likely to be a laborious and costly matter, and from a practical standpoint much will depend on using a necessarily limited range of special numerical and observational studies to identify the most severe modifications of the behaviour over flat terrain.

5.5 POLLUTION IN URBAN AREAS

In practice most of the concern with local air pollution naturally arises in urban areas where, as already discussed in 5.4, the airflow as a whole is modified significantly by the greater aerodynamic roughness of arrays of buildings and by the 'urban heat-island'. Essentially the modification amounts to a general increase in the mixing action of the air as compared with that in the same airstream upwind over relatively smooth open country. In particular the characteristic stabilization of surface air on clear nights is developed to a smaller extent and may be absent altogether. There are also important details in the airflow – not only the effect of an individual building but also the channelling and vertical circulations produced in streets. Finally, apart from the special conditions of airflow, the pollution itself is emitted from a complex array of sources ranging from the effluent stacks of power stations and factories to the flues of domestic heating systems. The widespread distribution of individual sources has important consequences on the general distribution of the concentration of pollution in the air. It also sets special problems in the procedures for estimating the concentration distribution from source inventories and meteorological data but the details of this aspect are considered separately in the next chapter.

The first systematic and thoroughly documented investigation of the distribution of smoke and sulphur dioxide in and around a large industrial city was conducted in Leicester in 1937–9, under the supervision of the Atmospheric Pollution Research Committee of the Department of Scientific and Industrial Research. Over the period of three years regular daily and monthly observations were carried out with a variety of instruments at twelve stations including one in the centre of the city and two in the surrounding country. For a description of these, and for full details of the results and analyses, the official report (Department of Scientific and Industrial Research, 1945) should be consulted; in the present context only a brief indication of the principal results and conclusions need be given.

The characteristic pattern of pollution was revealed in composite maps which were drawn of the concentrations of native Leicester smoke or sulphur dioxide. Native values were obtained by subtracting from the total pollution estimates of the country pollution, based on the measurements on the outskirts of the city. This process was carried out for summer and winter averages at each station, discriminated according to the eight standard wind directions, and according as the wind speed was in specified ranges. For each season and wind-speed range it was thus possible to draw eight contour maps representing the distribution, and the composite maps were obtained by, in effect, superimposing the individual maps with the wind directions coincident. This led to smoother distributions of pollution, which it appeared could be adequately represented by circular isopleths. With decrease of the level of pollution the centres of the isopleths were displaced systematically downwind, but only slightly, and indeed one of the conclusions which was particularly emphasized was that the point of maximum concentration was never more than about half a mile from the centre of the city. For an area with a uniform distribution of equal sources it would be expected that pollution would build up steadily to a maximum at the downwind edge, and the result at Leicester was evidently mainly a reflection of the concentration of sources of pollution in the central districts. Lucas (1958) has carried out calculations which demonstrate this effect for a hypothetical distribution of sources in which the effective rate of emission rises linearly from the perimeter to the centre of the city.

Another outstanding feature of the Leicester results is that the decrease of concentration with increase of wind speed was usually less than according to the theoretical inverse law, the departure from this being the more marked the greater the instability as indicated by temperature gradient, with the effect that in the most unstable conditions the variation with wind speed was of little practical consequence. In the light of the current ideas it seems that there were probably two main contributions to this effect: a decrease of wind speed would mean that the rise of effluent due to efflux velocity and buoyancy would be greater, as discussed in 5.1, and this elevating of the effective source would lead to lower concentrations; furthermore, the effect of stability on vertical mixing near the ground is known to be more clearly related to the Richardson number, rather than to lapse rate alone and this means that in unstable conditions any tendency for a reduction of wind speed to produce higher concentrations would be opposed by an increase in effective instability. In general the total effect of wind speed on averaged levels will be complicated by the distribution of the main sources with height, their variation with time of day, and their thermal rise. This may explain why other estimates of the effect of wind speed on urban pollution give widely conflicting results (see World Meteorological Organization, 1958).

Since the Leicester Survey much further effort has been devoted to detailed studies of urban pollution, especially in recent years following the increased interest in developing and testing 'mathematical models' to simulate the distribution of pollution. Analyses such as that undertaken in the Leicester Survey leave two basic features to be clarified. The first is the nature and magnitude of the detailed variation of concentration in space and time. The second is the extent to which the absolute levels of concentration can be reconciled with information on the amounts emitted and on the dispersive conditions.

Temporal and spatial variations of pollution concentration

Some useful inferences may be immediately drawn from a consideration of the properties of the distribution from an individual source as basically outlined in 5.3. The individual distribution from an elevated source (in practice most sources have some elevation) is strikingly peaked, as indicated in the idealized representation for an elevated continuous source in Fig. 5.17. Variations in the position and magnitude of the maximum concentration arise from changes in wind direction, effective height of source and rate of vertical spread. Resulting variations in the concentration at a position fixed relative to the source may be considerable, especially if this is in a region where there is a high probability of occurrence of the maximum value. Some examples are given in Table 5.X for the area between the source and the typical position of the maximum.

The changes of concentration arising largely in this way are reflected in the typically highly irregular record obtained with a sampler stationed relatively close to and generally downwind of a source. Overlapping of the several plumes from an array of point sources will introduce some smoothing of the overall concentration pattern and hence of the time-lapse variations at any position. However, overlapping in the downwind direction seems unlikely in view of the relative narrowness of plumes, and effective crosswind overlapping at the distance of maximum concentration requires a crosswind spacing of sources less than about five times their height. It seems unlikely therefore that prominent sources will in practice be sufficiently close (and sufficiently similar in output) for substantial smoothing of the concentration patterns. The examples in Table

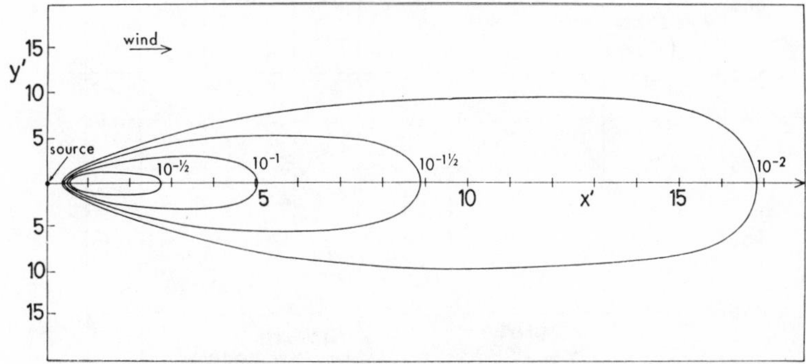

Fig. 5.17 – Idealized distribution from an elevated source. The isopleths are of $[\chi(x, y, 0)/\chi_m$, indicated as powers of 10, according to Eq (6.17) with $\sigma_y/\sigma_z = a$ irrespective of distance, and $\sigma_z \propto x^q$. The normalized alongwind distance x' is $\sigma_z\sqrt{2}/H\ [=(x/x_m)^q]$. The normalized crosswind distance y' is y/aH. For the simple case of $q = 1$ the apparent crosswind and alongwind proportions of the isopleths are correct if $x_m/aH = 5$, e.g. if $a = 2$ and $x_m/H = 10$, both of which are typical values.

5.X are for a particularly sensitive position in relation to a source but on the other hand the changes assumed in the controlling parameters are by no means extreme. Accordingly there is reason to expect order-of-magnitude changes in ground-level concentration, both in a spatial sense and in a time-lapse sense at a given position.

Table 5.X –Examples of modification of the ground-level concentration as a result of changes in (σ_z/H) and wind direction.
The following calculations are based on the distribution in Fig. 5.17, for a sampling position directly downwind of the source where initially $\sigma_z/H = 1/\sqrt{6}$, i.e. the 'position' at which $d\chi/d(\sigma_z/H)$ is a maximum

Change in (σ_z/H)	+20%	−20%
Modification of calculated χ if:		
(a) change is in σ_z alone	+75%	−73%
(b) as (a) with 10° change in wind direction	−12%	−87%
(c) change in H alone with 10° change in wind direction	+27%	−91%

Examples of statistics of variations which have been observed in systematic urban surveys of the concentration of sulphur dioxide are reproduced in Fig. 5.18. The first example (a), refers to 6-hr averages at three of the sampling stations used in Reading, England, in a 15-month survey during 1964 and 1965, while (b) refers to half-hour averages at one of the sampling stations in a survey

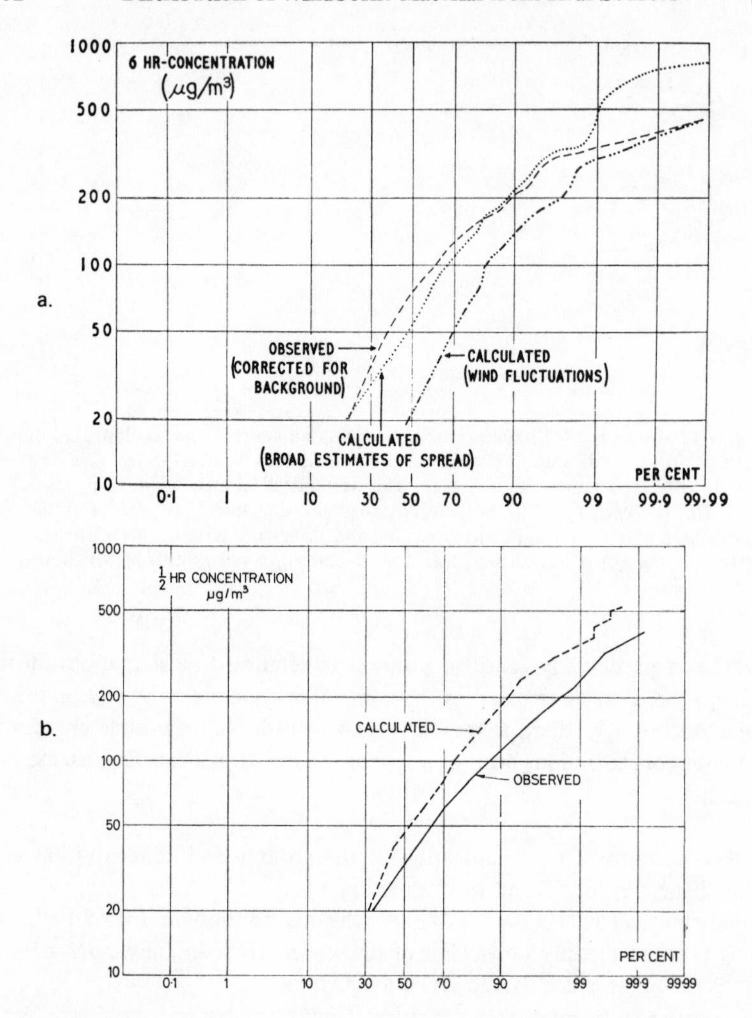

Fig. 5.18 – Examples of statistics of concentration of sulphur dioxide at urban sites.

(a) Cumulative frequency distribution for a combination of three sampling stations in Reading, England, 1964/65 (from an analysis by Marsh, reported by Pasquill (1970).

(b) Cumulative frequency distribution of ½-hr averages for Site No. 3 in the 1967/68 survey in Bremen (Fortak, 1970). (Reprinted in replotted form from Air Pollution Control Office Publication No. AP-86, U.S. Environmental Protection Agency, 1970.)

in Bremen, Germany, during the heating season 1967/68. The figures also include calculated distributions to which reference will be made later in this section. There is a considerable degree of similarity in the two independent sets of observations. Both distributions cover a range 10–1000 $\mu g/m^3$. Both show the same form and much the same degree of departure from log-normality.

The general question of the probability distribution of urban concentrations of pollution has received considerable attention, especially from Larsen. In his analysis Larsen (1971) concludes that the concentrations are log-normally distributed for all averaging times. As already noted the Reading and Bremen data show departures from log-normality but may be represented tolerably by this form over a substantial range. Daily average values in London (see Smith and Jeffrey, 1971) show a close approach to log-normality in both the spatial distribution and the time variation, but contain the same tendency as in the Reading and Bremen data for extremely high concentrations not to be realized.

Relation between urban pollution levels, emission inventory and dispersive conditions

The first serious attempt to calculate the general level of pollutant concentration built up in the airflow over a modern city appears to be that made by Lucas (1958). In this analysis use was made of estimates of annual national domestic usage of coal to arrive at a figure for the average rate of emission of sulphur dioxide, and factors allowing for population density, seasonal and diurnal variations of usage and low temperatures were applied to give a rate for Greater London in cold weather. This emission was assumed to be distributed uniformly over the whole area, and the concentration field was calculated by integrating along the wind direction the effects of a series of crosswind line sources, using for the latter the formula due to Sutton (1947b). On this basis Lucas calculated a mean winter concentration in Greater London, of 320 $\mu g\ m^{-3}$, which was compared with 1954–55 winter measurements at nine sites in London, averaging 300 $\mu g\ m^{-3}$ for the week-end and 400 $\mu g\ m^{-3}$ for week-days. Bearing in mind the crudity of the emission estimates and the questionable nature of the vertical dispersion estimates, the very close agreement is undoubtedly fortuitous, but the analysis is nevertheless a convincing demonstration that the broad magnitude of the long-term average pollution level over an urban area may be usefully estimated from emission estimates and general meteorological data. In such considerations the details of variation associated with the relatively narrow individual plumes are irrelevant and the principal factors in the dispersion are the average vertical spread and the wind speed. These factors can be estimated with a useful degree of precision and we shall see later in this section a more explicit demonstration of the success with which the long-term average level of concentration may be calculated.

A more detailed examination of the levels of sulphur dioxide pollution in a modern city in relation to the emissions (the source inventory) and the dispersive conditions was provided by Turner (1964). In this study Turner adopted the general system proposed by Pasquill (1961) for estimating dispersion from an individual source (see 6.6), with a modified system for prescribing the 'stability categories'. Essentially the modification consisted of replacing the qualitative description of insolation by the solar altitude and cloud cover, and defining seven stability categories by number in place of the original six defined by letters.

For the urban area concerned (Nashville, Tennessee) Turner was able to draw up source inventories for 2-hr intervals on a 1-mile square grid over a total area 17 × 16 mi. Each square was then assumed to produce a plume of pollution with Gaussian distributions acrosswind and vertically. As a practical means of allowing for the finite area of each source the σ_y in the plume was taken to be approximately 400 m at the centre of the source square, and to this was then added the point-source σ_y corresponding to the distance downwind of the centre. Concentrations produced by the whole array of sources were evaluated at points on a rectangular grid, for 2-hr periods in thirty-five 24-hr periods. Averages over each 24-hour period were then plotted, isopleths drawn, and values interpolated for the 32 sampling stations at which 24-hr measurements were available. The individual pairs of calculated and observed concentrations showed large discrepancies which were largely random in nature though with some tendency on the whole for the calculated values to be overestimated. Neglecting zero values 70 per cent of cases gave calculated values within a factor of two of those observed. Turner emphasizes a number of limitations unavoidably incorporated in the model, but his study gives the first realistic indication of the general level of correlation to be expected between the details of urban population and data on emissions and meteorological conditions. The experience has been largely confirmed and extended in several important ways in a study conducted in Reading, England, by the British Petroleum Research Centre, and a summary of the main features and results follows.

The Reading experiment included extensive measurements of the sulphur dioxide concentration distribution and detailed estimation of the source inventory over a 15-month period during 1964 and 1965. Aspects of the study have been discussed in a series of papers. The last of these publications [Marsh and Withers (1969)] compares the observations with estimates derived from the source inventory and meteorological data, using a number of methods.

The pollution data consisted of sulphur dioxide concentrations averaged over 6-hr sampling periods, for 40 sites in and around the town. Continuous records were available for wind direction, wind speed, air temperature, humidity, and the intensities of the vertical and lateral components of turbulence. The wind speed and turbulence measurements were made at a height of 14 m in a relatively open space, 100 m or more from buildings about 10 m high, and about 700 m from the taller buildings in the centre of the town. For the source distribution, three categories of fuel-burning installations were considered – domestic heating, industrial and commercial installations for space-heating, and processing plants. Domestic (space-heating) sources were divided into 251 housing areas with a diameter of roughly 400 m, and the strength of source in each area was based on official estimates of the average fuel consumption in private houses.

Installations burning more than 10 tons of fuel per year were considered individually.

For evaluation of the diffusion formulae, Marsh and Withers defined 67 area sources that were treated as individual sources if their centres were farther than 1000 m from the sampling site, and 7 major sources, that were treated as individual sources regardless of distance. If the centre of an area was less than 1000 m from the sampling site, the individual housing areas or installations were also treated as individual sources. In some of the calculations, the housing areas and source areas were treated as equivalent line sources. For domestic sources, a standard height of 15 m was adopted, but for individual sources the effective heights were estimated from the chimney height and the plume rise calculated from the CONCAWE empirical formula.

Marsh and Withers give an extensive series of comparisons of observed and calculated concentrations from which only a selection most directly reflecting on the present point of interest is reproduced here. The calculated concentrations were obtained in three ways:

(1) Using the familiar Gaussian form for the distribution from a point source with the standard deviations (of particle displacement) σ_y and σ_z obtained from wind fluctuation data using the method proposed by Hay and Pasquill (1959).

(2) As in (1) but with σ_y and σ_z obtained from the broad estimates of spread suggested by Pasquill (1961).

(3) From a regression of the observed values against a parameter involving the air temperature (as representing the variation of effective overall source strength) and the reciprocal of the wind speed.

The selection of comparisons, in which the observed concentrations have been corrected for 'background' concentration using measurements made upwind of the town is given in Table 5.XI.

It is seen that although the average level of concentration is usefully predicted by the dispersion estimates, especially when the emission and wind data are the most reliable, there are very large discrepancies in individual pairs of 6-hr values. Further analysis of these data subsequently, reported by Pasquill (1970) and included in Fig. 5.18(a) throws an interesting light on these discrepancies. It appears that the *frequency distribution* of the calculated 6-hr values is in quite good agreement with the observed distribution. A reasonable interpretation would now appear to be provided by the arguments given earlier in this section concerning the spatial and temporal variability of concentrations arising from an array of sources. Referring to the details in Table 5.X the 'changes' cited there for σ_z, H and wind direction may alternatively be regarded as errors which might realistically be expected to arise or indeed be exceeded in any dispersion calculation of this nature. In so far as these errors are random they cannot greatly distort the shape of the typically wide frequency distribution of concentration values, even though they produce individual deviations which, according to the calculations in Table 5.X, may occasionally be very large.

Table 5.XI – Comparison of observed and calculated SO_2 concentrations at six selected sites in Reading, England (Marsh and Withers, 1969)

Method of calculation	Mean of 6-hr concentration ($\mu g/m^3$)		R.m.s. difference between calculated and observed C	$\frac{\text{R.m.s}}{\text{Obs. } C}$
	Obs. $\bar{C}$†	Calc. $\bar{C}$		
All periods				
(a) Wind fluctuation	68	38	84	1.2
(b) Broad estimates of spread (stability categories)	68	88	132	1.94
(c) Regression on temperature and wind	68	–	68	1.0
119 selected periods‡				
(a) Wind fluctuation	66	53	63	0.95
(c) Regression on temperature and wind	66	–	57	0.86

Methods (a) and (b) follow Pasquill (1961).

†Corrected for background concentration as observed upwind of town.

‡Steady wind direction and speed, 6 a.m. to 12 noon and 12 noon to 6 p.m. only, for which most confident estimates of emission were made.

The general position therefore seems to be that the level of pollution in an urban area with a complex array of sources is demonstrably in accordance with the source inventory and the conditions of wind speed and dispersion in two respects:

(a) the long-term spatial average,
(b) the statistical frequency of occurrence of specified short-term local levels of concentration,

and in both these respects therefore the pollution levels may be predicted, probably within a factor of two or less, depending on the precise quality of the information available. However, there are likely to be larger unpredictable discrepancies in individual local values, of a magnitude which is broadly understandable in relation to practical uncertainties in the specification of the meteorological and dispersive conditions. It should be emphasized that these limitations apply for relatively ideal terrain and conditions of airflow, and it must be expected that additional complexities in the form of irregularities of terrain and rapid changes in the airflow characteristics will be reflected in even greater uncertainties in the prediction of urban pollution.

Magnitude of pollution in smog

Some interesting indications of the magnitude of sulphur dioxide pollution in smog, in relation to the amount of sulphur dioxide emitted, are provided by the case of the London smog of December 1952. A brief account of the general meteorological aspects of the situation has been given by Absalom (1954). The main feature was the maintenance over the country of the almost stationary central region of an anticyclone, from midnight 4–5 December to 9 December. With little cloud, and very light or calm surface winds, dense fog persisted over the London Basin throughout the period. No information was available on the local temperature inversion, but at Cardington 50 miles away, where there was little fog, the inversion extended to heights between 500 and 1000 ft. Wilkins (1954) has described the data of mean daily air pollution which were available from sites in the Greater London area. The apparatus used at these sites was the type in which air is drawn successively through an inverted funnel, a filter paper, a solution of hydrogen peroxide and a flow-meter, and is assumed to collect gaseous sulphur dioxide, and particles and droplets less than about 20 μ in diameter. Wilkins's figures show that the average sulphur dioxide concentration, for the eleven sites which operated regularly, rose from the normal figure of just over 290 $\mu g\ m^{-3}$ in the 24 hr ending noon on 4 December, to near 2000 $\mu g\ m^{-3}$ during the period ending noon on 8 December.

Assuming that the measured concentrations applied to the whole volume of air over the area of Greater London (500 square miles) to the top of the fog layer (approximately 300 ft), Wilkins pointed out that the rate of increase of approximately 600 $\mu g\ m^{-3}$ per day was equivalent to a total daily accumulation of sulphur dioxide of about 70 tons. This is only about 4 per cent of the estimated daily release of sulphur dioxide, implying a rate of removal of sulphur dioxide which may at first seem surprising. A rough analysis of the rate of sulphur dioxide on this occasion has also been made by Meetham (1954). Meetham assumes a steady state, in which a sulphur dioxide concentration of 2.1 mg/m^3 (equivalent to 0.74 p.p.m. by volume) is maintained, with a release which in the absence of decay would be equivalent to an increase of concentration of 8.0 mg/m^3 per day, in the air to a height of 500 ft over an area of 450 square miles. If the released sulphur dioxide decays exponentially with time after release, with a time constant t', these figures mean simply that $t' = 2.1/8.0$ days or approximately 6 hr. Alternatively, putting $t' = h/v_d$, with $h = 500$ ft, this corresponds to a deposition velocity v_d of 0.7 cm/sec. Wilkins's figures give a maximum average concentration (0.7 p.p.m.) only $\frac{1}{8}$ of the increase which would have occurred per day had there been no decay. The difference between this and the value of $\frac{1}{4}$ in Meetham's analysis is due mainly to taking $\bar{h} = 300$ ft, but also partly to taking a release of 2000 tons of sulphur dioxide per day and an area of 500 square miles, compared with respective figures of 1400 and 450 adopted by Meetham, again assuming a steady state $t' = 3$ hr and $v_d = 0.8$ cm/sec. In the model of diffusion and decay implied in these calculations the value of v_d is independent of the assumed value of h, and the small discrepancy between the values obtained from Meetham's and Wilkins's data is due to the differences in the estimates of the amount of sulphur dioxide released and of the area affected.

It is interesting to consider to what extent the apparent loss of sulphur dioxide could be due to a slow advection of air over the area. For rough calculation assume the material to be released into a box 20 miles square and 500 ft (0.1 mile) high. If the loss of sulphur dioxide is ascribed entirely to a flow out of the downwind side of the box, the wind speed required would be 20/0.1 times the deposition velocities evaluated above, i.e. approximately 1.5 m/sec or 5 ft/sec. According to data quoted by Lucas (1958) daily average surface wind speeds at Kew Observatory were 0.5, 0.5, 0.6 and 1.6 ft/sec for the period 5–8 December, i.e. an overall average of 0.8 ft/sec. Unless, therefore, the height of the layer into which the sulphur dioxide was mixed has been seriously underestimated, it seems that clearance by advection could not have been an important factor. A similar conclusion follows from the calculations made by Lucas (1958), using an integrated form of a simple line-source equation, in which vertical spread of the cloud is assumed to increase linearly with distance.

The considerable variations in space which are evident in the measurement of sulphur dioxide concentration, and the lack of precise knowledge about the air movement and mixing process, are such that the inferences drawn above must be regarded as very rough. There is, however, a fairly convincing indication that in the presence of smog there is a mechanism, other than advection, which removes sulphur dioxide as effectively as it appears to be removed by deposition in normal weather. Solution of the sulphur dioxide in fog droplets which are subsequently deposited is an obvious possibility to be considered, but the quantitative acceptability of this mechanism has yet to be demonstrated.

6

The estimation of local diffusion and air pollution from meteorological data

This chapter deals with the application of all the ideas, results and experience which have been discussed in the preceding chapters. Essentially the practical problem may be stated as follows – given a specification of the releases of an air pollutant, of the nature of the terrain and of the meteorological conditions, what is the expected distribution of pollutant concentration? In utilizing such general rules as can now be formulated for answering the foregoing question it should constantly be kept in mind that the applicability of these rules depends on the extent to which the actual circumstances approach the somewhat idealized conditions necessarily specified or selected in the basic theoretical and experimental studies. In particular this means that in general the methods cannot be assumed to give reliable estimates in the following circumstances:

(a) when the airflow is indefinite (calm conditions);
(b) when there are marked local disturbances of the airflow, e.g. in the immediate vicinity of buildings and obstacles, unless the diffusing cloud has already grown to a size considerably bigger than the disturbance;
(c) when the airflow is channelled or when it contains circulations or drainage set up by the heating or cooling of undulating or hilly terrain.

Even in these complex conditions however it may often be possible to speculate usefully on the sense and extent of the changes which the complexities are likely to impose on the diffusion, but clearly such cases need to be given special consideration. In any case, even if the conditions of terrain and weather approach the ideal of uniformity and steadiness, the estimates which may be made correspond to an ensemble average, from which there may be appreciable deviations on individual occasions.

The chapter has been written in a fairly self-contained form so that if necessary it may be studied and used without reference to the preceding chapters.

6.1 QUALITATIVE FEATURES OF DIFFUSION AND THEIR VARIATION WITH WEATHER CONDITIONS

There are certain aspects of diffusion which are obvious from the visible behaviour of smoke in the atmosphere, and an appreciation of these features forms a useful introduction to the more quantitative applications which follow. It is well known that the smoke trail from a source on the ground takes a variety of forms according to the weather conditions and the time of day, and that certain characteristic properties may be recognized. In open country, with at least a moderate wind speed and a thoroughly cloudy sky (or even with a clear or partly clear sky as long as the wind is strong), the smoke forms into a fairly straight well-defined trail which increases perceptibly and steadily in width and height as distance from the source increases. If, however, the wind is light, and there is sufficient sunshine to warm the ground surface, a much greater degree of irregularity appears in the form of the plume. Apart from the obvious effect that the smoke itself may be heated by absorption of solar radiation, it is clear that the movement of the air itself leads to a more rapid spread of the smoke in the vertical, and to an erratic variation in the direction of travel of successive sections of the smoke plume. The result is that the plume has a sinuous and sometimes even disconnected form, and rapidly reaches a stage when it is no longer visible. On the other hand, at night, if the sky is sufficiently clear to result in appreciable cooling of the ground, and the wind is light, bodily rise and vertical spread are considerably reduced and the smoke trails off downwind in a compact visible form for appreciable distances.

The three categories just described constitute the simplest classification of diffusive conditions. They are associated with characteristic vertical gradients of temperature and associated stabilities in the lower atmosphere, i.e.

near-zero gradient – neutral stability,
lapse (decrease of temperature with height) – unstable,
inversion (increase of temperature with height) – stable.

To be exact, because of the compressible nature of the atmosphere the thermal stability must be specified in terms of the difference between the actual vertical gradient of temperature and the *dry adiabatic lapse rate* (0.98°C fall per 100 m rise). The latter is the rate at which a parcel of air changes its temperature adiabatically as a result of the change of pressure associated with vertical displacement. Alternatively the specification may be in terms of the gradient of *potential temperature,* the latter being the temperature which the air would assume if brought adiabatically to a standard pressure. Differences between actual and potential temperature gradients are usually unimportant over very shallow layers close to the ground (i.e. within a metre or so) but they become progressively more important as the layer of interest deepens.

In accordance with the dimensions attained by a smoke plume the concentrations within the plume tend to be relatively low in unstable conditions and relatively high in stable conditions. In addition the irregularities which occur (to various extents), in the instantaneous appearance of a plume of smoke, have important consequences on the average crosswind distribution of smoke concentration. These irregularities arise from the fact that successive sections of the plume travel from the release point along different (approximately straight) trajectories (see Fig. 6.1). The result is that in the area downwind of the source the crosswind distribution of smoke concentration at any instant is characterized by relatively high values on a narrow front, while the distribution averaged over some period of time possesses lower values extending over a wider front. Accordingly the average concentration at any position will normally be made up of a series of fluctuations, the amplitude and rapidity of which will depend on the distance from the source and the diffusive conditions.

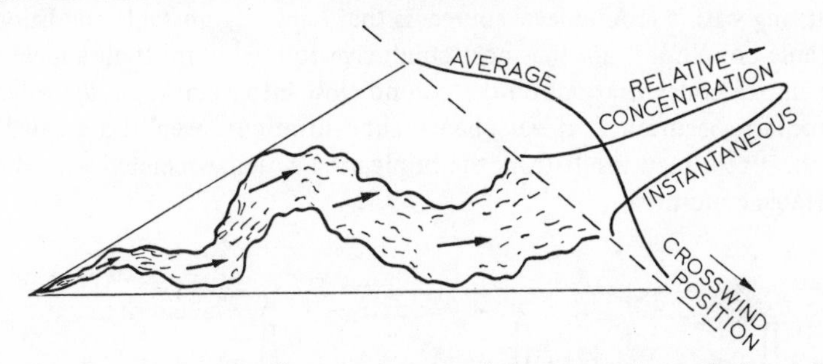

Fig. 6.1 – Instantaneous and average aspects of the crosswind spread of a smoke plume.

When the source of smoke is elevated above the ground, as in the case of effluent from an industrial chimney, the preceding modes again appear in the horizontal structure, and the smoke plume also assumes characteristic shapes in the vertical. These are easily identifiable when viewed from the side some distance away. The vertical displacements are undoubtedly amplified if the smoke is hot, as was clearly recognized in an early discussion of smoke plumes (Etkes and Brooks, 1918), but there are basic patterns which are essentially determined by the properties of turbulence and convection in the atmosphere. In the American literature on the behaviour of effluent from chimney stacks certain graphic terms have been coined to describe the modes of behaviour in the vertical (see Church (1949) and the United States Atomic Energy Commission publication edited by Slade (1968)). The three basic modes are:

looping – in unstable conditions
coning – in near-neutral conditions
fanning – in stable conditions

The looping form is produced by successive sections of plume travelling with different inclinations, not by the up and down motion of a given section. In coning, the successive sections follow trajectories which are not widely different, and the plume as a whole tends to assume a steady conical form. In fanning the vertical spread is greatly reduced and after some initial growth may even be completely halted, but the crosswind spread is maintained by the marked turning of wind with height which is characteristic of stably stratified flow. The schematic diagrams in Fig. 6.2 illustrate these three main forms, and also two transitional conditions. *Lofting,* in which vertical spreading occurs much more effectively on the upper side of the plume than on the lower side, occurs with the diurnal transition from unstable to stable conditions, near sunset in clear weather. *Fumigation* is the reverse of lofting, and occurs with the reverse transition in stability in the early morning. For a more detailed discussion of these properties, and of their implication as regards the ground-level distribution of effluent from a high stack, reference should be made to the U.S. A.E.C. publication. One of the most important implications, in obvious contrast to the situation arising with a ground-level source, is that relatively unstable conditions, and not stable conditions, are the most conducive to the intermittent appearance of heavy smoke concentrations on the ground close into a stack. On the other hand a temporary occurrence of very heavy concentrations, even at a great distance from the stack, may result from the fumigation action associated with the onset of unstable conditions.

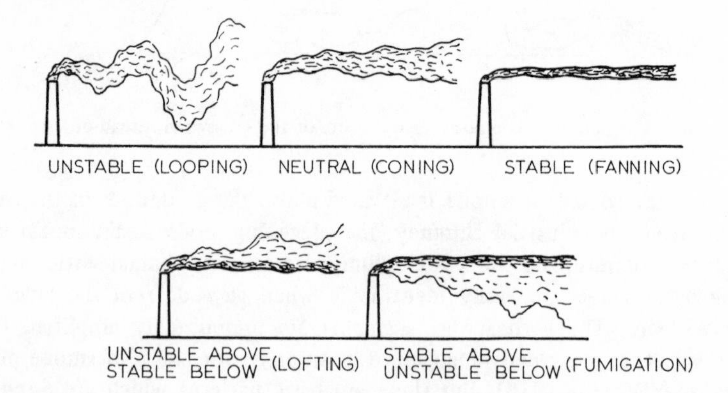

Fig. 6.2 – Characteristic forms of smoke plumes from chimneys (Church, 1949, and Slade, 1968).

Because of the clear association between diffusive action and the thermal stability of the atmosphere, temperature gradient was adopted from the beginning as the main indicator, and much effort has been expended in many countries towards obtaining statistics and maintaining current measurements of this quantity. Although it may be particularly effective in indicating the likelihood of extreme conditions such as fanning and fumigation, there has been a growing recognition of its inadequacy on its own as a general indicator. This arises partly from the observation that the influence of stability involves the vertical gradient of wind speed, as well as that of temperature, and partly from the observation that diffusion is manifestly affected by the roughness and topography of the terrain. Accordingly, there is a growing tendency, supported both by practical experience and by the development of the fundamental understanding of diffusion processes, either to adopt more sophisticated stability parameters such as the Richardson number and, more recently, the Monin–Obukhov length, or to use measurements of the intensity and scale of turbulence.

The widths of the ink traces obtained in routine instrumental records of wind speed and wind direction were used as a measure of turbulence at a very early stage. An application of this method in a qualitative way is implied in the classification of types of wind fluctuation by Giblett *et al.* (1932), based on records of wind speed and direction at Cardington. In this classification eddies were divided into four types, ranging from a Type I, which was characterized by gusts of large amplitude and long period, typical of well-developed convection and thunderstorms, to a Type IV represented by the fluctuations of very small amplitude which tend to occur on a clear night. This classification may be regarded as the fore-runner of a more detailed system established at the Brookhaven National Laboratory, U.S.A. In this case the range and appearance of the fluctuations of horizontal wind direction were used to define the type of airflow. An original division into four categories by M. E. Smith (1951), was later extended to five categories (Singer and Smith, 1953). There is an obvious association to be expected between wind-trace types and plume characteristics. The availability of wind direction records of this type also provides a useful basis for making reasonably realistic estimates of the standard deviation of the wind direction fluctuation, for use in the statistical theory approach to the calculation of crosswind spread.

6.2 THE SPECIAL METEOROLOGICAL FACTORS IN DIFFUSION CALCULATIONS

The parameters involved in the application of the most advanced methods of estimating dispersion in the atmospheric boundary layer are summarized in Table 6.I. The two basic parameters u_* and H_0 are in principle measurable in special observational programmes, but these are undertaken only in the most fundamental research studies and in practice the requirement is to estimate the parameters from regularly available meteorological data or from special measurements of a less demanding nature. Using the Monin–Obukhov theory u_* and L (hence H_0) may be evaluated from an estimate of z_0 (adequately made from the nature of the surface) coupled with measurements of the low-level wind speed and of the temperature difference between two low levels. An example of a suitable nomogram system with measurements of wind speed at a reference level

Table 6.I – The special meteorological factors.

Property and significance	Parameter	Definition – further details
Aerodynamic roughness. Determines vertical profile of wind and mechanical maintenance of turbulence in boundary layer	Roughness parameter z_0	$\bar{u}(z) = \frac{u_*}{k} \ln \frac{z}{z_0}$ in neutral flow, i.e. near-zero Ri
	Shearing stress and friction velocity	$u_* = (\tau_0/\rho)^{1/2}$
Thermal stratification. Represents magnitude of buoyancy forces which enhance or diminish turbulence and mixing	Richardson No.	$Ri = \frac{g}{T} \frac{\partial\bar{\theta}/\partial z}{(\partial\bar{u}/\partial z)^2}$
	Vertical flux of sensible heat, H, at the surface	$H = \rho c_p \overline{w'\theta'} = -\rho c_p K_H \frac{\partial\bar{\theta}}{\partial z}$ determined by surface heating and cooling
	Monin–Obhukov length scale	$L = -\frac{\rho c_p T}{kg} \frac{u_*^3}{H}$
	Mixing depth h	Determined by prevailing conditions of turbulence and convection. Development frequently limited by overhead inversion
	Deardorff convective velocity	$w_* = (gH\, h_c/\rho c_p T)^{\frac{1}{3}}$

Turbulence Directly represents mixing quality of the air. May be used more or less directly to prescribe rate of diffusion of windborne material	R.m.s. fluctuation of eddy velocity	$\sigma_w = (\overline{w'^2})^{1/2}$ etc.
	Intensity of turbulence	$i = \sigma/\bar{u}$
	Rate of dissipation of turbulent kinetic energy ϵ	Related to shearing stress and heat flux as determining feed of turbulent energy
	Integral scales of turbulence	$t_s = \int_0^\infty R(t)\,\mathrm{d}t$, $l_s = \int_0^\infty R(x)\,\mathrm{d}x$ etc., representing scale of mixing action
	Equivalent wavelength for peak $nS(n)$	$\lambda_m = \bar{u}/n_m = a\bar{u}t_s$, a depending on spectrum shape
	Lagrangian/Eulerian scale ratio	$\beta = t_L/t_E = \bar{u}t_L/l_E$
Diffusivity. Represents ratio of diffusion relative to gradient	Eddy diffusivity	K = flux/gradient. Representable in terms of intensity and scale of turbulence

z_1 and temperature at z_1 and $0.2\,z_1$ is reproduced here at Fig. 6.3. Some guidance on the qualitative estimation of z_0 is in Table 6.II.

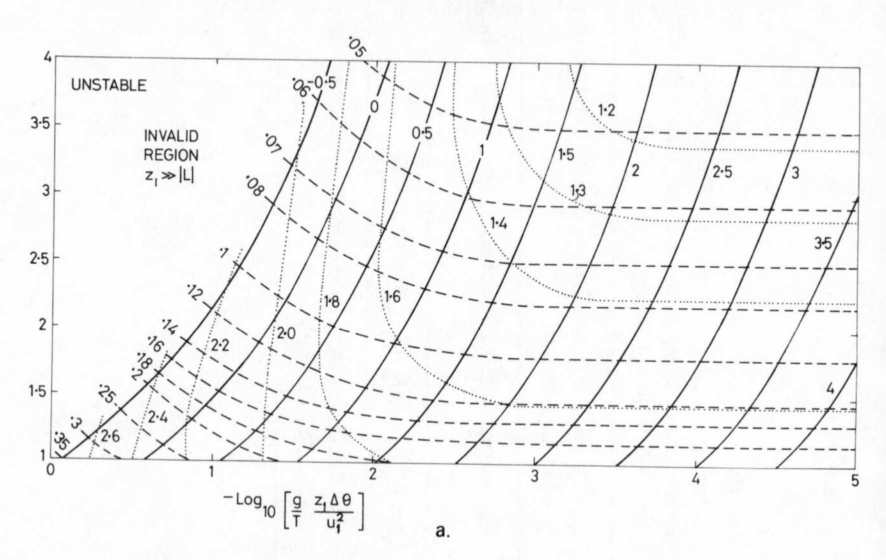

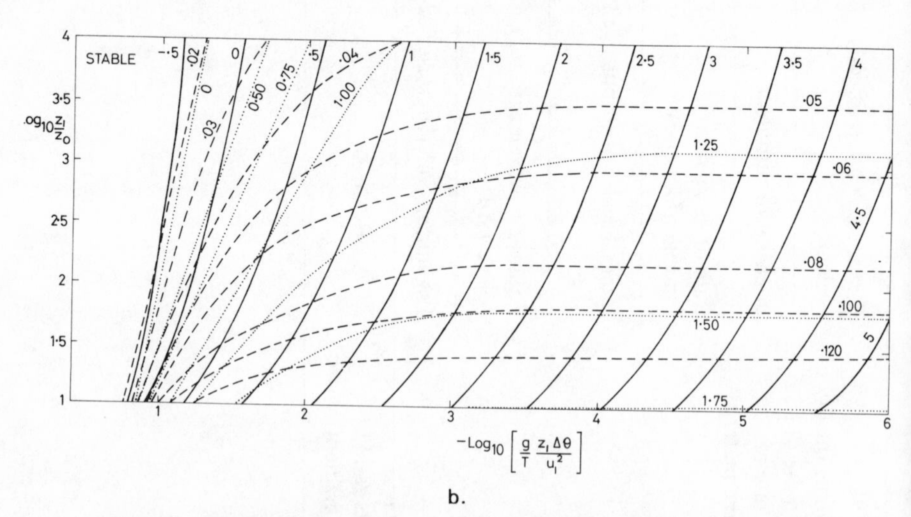

Fig. 6.3 – Nomograms for deducing the heat flux H_0, the friction velocity u_* and the Monin–Obukhov length L from data on the surface roughness z_0, wind speed u at height z_1 and temperature difference $\Delta\theta$ between z_1 and $0.2\,z_1$, (a) in unstable conditions, (b) in stable conditions.

Notes: The contours are ——— $\log_{10}(L/z_1)$, – – – – u_*/u_1, ... $\log_{10} |H_0|/u_1\Delta\theta$. Units are $\mathrm{ms^{-1}}$ for u, m for z, °C for $\Delta\theta$, $\mathrm{Wm^{-2}}$ for H_0. The Businger (1966) simple analytic forms of the M–O functions were assumed (i.e. $\phi_M = (1-16z/L)^{-\frac{1}{4}}$, $\phi_H = \phi_M^2$) for unstable conditions, and the Webb (1970) forms (i.e. $\phi_M = \phi_H = 1 + 5z/L$) in stable conditions.

Table 6.II – Typical values of z_0 and H_0 and formulae for L, w_* and h_s in practical units.

$z_0(m)$

Grass: closely mown 10^{-3}, short (*ca* 10 cm long) 10^{-2}, long 3×10^{-2}
Agricultural–rural complex 0.2
Towns, forests 1

$H_0(Wm^{-2})$ in S. England

Summer day 200 clear night 30

In units m, W and s

$L = 9.1 \times 10^4 u_*^3 / H_0$, $w_* = 3 \times 10^{-2} (H_0 h_c)^{\frac{1}{3}}$
h_s (empirical form in Fig. 2.9) $= 2.15 \times 10^4 u_*^2 H_0^{-\frac{1}{2}}$

The ideal site for such measurements is a uniformly smooth ($z_0 \cong 1$ cm) level area such as an airfield, with $z = 10$ m. Areas with much larger effective values of z_0 are more likely to be of interest in respect of pollution, with roughness elements which are large (hedges, trees, buildings) and probably of non-uniform size and distribution. This means that the foregoing choice of low reference levels (desirable both for operational convenience and the occurrence of relatively large vertical gradient of temperature) may be too small (in relation to z_0), and higher levels will be less convenient and will often involve smaller temperature gradients requiring greater accuracy of measurement. In either case the area-representativeness of measurements at one position is much more in doubt than for an ideal site. Theoretically this difficulty may be partly overcome by using Rossby-number similarity to link the properties z_0, u_* and H_0 in terms of the geostrophic wind speed G, as in the nomograms reproduced in Fig. 6.4.

Depth of mixing (h_c) in unstable conditions may be derived from synoptic (upper-air) data, or from soundings with a monostatic sodar or acoustic sounder, or calculated from a theoretical model referred to in Section 2.2. An example of a nomogram based on the latter is given here in Fig. 6.5. For stable conditions h_s may be calculated from estimates of u_* and H_0, using the formula fitted to available observational data in Fig. 2.9, and this formula is repeated here for convenience in Table 6.II together with other useful formulae and numerical data.

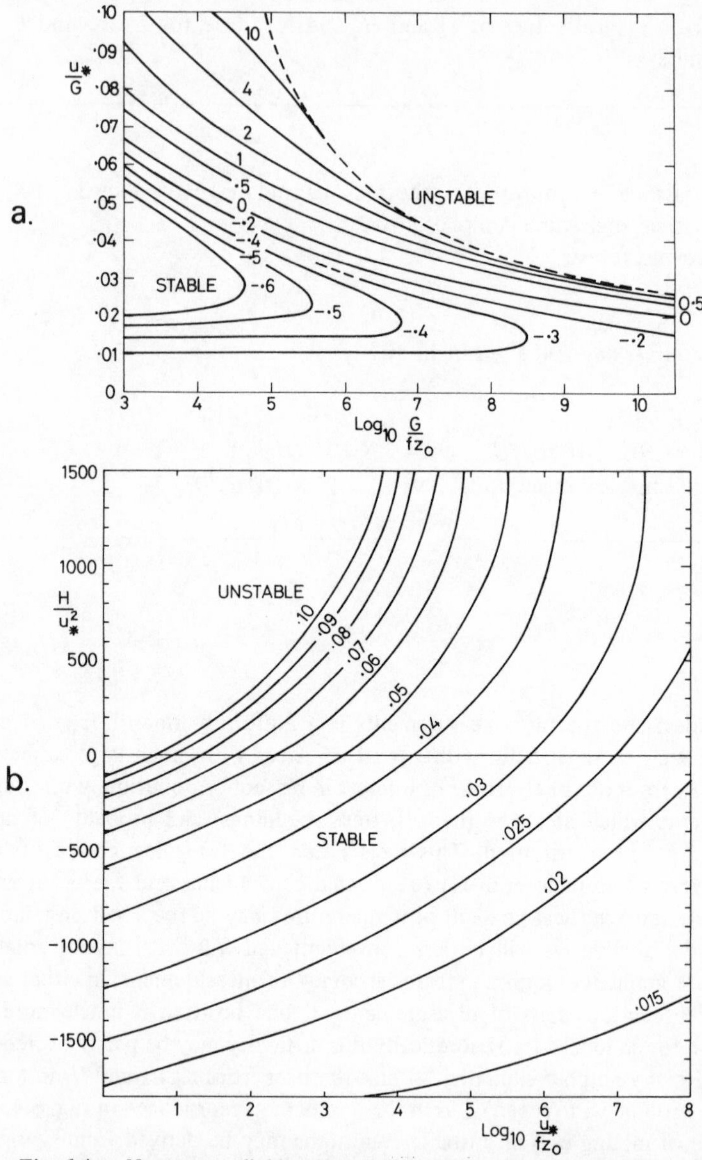

Fig. 6.4 – Nomograms linking 'external' boundary layer parameters G. f, z_0 to 'internal' parameters u_* and H_0, (a) Contours of H_0/G^2. (b) Contours of u_*/G.

Notes: Units are as in Fig. 6.3. The curves are constructed from Arya's (1975) Rossby-number similarity form for the wind profile. Nomogram (a) may be used to derive u_* (hence L) from G, H_0, z_0: Nomogram (b) may be used to derive G given z_0, u_* and H_0. This is a useful procedure if H_0 and u_* are available from 'ideal-site' measurements and Fig. 6.3, and it is desired to use the same value of H_0 for an adjoining 'non-ideal' site as an approximation, with an estimate of G, so as to obtain an estimate of u_* from nomogram (a).

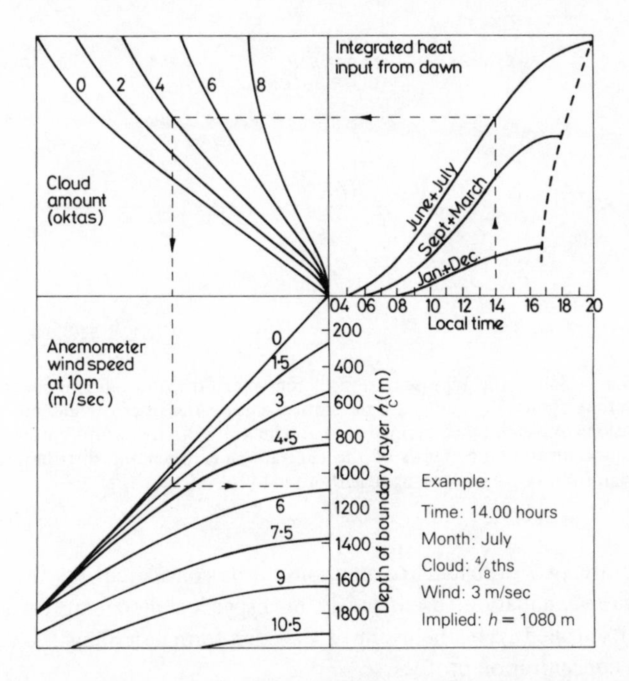

Fig. 6.5 – A nomogram for estimating the depth of the mixed layer in daytime conditions typical of the U.K., assuming no marked advective effects or basic changes in weather conditions. The broken line shows how the diagram is to be used.

6.3 DISPERSION FORMULAE FOR THE LOCAL DISTRIBUTION OF CONCENTRATION FROM SURFACE RELEASES

Consider first the mathematical formulation of the distribution in terms of the time-mean dimensions of the plume acrosswind and vertically for the schematic arrangement in Fig. 6.6. The dimensions are customarily represented by the quantities σ_y acrosswind and σ_z vertically, as defined in 4.1, and sometimes by $\bar{Z}$ for the vertical spread or by *outer* dimensions prescribing the height or overall width defined by a concentration say one-tenth of that at ground-level or on the x axis.

The spatial distribution of concentration is expressible in terms of the dimensions σ_y and σ_z, the shape of the distribution acrosswind and vertically, the wind speed and source strength, through the continuity condition

$$\int_0^{\infty}\int_{-\infty}^{+\infty} u(z)\chi(x, y, z)\,\mathrm{d}y\,\mathrm{d}z = Q \tag{6.1}$$

where Q is the quantity of material released in unit time.

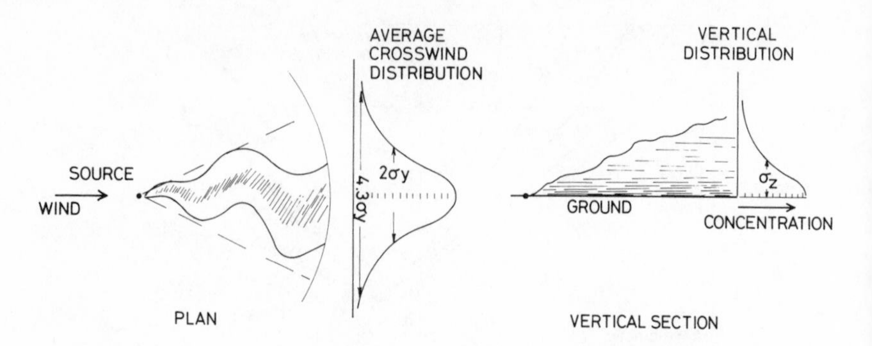

Fig. 6.6 – Schematic representation of time-mean distribution and spread for a continuous plume. σ_y and σ_z are the statistical measures of crosswind and vertical dimensions explained and defined in 4.1. The 4.3 σ_y is the width corresponding to a concentration one-tenth of the central value, when the distribution is of Gaussian form (a corresponding height of cloud is 2.15 σ_z).

There are two principal formulations to be considered, the first being the *Gaussian-plume* equation, used widely in dispersion modelling for several decades, and so-called from the assumed Gaussian form describing the vertical and crosswind concentration profiles,

$$\chi(x, y, z) = \frac{Q}{\pi \bar{u} \sigma_y \sigma_z} \exp\left[-\tfrac{1}{2}\left(\frac{y^2}{\sigma_y{}^2} + \frac{z^2}{\sigma_z{}^2}\right)\right] \tag{6.2}$$

This equation satisfies Eq. (6.1) when the variation of wind speed with height is neglected, but may be adjusted to allow for that variation by using $\bar{u}$ at an appropriate height increasing with distance downwind of the source. Of course Eq. (6.2) could be made formally consistent with Eq. (6.1) by replacing u by $u(z)$, though this would distort the Gaussian profile and lead to invalid concentrations near the ground where u becomes small.

Observations of passive plumes have confirmed that the Gaussian form is a satisfactory description for the crosswind distribution, in an ensemble average sense. The argument that such a distribution follows from the classical parabolic equation of diffusion for Fickian diffusion (K constant) is inadequate, since the latter condition implies $\sigma_y{}^2 = 2Kx/\bar{u}$ which is not generally observed though it may be regarded as a limiting form approached with increasing distance (see Section 3.6 for further reference to the theoretical basis for the Gaussian form). For the vertical distribution it has been found that the shape is not generally Gaussian (except when the plume is elevated, as will be considered in a later section), but this deviation has tended to be disregarded as the practical consequence is not very significant.

The foregoing deficiencies in the Gaussian-plume equation may be eliminated by incorporating vertical diffusion terms in accordance with the solution of the two-dimensional (x, z) parabolic equation of diffusion, with u and K regarded as functions of height, for details of which see 3.2 and Eqs. (3.20) and (3.25) in particular. The equation defining $\chi(x, z)$ from a line-source of infinite extent acrosswind, releasing Q units of material in unit time from unit length of the line, is exactly the same as that defining $\int_{-\infty}^{\infty} \chi(x, y, z)\, dy$, i.e. the crosswind integrated concentration or *CIC* from a point source releasing Q units in unit time. Also, assuming the crosswind and vertical distributions to be independent of z and y respectively we may write

$$\chi(x, y, z) = \chi(x, 0, 0) f_1(y, x) f_2(z, x) \tag{6.3}$$

which, using the identity

$$\int_{-\infty}^{+\infty} \chi(x, y, z)\, dy = \chi(x, z) \tag{6.4}$$

may be transformed into

$$\chi(x, y, z) = \chi(x, z) f_1(y, x) \Big/ \int_{-\infty}^{+\infty} f_1(y, x)\, dy \tag{6.5}$$

For $\chi(x, z)$ or *CIC* the solution in Eq. (3.20) may be used. Rearranged in terms of vertical spread σ_z, in a form corresponding to that (Eq. (3.25)) of van Ulden (1978), this becomes

$$\chi(x, z) = \frac{Q}{A'(s) u_{px} \sigma_z} \exp\left[-(z/B'(s)\sigma_z)^s\right] \tag{6.6}$$

and substitution in Eq. (6.5) with $f_1(y,x)$ taken as $\exp[-(y^2/2\sigma_y^2)]$ (the Gaussian form) gives

$$\chi(x, y, z) = \frac{Q}{(2\pi)^{\frac{1}{2}} A'(s) u_{px} \sigma_y \sigma_z} \exp\left[-(y^2/2\sigma_y^2)\right] \exp\left[-(z/B'(s)\sigma_z)^s\right] \tag{6.7}$$

the parameters $A'(s)$, $B'(s)$ and u_{px} being defined, with certain tabulated values, in Table 6.III.

The two equations are identical, except for the interpretation of the u terms, when $s = 2$, i.e. when the exponents m and n satisfy the condition $m - n = 0$. In the special (Fickian) case of this condition when $m = n = 0$ the wind representations are also identical. Otherwise (for $s \neq 2$) the concentrations provided by the two equations differ, to an extent that can be seen from Table 6.III. For example, if $s = 1$ the ground-level concentration for given σ_z and u set equal to u_{px} is 1.77 times that provided by the 'wholly Gaussian' form of equation.

Theoretically the shape exponent s is prescribed in terms of the exponents in the stability-dependent power-law forms of u and K, and the observations of

Table 6.III – Parameters in Eq. (6.7) for the concentration distribution $\chi(x, z)$ from a crosswind line source of infinite extent at ground level, or the integrated crosswind concentration *CIC*(x, z) from a continuous point source, and certain other functions of interest.

$s =$	1.0	1.25	1.5	1.75	2.0
A'	0.707	0.902	1.050	1.165	1.253
B'	0.707	0.968	1.164	1.308	1.414
$\frac{CIC(x, 0)}{CIC(x, 0) \text{ for } s=2}$	1.77	1.39	1.19	1.08	1.0
$\sigma_z/\bar{Z}$	1.414	1.346	1.303	1.274	1.253

$s = 2 + m - n$ where m and n are the z-exponents in the power law forms for the variation of u and K with height.

$A'(s) = s^{-1}(\Gamma(1/s))^{3/2}\,(\Gamma(3/s))^{-\frac{1}{2}}$

$B'(s) = (\Gamma(1/s))^{\frac{1}{2}}\,(\Gamma(3/s))^{-\frac{1}{2}}$

$\sigma_z/\bar{Z} = (\Gamma(1/s))^{\frac{1}{2}}\,(\Gamma(3/s))^{\frac{1}{2}}\,(\Gamma(2/s))^{-1}$

u_{px} = u at height $c\bar{Z}$, where c is a function of stability, i.e. of s. As noted in Table 3.II, however, the effect of the variation on u_{px} is slight, and for practical purposes (e.g. as in the construction of Fig. 6.7) the use of a constant (average) value of c appears to be acceptable.

vertical distribution have shown values in the range 1.0 to 2.0, broadly consistent with theory. In view of the relative insensitivity of $\chi(x, 0)$ to the value of s the present tendency is to adopt an average value of 1.5 irrespective of stability.

Evaluation of σ_y and σ_z or CIC

For the crosswind spread σ_y the use of data on the wind direction fluctuation is now widely advocated (see Hanna *et al.*, 1977), in terms of the simple expression

$$\sigma_y = \sigma_\theta x f(x) \tag{6.8}$$

the function $f(x)$, which decreases from unity as x increases, being prescribed semi-empirically. The properties *CIC* at ground level (which is the property of practical interest) and σ_z (which by substitution in standard formulae will give the *CIC*) are derivable as a useful approximation from gradient-transfer/similarity-theory and observed parameters of the vertical transfer in the surface layer of at atmosphere. Various alternative procedures are summarized in Table 6.IV [See pages 324–5].

For the croswind-integrated-concentration the most promising procedure appears to be that developed recently and independently by van Ulden (1978) and Horst (1979), leading to a specification of *CIC* as a function of x/z_0 and

z_0/L. Prediction of *CIC* may thus be achieved given a specification of the boundary-layer parameters z_0, u_* and L either directly from observations or from a climatological specification in terms of routine meteorological data. Graphs of van Ulden's calculations are reproduced in Fig. 6.7 with explanatory notes. Horst's calculations differ from van Ulden's only in the interpretation of the distance of travel of the dispersed material – Horst uses $\bar{Z}$ as a function of travel T and hence of $\bar{X}$ the mean distance of travel of particles, whereas van Ulden's procedure correctly uses $\bar{Z}$ as function of a specified distance x. The result is that for equality in $\bar{X}$ and x Horst's values of $1/\bar{Z}$ and *CIC* are less than van Ulden's, by roughly 30 per cent (stable flow), 15 per cent (neutral) and 20 per cent (unstable). In the latter condition the discrepancy decreases and ultimately reverses with increasing range, but then the range of $\bar{Z}/L$ is evidently outside the z/L range covered by the boundary layer observations used in specifying the formula used for ϕ_H.

Both van Ulden and Horst demonstrate the success of their calculations in graphical comparisons of predicted values of *CIC* with those observed in the *Prairie Grass* study – both show good agreement on average but considerable

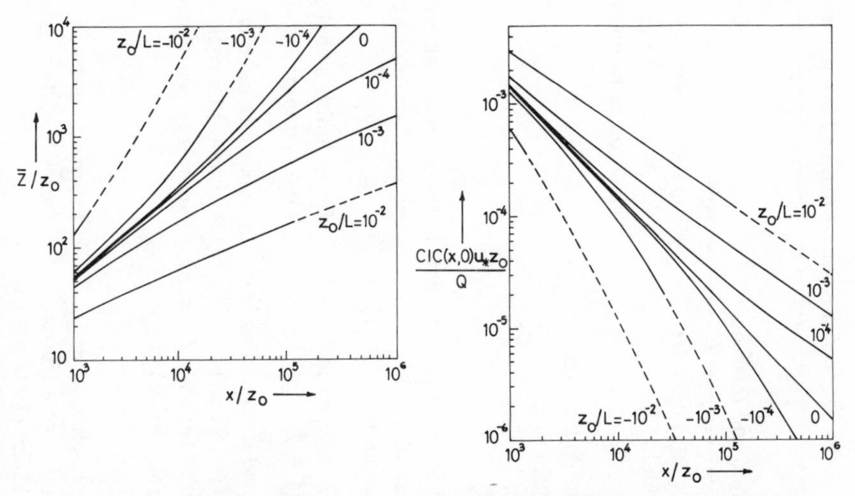

Fig. 6.7 – Simple estimates of vertical spread $\bar{Z}$ and ground-level crosswind-integrated-concentration *CIC* from a continuous point source of passive material at ground level (after van Ulden, 1978). (i) The curves for $\bar{Z}$ were derived by integrating (with some approximations) the form for $d\bar{Z}/dx$ in Table 3.II, with $p = 1.55$, $K = ku_*z/\phi_H$, $u(z) = (u_*/k)\ [\ln(z/z_0) - \psi(z/L)]$ and with $k(0.35)$, ϕ and ψ as advocated by Businger (1973). (ii) The curves for *CIC* are according to the relation

$$CIC(x, 0) = Q/1.37\,\bar{Z}\,u_{px}$$

(which is equivalent to Eq. (6.6) with $s = 1.5$) with $u_{px} = u$ at $0.6\,\bar{Z}$ as a general approximation independent of L.

Note that the curves are continuous only up to $\bar{Z}/z_0$ corresponding to the upper limits of $z/|L|$ represented in the empirical specification of ϕ and ψ.

Table 6.IV – Procedures for estimating $f(x)$, σ_y, σ_z, CIC for use in Eqs. (6.6)–(6.8).

Turbulence data available	Method and reference	Explanations: advantages or limitations
v or θ time-lapse record	$f(x) = \sigma_\theta(\tau, x/\beta u_e)/\sigma_\theta(\tau, 0)$ Hay and Pasquill (1959) $\beta \cong 0.6/\sigma_\theta$	Implicit allowance for scale of turbulence, but β only roughly specified. Applicable for travel time less than sampling duration τ. No allowance for variation of turbulence with height.
σ_θ and t_E	Use $t_L = t_E$ in Phillips and Panofsky (1982) form $T'^2 f(x)^2 = 2T' - 2\ln(1 + T')$ where $T' = T/t_L$	
σ_θ only	Use empirical forms of $f(x)$ tabulated in Note (i) Pasquill (1978) Note (ii) Doran *et al.* (1978a)	These are rough values of $f(x)$ derived from averaged miscellaneous data on σ_y. Those in Note (i) include data in urban airflow.
σ_θ and h_c	Use $t_L = 0.275\, z_i/u_e\sigma_\theta$ in $f(x/u_e t_L) = (1 + T/2t_L)^{-\frac{1}{2}}$	From Smith and Blackall's (1979) rearrangement of Deardorff and Willis's (1975) interpolation form. (u_e is effective wind speed ($=u_{px}$ in Eq. 6.6)).
H_0, h_c (or w_*)	Deardorff mixed-layer scaling law $\sigma_y/h_c = f(w_* x/u_e h)$ with estimates of function f, e.g. as in Eqs. (6.14) and (6.16) in Section 6.4.	Applicable in convective mixing, especially in light winds, but may under-estimate in strong winds and on account of horizontal limitations in the laboratory convective tank.
Surface-stress layer statistics z_0, ϕ_M, ϕ_H	CIC (or σ_z) from analytic or numerical solution of diffusion equation van Ulden (1978), Horst (1979) Nieuwstadt and van Ulden	Provides universal relation in terms of z_0, u_* and L (see Fig. 6.7). Basically and empirically questionable for very unstable flow.

u_* and L	Similarity analysis in terms of scale height $Q/u\ CIC$, Briggs and McDonald (1978), with fitting to dispersion data, giving notably $CICu_*x/Q = \begin{matrix} (1+X^{\frac{1}{2}}) & \text{stable} \\ (1+5X^2)^{-\frac{1}{2}} & \text{unstable} \end{matrix}$ where $X = u_*x/uL$	Provides rational universal representation of research data. Fitting at present confined to one set of data (*Priarie Grass*). Fitting fails in most stable conditions, when $X > 1$.
H_0, h_c (or w_*)	Mixed-layer similarity analysis and fitting to *Prairie Grass* data in convective conditions, Nieuwstadt (1980) gives $CICux/Q = 0.9\ x(gHt^3/\rho c_p T)^{-\frac{1}{2}}$	Fitting confined to intermediate range $0.03 < w_*x/uh_c < 0.23$ but laboratory and numerical simulation studies point to dependence on w_*x/uh_c at longer range.

Notes

(i)	x (km)	0.1	0.2	0.4	1	2	4	10	10
	$f(x)$	0.8	0.7	0.65	0.6	0.5	0.4	0.33	$0.33\ (10/x)^{\frac{1}{2}}$
(ii)	x (km)	0.1	0.2	0.4	0.8	1.6	3.2	10	
$f(x)$ for τ =	1800 s	0.95	0.85	0.76	0.70	0.64	0.58	0.52	
=	3600 s	1.04	0.98	0.92	0.85	0.77	0.67	0.54	

In (i) it is implicit that σ_θ is effectively for zero averaging (smoothing) time, whereas in (ii) an averaging time of 5 sec is implied.

deviation in individual values. Horst *et al.* (1979) also provide a table of statistics of the agreement of individual values grouped in stability classes. This shows the predicted values to be on average 8 per cent higher than observed and to have root-mean-square deviations from observed of 33 per cent, though with some dependence of the statistics on distance and stability, notably a tendency for the 'over prediction' to be considerably higher than the overall average in unstable conditions at the larger distances. For σ_y Horst *et al.* (1979) show impressive agreement with several sets of observational data (including *Prairie Grass*) on average, but with considerable scatter, using the empirical value of $f(x)$ recommended by Doran *et al.* (1978a).

Table 6.IV also contains reference to empirical modelling as rationally guided by dimensional and similarity considerations. Both Briggs and McDonald (1978) and Nieuwstadt (1980) demonstrate an impressive ordering of *Prairie Grass* data in terms of boundary-layer vertical-transfer properties. Then in regard to crosswind spread Deardorff and Willis (1975) have provided laboratory demonstration of a useful similarity relation for a convectively mixed layer. Clearly these developments deserve careful watching.

Although the qualifications made in the foregoing paragraphs and brought out in more detail in the basic discussions of Chapters 3 and 4 need to be kept in mind, it appears that useful prediction of dispersion from a surface release of passive material is now achievable for near-ideal boundary-layer flow with thermal stratification, at least for dispersion through the depth of the surface-stress layer, which may be up to about 100 m in daytime conditions, though much less in nocturnal stable flow. On the basis of available empirical confirmation there does not appear to be any better procedure than that advocated by van Ulden and Horst *et al.* The former is the more rigorous, but on the present evidence the latter is the more successful in providing a smaller degree of overestimation of the crosswind integrated concentration observed in the *Prairie Grass* study. There is some reservation to be made on this point, however, in that earlier observations in neutral conditions, with a chemical-smoke tracer instead of sulphur dioxide as used in the *Prairie Grass* study, are fitted better by the van Ulden (higher) calculated values of *CIC* (see 4.6 and Fig. 4.12)

Note that the progressive widening of the crosswind distribution as the release continues, or as the time of exposure to the effects of a continuous plume increases, is broadly taken into account by the formulation of σ_y in terms of σ_θ measured over a corresponding sampling duration τ. It should however be kept in mind that this disregards more subtle aspects, which are not fully resolved, especially when τ is smaller than the time of travel.

6.4 ESTIMATION OF VERTICAL DIFFUSION, FROM A SURFACE RELEASE, THROUGH THE DEPTH OF THE BOUNDARY LAYER

Extension of the theoretical procedures of the previous section to take into account the further vertical dispersion into the upper part of the boundary layer may be made on two main lines:

(a) the continued adoption of the gradient transfer approach, but with certain well-recognized limitations,
(b) for the daytime mixed layer the application of the recent developments in the modelling of convective mixing.

More advanced modelling by second-order closure treatments and by numerical statistical methods is under consideration (and the latter method seems likely to be helpful for the difficult problem of elevated sources). An additional bar to progress is the lack of any body of observational data on vertical diffusion comparable with that available for the surface layer.

The gradient transfer method

The main additional difficulties that arise in the extension of this method above the surface-stress layer are:

(i) the $K(z)$ properties are not well defined,
(ii) the vertical profiles of K may be expected to have more complicated form than those of the surface layer and may not be convenient for use in analytic solutions of the diffusion equation,
(iii) the gradient-transfer assumption, traditionally open to suspicion in general, is now widely rejected for strong convective mixing such as frequently occurs in daytime.

For the analytic specification of the K profile the traditional device is to combine the surface-layer increase with height and a constant K thereafter or, more realistically, a K reaching a maximum at some height and then falling off to smaller or zero value at the top of the mixing layer h'. An interpolation form $K \propto z(h'-z)$ for $0 \leqslant z \leqslant h'$ was used by F. B. Smith (1957a) to obtain solutions for the time-dependent diffusion equation (equivalent to the two-dimensional x, z form with wind independent of height). These are in the form of Legendre polynomials and have been used recently by Nieuwstadt (1980b) with K taking the explicit form $0.4\, u_* z\, (1-z/h')$. Numerical solutions of the diffusion equation are now more readily undertaken and so analytically inconvenient empirical profiles of K may also be considered, for example those emerging from the studies of the transport of momentum etc. over the whole depth of the boundary layer or from the detailed studies of the statistical properties of turbulence. In the latter context a form which has been argued to follow from the Taylor statistical theory and which has been independently suggested by Hanna (1968) is

$$K \cong \mathrm{a}\, \sigma_w \lambda_m \tag{6.9}$$

involving the measurable intensity and scale of the vertical component of turbulence (though with a subsequently recognized as a function of stability).

The immediately foregoing form of K with constant **a**, evaluated from the Minnesota boundary-layer study (Izumi and Caughey (1976), Kaimal *et al.* (1976) and Caughey and Kaimal (1977)), has been used by Smith and Blackall (1979) in an experimental series of calculations of dispersion. Test of these calculations against the field measurements of dispersion has not yet been provided, but the lack of physical validity in the gradient transfer assumption for convective mixing has been demonstrated by the laboratory work of Willis and Deardorff (1976). For near-neutral and stable flow however the calculations are theoretically no less valid than those for the surface layer, and for the time being they are a basis for useful extension of the procedure of the previous section. Fig. 6.8 (see later) includes σ_z, x curves calculated by Smith and Blackall and also demonstrates a reasonable matching with the van Ulden/Horst calculations at short range. At the extreme range the levelling off in σ_z represents the effect of the upper boundary of the mixed layer and the attainment of a concentration uniform with height, assuming there is no removal of material by physical or chemical processes.

The effect of finite mixing depth in the conventional plume model

When a uniform vertical distribution is achieved Eq. (6.2) must be replaced by

$$\chi(x, y) = \frac{Q}{uh_c(2\pi)^{\frac{1}{2}}\sigma_y} \exp(-y^2/2\sigma_y^2) \tag{6.10}$$

in which u must now imply an average value through the depth h_c of the mixed layer. Prior to this stage the precise vertical distribution will depend on the stage reached in the redistribution consequent on the downward diffusion of material from the top of the layer. An important practical question is the specification of the stage at which Eq. (6.10) is correct, at which stage the equivalent value of σ_z which would give the correct ground-level concentration on substitution in Eq. (6.2) is $(2/\pi)^{\frac{1}{2}}h_c$ i.e. 0.80 h_c. Applying the 'reflection' or 'image source' model as conventionally adopted for an elevated source (see later) and allowing for multiple 'reflections' between the ground and the top of the mixed layer, it is easily seen that the ground-level concentration from a surface release may be written, using the Gaussian model, as

$$\chi(x, y, 0) = \frac{Q}{\pi u\sigma_y\sigma_z} \exp(-y^2/2\sigma_y^2) \sum_n \exp(-Z_n^2/2\sigma_z^2) \tag{6.11}$$

where $Z_n = \pm 2nh_c$, $n = 0, 1, 2, \ldots.$ and σ_z is given an effective value which implies continuation of its early-stage growth in the absence of a limit to vertical spread. Examination of the term by term contribution to the summation part of Eq. (6.11) shows that the first three terms are always adequate to give the final value. Indeed the $n = 0$ term alone (which of course corresponds to Eq. (6.2) for unlimited vertical diffusion) gives 0.99 of the total value even when

σ_z/h_c is as large as 0.6, and 0.92 when σ_z/h_c is 0.8. At this latter stage the total value is already close to the limiting value given by Eq. (6.10), being in fact 1.085 times that value, and with further increase in σ_z/h_c there is a rapid levelling off to the limiting value. For practical purposes therefore Eq. (6.11) may be replaced by a two-stage application of Eq. (6.2) in which

(a) $\sigma_z(x)$ is given the value estimated for unlimited vertical diffusion until it reaches $0.8\ h_c$,

(b) σ_z is taken to be constant thereafter at the value $0.8\ h_c$.

Application of convective-mixing similarity

Deardorff and Willis's computer modelling and laboratory measurements of dispersion of passive material by convective mixing indicate the form of a universal similarity relation

$$\sigma_z/h_c = f(w_* T/h_c) \tag{6.12}$$

where T is the time of travel. They also include the important result that the time for attainment of concentration virtually uniform with height is about $3h_c/w_*$. However, a ground-level concentration *below* that of the ultimate uniform vertical distribution appears over an intermediate range of T, associated with the maximum in the vertical profile occurring not at ground level but on an ascending path, a phenomenon which has yet to be confirmed in the atmospheric boundary layer. In the meantime these modelling results are the best available guide to the estimation of vertical dispersion above the surface-stress layer in unstable conditions. As pointed out by Pasquill (1978) a useful simplification of the laboratory curve over a limited range is provided by the linear form

$$\sigma_z = 0.6\, w_* T \quad 0.2 < w_* T/h_c < 0.8 \tag{6.13}$$

to within a few per cent. These results are actually for a source at a height $h_c/15$ but it is reasonable to expect the vertical distribution to become independent of height of release h_r when σ_z/h_r is sufficiently large, which according to Willis and Deardorff probably applied at $\sigma_z/h_c = 0.3$. This means that for a surface release the range of applicability of Eq. (6.13) is reduced to $0.3 < \sigma_z/h_c < 0.48$, but even this provides a useful fixing of the position of the σ_z, x curve, facilitating interpolation between the section in the lower part of the boundary layer and the levelling off in σ_z.

The numerical simulations by Lamb (1979) provide a more comprehensive approximation for the equivalent σ_z to be used in Eq. (6.2)

$$\left.\begin{aligned} \sigma_z/h_c &= 0.5(w_* T/h_c)^{1.2}, \quad && w_* T/h_c \leqslant 1.2 \\ &= 0.6\,, && w_* T/h_c > 1.2 \end{aligned}\right\} \tag{6.14}$$

and this gives values 20–27 per cent lower than Eq. (6.13) over the reduced applicable range. The limiting value of 0.6 represents approximately the true σ_z for a uniform vertical distribution over $0 \leqslant z \leqslant h_c$, as distinct from the effective value of 0.8 which gives the corresponding $\chi(x, 0)$ on substitution in Eq. (6.2). Use of the Eq. (6.14) values in Eq. (6.2) requires imposition of a virtual source height, not only at large T but also over a lower range of T to allow for ascent of the locus of the maximum as noted above.

Estimation of crosswind spread

As diffusion proceeds above the surface layer several aspects have a progressively more important bearing on the crosswind dispersion, as follows:

(a) the full effects of convective mixing or of reduction of turbulence with height come into play and may be relevant even at quite short range,
(b) the interaction between the increasing vertical spread and the progressive turning of wind with height becomes increasingly effective, and
(c) larger scale components of the horizontal field of turbulence become effective in a complex way.

Although the representation of σ_y in terms of the observed or climatological estimates of σ_θ as advocated in Table 6.IV has some theoretical and empirical support there are serious uncertainties arising from aspects (b) and (c). In respect of (b) a rough empirical correction (Pasquill, 1978) is provided by adding to the σ_y^2 calculated as in Table 6.IV a contribution $0.03\ \Delta\theta^2 x^2$ where $\Delta\theta$ (radians) is the total change of mean wind direction over the whole depth of the plume. For (c) the problem is least serious when the release-time or sampling duration is sufficiently long – say $\tau > x/u$ – in which case the procedures of Table 6.IV may be expected to be a useful guide. The problem does however become increasingly difficult as $\tau u/x$ is reduced. For both aspects several points need to be resolved if the still substantial reliance on empirical guidance is to be reduced.

For the case of convective conditions the recent experience of laboratory and numerical simulations encourages the adoption of formulations in terms of w_* and h_c. For low-level releases Lamb (1979, 1982) offers the following generalization

$$\begin{aligned} \sigma_y/x &= 0.33\, w_*/u \quad , \qquad w_* T/h_c \leqslant 1 \ , \\ &= (h_c/3x)\,(w_* T/h_c)^{\frac{2}{3}} \ , \qquad 1 < w_* T/h_c < 3 \ , \end{aligned} \tag{6.15}$$

on the basis of his numerical simulations. With some empirical guidance Venkatram (1980) has suggested the following irrespective of distance

$$\sigma_y/x = 0.45\, w_*/u \tag{6.16}$$

It is emphasized that these formulae do not include possibly important contributions from large eddies not generated by convection.

Testing of the foregoing formulae

Satisfactory testing of the formulae and procedures outlined in this and the previous section has not been fully achieved. Vertical profiles of concentration from which to evaluate the vertical spread σ_z are not easily measured further than about 100 m from a source except in stable conditions. On the other hand measurements of the concentration at ground level provide a test of theoretical representation of vertical diffusion only if the removal of material by deposition

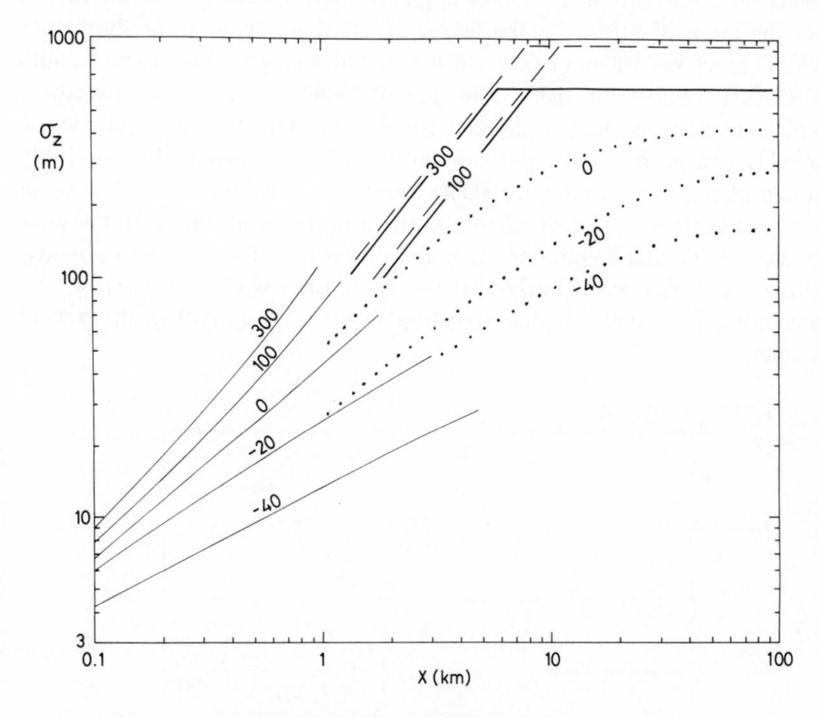

Fig. 6.8 – Examples of calculated σ_z (surface release), over the whole depth (h_c or h_m) of the boundary layer, as a function of surface heat flux H_0.

——— interpolation from the van Ulden (1978) curves

——— a, – – – b } Eq. (6.14) (Lamb's 1979, 1982 formulae).

. . . . Smith and Blackall's (1979) calculations.

Boundary layer parameters

z_0 0.1 m

G 10 ms^{-1}

h_c (a) 1000 m (b) 1500 m

$H_0\ Wm^{-2}$		−40	−20	0	100	300
w_* ms^{-1}	(a)				1.41	2.03
	(b)				1.61	2.33
h_m m		280	470	700		
u_* ms^{-1}		0.24	0.33	0.36	0.47	0.54
L m		31	165		−97	−50

is negligible or otherwise taken into account. The most exhaustive testing has been provided by the *Prairie Grass* study, in which sulphur dioxide was the tracer. Even here there is still some question of the possible magnitude of deposition, and in any case the data are confined to a distance within 800 m from the source.

The general quality of the estimates which may currently be made of the dispersion from a surface release, or from an elevated release at long enough distance, is evident from the two examples in Figs. 6.8 and 6.9. In the first of these the point of interest is the degree of matching between the short-range calculations of van Ulden (1978) and the extended-range calculations of Smith and Blackall (1979) for stable flow and of Lamb (1982) for a convectively mixed layer. The matching at intermediate range is imperfect by a factor which is roughly two in the most stable case but less in the remainder. It seems likely that the discrepancy in the most stable case is associated with the adoption of too high a value for the coefficient **a** in the form for K in Eq. (6.9), the value estimated for neutral conditions being used throughout irrespective of stability. If this is so it also seems likely that the discrepancy will be even greater in a lighter wind than that adopted (geostrophic value 10 m/sec) in the present examples.

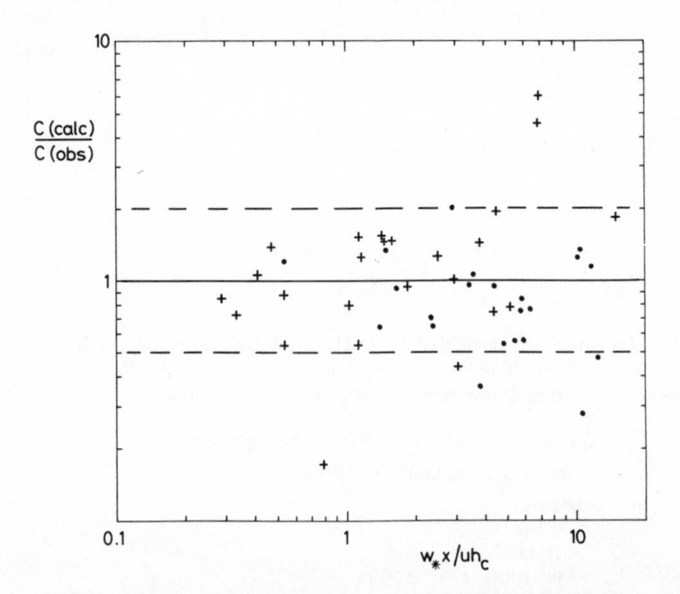

Fig. 6.9 – Ratios of calculated and observed ground-level plume-centre-line concentrations of sulphur dioxide at distances up to 30 km from two power plants in U.S.A. (Venkatram, 1980) in convectively mixed conditions. Calculated values are according to the simple formula in Eq. (6.10) with $y = 0$ and $\sigma_y/x = 0.45\ w^*/u$. Observed values: •, Morgantown plant, $4.6 \leqslant x \leqslant 31.8$ km; + Dickerson plant, $1.7 \leqslant x \leqslant 18.1$ km.

Fig. 6.9 shows predictions of sulphur dioxide concentrations in convectively mixed conditions using the simple 'vertical box' model represented in Eq. (6.10), with the simple linear form of $\sigma_y(x)$ advocated by Venkatram (1980), in relation to observations of ground-level plume-centre-line concentrations from two power plants in the U.S.A. tabulated by the same author. The scatter is wide but not markedly greater than that found in the testing of the much more sophisticated model advanced by Venkatram and modified by Venkatram and Vet (1981) (see discussion in 5.3). In the simple model there is no dependence on stack height and plume rise, the ultimate uniform vertical distribution being assumed irrespective of distance from the source. Purely from the standpoint of vertical diffusion the applicability of this simple model must deteriorate near the source, but there is no sign of such deterioration at distances of a few kilometres, and overall the agreement is within a factor of two for 80 per cent of the observations.

6.5 ALLOWANCE FOR ELEVATION OF SOURCE AND FOR PLUME RISE

Representation of the effect of elevation of the source of pollution was originally based on the simple idea of 'reflection' of the plume from the ground surface (in the absence of any deposition or absorption). Mathematically this condition is prescribed by adding to the distribution of the plume from the real source that from an imaginary (image) source situated h_s below the surface, giving a total $C(x, y, z)$ as follows for the customarily assumed Gaussian shape

$$\chi(x, y, z) = \frac{Q}{2\pi u \sigma_y \sigma_z} \exp\left(-\frac{y^2}{2\sigma_y^2}\right)\left[\exp\left(-\frac{(h_s - z)^2}{2\sigma_z^2}\right) + \exp\left(-\frac{(h_s + z)^2}{2\sigma_z^2}\right)\right] \qquad (6.17)$$

$C(x, 0, 0)$ is zero at $x = 0$, remains effectively zero for some distance downwind, then rises fairly sharply to a maximum $C(\max)$, and thereafter falls off and asymptotically approaches the curve for a surface release. The simplest assumption about σ_y and σ_z, i.e. that σ_y/σ_z is constant independent of x, leads to the familiar results for $C(\max)$ and for the distance $x(\max)$ at which this maximum concentration occurs

$$C(\max) = (2Q/e\pi u h_s^2)\,(\sigma_z/\sigma_y) \qquad (6.18)$$

$$x(\max) = x \text{ for } \sigma_z = h_s/2^{\frac{1}{2}} \qquad (6.19)$$

If σ_y and σ_z are more realistically assumed to have different growth functions the expressions for $C(\max)$ and $x(\max)$ are modified, but in all cases $C(\max)$ is determined by $1/h_s\sigma_y$ at the distance $x(\max)$.

Solutions of the diffusion equation for an elevated source do not generally give a vertical distribution exactly in accordance with the Gaussian model, though for practical purposes the difference is probably of secondary importance. Moreover these solutions give σ_z behaving as $x^{\frac{1}{2}}$ as $x \to 0$, whereas the initial behaviour should be a linear growth. This difficulty is avoidable by regarding the diffusion equation solution as the asymptote for large x and adopting the statistical theory form as the asymptote for $x \to 0$, the whole curve being completed by judicious interpolation. Smith and Blackall (1979) provide this interpolation by assigning the diffusion-equation results to an adjusted distance $x + ut_L\ (1 - \exp(-x/ut_L))$, where t_L is given its value at h_s. However, the most serious objection to these solutions is that in strongly convective mixing both theoretical modelling and laboratory measurements indicate failure of the gradient-transfer assumption.

Recently the progress toward a more realistic solution of the elevated source problem has been given new impetus, not only by the emergence of a successful similarity treatment of turbulence and dispersion in convective mixing but also by the development of numerical statistical solutions which amount to the construction of ensembles of individual particle trajectories in accordance with a specified field of turbulence. This specification may contain a height dependence, in accordance with observation or other modelling procedures, without invalidating the theory as would be the case in application of the Taylor form of analytical statistical solution. For practical purposes the most significant progress has so far been made for convective conditions, notably by Lamb (1979, 1982) and Hanna (1981). In an attempt to provide more realistic predictions in terms of the conventional Gaussian-plume model represented in Eq. (6.17) Lamb has offered a generalization for passive releases in the upper nine-tenths of a convectively mixed layer. This combines his numerically simulated values of σ_y and σ_z, expressed as simple functions of $w_* x/uh_c$, with virtual source heights also a function of $w_* x/uh_c$. As Lamb points out, this forcing of the more realistic plume behaviour into the over-simplified mould of the old conventional model does inevitably contain imperfections, though the main effects of substantial changes in source height are reasonably well described.

For the important practical case of a buoyant plume there is also the question of the significance of the spread of the instantaneous plume by turbulent entrainment generated as the plume rises. Clearly this process must be the dominant one when ambient turbulence is low, determining the instantaneous concentration in a plume. It must also be important even in the presence of convective mixing when the concern is with *instantaneous* concentrations experienced close in to a stack when sections of plume are brought to ground in the 'looping' behaviour. However, for *time-mean* concentrations which are the objective of most of the modelling procedures the controlling feature would appear to be the range of downdraft velocities, determining the spread of distances over which plume elements reach the ground. A tentative estimate of the σ_z (and σ_y) equivalent to the induced growth of a rising plume (see 5.3) is

$$\sigma_z = a\Delta H \tag{6.20}$$

with the coefficient a taking the value $1/2\sqrt{3}$ (roughly 0.3) if the concentration across the instantaneous plume is assumed to be uniform. The magnitude of the plume-rise ΔH is in any case of prime importance in setting the effective plume-height. Briggs's latest recommendations for calculation of ΔH are repeated below.

$$\text{Neutral windy atmosphere} \quad \Delta H = 1.6\, F^{\frac{1}{3}}\, u^{-1}\, x^{2/3} \tag{6.21a}$$

$$\text{Stable windy atmosphere} \quad \Delta H = 2.6\, (F/us)^{1/3} \tag{6.21b}$$

where F is the stack parameter $gQ_H/\pi\rho c_p T_0$ and s the stability parameter $(g/T)/d\theta/dz)$. For large plants ($Q_H = 20$ MW or more) Eq. (6.21a) is advocated up to a distance of 10 stack heights, with no further rise. However, a more satisfactory allowance for the combined effects of plume buoyancy and ambient turbulence is an outstanding requirement, on which preliminary ideas have recently been advanced by Lamb (1982).

Proper allowance for the behaviour of an elevated plume, as regards both its trajectory and its dispersion, is particularly important in estimating the pollution experienced in hilly terrain, where impingement of the plume on ground elevated with reference to the plant site may result in much higher concentration than would have occurred at the same distance over level ground, (see 5.4). When the impingement is at short range, or even at long range if the airflow is stably stratified, the growth of a power-plant plume will be predominantly due to the plume-rise entrainment process. In this case a pessimistic estimate of the pollution would follow from use of the rule in Eq. (6.20) for σ_y and σ_z, but for time-mean estimates the effective values of these dispersion parameters may be expected to be greatly increased by the fluctuations in the trajectory of the plume.

6.6 PRESENTATION OF DISPERSION ESTIMATES IN TERMS OF ROUTINE METEOROLOGICAL DATA

In the foregoing sections we have considered the latest developments in the calculation of dispersion given a specification of the basic parameters representing the turbulent and convective mixing in the boundary layer. Most of these parameters (listed in Table 6.I) are not routinely measured by meteorological services and so from an early stage there has always been a requirement for presenting 'dispersion formulae' in forms that could be evaluated with the simpler meteorological data regularly available. Such was the situation in 1957 when, following the accidental release of radioactive material from the reactor at Windscale in Cumbria, the British Meteorological Office was asked to provide an up-to-date readily applied procedure for calculating the concentrations downwind of a specified release of windborne material. At that time, as now, the problem naturally divided into two questions – on the one hand what basic relations between dispersion and boundary-layer flow conditions could be offered; and on the other hand how could these be presented in terms of available meteorological data?

The Meteorological Office 1958 system, issued first as a departmental note until its appearance in publication three years later (Pasquill, 1961), emphasized the desirability of utilizing data on wind-direction fluctuation (σ_θ) and also, if possible, data on wind-inclination fluctuation (σ_ϕ), a preference that was also being advocated in the U.S.A. by Cramer (1957). This preference is still being urged, though now with the greater consensus provided by an American Meteorological Society workshop on the matter (Hanna *et al.*, 1977). However, to meet the requirement for application in terms of routine data the Meteorological Office method included a diagram entitled 'Tentative estimates of vertical and lateral spread', containing curves for the vertical spread and a concise tabulation for the lateral spread for distances up to 100 km in terms of broad stability categories semi-quantitatively specified in terms of wind speed, insolation and night-time state of sky (see Table 6.V), for use in the Gaussian plume equations. These represented the most realistic, albeit rough, estimates that it seemed possible to offer in the current state of knowledge.

Table 6.V – Stability categories in terms of wind speed, insolation and state of sky.

Surface wind speed (m/sec)	Insolation			Night	
	Strong	Moderate	Slight	Thinly overcast or $\geqslant 4/8$ low cloud	$\leqslant 3/8$ cloud
<2	A	A–B	B	–	–
2–3	A–B	B	C	E	F
3–5	B	B–C	C	D	E
5–6	C	C–D	D	D	D
>6	C	D	D	D	D

(for A–B take average of values for A and B etc.)

Strong insolation corresponds to sunny midday in midsummer in England, slight insolation to similar conditions in midwinter. Night refers to the period from 1 hr before sunset to 1 hr after dawn. The neutral category D should also be used, regardless of wind speed, for overcast conditions during day or night, and for any sky conditions during the hour preceding or following the night as defined above. (Pasquill, 1961, from *The Meteorological Magazine,* February 1961, H.M.S.O. Crown Copyright Reserved).

The system was taken up in the U.S.A. with some rearrangement (Gifford, 1961) into a format widely known as the Pasquill–Gifford curves, and these were used to calculate the numerous concentration–distance curves which appear in the *Turner Workbook* (Turner, 1970). In his use of the method Turner introduced a more quantitative specification of insolation in terms of the sun's altitude. The first attempt to improve the fundamental character of the estimates was provided by F. B. Smith (1972) on the basis of a limited number of numerical solutions of the two-dimensional diffusion equation using a diffusivity in terms of ϵ and λ_m and a limited quantity of data on the vertical profiles of those properties. Smith's revision of the system included allowance for the surface roughness z_0 (the original estimates were for a rather small roughness, $z_0 = 3$ cm) and for thermal stratification in terms of the basic parameter H_0, arranged in nomograms providing:

(a) stability categories (designated alphabetically as originally and also as a number P) as a function of upward heat flux or incoming solar radiation, cloud amount at night, and surface wind speed (see Fig. 6.10),

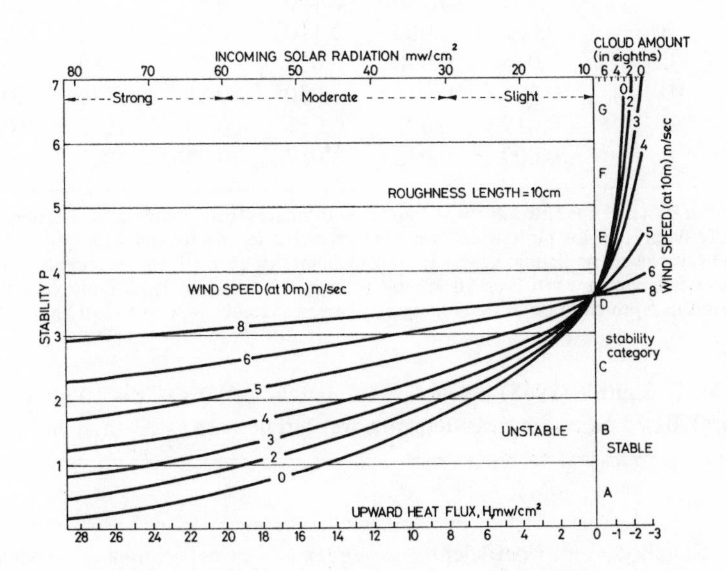

Fig. 6.10 – Revised system for estimating σ_z from a ground-level source (Smith, 1973). Stability number P and stability categories A–G prescribed in terms of upward heat flux (daytime) or cloud amount (night time) together with surface wind speed, for $z_0 = 10$ cm. For use in the absence of heat flux estimates a solar radiation scale is included which is roughly in accordance with typical conditions of terrain and weather in England.

(b) $\sigma_z(x)$ for $z_0 = 10$ cm and neutral flow ($P = 3.6$),
(c) isopleths of $\sigma_z(P)/\sigma_z$ (neutral) for specified P and $x = 0.1$ to 100 km,
(d) isopleths of $\sigma_z(z_0)/\sigma_z(z_0 = 0 \text{ cm})$ for a wide range of z_0.

Table 6.VI gives at A simple power-law approximations to Smith's $\sigma_z(x)$ for use over $0.1 \leqslant x \leqslant 10.0$ km.

Table 6.VI – Various general estimates of σ_y and σ_z.

A. Power-law approximations $\sigma_z = ax^s$ in F. B. Smith's (1972) revision of Pasquill's (1961) values (surface release). σ and x in km.(s as used here and in Section 6.7 is to be distinguished from s in the vertical distribution of concentration in Eqs. (3.25), (4.4), (6.6)).

Stability category	Coefficient a			Index s		
z_0	1 cm	10 cm	1 m	1 cm	10 cm	1 m
A	0.102	0.140	0.190	0.94	0.90	0.83
B	0.062	0.080	0.110	0.89	0.85	0.77
C	0.043	0.056	0.077	0.85	0.80	0.72
D	0.029	0.038	0.050	0.81	0.76	0.68
E	0.017	0.023	0.031	0.78	0.73	0.65
F	0.009	0.012	0.017	0.72	0.67	0.58

The form $\sigma_z = ax^s$ was fitted at $x = 0.3$ and 3 to estimates from nomograms. These estimates actually lie on log–log plots which are concave to the log x axis, and with respect to these curves the power-law fittings generally underestimate σ_z at $x = 1$ km by roughly 5 per cent and overestimate at $x = 0.1$ or 10 km by roughly 10 per cent. It is emphasized that the error in this representation of the nomogram increases rapidly beyond $x = 10$ km.

B. M. E. Smith's (1968) power-law forms $\sigma = ax^s$, $0.1 \leqslant x \leqslant 10$ km (elevated release). Based on tracer measurements over terrain with $z_0 \cong 1$ m. σ and x in km.

	σ_y		σ_z	
Stability	Coefficient a	Index s	Coefficient a	Index s
Very unstable	0.215	0.91	0.215	0.91
Unstable	0.137	0.86	0.125	0.86
Neutral	0.070	0.76	0.048	0.76
Stable	0.042	0.71	0.008	0.71

C. Briggs's $\sigma_y(x)$ and $\sigma_z(x)$ formulae for elevated small releases (see Gifford, 1975), $0.1 \leqslant x \leqslant 10$ km.

	Stability category	σ_y	σ_z
Open country	A	$0.22x(1+0.1x)^{-\frac{1}{2}}$	$0.20x$
	B	$0.16x(1+0.1x)^{-\frac{1}{2}}$	$0.12x$
	C	$0.11x(1+0.1x)^{-\frac{1}{2}}$	$0.08x(1+0.2x)^{-\frac{1}{2}}$
	D	$0.08x(1+0.1x)^{-\frac{1}{2}}$	$0.06x(1+1.5x)^{-\frac{1}{2}}$
	E	$0.06x(1+0.1x)^{-\frac{1}{2}}$	$0.03x(1+0.3x)^{-1}$
	F	$0.04x(1+0.1x)^{-\frac{1}{2}}$	$0.016x(1+0.3x)^{-1}$
Urban areas	A–B	$0.32x(1+0.4x)^{-\frac{1}{2}}$	$0.24x(1+0.1x)^{\frac{1}{2}}$
	C	$0.22x(1+0.4x)^{-\frac{1}{2}}$	$0.20x$
	D	$0.16x(1+0.4x)^{-\frac{1}{2}}$	$0.14x(1+0.3x)^{-\frac{1}{2}}$
	E–F	$0.11x(1+0.4x)^{-\frac{1}{2}}$	$0.08x(1+0.15x)^{-\frac{1}{2}}$

An alternative description of $\sigma_z(x)$ and $\sigma_y(x)$ summarized at B in Table 6.VI has been advocated in the U.S.A. by M. E. Smith (1968) on the basis of measurements of passive tracer concentrations from an elevated release at the Brookhaven National Laboratory (the U.K. systems referred to above are strictly for a surface release, but in the lack of other information were originally offered for use irrespective of elevation of source). Briggs has recently offered a revision of the Pasquill–Gifford σ_y, σ_z estimates (see Gifford, 1975) with elevated releases specially in mind (C in Table 6.VI). Comparing the various estimates, the difference between the F. B. Smith and M. E. Smith figures is wide only in the most stable condition, especially at short range, where the respective figures are in the ratio 2:1. The difference between the F. B. Smith and the Briggs figures is widest in the most unstable condition, the ratio being roughly 1:2, if Briggs's open country values are taken for $z_0 = 10$ cm.

Under the stimulus of the air pollution regulatory policies developed especially in the U.S.A. the demand for generalized but easily applied dispersion-prediction systems continues. Within the U.S. Environmental Protection Agency's programme new proposals have been put forward by Irwin (1979) taking into account much of the recent basic research and with tall stacks especially in mind, and the examination and development of these proposals continues (see the report of the American Meteorological Society's Review Panel, Randerson, 1979).

Categorization of dispersion conditions

Practical dispersion-prediction systems of the Meteorological Office 1958 type need two basic ingredients:

(a) Specification of the relation between dispersion and flow conditions in the most fundamental form available and with a minimum of dependence on specific pollution surveys.
(b) Specification of the flow conditions in terms of routinely available meteorological data.

The stability categories of Table 6.V were constructed rather subjectively with two basic properties of surface boundary-layer flow in mind:

(c) The restriction of large numerical values of the low-level Richardson number (or the equivalent z/L) to very light winds.
(d) The well-defined diurnal cycle of the vertical profile of temperature in the first few metres and its dependence on the state of sky.

A serious weakness of the system, which would apply to any system in terms of radiation properties, is the fact that the relatively unstable layer formed some time after sunrise grows in depth in accordance with the stable temperature gradient formed near the surface during the pre-dawn hours and the increasing intensity of surface heating, and ultimately approaches the limit of maximum mixing depth set by overlying stable air. This means that the upper boundary of the plume from a surface release may reach the top of the mixed layer while the latter is still in the early stages of development, and further increase of vertical spread of the plume will then be determined by further rise of the top of the mixed layer. On the other hand the vertical spread from an elevated source will obviously not be rapid until the mixed layer has engulfed the plume (as in fumigation).

Lack of a clear-cut orderly relation between the stability categories and observations of the turbulent fluctuations of the wind has been demonstrated by Luna and Church (1972), and it seems likely that to some extent this was a consequence of the evolutionary aspect just noted. Also another disturbing feature of Luna and Church's results which needs careful interpretation is the demonstration that the D-category does not isolate cases of near-zero lapse rate of potential temperature. This feature is, however, not surprising in view of the recognition that such lapse rates also occur, even at relatively low level, in well-developed convective conditions which would be appropriately classified as A or B. Nevertheless it is clear that stability classification of this type needs very careful application, and that especially in the diurnal evolutionary conditions a modified procedure is called for, on the lines for example of that incorporated by Smith (1973) on the basis of Carson's (1973) developing boundary layer model. For distances of practical interest it appears that for substantial periods of the day, depending on the distance and the meteorological conditions, the effective vertical spread is determined by the developing mixing depth and is less than would be calculated on the assumption that vertical spread is limited only by higher-level inversions associated with the large-scale weather pattern.

The whole subject of the choice of parameters in specifying dispersive conditions has recently been aired in two conferences in the U.S.A., one held by the American Meteorological Society (Hanna *et al.*, 1977) and another sponsored by the Environmental Protection Agency (Hoffnagle *et al.*, 1981). These discussions have reaffirmed the preference for the use of the primary and derived turbulence parameters (listed in Table 6.I) over the more qualitative stability categories and, in the second case, include recommendations on choice of instruments and on the quality-control and processing of the data thereby obtained. The following up of these recommendations, in association with the continuing improvements in the theoretical bases for the dispersion estimates, may well in due course eliminate the need for prediction systems applicable in terms of routine data, and there would be the advantage of using parameters specific to the site. In the meantime 'routine data' systems could be improved and brought up to date in the light of the developments referred to in the foregoing sections of this chapter. The most urgent requirements are:

(a) abandoning of the qualitative stability categories in Table 6.V in favour of the parameters H_0, u_*, L and w_*, for which there is now a reasonable prospect of specification in terms of routine meteorological data (see Section 6.2);
(b) adoption of estimates of σ_y and C/C in accordance with the relations summarized in 6.3 and 6.4.

An example of an improved practical scheme for estimating H_0 has been provided by Smith and Blackall (1979). Although this is acknowledged to be an interim solution, it nevertheless represents a reasonably accurate and theoretically more acceptable approach.

6.7 MATHEMATICAL MODELLING FOR MULTIPLE SOURCES

Practical problems of pollution distribution from an array of sources, with an arbitrary distribution of position and strength of emission, may be dealt with by integrating the specifiable distributions from individual point sources in some acceptable and convenient way. The process is relatively straightforward only for a pollutant which is effectively chemically inert. Otherwise, and especially if the particular pollutant reacts with other pollutants, there are additional complications which have not yet been overcome.

Integration of the effects of a number of continuous emissions may be carried out either algebraically, by carrying out tractable integrations of the theoretical distributions from idealized arrays of sources, or numerically merely by superimposing the patterns of pollution from the sources and adding the contributions at any point of interest. The algebraic method is more elegant and economical in numerical effort but, as will be seen, is mathematically possible only in special cases, as well as requiring some approach to uniformity in the source distribution. Much greater universality is attached to the numerical method, which does not require this condition of uniformity, but which may impose enormous computing demands requiring high-speed computer facilities. In the application of either method it is an implicit requirement that there be a sufficiently close approach to ideal conditions in respect of

steady emission rates
straight-line mean flow
steady and spatially uniform conditions of dispersion.

Variations from the ideal in either emission or dispersive conditions may be allowed by considering sampling or exposure times and times of travel which are both small compared with the characteristic time-scales of the variations. The restriction on sampling time merely adds to the numerical labour, as any period of variable conditions may be treated by adequate sub-division but that on time of travel is more troublesome in that it sets a limit to the downwind extent of a multi-source array which may be considered. The process of sub-dividing the sampling or exposure time may of course be carried to the extreme of virtually instantaneous effects, regarding these as determined by isolated 'puffs' of pollutant released sequentially from each source. This 'puff' model, as distinct from the 'steady plume' model which has so far been implied, is clearly only to be considered for numerical integration, but with this limitation it is in principle applicable with any degree of variability of the dispersive conditions and airflow pattern in either time or space. On the other hand such an advantage is secured only at the expense of greatly increased numerical effort. Additionally, the random errors in specifying the dispersion and trajectories of individual puffs may be substantial – the plume model is essentially a composition of puffs in which these errors have been largely smoothed out – and this must to some extent reduce the benefits of the puff model in respect of variable conditions.

Superimposition of individual straight plumes is currently a popular approach, as can be seen from the considerable number of reports and papers which have appeared, especially in the U.S.A. The first step is the adoption of a system, such as that outlined in the previous section, for prescribing the individual pattern of concentration in terms of data on emission and dispersive conditions. Much stress is often laid on the assumption of Gaussian distribution in the plume, though for practical purposes this is not a crucial assumption but rather a matter of convention and convenience. Substantial departures from Gaussian shape do not have an important effect on the ground-level concentration. The more important requirement is to avoid large errors in the magnitude of the spreads.

Even with modern high-speed computing facilities it has been generally accepted as impracticable to treat numerous small sources (e.g. domestic sources) individually. It is customary to group these in areas and to use a composite plume, derived either from a previous numerical summation of an idealized distribution of point sources within the area, or by replacing the area source by an equivalent point source at a position upwind such that its crosswind width matches the crosswind size of the area. Provision and organization of the meteorological data are of course important components, otherwise the operational practice is essentially a matter of numerical processing. There has been a tendency for preoccupation with this aspect to lead to a numerical elaboration which is not always obviously justifiable and to divert attention from more significant basic aspects of modelling.

The guidance which at present may be expected from mathematical modelling of multiple-source distributions of pollution has already been brought out in the discussion of urban pollution (5.5), particularly in the studies by Turner and others in the U.S.A., Marsh and Withers in England and Fortak in Germany.

It is noteworthy that in the plume (or puff) model conservation of pollution is already implied. In considering pollution distribution on the meso-scale or on even greater scale, when the times of travel are such as to preclude the use of the 'steady plume' model, there is advantage in a primitive application of the conservation principle for pollution, in essentially the same way as it is applied to meteorological properties of the atmosphere in numerical solutions of the equations of motion.

Some of the foregoing features will now be considered in greater detail.

Area source solution by algebraic integration

Consider first the case of emissions from an area at a uniform rate of q per unit area at ground level and, as in Fig. 6.11(a), consider a crosswind element of downwind extent δx. This constitutes an elementary crosswind line source of strength $Q=q\ \delta x$ per unit length. At downwind distances which are not large compared with the crosswind extent of the element in either direction the effect of the elementary line source may be taken for many practical situations as approximating to that from an idealized line source of infinite crosswind extent.

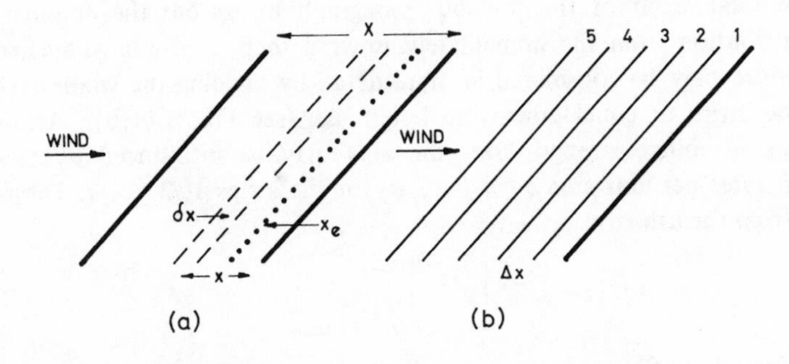

Fig. 6.11 – Representation of an area source.

Solutions for this (which have been considered in more detail in Section 3.2 and 6.3) take the basic form

$$\chi = AQx^{-s} \tag{6.22}$$

where A depends primarily on the wind speed and the magnitude of the vertical spread (at a given distance downwind) but also on the shape of the vertical distribution, while s is determined essentially by the growth of vertical spread with distance. Thus in the simple case of wind constant with height the above variation of concentration would follow from a variation of σ_z with x^s.

Referring again to Fig. 6.11(a) the contribution $\delta\chi$ on the downwind boundary of the area source from an element distant x upwind is $Aq\ \delta x\ x^{-s}$. Thus the total concentration at ground level on the downwind boundary, from all sources within a distance X upwind is

$$\chi = Aq \int_0^X x^{-s}\, dx \tag{6.23}$$

which for $s < 1$ is

$$\chi = Aq(1-s)^{-1}\, X^{1-s} \tag{6.24}$$

Eq. (6.24) may be rewritten in the form

$$\chi = AQx_e^{-s} \tag{6.25}$$

where Q is the total source strength per unit crosswind width from the area source of extent X upwind, and x_e is the upwind distance of the equivalent crosswind line source of strength Q. From (6.24) and (6.25)

$$x_e = (1-s)^{1/s}\, X \tag{6.26}$$

and so for example with $s = 0.8$, which is typical of neutral flow, x_e is a fraction 0.13 of the total downwind extent X.

The final result of the previous paragraph brings out the dominance of the contribution from the immediately upwind section of a large area source. This feature may be considered in more detail by dividing the whole area into crosswind strips of equal downwind length Δx (see Fig. 6.11(b)). Also broad variations in source strength over the area may be introduced by assigning emission rates per unit area $q_1, q_2, \ldots, q_i$ for the strips $1, 2, \ldots, i$. The contribution from the nth strip is then

$$\chi_n = Aq_n(1-s)^{-1} \left[x^{1-s}\right]_{(n-1)\,\Delta x}^{n\,\Delta x} \tag{6.27}$$

Apart from the differences in q_1, q_2 etc. the relative importance of the separate contribution from the strips is represented by the weighting factor $n^{(1-s)}$ –

$(n-1)^{(1-s)}$, which falls off with n with a rapidity sensitively dependent on the choice of s. For $s = 0.8$ it has values as follows:

n =	1	2	3	4	5
weighting factor =	1	0.149	0.097	0.073	0.061

Such a rapid fall-off in the contribution from the more distant strips means that a useful economy may often be possible by the foregoing combination of algebraic and numerical summation. The general principle has been strongly advocated and developed by Gifford and Hanna (1971) for application with a two-dimensional square grid representation of the source distribution.

Note that a sustained build-up of concentration as the air flows over an area source requires $d\sigma_z/dx$ to be small, i.e. in the foregoing power-law description of the growth of σ_z, s must be small. In practice this condition can exist from the beginning when stable conditions prevail *near* the ground (conditions E and F of Table 6.V) or it can set in after some distance of travel as a result of the limitation of vertical mixing by an overhead stable layer.

The foregoing analysis has neglected lateral (crosswind) dispersion and this is justifiable with the limitation imposed at the outset, namely that the distance downwind of the line source element of interest should not be large compared with the crosswind extent of the line source. This criterion arises from the property that in general the width of plume from an elementary point source is a small fraction of the distance downwind. For an area source which has a sharply defined crosswind edge the approximation adopted in neglecting crosswind spread must become doubtful for concentrations near that edge. Likewise, if the area source contains isolated sources with large emission rates, crosswind spread will be important and the effects of such sources must be considered separately.

Other limitations in the foregoing algebraic integration technique arise from the restriction of Eq. (6.22) to ground-level sources and to values of the exponent s less than 1. The latter restriction is fortunately unlikely to be important though the possibility of $s > 1$ exists at shorter ranges in very unstable conditions. For elevated sources over an area, integration into a series which converges fairly rapidly, may be carried out for the special case of linear growth of vertical spread ($s = 1$), as first pointed out and applied by Lucas (1958).

Lucas's solution may be presented in the following general terms (and is not restricted to the special Sutton form of diffusion formula originally used by Lucas). For an elevated crosswind line source at height H, neglecting crosswind spread as before, the solution for Gaussian distribution, wind constant with height, and assuming complete reflection at the ground, is

$$\chi(x, 0) = \frac{2^{1/2} Q}{\pi^{1/2} \bar{u} \sigma_z} \exp\left[-\frac{H^2}{2\sigma_z{}^2}\right] \tag{6.28}$$

This result can be seen to follow from integrating w.r.t y the point source expression in Eq. (6.17). Considering an element δx in an area source as before [Fig. 6.11(a)] and taking $\sigma_z = ax$ say,

$$\delta\chi = A_e q \frac{\delta x}{x} \exp\left[-\frac{H^2}{2a^2x^2}\right] \tag{6.29}$$

where $A_e = 2^{1/2}/\pi^{1/2}\bar{u}a$ (to be distinguished from A in Eq. (6.22). Eq. (6.29) integrates to give the ground-level concentration arising from the area source extending X upwind as

$$\chi = \frac{A_e q}{2}\left[\ln r^2 + \frac{1}{r^2} - \frac{1}{r^4 \times 2 \times 2!} + \frac{1}{r^6 \times 3 \times 3!} + \dots\right] \tag{6.30}$$

where $r^2 = 2a^2 X^2/H^2$.

Reference has already been made (see 5.5) to the early use of this form by Lucas in estimating the general build-up of pollution in the air flowing over a city, but no further application has so far been made. The method has the attraction of allowing for the fact that in practice the emissions of pollution from the multiple small sources of heating and industrial processes are mostly elevated, and the validity of neglecting this feature in using the *ground-level* source solutions for an area source has not yet been made clear.

Box models and their limitations

A rough calculation of the effects of a large area source may be made by using the *box model*, in which it is assumed that the pollution from all sources upwind is uniformly distributed with height up to an effective lid at height h'. For an idealized area source distribution as in Fig. 6.11(b) conservation requires that the concentration χ_a accumulated during a downwind traverse of distance X containing i crosswind strips of downwind length Δx, is

$$\chi_a(X) = \Delta x \sum_{n=1}^{n=i} \frac{q_n}{uh'} \tag{6.31}$$

or for the simple case of a uniform rate of emission of q per unit area

$$\chi_a(X) = qX/uh' \tag{6.32}$$

As before these results are valid for areas of sufficient crosswind width and for wind speed constant with height and they represent the steady-state which would be attained after a time X/u.

The effectiveness of these simple results may be assessed by comparing them with a more realistic development of concentration in which allowance is made for the relatively small vertical spread attained by the pollution from the sources immediately upwind. For the practical case when the source is at height H above ground and when vertical spread is ultimately limited by an effective lid at height h', integration of Eq. (6.17) w.r.t. y yields the following expression for the contribution $\delta\chi$ from an element δx at distance x upwind

$$\delta\chi(x, 0) = \left(\frac{2}{\pi}\right)^{1/2} \frac{q\delta x}{\bar{u}\sigma_z} \sum \exp\left[-\frac{Z_n^2}{2\sigma_z^2}\right] \tag{6.33}$$

with $Z_n = H \pm 2nh'$, $n = 0, 1, 2 \ldots$

For $n=0$, $H=0$ Eq. (6.33) reduces to the ground-level source case with no overhead limitation in vertical spreading and for that case the progressive build-up of ground-level concentration with fetch over an area source may be derived algebraically as in Eq. (6.23) *et seq.* For the model applicable in Eq. (6.33) A in Eq. (6.23) *et seq.* is defined by

$$A = \left(\frac{2}{\pi}\right)^{1/2} \frac{1}{a\bar{u}}, \quad \sigma_z = ax^s \tag{6.34}$$

and Eq. (6.24) may be re-arranged to

$$\frac{\chi_a(X)\bar{u}}{q} = \left(\frac{2}{\pi}\right)^{1/2} \frac{X^{(1-s)}}{a(1-s)}, \quad 0<s<1, \tag{6.35}$$

which specifies the accumulated magnitude of the ground-level concentration arising in a distance X over an area source with uniform rate of emission per unit area.

For the general case the evaluation of Eq. (6.33) and the subsequent integration w.r.t. x must be carried out numerically. However, inspection of the terms in Eq. (6.33) readily shows that although the required number of terms in the summation increases as σ_z increases (with x), in practice summation to $n=2$ is adequate. This value of n is required when σ_z (as it would apply in the absence of the overhead lid) reaches values near h', but at the distance then concerned the contribution from an elementary line source effectively reaches the asymptotic limit of uniformity with height, i.e. at that distance from an element δx the contribution therefrom is $q\ \delta x/\bar{u}h'$.

On the other hand there is an appreciable range of σ_z (hence of x) for which only the first term ($n=0$) is required, and then for $H=0$ the algebraic solution in Eq. (6.35) may be used to specify the 'correct' accumulated concentration $\chi_c(X)$ for comparison with that $[\chi_{bm}(X)]$ given by the simple box model in Eq. (6.32). Hence taking the ratio of the two and putting $aX^s = \sigma_z(X)$,

$$\frac{\chi_{bm}}{\chi_c} = \left(\frac{\pi}{2}\right)^{1/2} \frac{(1-s)\sigma_z(X)}{h'} \quad \text{for } H=0 \text{ and } \frac{\sigma_z(X)}{h'} \text{ small} \tag{6.36}$$

For example, when $\sigma_z(X)/h' = \frac{1}{2}$ (for which the neglect of all terms other than $n=0$ in Eq. (6.33) is justifiable) and when $s=0.85$ (a typical value in daytime mixing), χ_{bm}/χ_c is 0.09. Thus, as would be expected, the simple box model is seen to give a marked underestimate for fetches such that the effect of the overhead lid is insignificant at the ground. Full evaluation of Eq. (6.33) and graphical integration w.r.t. x has been carried out for a typical situation, with $H=0$ or 10 m, $h'=500$ m, and with vertical spread characteristics appropriate to $z_0=10$ cm, stability category B, as in Table 6.VIA. The results are given in Fig. 6.12. They bring out further the inadequacy of the simple box model, even when allowance is made for elevation of the area source to a height probably representative of many of the so-called 'low-level' sources of an urban-industrial complex. As already noted, the contribution reaches the 'uniformly mixed' limit when $\sigma_z \cong h'$ (i.e. when $x=8.8$ km in the example) and thereafter the accumulated ground-level concentration increases linearly at a rate q/uh' per unit increase of fetch X. It is evident (see the example in the notes below Fig. 6.12) that a very large fetch may be required for the simple box model to give a close approximation to the more realistic estimate of accumulated concentration.

It is emphasized that the reference here is to the ground-level concentration at the *downwind edge* of an area source. For the *average* concentration

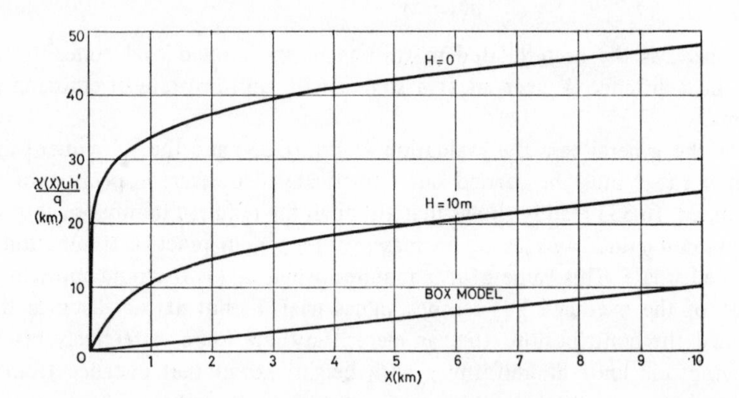

Fig. 6.12 – Theoretical ground-level concentration over an area with uniform rate of emission. Mixing depth $h'=500$ m, $z_0=10$ cm, stability category B (see Table 6.VIA).

$\chi(X)$ is the accumulated ground-level concentration at distance X from the upwind edge, obtained from the algebraic integration as in Eq. (6.35) for the surface source ($H=0$) and $X \leqslant 3$ km, and from Eq. (6.33) and graphical integration for $H=0$, $X>3$ km, and for the elevated source $H=10$ m, $X>0$. The 'box model' line represents Eq. (6.32).

Designating the 'box model' value by χ_{bm}, and the 'correct' value (in which allowance has been made for progressive vertical spread) by χ_c, then for example

$$X = 8.8.\ \text{km} \qquad \chi_{bm}/\chi_c = 0.37 \text{ for } H=10 \text{ m}$$
$$\text{or } 0.19 \text{ for } H=0$$

For low-level emission in practice $H=10$ m is probably more realistic, and for this case the increase in fetch ΔX beyond 8.8 km required to give $\chi_{bm}/\chi_c=0.8$ say is approximately 50 km.

over the area it is evident that the discrepancy between box model estimates and those made with recognition of the progressive vertical spread will be even greater, and the required fetch for adequate approximation will also be even greater. Thus for *low-level* sources the simple box model should not generally be applied for areas less than about 100 km in downwind extent unless it is clear that the mixing depth is much less than the typical value of 500 m assumed in the example above.

The results of Fig. 6.12 also indicate the discrepancy which may result from assuming (for mathematical convenience) that the sources are strictly at the surface rather than at say 10 m. At 8.8. km, the ratio $\chi_{H=0}/\chi_{H=10\,\mathrm{m}}$ is 46.5/23.6 and reduction of this to say 1.2 would require an extra fetch of 90 km. The point was made earlier in this section that the effect of elevation of source required further consideration – the foregoing example indicates that the effect may often be significant. A practical method of eliminating this difficulty, while retaining the simplicity of the 'surface-source' method, would be to neglect all sources immediately upwind within the distance at which $\sigma_z = H$. This would reduce the accumulated concentration by a constant amount at all values of X. Thus, in the example of Fig. 6.12, for which $\sigma_z = 10$ m at $x = 86$ m the ordinate of the uppermost curve would be reduced by about 23 and would then be in close correspondence with the $H = 10$ m curve.

Grid-cell systems

The working methods discussed so far are specially suitable for prescribing the separate effects of localized sources or for integrating the effect of an upwind area of virtually uniform and steady emission. For the more general case of an area in which there are marked spatial and temporal variations in the density of emission there is now a growing tendency to advocate grid-cell systems. In principle this is a return to the primitive conservation equation [Eq. (3.10)]. In practice the equation is approximated by a finite-difference scheme based on a grid of points with convenient spacing horizontally and vertically. Effectively the whole domain is divided into an array of boxes or cells and the numerical analysis has to be designed to represent the advection through the boxes and the mixing between adjoining boxes. As in the meteorologically more familiar finite-difference schemes for numerical forecasting there are many practical problems to be overcome and these have received much attention in the literature on air pollution meteorology.

The difficulties arise mainly in the following respects:

(i) the process by which the effects of simple advection through the cells are adequately represented,

(ii) the inadequacy of the finite-difference representation of the turbulent flux in prescribing inter-cell mixing,

(iii) the realism of the eddy-diffusion coefficients used in determining eddy flux.

The requirement is to represent for each cell the full 'budget' of (a) import of material through the upwind boundaries, (b) introduction (sources) and extraction (sinks) of material within the cell and (c) export of material through the downwind boundaries. All these terms are specified for a time increment Δt. Each import term is expressed as the product of the airflow normal to the upwind boundary and the average concentration existing in the upwind adjoining cells during the interval Δt. Likewise the export terms are determined by the appropriate wind components on the downwind boundaries and the concentration in the cell itself. Various statements of the detail of this 'budget' have been set out (e.g. by Reiquam, 1970).

Even if the advective speeds and directions are satisfactorily specified there is a difficulty in that material released in a cell over the time-step Δt will be displaced according to the speed and direction of the flow, into an area overlapping several cells. In applying the finite-differences scheme any material imported into a cell is conventionally supposed uniformly spread throughout the cell. This produces a spurious horizontal advection which may be specially important when considering the progress of an isolated dense patch of pollution which is not large compared with the cell area. In the latter case the problem may of course be avoided altogether simply by constructing the complete trajectory of the individual patch of material and treating it as an isolated source. However, if it is desired to incorporate such patches in a general grid-cell treatment some way of avoiding this gross exaggeration of the horizontal spread will be required. A method in which at the end of each time-step the moments of the distribution of the material within each cell are taken into account before the advection in the next time-step is computed has been proposed by Egan and Mahoney (1972).

The essential point in item (ii) is that even if the finite-difference estimate of concentration gradient and the computed turbulent flux of material are realistic, the step of uniformly distributing any material which has penetrated into a cell again imposes a spurious horizontal spread. With the object of avoiding this difficulty Sklarew *et al.* (1971) have designed a numerical procedure described as a 'particle-in-cell' method. The procedure is a combination of 'particle trajectory' and 'inter-cell exchange' treatments, and incorporates the novel feature of representing the eddy flux of particles by the product of the mean concentration and an equivalent diffusion velocity. This is analogous to the specification of a vertical deposition in 5.2, but is now applied to the horizontal fluxes as well. In effect the diffusion is represented as an extra advection velocity added to the true advection, and the total velocity assigned to each one of a specified initial distribution of particles is interpolated according to the position of the particle between cell centres. Both the advective and diffusive movements of the particles are thereby represented without incurring the spurious effects of distributing particles uniformly over each cell.

Egan and Mahoney's suppression of spurious advection by constraining the movement so as to maintain the true moments of the distribution of material is of course essentially equivalent to the use of actual particle trajectories in Slarew *et al.*'s system.

Item (iii) is a difficulty which applies equally to algebraic solutions of the diffusion equation. There are arguments which may be advanced against the use of the K concept for horizontal spread of individual plumes but for sub-grid scale horizontal mixing within a large area distribution of pollution these objections are less important. More important probably is the problem of giving a numerical specification of the horizontal K, consistent with the intensity and scale of the horizontal components of turbulence. In this context of the large area distribution of pollution, the statistical theory form of K (see Table 3.I), now with eddy velocity and time-scale appropriate to the horizontal, may form a useful starting point.

The difficulty arising from imposing a uniform concentration throughout a cell also applies in respect of vertical advection and diffusion. It is particularly important that the height/length ratio of the cells over an area source should be small enough for the simple box-model relation of Eq. (6.32) to be a suitable approximation, otherwise the early stages of vertical spread of pollution may be grossly exaggerated. In the foregoing considerations of box models, the calculations for a box of height 500 m indicated that for low-level sources a length in the region of at least 100 times the height would be required for the average concentration over the box to be close to that given by the model. This assumed relatively rapid vertical mixing and the necessary ratio will be even larger for slower mixing. For the particular condition of vertical mixing it is however a conservative estimate for shallower boxes (it would apply exactly to all depths of box if σ_z increased linearly with x, but as previously seen the growth is typically somewhat slower than linear).

The implication of the previous paragraph is especially significant if the simplest form of grid-cell system is adopted – namely a single layer of cells of height corresponding to the potential mixing layer. With the latter typically 1000 m the horizontal dimensions of the cells over an area source should preferably be 100 km or more, so precluding any useful resolution in practice over an urban area, except at the expense of introducing multiple layers or taking into account the incomplete vertical spread as in the early part of this section.

6.8 ESTIMATES USING ROUTINE METEOROLOGICAL DATA AND HISTORICAL DATA ON POLLUTION

The most detailed application of the foregoing methods for estimating the level of air pollution requires a combination of emission data and special meteorological data of the type summarized in 6.2. This may be feasible in special studies, but for many of the practical requirements which arise in air pollution studies the methods have to be simplified into a form which may be operated in terms of routine meteorological data and possibly without detailed information on the rates of emission.

In the regular programme of synoptic and climatological observations carried out by meteorological services the most relevant elements are:

(a) 'surface' wind speed and direction (conventionally referring to a height of 10 m),
(b) amount and nature of cloud cover,
(c) the broad variations with height of wind and temperature (from radiosonde ascents),
(d) 'surface' air temperature (conventionally referring to about 1 m above ground),
(e) the surface pressure distribution (providing a broad picture of the airflow pattern),
(f) duration of sunshine,
(g) intensity of incoming solar radiation.

Correlation of the special meteorological factors of 6.2 with some suitable combination of the above routine data is an important step in the chain of operations. The special data are being gradually accumulated in the course of research studies of turbulence and transfer processes in the atmospheric boundary layer and also in the more elaborate observational studies of dispersion of tracers or of actual pollutants. However, this development of a dispersion climatology is inevitably slow, especially in respect of the most fundamental of the dispersion parameters, such as the intensity and scale of turbulence and the turbulent heat flux. In the meantime the simplest starting point for a dispersion climatology is provided by the well-established general character of the vertical profile of temperature in the lowest layers of the atmosphere. The development on this basis of the system of *stability categories*, requiring as a minimum a knowledge of wind speed and cloudiness in relation to time of day, has already been considered in 6.6 and some of the difficulties noted.

Climatological summaries of surface wind direction and speed are already fairly readily available. Useful summaries of the incidence of suitably defined stability categories and of potential mixing depths are therefore now derivable from the general background of routine observations. Examples of summaries compiled so far are:

stabilities in the surface layer over the British Isles (Bannon, Dods and Meade, 1962),
incidence of inversions over the British Isles (Best and Meade, 1956),
mixing depths over the U.S.A. (Holzworth, 1967).

Statistics of this kind must be used with care and with regard for the difficulties noted in 6.6. In particular it should be kept in mind that potential mixing depths as designated by an overhead inversion will be realized only for some period of the day, and may not be realized if the foregoing nocturnal stability is strong and the daytime convection weak. Also, statistics of stability categories should not be combined with those of other parameters (such as wind direction) without considering the possible existence of special correlations between the two.

Given a specification of stability categories either from current meteorological data or from climatological statistics mathematical modelling of air pollution distributions may be accomplished on the lines already described. However, even this degree of simplification of procedure and meteorological data requirements may be insufficient for many practical purposes, and in any case it may be difficult or even impossible to obtain emission data of a quality which would justify carrying out the detailed calculations of the model. These complications arise in their most extreme form when estimates of air pollution are required in a warning or forecasting procedure and this has led to a demand for even further simplifications.

The idea of forecasting the potential for high levels of air pollution, in a purely meteorological analysis of the prospects of conditions of relatively ineffective dispersion, has been given great attention in the U.S.A. following the the specification by Niemeyer (1960) of appropriate wind and subsidence criteria associated with stagnant anticyclonic conditions. More recently in the U.S.A. this qualitative approach has been extended progressively in a quantitative sense, by including the mixing depth and by using the product of mixing depth and wind speed which appears in the 'box model' specification of the build-up of air pollution over an area [see Eq. (6.31)]. A summary of this development is given by McCormick (1970).

For the case in which there is an established pattern of sources of pollution effective use may be made of historical data on air pollution, by examining these in relation to the general meteorological conditions. The rules which can emerge from correlations of this kind reflect the net result of a whole complex of weather factors, including their effects on the rate of emission and thermal rise of the pollution, as well as on its subsequent dispersion. One obvious correlation which has a useful bearing is that between consumption of fuel for heating purposes and the prevailing temperature conditions. The practical significance of this concept in an urban pollution context was demonstrated by Marsh and Withers (1969) in their derivation of a regression relating sulphur dioxide concentrations to a parameter incorporating air temperature and the reciprocal of the wind speed.

In the context of providing warning of the onset of very high concentrations of pollution it is obvious that incidence of very light winds is a dominant criterion. A simple application of this principle has been designed by Benairie (1971) on the basis of limited experience of trends of air pollution in the Rouen area in France. From an inspection of the data factors were derived relating the average concentration over selected days (with specified low wind speeds) to the preceding long-term average concentration, represented in practice by an average over 30 days, terminating shortly (i.e. a few days at most) before the selected day. The factors were given mainly in terms of wind speed but also included some allowance for temperature and wind direction. Clearly the success of this procedure depends on the extent to which the adopted long-term average is representative of the general amount of pollution released and of a typical range of wind speeds and dispersive conditions.

The technique of correlation with the past history of the concentration in terms of routine meteorological data has been explored in more detail in a study of the daily average levels of sulphur dioxide in London and Manchester, by Smith and Jeffrey (1971). In this study the following parameters were included in a multiple regression analysis of the daily average concentrations for four selected sites in Central London, in relation to the long-term mean and the value for the previous 24 hr:

(a) the minimum temperature – which reflects both the emission rates and the nocturnal stability in the lower atmosphere,
(b) the wind speed – which represents the initial dilution near a source and also to some extent the stability condition as it affects dispersion,
(c) wind direction,
(d) mixing depth.

The air pollution data were obtained from the Department of Trade and Industry National Survey for two winter-half-years in the period 1968–70. The analysis led to the the two prediction equations below, the second being a simplified version in which certain parameters are omitted.

$$C_{\text{est}} = \left[1 + \frac{\delta_d}{3}\right]\left[1 + \frac{\delta_m}{6}\right]\left[1 - \frac{T}{28} + \frac{t}{20}\right](a\bar{C} + bC_p) \qquad (6.37)$$

where $\bar{C}$ = long term mean concentration
C_{est} = estimated concentration (24-hr average)
C_p = concentration for the previous 24 hr
T = minimum temperature (°C) expected up to midnight
t = number of hours of mean wind less than 3 knots during the 24 hr
$a = \frac{2}{3}, b = \frac{2}{9}$ if the mean wind for the day exceeds 6 knots
$a = \frac{3}{7}, b = \frac{8}{21}$ otherwise

$$\delta_d = \begin{cases} 1 \text{ if the mean wind comes from the 'dirty' sector, 060° to 120° in London only} \\ 0 \text{ otherwise} \end{cases}$$

$$\delta_m = \begin{cases} 1 \text{ if the mixing depth is low} \\ 0 \text{ otherwise} \end{cases}$$

$$C_{\text{est}} = 0.085\left(1 + \frac{\delta_m}{6}\right)\left(1 - \frac{T-t}{28}\right)(5\bar{C} + 4C_p) + 0.15\bar{C} \qquad (6.38)$$

where δ_m = 1 if the mixing depth is low for London only and is 0 otherwise
T = minimum temperature from 0900 to 0900
t = hours when mean wind falls below 5 knots

Low mixing depth is defined by the following criteria:

(i) *day*: lowest inversion, or cloud base, below 500 m
following night: surface inversion sets in before 1800 because the 10 m wind $V \leqslant 8$ knots and cloud $\leqslant \frac{5}{8}$ths

or (ii) *day*: lowest inversion, or cloud base, below 300 m
following night: surface inversion sets in before 2100 because $V \leqslant 8$ knots and cloud $\leqslant \frac{7}{8}$ths.

Figure 6.13 is a nomogram representation of Eq. (6.38) and Table 6.VII summarizes the correlations and root-mean-square differences between the observed values and those calculated from the two regressions using actual meteorological data.

Examination of the graphs in Fig. 6.13 and the statistics in Table 6.VII brings out the following points:

(i) the depth of the mixing layer is of secondary importance,
(ii) a substantial improvement over a 'persistence' estimate is provided by using simply the long-term mean concentration for the 'previous day' concentration,
(iii) use of an actual 'previous day' concentration is only marginally better than use of a value estimated by the formula

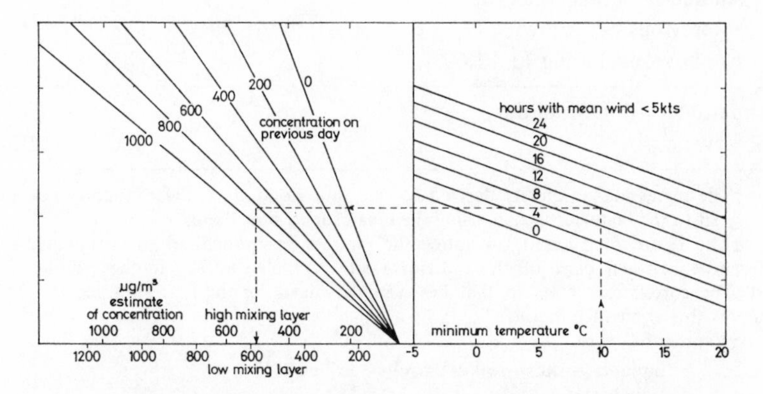

Fig. 6.13 – Nomogram of Eq. (6.38) for estimating the concentration of sulphur dioxide in Central London averaged over 24 hr. An example is represented in the dashed line. (Smith and Jeffrey, 1971). Note that this nomogram applies to years when the average winter concentration $\bar{C}$ was approximately 350 $\mu g/m^3$. As a result of the Clean Air Acts and the increased use of natural gas the value of $\bar{C}$ was virtually halved by 1980, and in that case the abscissae on the left of the nomogram would have to be modified accordingly.

Table 6.VII – Results of applying the prediction formulae (6.37) and (6.38) to London and Manchester SO_2 data for two winters (Smith and Jeffrey, 1971).

	Correlation between actual and estimated concentrations		Root-mean-square difference ($\mu g/m^3$)	
Detailed Scheme Eq. (6.37)	London	Manchester†	London	Manchester
Inherent error from taking only 4 sites‡	0.94	–	50	–
All data – actual conc. for previous day known	0.87	0.80	76	65
As above, excluding 11.12.69 (an unexplained bad day)	0.90	–	66	–
Previous day's concentration replaced by mean, $\bar{C}$ §	0.80	–	95	–
Previous day's conc. replaced by estimated value for that day	0.85	–	79	–
Persistence estimate – previous day's conc. used as estimate for today	0.55	–	142	–
Simplified Scheme Eq. (6.38)				
All data – actual conc. for previous day known	0.81	0.79	88	66
As above, excluding 11.12.69	0.84	–	83	–
Standard deviation of all the observations	–	–	150	106

†The regressions originally derived for the four London sites were found to be a satisfactory fit to the Manchester data using the mean for *all* sites there.

‡The figure of 50 is $1/\sqrt{4}$ times the random component of the root-mean-square differences between pairs of the individual sites, which was 100 in the sample of data used. The correlation refers to that between the mean for the four sites and the implied mean for the whole of London.

§Values of $\bar{C}$ were

London (4 sites) weekdays only	352 $\mu g/m^3$
Manchester	252

A statistical estimate of the discrepancy E_t of an estimate for the average over any other number of stations n may be derived by writing

$$E_t^2 = E_i^2 + E_s^2$$

where E_i is the random discrepancy (i.e. in the present case for London $100/n^{1/2}$) and E_s is the basic error of the relation for C_{est}. Thus if E_t for 4 sites is 95 $\mu g/m^3$ then E_s is 81 $\mu g/m^3$, and the magnitude of E_t for 1 and 100 sites will be approximately 129 and 87 respectively.

(iv) the simplified form in Eq. (6.38) and Fig. 6.13 is not greatly inferior to the more detailed equation, indicating that of the meteorological parameters the duration of light wind and the minimum temperature alone represent most of the effect of changes in meteorological conditions.

6.9 LIMITATIONS AND UNCERTAINTIES IN METEOROLOGICAL ESTIMATES OF POLLUTION

The aim of this chapter has been to outline the methods which have evolved for estimating incident levels of pollution in terms of meteorological data. In practice the final interest lies in the prospects which these methods offer for providing two main types of information:

(a) estimation in advance of the effects likely to be produced by new sources of pollution,

(b) warning of the likelihood of a serious build-up in the pollution from an existing complex of sources

Limitations in the provision of this information are encountered in the following respects:

(c) the special nature of the meteorological conditions and other circumstances which occasionally combine to form the worst pollution episodes,

(d) the availability in required form of the essential meteorological information,

(e) the basic inaccuracy and unrepresentativeness of the calculations of dispersion, even in the circumstances for which the dispersive action of the atmosphere is most clearly understood.

Association of the most disastrous cases of air pollution with stagnant air and confining effects of topography is well-known, and it may fairly be said that the more elaborate aspects of the theoretical and observational study which have been the subject of the first four chapters of this book have added little to the prospects of usefully anticipating the incidence and severity of such episodes. For this requirement the over-riding problem is the forecasting of the specially stagnant anticyclonic conditions in which the episodes are possible. Beyond this the appreciation of the intensity of the build-up of pollution rests on the experience of a limited number of acute episodes, and on the inevitably slow progress in formulating useful rules on the basis of experience of sub-acute levels of pollution in conditions of light wind.

The lack of meteorological information is of course particularly obvious in respect of the special factors, which are required in the fullest application of the dispersion relations and mathematical models, and which are beyond those listed at the beginning of 6.8. In considering the need for such information it is clearly desirable to bear in mind the cost–benefit implications in the mounting of the required observational programmes and in the reliability of the pollution estimates which are thereby achieved. The procedure may be criticized on the score that dispersion relations apply at their best only to idealized situations of airflow and site topography, which are rarely met in important pollution problems. Also it is often asserted that estimates of air pollution based for convenience on a climatology of dispersion conditions may be of little value, and that more attention should be devoted to the special features of sites and to specially adverse conditions of dispersion.

There have been several published discussions and comments on the accuracy attainable in the prediction of levels of air pollution. The essential basis for judgement on this issue is in the background of tracer study, pollution surveys and special meteorological data which has formed a major part of this book. In concluding this chapter it is appropriate to be reminded of the overall indications. An accuracy as good as ±10 per cent in ensemble-average prediction is evidently attainable only in the most ideal combination of circumstances as regards the release of the pollutant and the properties of the dispersing airflow. In most circumstances of practical interest however there is no reason to expect the uncertainties to be less than several tens per cent in ensemble averages or factors of two or more in individual examples of time-mean concentration even over a time as long as one hour. It remains to be seen whether any significant improvement in this capability will ensue as a consequence either of the ever-increasing sophistication in the studies of atmospheric flow or of the continuing elaboration in mathematical modelling techniques.

7

Dispersion over distances dominated by Mesoscale and Synoptic scale motions

7.1 THE ATMOSPHERE ON A LARGE SCALE

Most attention so far in this book has been concentrated on diffusion out to, at most, a few tens of kilometres, and it is indeed true that this is the range of most general practical interest. Beyond such distances concentrations are usually so low that resulting hazards are negligible. However, this is not always so. Airborne animal viruses, pollens, volcanic dust, industrial pollutants and radioactive material released in nuclear power plant disasters are but a few instances where dispersion is important on much larger scales. Sometimes the release duration is short, perhaps less than an hour, but more generally these problems are associated with long period releases, perhaps even never-ending emissions, and in consequence some of the simplifying assumptions such as steady-state boundary layer conditions applied earlier to short-range problems cease to be relevant. Long-term average concentrations, dosages and depositions now have to take into account the complex inter-related climatologies of wind speed and direction, stability, boundary-layer depth, boundary layer 'break-down' processes and so on. The first and perhaps most important question to be faced is where does the effluent go to. Previously in short range studies it has been sufficient to consider mean wind fields with a superimposed diffusion in the vertical and the horizontal characterized by turbulence spectra that depend on the micrometeorological inputs of energy associated with the surface fluxes of momentum, sensible and latent heat. Now, at longer scale and over longer release durations, the mean wind itself becomes variable and is part of the spectrum of turbulence. Whether it is treated as such or whether it is determined from actual synoptic observations of wind depends largely on the nature of the problem, for example whether we are attempting to understand an actual event that has happened or whether we are attempting to predict probabilities of possible events in the future. Either way recognition of the wind's spectral character is important. Fig. 7.1 shows an example of spectra due to Hess and Clarke (1973) for the eastward component of wind derived from hourly observations of wind deduced from measurements of pilot balloon motions at various heights made during the

Wangara experiment in Australia. The spectra have maximum energy density around 10 days, corresponding to associated changes in the general synoptic pattern. Hess and Clarke show in their paper that in these measurements there was comparatively little energy in the mesoscale (a few kilometres up to a few

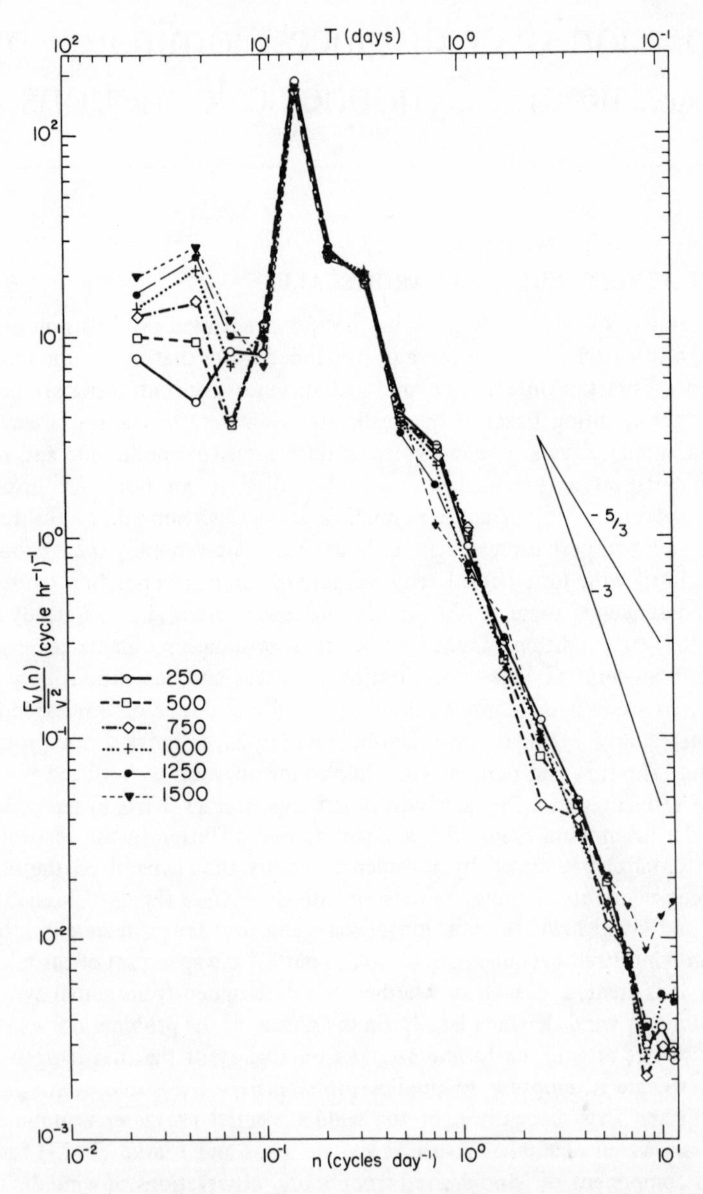

Fig. 7.1 – Normalized *u*-spectra at various heights averaged over bandwidths.

tens of kilometres) and this was probably due to the uniformity of the terrain around the site of Hay in New South Wales (34°30′S, 144°56′E). The $S_u(n)$ spectra also peak at a somewhat lower frequency than that due to Van der Hoven (1957) which has been widely quoted and appeared in the second edition of this book (shown as an $nS(n)$ spectrum). Note that in earlier sections we have been concerned with the so-called micrometeorological part of the spectrum in which the peak occurs at a frequency at least three orders of magnitude greater than that shown in Fig. 7.1.

Lagrangian trajectories and time-scales

The advection of pollution on a large scale can be determined approximately by making estimates of wind speed and direction from computer-stored synoptic fields of pressure or from successive charts and using these to construct the trajectories of imaginary air particles. The method of Sykes and Hatton (1976) is probably the best of the computer techniques. Trajectories can also be derived by tracking constant-level balloons over long distances. Durst *et al.* (1957) derived trajectories from synoptic charts, 3 or 6 hours apart, assuming geostrophic motion. 'Particles' were released at specified pressure levels (either 500 mb, 700 mb or surface) and followed at the same level using time-centred steps for up to 72 hours. Lagrangian auto-correlation coefficients were calculated from the serial values of velocity, and these were a close fit to an exponential form $R(t) = \exp(-t/t_L)$. The average time scale t_L was about 22 hours for the 500 mb and 700 mb trajectories, and about 5 hours for the surface trajectories. A value of 8 hours was also obtained using balloon data collected by the U.S. Navy in 1953 using trans-oceanic-sondes flying at 300 mb (see Angell (1958) and (1960) for a report on these and later flights). The differences in these upper pressure level values is possibly not very meaningful, perhaps all that can be deduced is that the time-scale is of the order of 10 hours. To a significant extent these time-scales depend on the lengths of the trajectories being analysed, since the mean velocity in each is removed. Taking the whole ensemble of trajectories at one level together over a long period like a year would yield a much larger time-scale; trajectories on a continental scale tend to curve only slowly, except when the wind speeds are low. In general forward-trajectories no matter in what direction they start all tend to move eastwards eventually in response to the overall flow of synoptic features from west to east in temperate latitudes.

Tetroons have also been flown at heights of about 100 metres within the boundary layer out to distances of the order of 200 km, and have been tracked by radar or have been returned to the launchers with details of the ultimate recovery point by members of the public, and have provided useful information on the behaviour of low-level winds. Pack *et al.* (1978) have summarized many of these experiments. Whilst remembering that these are one-level motions and do not fully represent the behaviour of 'free' particles of pollution, the following are just two of Pack's conclusions:

(i) the backing in southerly flows is on average 20° from the surface geostrophic direction and about 40° for northerly flows.
(ii) the best fits to the trajectories were obtained using twice the 10-metre wind speed veered by 10°, although even here the standard errors were not insignificant.

7.2 DISPERSION INTO THE WHOLE ATMOSPHERE AND ITS EFFECT ON CONCENTRATIONS

A brief description of the structure of the atmosphere

Fig. 7.2 shows a simplified picture of the temperature structure of the atmosphere. Temperature is plotted against height on a non-linear scale. The ordinate actually represents the fraction of total mass of the atmosphere up to that height. The atmosphere can be divided into three parts: the troposphere, where

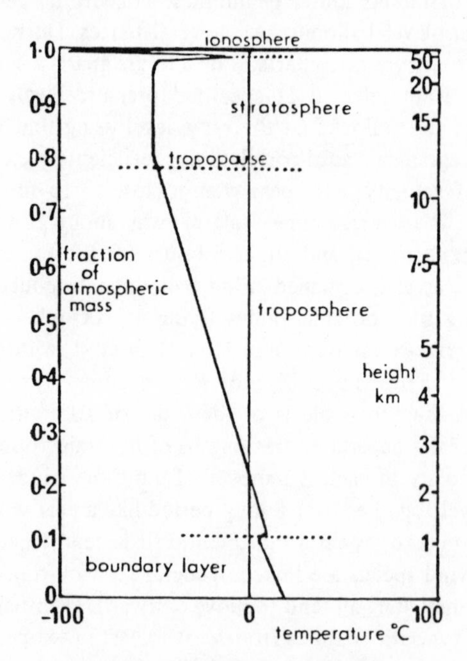

Fig. 7.2 – The standard atmosphere.

the temperature falls on average with height by about 0.65°C/100 m, the stratosphere, where the temperature rises and the ionosphere (containing a very small fraction of the total mass), where the temperature first falls and then rises until the top of the atmosphere is reached. The reasons for these temperature variations are rather complex but at their simplest can be thought of in the following terms:

(i) In the troposphere, vertical mixing is important. Since a vertically displaced element of air experiences falling pressure as it rises (pressure = mass of air above), it expands and cools. Hence temperature falls with height.

(ii) In the stratosphere, ozone is generated and absorbs ultraviolet light. This absorption causes warming and energy is transferred between layers by the radiational exchange. The generation of ozone and hence the warming is greatest above 30 km.

Note that the boundary layer typically contains some 10 per cent of the total atmospheric mass, the troposphere some 75 per cent of the total and the stratosphere only some 25 per cent. Penetrations of pollution into the stratosphere will therefore only have a relatively small effect on ground-level concentrations. Tropospheric air tends to penetrate into the stratosphere through a break in the tropopause at a latitude near 40° or by the action of large tropical cumuli at lower latitudes. Stratospheric air on the other hand tends to be drawn down into the troposphere in regions of active deep convection (i.e. vigorous storm areas).

Time-scales of dispersion and resulting concentrations

Table 7.I gives a very approximate idea of how far dispersion has proceeded on different time-scales for material released near the ground.

Table 7.I

Time of travel	Typical distance travelled	Dispersive eddies	Area affected
minutes	hundreds of metres	small-scale	plume within boundary layer, and growing vertically and sideways linearly with time
hours	tens of kilometres	horizontal spread by mesoscale eddies	plume reached top of boundary layer; concentration determined by lateral spread
days	thousands of kilometres	synoptic scale	escape of pollution from b.l. into remainder of troposphere becoming significant

Table 7.1 – continued

Time of travel	Typical distance travelled	Dispersive eddies	Area affected
weeks	right round earth	global circulation	material anywhere throughout troposphere in one hemisphere. Transport to other hemisphere beginning
months	right round earth	global circulation	all global troposphere affected. Some penetration into lower stratosphere, mainly through tropospheric gap
years	right round earth	global circulation	whole atmosphere affected

If we were to suppose that 10 tonnes of passive airborne pollution were to be emitted within the boundary layer, uniformly over 24 hours, then we may interpret the rough values given in Table 7.I in terms of maximum statistical-average concentrations within the resulting plume. We suppose the concentrations within the resulting plume refer to a sampling time of 1 hour and that the surface geostrophic wind at the time of release is 10 ms^{-1}. No marked stabilization of the lowest layers is included. The values quoted in the following table (Table 7.II) assume no loss processes and may be at least one order of magnitude lowere than could be experienced locally over short sampling times.

Table 7.II – Approximate maximum average concentrations resulting from the release of 10 tonnes of pollution in 1 day.

Time of travel	Typical distance of travel	concentration (μgm^{-3})
1 min	300 m	10^4
1 hour	30 km	5
1 day	900 km	3×10^{-2}
1 week	5000 km	5×10^{-4}
1 month	–	3×10^{-5}
1 year	–	4×10^{-6}
10 years	–	2×10^{-6}

Table 7.III – Table of typical average concentrations of specified substances (for a release rate of 10 tonnes per day) after given times of travel, having been subjected, where appropriate, to dry deposition, removal by precipitation, chemical change and radioactive decay. Concentrations are in units of μ gm^{-3}.

Substance	Loss rate (% h^{-1})	Time of travel				
		1 minute	1 hour	1 day	1 week	1 month
Noble gases (no loss processes)	0	10^4	5	3×10^{-2}	5×10^{-4}	3×10^{-5}
Inert small airborne particles (diameter ~ 1 μm)	2	10^4	4.9	2×10^{-2}	2×10^{-5}	4×10^{-11}
Inert large airborne particles (diameter ~ 10 μm)	4	10^4	4.8	1×10^{-2}	5×10^{-7}	4×10^{-17}
Reactive gases (e.g. SO_2)	6	10^4	4.7	7×10^{-3}	1.5×10^{-8}	0
Airborne material undergoing radioactive decay *only*:						
half life: 1 hour	50	10^4	2.5	1×10^{-9}	0	0
1 day	3	10^4	4.85	1.4×10^{-2}	3×10^{-6}	4×10^{-14}
1 week	0.4	10^4	5	2.7×10^{-2}	2.5×10^{-4}	2×10^{-6}

Loss–processes resulting from chemical decay, dry deposition and wet deposition (in precipitation) are often very important and could have a profound effect on the concentrations given in Table 7.II.

Some examples are given in Table 7.III which is really an extension of Table 7.II. The given concentrations resulting from the combined effect of these processes are typical for the substances listed, although in reality quite a lot of scatter must be expected. A simple source-depletion model has been used. The loss rates are typical annual-averaged values appropriate to temperate climates.

Concentrations for other releases can be obtained by scaling the values in Tables 7.II and 7.III linearly according to the release rate.

7.3 MESOSCALE DISPERSION

Mesoscale motions are not readily quantified either in detail or statistically because they depend on terrain characteristics which themselves are not easily quantified, nor are they distinguished by meteorological observing stations which for this purpose are too widely spaced. Consequently the understanding of these motions is still rather fragmentary although in recent years various field experiments have been conducted to study some of them.

Satellite pictures have revealed boundary layer mesoscale structures unsuspected previously. For example cumulus clouds tend to form in patches or in lines of scale 100–1000 kilometres rather than in uniform sheets, and these have been the subject of such major investigations as the GARP Atlantic Tropical Experiment (GATE) described in WMO GATE Reports Nos. 5 and 14. Another experiment off the German coast in the North Sea has also studied the same phenomena (Kontur Expt., see Hoeber, 1982) in temperate latitudes. Such well-organized motions are bound to have an influence on dispersion and one unanswered question is to what degree they enhance or suppress dispersion beyond their own scale. It is conceivable pollution is largely trapped within such rolls and is passed only slowly to neighbouring rolls, thus being essentially non-diffusive on the large scale in spite of the appreciable velocities associated with the circulation.

Sea Breeze circulations

Other examples of mesoscale circulations are sea and land breezes, flows associated with large convective clouds, urban circulations and topographically induced flows. Although this is not the place for a detailed description of these since they can be found in modern meteorological handbooks and journals, very brief summaries can be given.

A sea breeze can occur when a significant temperature difference, greater than about 2°C, between the air over land and over the sea is established during the day and other factors such as the wind and the local topography are right. Convection over the land is important but should be limited in vertical extent and should not begin near the coast before the sea breeze has been established. Sea breezes also appear not to occur if the undisturbed 900-metre wind exceeds about 7 m s^{-1}.

Once established the depth of the inflow (sea to land) is typically some few hundred metres deep (usually about 500 metres) and is some $\frac{1}{3}$ to $\frac{1}{4}$ of the total atmospheric boundary layer depth; the remainder consisting of the outflow back to the sea. The two sections are separated by a stable interface. The onset of the sea breeze as it moves inland is characterized by a turbulent frontal zone in which the air on average moves upward to become part of the outflow section. The front moves forward at a speed proportional to $\sqrt{\Delta\theta h}$, where $\Delta\theta$ is the land–sea temperature difference and h is the height of the stable interface. Speeds of 3–5 m s^{-1} are typical. Sea breezes, although not un-common at most temperate-latitude coastlines, see Table 7.IV, occur frequently in lower latitudes and are potentially important for the following reasons:

(i) they sometimes appear to be associated with a largely unexplained suppression of plume rise associated with industrial sources (Palumbo, private communication);
(ii) strong fumigation usually occurs at the sea-breeze front which can bring down high concentrations to ground level several tens of kilometres downwind of the source;
(iii) the total circulation can mean the same air is being cyclically and additively affected by emissions from coastal sources.

Table 7.IV – The estimated frequency of sea breezes inland from the east coast of England (based on observations by O. W. Brittain, given in Internal Meteorological Office Memoranda).

Distance inland (km)	Estimated number per year
0	~ 80
10	20
30	10
60	6
100	3

Land breeze circulations

Land breezes at night are generally more feeble and rarely exceed about 60 metres in depth. Even a light onshore component of the gradient wind will often completely suppress it. If, however, high ground exists only a few kilometres inland downslope flows may reinforce the land breeze provided the pressure gradient is slack and the skies are virtually cloudless. Generally land breezes are 'good' for pollution since they carry it out to sea where it can do little harm.

Storms

Large storms have a very dramatic effect on the boundary layer and its contents. They represent a very real breakdown of the normal structure. Fig. 7.2a which is due to Thorpe (1981) and is based on the work of Browning and Ludlam (1962) shows the strong exchange of air between the boundary layer and the upper and middle troposphere. Not only will this imply a rapid loss of boundary layer pollution to other levels but presumably the associated precipitation will remove much of the soluble pollution, giving localized episodic depositions. In

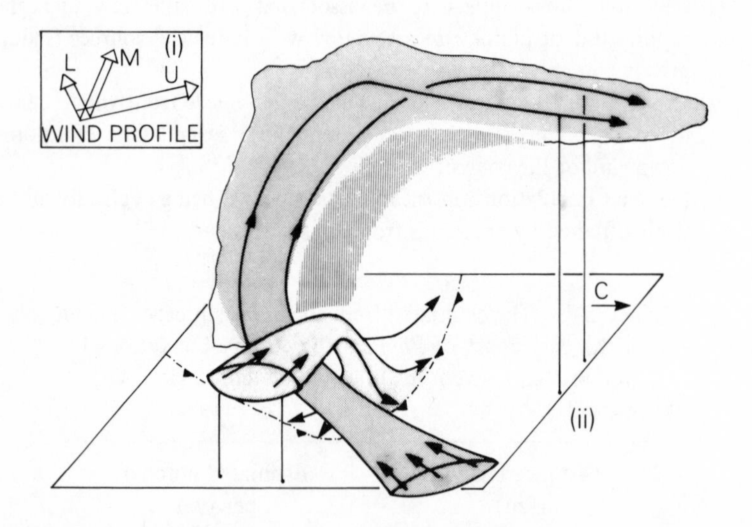

Fig. 7.2(i) – A schematic but typical, wind profile in terms of a lower (L) (0–2 km), middle (M) (2–5 km), and upper (U) (5–10 km) layer. (ii) A conceptual model of the structure of a storm existing in the wind profile of (i). Also indicated is the rain (vertical lines), the surface gust front, and the storm propagation vector C. (Thorpe, 1981).

some cases the outflow air from such storms becomes the inflow air to neighbouring storms and pollution contained in a single plume within the boundary layer may be drawn into one such storm and wet deposited not only beneath the storm of entry but also below neighbouring storms which may lie across the direction of the low-level winds.

Urban circulations

Cities generate their own modified climates and these have been well described in the literature, for example by Oke (1978). Essentially the increased surface roughness and the significant anthropogenic input of heat together with the different sub-division of naturally available heat between sensible and latent heat cause increasing atmospheric instability over the area and mesoscale wind circulations that have consequences for diffusion some of which are discussed in 5.4. These may be summarized as (a) an increase in vertical diffusion, (b) an increase in mixing depth, (c) an increase in the probability of convective precipitation, especially in summer, (d) difficulty in prescribing appropriate wind speeds and turbulence intensities, and (e) a breakdown in the simple scaling laws because the wakes of individual buildings retain their identity through too large a fraction of the total mixing layer.

Topographically induced circulations

The effect of hilly or mountainous terrain on airflow and on the dispersion of plumes is equally complex, and is a matter of continuing research employing theory, wind-tunnel modelling and major field experiments. Effects are evident on a wide range of scales but a satisfactory understanding of them is still far off in the future. Suffice it to mention just a few qualitative points relevant to dispersion:

(i) With simple 'Gaussian'-shaped isolated hills, the speed-up factor of the wind at the crest, $\Delta u/u$, is approximately $2h/l$, where h is the height of the hill and l is the half-width of the hill at a height $h/2$.

(ii) Turbulence levels are also increased in proportion:

$$\text{i.e.} \quad \frac{\sigma_u}{\sigma_{u_0}} \approx 1 + \frac{2h}{l} \tag{7.1}$$

where the suffix 'o' refers to values over upwind open country.

(iii) When the slopes of the hill are greater than about 1 in 3 near-neutral airflows tend to separate from the surface and closed eddies are set up. Stability and the existence (or otherwise) of sharp edges tend to have a pronounced effect on the occurrence and size of these recirculating eddies. If a source of pollution lies inside one of these eddies then high concentrations can be found within it with a relatively slow escape to the main flow outside.

(iv) For a source upwind of the hill the behaviour of the plume and the resulting ground-level concentrations depend very much on stability. In unstable and neutral conditions the air flows easily over the hill, but in stable conditions work has to be done to lift the airflow and this has to come from the kinetic energy of the flow itself. Consequently there is a critical height h_c above which this is possible, see Table 7.V, but

Table 7.V – The height below the hill crest at which the airflow has just sufficient kinetic energy to rise over the hill in stable conditions.

$h-h_c$		$\Delta\theta$ Increase in potential temperature over 100 m				
		0.1	0.3	0.5	1	3
	1	172	100	77	55	32
wind	2	344	199	154	109	63
speed	4	688	397	308	217	126
u	6	1032	596	461	326	188
(m s^{-1})	8	1375	794	615	435	251
	10	1719	993	769	544	314

below which the air will be forced either to find a way round the hill or to stagnate. h_c is given in terms of the Froude number:

$$Fr = U h^{-1} \left(\frac{g}{T} \frac{\partial \theta}{\partial z} \right)^{-\frac{1}{2}}$$

$$h_c = (1 - Fr) \qquad (7.2)$$

Increased stability with $Fr < 1$, say, also inhibits the occurrence of separation on all but the steepest slopes.

If the plume is advected close to the hill then high surface concentrations may be experienced, as large as the maximum concentration that would have been found at the centre of the plume in the absence of the hill. The track of the plume on the forward side of the hill is likely to be quite variable as it responds to upstream variations in the flow, and a given point will sometimes be affected and at other times not affected by the plume, whereas if complex separated flows are present at the rear of the hill a point there may be almost continuously affected and may experience an even higher dosage.

(v) If the hill is fairly isolated, a persistent single vortex may form in the lee stretching downwind for many kilometres. The vortex will have its axis orientated close to the upstream wind direction with the sense of the rotation appearing to depend on the precise nature of the asymmetries of the shape of the hill. An example of this has been studied at Ailsa Craig, an island off the Scottish coast, by Jenkins *et al.* (1981). The consequences of this for vertical and lateral dispersion of a plume entrained into the vortex have not been explored but it is possible the stability of such vortices may inhibit such dispersion beyond its own scale.

(vi) On a larger scale there is evidence from aircraft-measured winds that areas of high ground may generate mesoscale horizontal eddies on a sub-synoptic scale. Fig. 7.3 shows an example of this downwind from the North Yorkshire Moors off the east coast of England observed on 18 June 1980 in which there appears to be a closed circulation embedded in the overall westerly flow. Similar eddies were observed on at least two other occasions by aircraft during experiments in which the

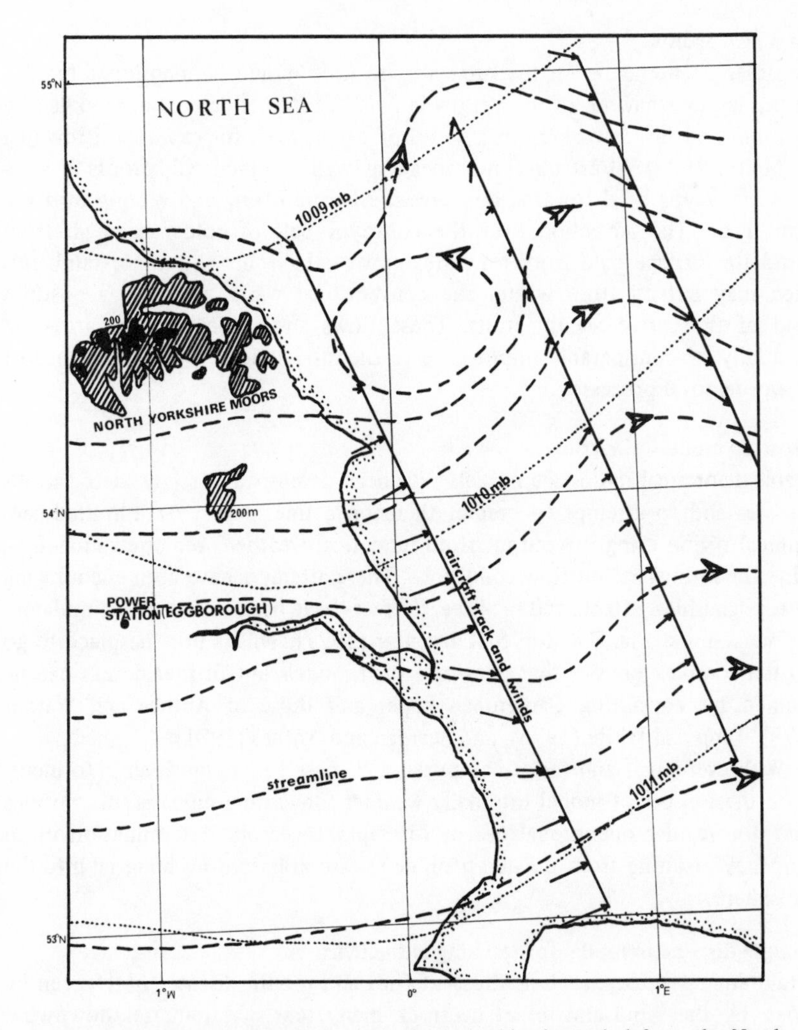

Fig. 7.3 – An apparent mesoscale horizontal eddy downwind from the North Yorkshire Moors implied by winds measured by the Meteorological Office's Hercules aircraft on 18 June 1980 during an experiment to study the plume from Eggborough Power Station (Crabtree, 1982).

plume from Eggborough Power Station was being studied as it was advected across the North Sea (Crabree, 1982).

Theoretical work on the effect of hills on dispersion is proceeding and useful understanding is being gained but understandably the complexity of the subject prohibits easy answers. Reference should be made to Hunt and Mulhearn (1973), Mason and Sykes (1979, 1981) and Puttock and Hunt (1979). There is further discussion of some of these issues in Section 5.4.

Flows near fronts

One other phenomenon on the sub-synoptic scale should be mentioned briefly, and this is the behaviour of the airflow in frontal zones of depressions studied by Browning and his coworkers in a series of papers (see, for example, Browning and Monk, 1982). Moist air flows ahead of well defined cold fronts in a so-called 'conveyor belt' towards low pressure before lifting and veering over the warm front. The advection over the conveyor belt of colder drier air from behind the surface cold front generates upper cold fronts and major rainbands which may extend from within the conventional warm sector to a position ahead of the surface warm front. These flows and associated rain areas are potentially of considerable importance to the modelling of pollution transport and wet removal processes.

Mesoscale models of airflow

Several theoretical mesoscale models have been developed to study these various processes and to attempt to predict them on a fine mesh over a limited geographical region using the output from numerical weather prediction models to define the inflow and outflow conditions. These attempts have been encouraging and topographical effects and sea-breeze effects have been realistically simulated, see for example Fig. 7.4 due to Carpenter (1979). This is not the place to go into the complex physics that goes into such models but further details can be obtained by consulting Carpenter's paper and those of Anthes and Warner (1978), Tapp and White (1976), and Seaman and Anthes (1981).

With ever-larger and faster computers it is now becoming feasible to incorporate these types of model into daily weather forecasting suites as sub-routines either for regular operational use or for rapid trajectory determination in an emergency resulting from the catastrophic release of hazardous material into the atmosphere.

A simple dispersion model for tracking animal viruses

Certain animal viruses, such as those of foot and mouth disease (FMD), can be carried by the wind and infect livestock many tens of kilometres downwind from the source. Epidemics, if not brought under control, can result and have resulted in vast numbers of farm and wild animals becoming infected and perhaps dying as a direct result or having to be intentionally slaughtered. Taking

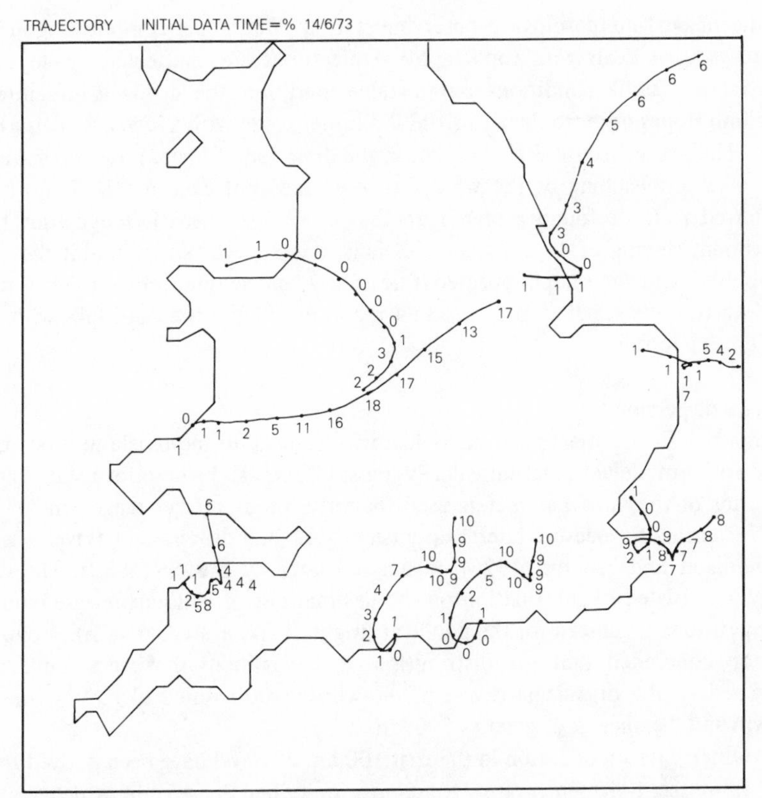

Fig. 7.4 – Calculated trajectories of fluid particles for a 24-hour forecast period. Particle heights are shown to the nearest 100 m every two hours (Carpenter, 1979).

FMD as an example, the virus multiplies inside the animal after the initial infection followed by a period of up to four days when the virus is exhaled and carried downwind possibly infecting other livestock. The survival time of the virus once airborne depends chiefly on the relative humidity of the air, and should this fall below about 60 per cent the virus soon dies, but otherwise it can survive for several days. Investigations of outbreaks in the U.K. revealed the importance of local topography on the areas of secondary infection and hence presumably on the low-level airflows that carried the virus. Blackall and Gloster (1981) have developed a conceptually simple model which appears in the light of the few outbreaks analysed in the U.K. to be successful in predicting areas of potential secondary infection. The model is a Gaussian plume model in which the σ_y and σ_z are functions of time-evolving stability obtained from knowledge of the solar elevation and from measurements of cloud amount and wind speed at the nearest meteorological observing station. The model is interesting in the simple way it deals with topography. Trajectories are unaffected by slopes in neutral or unstable conditions, but in stable conditions the airflow is not allowed to climb slopes of more than 1 in 100 if another route with a lower slope is available. The 'easier' route is first sought in the direction of lower pressure provided the resultant backing of the wind direction does not exceed 90°. If no such preferred route is found a search up the pressure gradient is sought out to a maximum veering of 45°. If again this fails, the slope taken to inhibit the flow is doubled and the search repeated if necessary, and so on until a route is found. Once a route is selected, it is assumed the entire plume rises and falls with the underlying terrain.

Lateral dispersion

Of course, by no means are all trajectories subject to mesoscale motions that have to be modelled deterministically; most effects can be described statistically in terms of the same parameters used for diffusion at shorter range. One of the first attempts to measure lateral dispersion over long distances of travel was by Richardson and Proctor (1925), reanalysed later by Sutton (1932). The data largely consisted of information on the destinations of balloons released during competitions organized for the public at Brighton and in Regent's Park, London. Sutton concluded that the distribution of the balloons that were found and returned to the organizers gave a plume whose width varied like $x^{0.875}$ out to downwind distances x as great as 500 km.

Other data on diffusion in the first 100 km of travel have been derived from the fluorescent-pigment tracer techniques described in 4.1, in which greater sensitivity is achieved by the wearisome process of actually counting the very small individual particles. The first results to be reported in the literature are due to Braham, Seely and Cozier (1952), who carried out their studies in 1951 from the New Mexico Institute of Mining and Technology. An effectively continuous release of the material at a rate of about 3×10^{10} particles per second was achieved, either by mixing it into oil fed into a military smoke generator, or by means of a mechanical type of dispenser. Air-sampling downwind of the point of release was carried out with a drum impactor mounted on an aircraft, which flew successive traverses through the plume of airborne material at various distances and heights. By turning the drum of the sampler at prescribed intervals of time samples corresponding to successive sections of a traverse were separately collected, so enabling the crosswind distribution to be determined. A continuation of the experiments in New Mexico in 1952, and similar experiments in Australia in 1953, have been described by Crozier and Seely (1955). Since 1955 (see Pasquill, 1955, 1956) the techniques have been used at Porton in connection with Meteorological Office studies of the diffusion of cloud-seeding agents for the artificial stimulation of rain. In the latter work, in order more effectively to study the vertical as distinct from the crosswind distribution, the tracer experiments have been carried out in a converse manner, that is, with release of the particles from a dispenser carried rapidly acrosswind on a motor vehicle or aircraft, the sampling being carried out with apparatus at various heights on the cable of a captive balloon flying at a fixed position downwind of the release-line. An additional advantage of this arrangement is that suitable meteorological instruments can also be carried on the tethering cable.

Both the Richardson and Proctor data and the Porton data are included in Fig. 7.5 which also includes data from other sources. The data collected from the Mt. Isa smelter plume in central Australia (Bigg *et al.*, 1978) covers an enormous downwind range from about 15 km to nearly 1000 km and is particularly valuable for this reason. The data are currently being analysed in greater detail by F. Gifford in an investigation of whether or not larger mesoscale and synoptic scale turbulence can lead to an acceleration phase in the development of σ_y as a function of distance, that is σ_y growing faster than linearly. Further comment on these points is made in Section 4.8. The data collection by Crabtree (1982) is also very interesting, representing as it does measurements made over the sea of the width of the Eggborough Power Station plume by the Meteo-

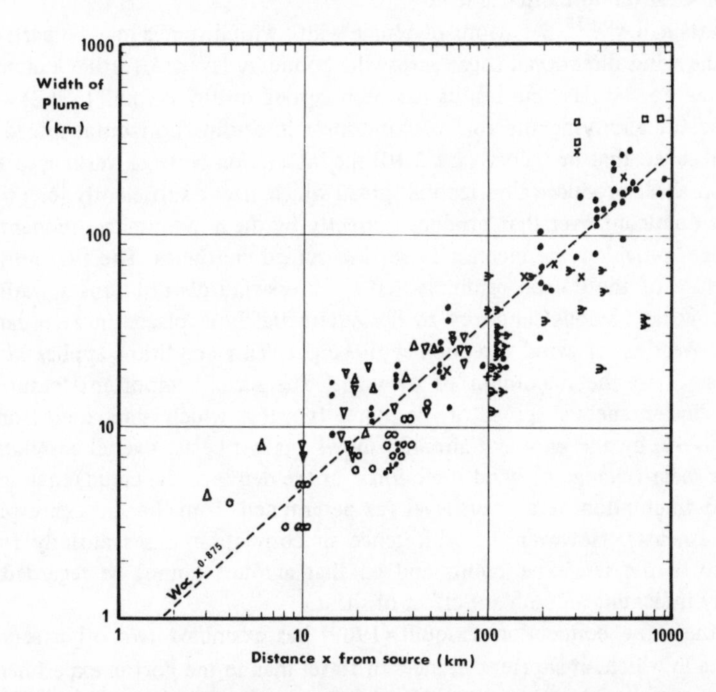

Fig. 7.5 – Data on the width of plumes as a function of distance from the source. × Richardson and Proctor (1925); ∇ Porton data (Pasquill, 1974); + Gifford (see Slade, 1968); □ Classified project (see Slade, 1968); △ Braham *et al.* (1952); ○ Smith and Heffernan (1956); ● Mt. Isa data (Bigg *et al.*, 1978; Carras and Williams, 1981); ⋡ Crabtree (1982): data collected over the sea.

rological Office's Hercules aircraft. As one might expect these widths are generally smaller than the corresponding widths measured over land due to the relative absence of mesoscale sources of energy.

It should be noted that in almost all the measurements displayed in Fig. 7.5 the sampling time is fairly short: often the time taken for the aircraft to fly through the plume. The Richardson and Proctor data however refer to balloons released over a period of about 10 hours and this should bias them, to some limited degree, towards higher widths compared to the other data. Fig. 7.5 suggests however that this bias must be quite small.

Fig. 7.5 confirms that Sutton's original estimate of the rate of lateral growth of the plume with distance appears to be valid out to 1000 km. The reduction of this rate to an $x^{\frac{1}{2}}$ – rate theoretically expected when x becomes large compared to the length-scale of turbulence is not evident in the land data but may be supported by Crabtree's data over the sea.

It is also noteworthy that an ensemble of the crosswind distribution from individual Porton flights is almost exactly Gaussian in shape.

Effect of shear on horizontal spread

The maintained $x^{0.875}$ behaviour of plume width with distance may be partially due to the wind directional shear across the boundary layer. A further examination of the Porton daytime results has been carried out by Pasquill (1962) with the object of clarifying the role of wind-shear in causing horizontal spread. It will be recalled that in theory (see 3.10) the interaction between vertical spread and mean shear produces a horizontal spread which after a sufficiently long time becomes dominant over that produced directly by the horizontal component of turbulence, provided the increase in *vertical* spread continues. The first noticeable feature of the Porton results is that the crosswind distributions at various levels show no obvious tendency to be systematically displaced in accordance with the veering of wind direction with height. This condition applies at all distances up to the maximum of 47 miles. The second important feature is that the 'instantaneous' spread of the plume (i.e. that which is apparent from a single traverse by the sampling aircraft) shows a rather better overall correlation with the shear (change of wind direction over the depth of the cloud) than with the wind fluctuation at a given level (as determined from the successive pilot balloon ascents). However, the difference in correlation arises entirely from only two of the ten experiments and on that amount cannot be regarded as decisively indicating a dominant effect of shear.

In the same connection Pasquill (1969) has examined two other sets of field data in which, at shorter distances of travel than in the Porton experiments, there was visual evidence for distortion of the clouds of material in accordance with the turning of wind with height. Both series were conducted in evening stable conditions – one being Högström's (1964) study of smoke puffs from an elevated source, the other being a tracer study using a continuous ground-level source of fluorescent pigment (Fuquay, Simpson and Hinds, 1964). The main point to be ascertained was whether or not the distortion of the puffs or plumes had in fact been effectively transferred from one level to another so as to produce a significant enhancement of the crosswind spread at a *given level*, over and above that associated with the crosswind component of turbulence. By considering the data in relation to the statistical treatments of 3.4 and 3.10 the conclusion was reached that the effect of shear was not relatively important within about 5 km of the elevated sources or within about 12 km of the ground-level source. Unfortunately it did not prove possible from these data to make a more critical test of the theoretical estimates of the effect of shear.

The Windscale accident – October 1957

The outstanding aspect of problems associated with the release of radionuclides into the atmosphere is that minute quantities of material are capable of producing significant levels of contamination over enormous distances and areas. This was exemplified by the accident at the Windscale Works on 10 October 1957, when some of the uranium fuel became white hot and fission products, notably iodine-131, escaped from the 12-m chimney stack into the atmosphere during the subsequent 24 hours. The resulting total ground contamination of north-west England amounted to only a few grams (amounting to less than a monolayer) but was sufficient nevertheless to be identified in normal air filters, exposed operationally on a daily basis, as part of an on-going survey of urban air pollution across England and Wales. The analysis of the accident and the deposition data have been described by Chamberlain (1959) and by Crabtree (1959). During the first half of the release the wind was from the southwest but later veered to the northwest as a cold front passed over Windscale, and this sequence of events was shown up by the V-shape of the reconstructed plume. Subsequently this plume moved southeast across England before being caught up in a developing high pressure cell which turned it westward across southern England and Wales. Crabtree showed that this history could be simulated if advection was made using observed surface winds, spatially and temporally smoothed, and increased by 25–30 per cent, implying a vertical spread of the plume of at least some hundreds of metres.

One of the more curious aspects of the plume was the magnitude of its width within 34 km of Windscale as it ran along the Cumbrian coast in the northwesterly airstream following the passage of the cold front. Chamberlain examined the resulting deposition field and the implied total width of the plume (obtained by fitting a best-fit Gaussian distribution to the measurements and estimating the distance between points at which the deposition fell to one-tenth of the peak value) was only 2.3 km at 10 km range and was increasing at only an $x^{\frac{1}{2}}$ rate with distance x, both considerably less than the values given in Fig. 7.5. Whether or not the coastline and the moutains immediately inland were the cause of this, by restraining lateral motions, is not clear.

7.4 LONG RANGE TRANSPORT AND THE ACID RAIN PROBLEM

Studies of air pollution are not new, the dangers of air pollution to human health was appreciated in medieval times when legislation was introduced in England in 1273 to control the sale of soft coal and its use in London.

Furthermore the threat to the countryside far away from the growing industrialized centres of England and Scotland was being investigated well over a hundred years ago (R. A. Smith, 1872) when regular samples of rain were being analysed to determine the concentrations of various acidic components, and quite acid they were too! However, the suggestion that man-made airborne pollutants deposited on the ground either by 'dry deposition' processes or by precipitation (i.e. 'wet deposition') hundreds or thousands of kilometres away from their points of origin might lead to ecological changes and damage were first made in Scandinavia in the late 1960s (Oden, 1968). Similar fears based on observed reductions in fresh-water fish populations have since been expressed in other parts of Europe (e.g. Scotland – Harriman and Morrison, 1980), and the Appalachian mountains of the U.S.A. and in large areas of Canada (Harvey, 1980). In fact there are many other sensitive areas of the world at equally large distances from major source areas of pollution which might be at risk. Although the best documented damage claimed to be caused by industrial air pollutants at long range is to fresh-water ecosystems, speculation exists that forests, agriculture and soils may be adversely affected in some critical areas, although hard evidence for this and the magnitudes of any effects have still to be ascertained.

Since the original suggestion that ecological damage was being caused by man-made airborne pollutants, two major international collaborative experiments have taken place in Europe. The first (1971–1976) involved most of the western European countries and was carried out under the auspices of the Organization for European Co-operation and Development, OECD (OECD, 1977). The second (1977) brought in many of the eastern European countries, including the USSR, recognizing the truly international character of the problem, and was mounted jointly under the auspices of ECE, UNEP and WMO (Eliassen and Saltbones, 1982). Both these experiments set out to monitor the deposition over their respective regions by making appropriate measurements on a daily basis at a large number of sites remote from local sources. They also attempted to quantify the emission fields using statistical data supplied by the governments of the countries involved, and to develop mathematical models which could use meteorological observations of low-level winds, rainfall and mixing depths to simulate the atmospheric transport and loss processes that should link the emissions to the observed depositions (Fig. 7.6 outlines the various processes that occur). Similar monitoring and modelling programmes have been under way in Canada and the U.S.A. since about 1977 and are now being co-ordinated following the Memorandum of Intent on Transboundary Air Pollution agreed by the two countries on 5 August 1980.

In addition to this work, aircraft sampling flights (sometimes using specially released tracers) have been mounted in the U.K. (see Smith and Jeffrey, 1975, and Crabtree, 1982, already referred to), in Norway and in the U.S.A. to answer very specific questions concerning the physical and chemical behaviour of the pollutants involved in the acid rain problem.

The flights documented by Smith and Jeffrey were the first to attempt budget studies for sulphur dioxide and sulphate, and to deduce the various loss and conversion rate constants. The flights were carried out along specified tracks and at three or more heights off the east coast of England in 1971 and 1973 in relatively simple westerly meteorological conditions. The implied fluxes were related to estimated emissions of sulphur dioxide into the same air as it passed earlier over upwind U.K. source regions. On average about 30 per cent was lost by dry deposition to the surface before reaching the flight tracks, corresponding to a velocity of deposition about 0.8 cm s^{-1}. A further 50–60 per cent typically remained as SO_2 in dry conditions and the remainder had already been oxidized to sulphate. The latter depended sensitively on the relative humidity of the air: the converted fraction was only about one-tenth in relative humidities near 75 per cent but about four-tenths when the humidity was as high as about 90 per cent, although, since the fraction seemed to depend only weakly on distance or time of travel, the implication was that much of the conversion took place soon after emission.

The findings of all the studies must be linked to a genuine understanding of how acid rain causes ecological damage. Unfortunately this understanding has only been partial. For example the relative importance of

(i) long-term depositions and average concentrations of the various acid components in the precipitation (these depositions will be referred to as LTD in later discussions);
(ii) short-term episodic depositions (STED) containing high concentrations of acidity which may in some circumstances be more hazardous to sensitive living species than LTD.

is still not clear. Most effort has been devoted to the development of models designed to predict LTD, but which are of rather limited use for STED. About all these models can be tuned to produce in this respect is a broadly correct statistical distribution of daily deposition magnitudes. This bias in modelling only partially reflects physical priorities but also reflects the fact that these particular models are conceptually easier to develop and run in terms of standard meteorological observations. Models for STED are much more difficult to develop from basic physical principles and would probably have to have a much greater degree of empirical content.

One thing is clear however, and that is that the countries or regions producing the largest contributions to LTD within the sensitive regions may not be the same countries or regions as are responsible for the largest and potentially most damaging contributions to STED in the same areas.

Up to the present time, both in Europe and in America, the greatest emphasis has been placed on modelling, monitoring and emission specification of the major component of acid rain, namely sulphur dioxide and its daughter products. Whilst other components such as the oxides of nitrogen, ammonia, hydrocarbons, chlorides etc. all have a role in the oxidation processes that convert sulphur dioxide to sulphate, and in the acidity of precipitation, as far as effects at long-range are concerned there seems to be growing agreement that:

(i) it is the acidic sulphates (H_2SO_4 and $(NH_4)_2SO_4$ in particular) in rain that are largely capable of affecting lake and river waters,
(ii) other components in rain undergo strong ionic exchange in the soil but cause little ecological damage at least in the short term, and
(iii) the conversion of sulphur dioxide to sulphate is usually dominated by wet-phase processes in cloud droplets and these are sufficiently rapid that even if they depend in detail on the concentrations of other pollutant species the LTD wet deposition fields are believed to be only marginally affected. Areas of strong topographically induced rainfall may be an exception to this general supposition.

Moreover determination of the magnitudes and spatial distributions of sources of these other pollutants is often very difficult, consequently there is a natural reluctance on the part of some workers to develop large and expensive models incorporating all the other components, given their marginal importance.

Some overall caution in this whole subject is warranted however because the processes involved in (i) and (ii) are not completely understood by any means and many important questions remain unanswered. The relationship between deposited sulphur and fish-kill, for example, is far from being a simple one and depends on many other factors, only partially understood, such as the availability of toxic metals like aluminium in the soil. Meaningful data in this area are consequently hard won and will take many further years of careful work to amass.

7.5 LONG-RANGE TRANSPORT MODELS

Fig. 7.6 illustrates the fate of pollutants within the atmosphere. The following aspects are involved:

(i) source characteristics: the mean rate of emission and any variations which can be prescribed, the source height, any buoyancy in the plume which may result in plume rise or fall, and its dependence on stability and mixing height;
(ii) advection: the definition of the appropriate advecting winds and the problem of prescribing these in terms of available data;

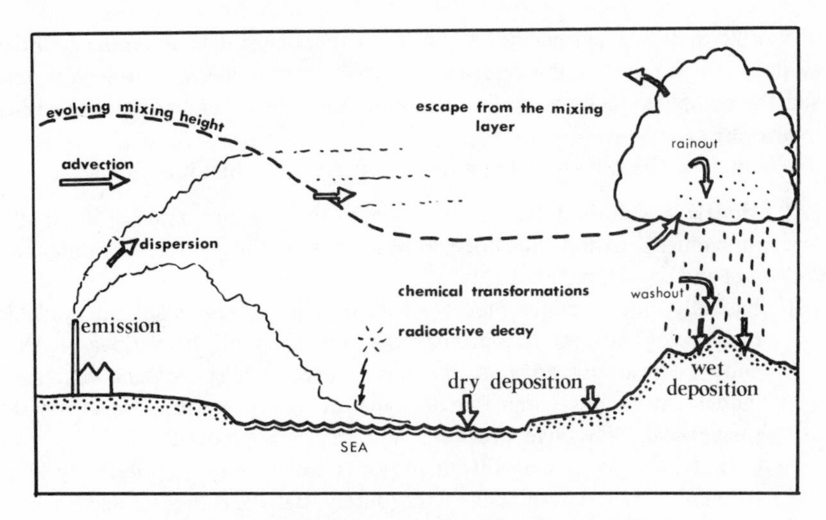

Fig. 7.6 – Some of the processes affecting airborne pollution.

(iii) dispersion: the parameterization of both lateral and vertical dispersion in terms of available data, and the influence of mixing height;
(iv) composition changes: chemical transformations that may take place, the dependence on other chemical species, and radioactive decay;
(v) loss processes out of the mixing layer: what these processes are and how to parameterize them; the ultimate return of material back into the boundary layer;
(vi) dry-deposition processes: the loss of pollution by absorption, sedimentation and impaction to the underlying surface; and the problem of parameterization; see 5.2 for some details;
(vii) wet-deposition processes: the removal of pollution by precipitation either by uptake into cloud droplets (rain-out) or into the precipitation during its fall through the sub-cloud mixing layer (wash-out);
(viii) and finally the fate of the pollution once it has reached the surface, and the damage it may cause there. These particular aspects are beyond the scope of this book and readers should refer to other texts, for example Drabløs and Tollan (1980), Record *et al.* (1982).

All these various aspects have been the subject of considerable study, especially since 1970, but inevitably, owing to the complexity of the subject and the relative sparsity of appropriate data, the models tend to involve a considerable degree of simplification and parameterization. For this reason many models have been developed which display quite a wide spectrum of methods and degrees of complexity. Ultimately the user will need to ask which model is best. The answer of course is not easily given, partly because no really exhaustive intercomparison with sufficient reliable field data has yet been made (at the time of writing in 1982) and partly because the answer must depend on the facilities available to the user – the ease with which he can obtain meteorological data and the computer facilities available to him – and the exact questions he wishes to answer.

Models can be sub-divided in a variety of ways. The first is as follows:

[A] Models that are statistical in concept, that do not attempt to model individual situations but hope to have some validity in terms of long-term wet and dry deposition fields.

[B] Basically quite simple models that attempt to use standard available meteorological data to optimize the modelling of the various physical and chemical processes in time and space. Whilst realizing that such models inevitably parameterize some of the processes, especially the smaller-scale advective processes and the sink processes, and therefore cannot hope to model individual events totally satisfactorily, they hope to achieve better results than the purely statistical models and to have validity on a shorter time-scale, say on a seasonal basis.

[C] Models that are complex in character: that follow air motions in three dimensions on a finer grid than the models in [B] and are therefore essentially non-hydrostatic baroclinic mesoscale models extended to cover as large an area as the available computer facilities permit. They attempt to allow for the effects of topography and use non-standard meteorological data such as highly resolved rainfall data from modern weather-radar facilities. These models will be expensive to run and will be aimed at individual situations, either to explain measured episodic depositions, or as noted later in this chapter to follow the accidental release of a large amount of hazardous material into the air, having potential consequences both at short range and out to perhaps a few thousand kilometres.

Alternatively models can be sub-divided as to whether they are Eulerian in character, where the concentrations and depositions are estimated at each time-step on a grid basis, or whether they are Lagrangian, where a succession of trajectories are followed and the development of individual 'masses' of pollution are simulated in time. Both approaches have their attractions and disadvantages. Eulerian models are useful when many of the parameters (e.g. the emission rates and the velocity of deposition) are specified on a grid basis. They also tend to be somewhat more economical in terms of computer storage space. Their disadvantages are (i) that special measures have to be taken to avoid spurious 'numerical' diffusion which arises from modelling spatial gradients on a finite grid, (ii) it is slightly harder to follow the fate of pollution arising from individual source regions, especially when non-linear processes are involved. Lagrangian models are generally simpler in concept and are not subject to spurious diffusion, although errors in trajectory determination do tend to grow exponentially in time. It is also comparatively easy to determine inter-country or inter-regional emission–deposition budgets. Disadvantages are that after some time the trajectories are advected out of the area of calculation (the same of course implicitly occurs in the Eulerian models) and that trajectories tend to converge into low pressure areas and diverge out of high pressure areas, which, although reflecting a valid physical process, makes the assessment of concentrations and depositions less certain in the latter.

Another sub-division could be in terms of the vertical resolution. Many models are one-layered, either being infinite in depth or are limited to a constant or evolving mixing depth. The former usually allow the pollutant to diffuse upwards according to some specified eddy diffusivity profile which often falls off in magnitude above the typical height of the top of the mixing layer to some non-zero value which in a very empirical sense parameterizes the escape processes that have been referred to earlier. The latter models either also permit diffusion in the vertical through a height-limited diffusivity profile or allow the plume to deepen according to the stability classification scheme described in Section 6.6 or assume uniform concentration within the mixing layer. Advection is by a single wind chosen for its appropriateness.

Some other models are multi-layered however and, whilst simulating exchange between layers, permit the pollution in each layer to be advected according to some prescribed or objectively determined wind appropriate to that layer alone, thus modelling the potentially important consequences of wind shear.

Many models do not attempt to simulate crosswind horizontal diffusion on the grounds that the effects on long-term deposition fields are largely self-cancelling. Others, especially those with non-linear air chemistry, are forced to introduce some diffusion and this can be done either empirically or by using the implied spatial variation in the wind field or by introducing a so-called pseudo-diffusion velocity $\mathrm{V} = \frac{1}{C}\, \mathrm{K} \cdot \nabla C$ where C is the concentration and K is the prescribed eddy diffusivity tensor. This last method is employed in the Eulerian models of Sklarew *et al.* (1971) and Lange (1978).

Models also differ in the way that pollution is advected. The statistical models (Type [A]) tend to use climatological wind roses to determine the frequency of different wind directions and let the pollution follow straight trajectories thereby determined. This assumption is not as bad as it might at first seem, air trajectories tend to be much less curved on the whole than the one-moment-in-time Eulerian wind fields or pressure patterns would suggest. This is due to the mobility of these patterns, the inertia of the air, and the magnitude of the inverse of the Coriolis parameter which determines the rate at which air velocities can respond to changes in the pressure field. Other models either use observed winds (usually taken from routine radiosonde observations) or modified geostrophic winds determined from the pressure fields at one or more heights. For the latter geostrophic winds at about 700 metres (around 925 mb) would be ideal for this purpose. However, these are not yet reported in conventional meteorological codes and either the surface geostrophic or the 850 mb geostrophic has to be used. These winds are then empirically modified by some constant amounts or according to the nature of the underlying surface and the stability of the air, usually by reducing the magnitude by an amount in the range 0–10 per cent and by backing the wind somewhere in the range 0–15°.

The dry deposition rate is assumed proportional to the concentration in the whole layer or to some implied surface concentration, and the constant of proportionality is called the velocity of deposition v_d. The velocity is usually of the order of 1 cm s^{-1} for reactive gases like sulphur dioxide, but is relatively much smaller for fine aerosols (radii near 1 μm) like sulphate aerosol and increases again for large aerosols (radii > 10 μm). It also varies with stability, becoming relatively small in very stable conditions because of the high aerodynamic resistance of the near-surface air. The nature of the surface cover also has an important effect on the velocity. Nevertheless attempts to determine the loss of sulphur dioxide by dry deposition using the budget technique, whereby the loss is equated to the difference in the emission rate and the total integrated flux some reasonable distance downwind (e.g. some hundred kilometres or more), rather suggest that local variations are relatively unimportant and that larger-scale differences (say between land and sea) are rather small. One source of evidence for this rests on the fact that values of v_d obtained by optimizing models in terms of observed concentrations (e.g. Eliassen, 1978) agree with total-flux estimates obtained using aircraft (e.g. Smith and Jeffrey, 1975).

On the other hand estimation of wet deposition is more difficult because precipitation is much more intermittent in time and space and variable in intensity and in its ability to remove pollution: the response of the load of pollution being carried by the air to precipitation depends on many complex factors which are only partially understood. It depends on the nature of the precipitating cloud and whether or not it is drawing polluted air from the boundary layer. It depends on the mix of nucleating particles and aerosols and the presence of other chemical species that may affect the uptake of the pollution into the droplets. It depends on the wash-out process below the cloud and the evaporation rate, and the possible release of droplet-borne pollution as the precipitation is falling to the ground.

Models take one of four courses: they may assume it is raining lightly everywhere all the time, at a rate which is constant or may vary according to known annual precipitation amounts, or they may attempt to interpolate rainfall from observed rates at nearby meteorological observing stations, or they may introduce rainfall stochastically depending on the existence of rain or no rain at nearby stations. Alternatively they may recognize the existence of wet regions (like frontal zones and areas of unstable polar outbreaks) and dry regions (like summer anticyclones) that occur along the trajectory with prescribed probabilities and time-scales: no rain occurring in the dry regions and rain occurring in the wet regions either all the time or with a comparatively high probability. These latter methods developed from the recognition by Rodhe and Grandell (1972) that the patchy nature of precipitation has a pronounced effect on the average amount of aerosol that can reach distant receptors. Fisher (1978), Smith (1980 and 1981) and Venkatram (1981) developed these ideas to the transport of sulphur and included the oxidation of sulphur dioxide to sulphate, with their different responses to wet and dry removal processes.

Smith's formulation, for example, differentiates between the relative fractions of SO_2 and sulphate that exist in a statistical sense after a given time of travel t in dry regions and in wet regions. If q refers to SO_2, Q to sulphate, suffix D to dry regions and W to wet regions then four Lagrangian rate of change equations are required to describe the statistical fate of the original emission:

$$
\begin{aligned}
\frac{\mathrm{d}}{\mathrm{d}t} q_D &= s_W q_W - \left(s_D + \alpha_D + \frac{v_d}{h}\right) q_D \\
\frac{\mathrm{d}}{\mathrm{d}t} q_W &= s_D q_D - \left(s_W + \alpha_W + \frac{v_d}{h} + A\right) q_W \\
\frac{\mathrm{d}}{\mathrm{d}t} Q_D &= s_W Q_W - s_D Q_D + \alpha_D q_D \\
\frac{\mathrm{d}}{\mathrm{d}t} Q_W &= s_D Q_D - (s_W + A)\, Q_W + \alpha_W q_W
\end{aligned}
\qquad (7.3)
$$

where s_W and s_D are related to the probability that a followed air-parcel will flow from a wet region into a dry region, or vice versa, recognizing that boundary layer air moves with a different velocity to that moving the main synoptic features. α_D and α_W represent the rates of oxidation of SO_2 to sulphate, h is the mixing depth which may be different or the same for wet and dry regions, and A is the wet removal rate applying to wet regions only. Eq. (7.3) can be solved analytically if all the coefficients are constant and prescribed, otherwise they may be solved numerically. Solutions appear to be in rather better accord with observed deposition fields than results of models that assume constant rainfall probabilities.

At the beginning of this section the question was asked 'which model is best?', and a partial answer was given. We can supplement that answer now by saying that in so far as the determination of long-term deposition fields are concerned, the very simple statistical models of type [A] appear to do very well indeed. Smith's model described later is just about the simplest and achieves correlations with observed average depositions around 0.9. More complex models have to work very hard indeed to improve significantly on this, although in other respects they may be better, for example their accuracy must fall off more slowly as the time of averaging is reduced to months or weeks when compared with that of the statistitical models. In this respect the situation is rather similar to that of weather forecasting models where the simplest models achieved respectable success which was only bettered after considerable effort and sophistication.

Twenty-eight models in current use are summarized in Tables 7.VI , 7.VII and 7.VIII. Table 7.VI lists five Eulerian models which are all event orientated; that is they are of type [B]. Table 7.VII lists fourteen Lagrangian trajectory models of type [B], and Table 7.VIII gives eight statistical Lagrangian models of type [A]. Some very brief additional details of these models are given below in order.

Eulerian Models

A. A pseudo-spectral model by Prahm and Christensen (1977). The model controls spurious grid-diffusion by Fourier-transforming the concentration conservation equation into spectral space. It uses wind fields derived for the operational model F described below.

B. A moment-conservation model by Nordø (1974) using a technique due to Egan and Mahoney (1972). After the advection of pollution over a time step, it calculates new zero, first and second moments of concentration in each grid element and derives equivalent rectangular distributions having the same moments. These distributions are the starting points for the next advective step.

Table 7.VI – Characteristics of the Eulerian models. Key to the models A to E is given in the text.

Eulerian Models		A	B	C	D	E
	Pseudo-spectral	●				
	Pseudo-diffusion coefficients		●			
	Turbulent flux velocity				●	●
	Conservation of moments			●		
Advecting wind	850 mb geostrophic	●	●	●		
	Non-divergent wind fit to data				●	●
Mixing depth	Constant	●	●	●	●	●
Vertical mixing	Assumed constant concentration			●	●	●
	Uses *K* theory	●	●			
Lateral diffusion	None	●			●	●
	Synoptic scale diffusion		●			
	Small-scale diffusion			●		
Dry deposition	Constant rate	●	●	●		
Wet deposition	Constant rate		●	●		
	Proportional to interpolated rain	●				
Number of layers	One	●	●	●	●	●
SO_2 to SO_4 conversion	Constant rate	●	●	●		

C. A particle-in-cell model by Nordlund (1973). It also uses Egan and Mahoney's (1972) method. The model also included mesoscale Fickian diffusion.

D. Another particle-in-cell model, by Lange (1978). This is a regional scale model using Sklarew *et al.*'s (1971 method, model E, and three-dimensional non-divergent wind fields, applicable over complex terrain, derived by a technique due to Sherman (1978).

E. The original particle-in-cell model by Sklarew *et al.* (1971). The model was not originally designed specifically for long-range transport. The area is covered by a grid containing a large number of Lagrangian unit-mass particles,

ble 7.VII – Characteristics of the trajectory Lagrangian models. Key to the models F T is given in the text

grangian models pe B		F	G	H	I	J	K	L	M	N	O	P	Q	R	S	T
dvecting wind	Surf. geostrophic			●							●			●	●	●
	850 mb geostrophic	●	●			●			●				●			
	Obj. analysed winds				●		●	●		●		●				
	Straight trajectories			●												
xing depth	Constant concentration			●		●		●	●				●	●		
	Diurnal/seasonal var.	●			●		●				●					●
	Interpolated									●		●			*●	
	Unlimited		●													
rtical mixing	Constant	●		●	●	●		●	●	●	●		●	●		●
	Uses *K*-theory		●				●									
	Layered											●				
	Stability dependent														*●	
teral diffusion	None	●			●		●		●					●		●
	Synoptic scale		●	●												
	Small scale					●		●		●	●	●	●		●	
	Stochastic							●								
y deposition	Constant rate	●	●	●	●	●			●	●	●		●	●		●
	Grid dependent											●				
	Stability dependent														*●	
	Diurnal/seasonal var.						●					●				
	Stochastic							●								
t deposition	Constant rate		●							●						
	Annual rainfall dep.			●												
	Uses interpolated rain	●			●	●	●	●	●		●	●	●	●	*●	●
$_2$ to SO_4 conversion	Constant rate	●	●	●	●	●			●	●	●		●	●		
	Larger near source						●									●
	Diurnal/seasonal var.						●					●				
	Stochastic							●								
	Stability dependent														*●	
ecies	Sulphur alone	●	●	●	●		●	●	●	●		●	●	●	●	●
	Sulphur + nitrogen					●					●					

ecial remarks: *inferred from standard surface meteorological observations.

Table 7.VIII – Characteristics of the statistical Lagrangian models. Key to the models is given in the text

Statistical Lagrangian Models: Type A		U	V	W	X	Y	Z	α	β
Advection	Wind rose	●	●	●	●	●			
	Trajectory rose						●		●
	Horiz. eddy diffusion							●	
Mixing depth	Constant	●	●	●	●	●	●	●	●
Vertical mixing	Uniform concentration			●				●	●
	K-theory: constant *K*				●	●			
	Stability dep.	●							
	Stochastic						●		
	Stability dependent		●						
Lateral diffusion									
single layer:	None	●	●	●	●	●			
	Synoptic scale diffusion							●	●
multi-layer:	Wind shear diffusion						●		
Dry deposition	Constant rate	●	●	●	●	●	●	●	●
Wet deposition	Constant rate				●	●	●		●
	Annual rainfall dep.	●		●					
	Wind direction dep.			●					
	Wet and dry periods	●	●	●				●	
SO_2 to SO_4 conversion	Constant rate				●	●	●		●
	Wet/dry period dependent	●	●	●				●	

which are advected by a wind V defined as the sum of the mean wind V and a turbulent velocity $V' = \frac{1}{C} K \cdot \nabla C$ discussed later.

Lagrangian models of Type [B]

F. Eliassen's trajectory model used during the OECD experiment (Eliassen, 1978) and during the ECE–EMEP experiment at the so-called Meteorological Synthesizing Centre–West in Oslo. Simple 850 mb winds are applied to a single layer trajectory model, interpolating rainfall from nearby meteorological stations to obtain wet deposition. This represents a successful operational model of acceptable accuracy in obtaining long-term deposition fields and budgets.

G. Pressman's model (see WMO Report, March 1981) used at the Meteorological Synthesizing Centre–East in Moscow during the EMEP experiment. The model was designed to evaluate transboundary fluxes and to this end includes lateral diffusion using an empirical law. In spite of the diffusion the plume is advected forward using a single wind.

H. A model of a different kind. It is the U.K. National Radiological Protection Board's model for long-range dispersion of radionuclides released over a relatively short duration (Jones, 1981b). It uses Smith's probability technique described later in this Chapter to predict areas of possible risk in future accidental release scenarios.

I. The Atmospheric Environment Services (Canada) model, conveniently described in a Working Paper of the Modelling Subgroup set up in response to the U.S.–Canada Memorandum of Intent on Transboundary Air Pollution, as are some of the other models. The reference is to Niemann and Young (1981). Model I uses three-dimensional objectively analysed wind fields at four pressure levels over a 381 km horizontal grid; it also uses monthly average mixing heights and daily precipitation data.

J. The ENAMAP–1 model by Johnson (see Niemann and Young, 1981). This is a development of model Q below to include chemistry. It uses objectively analysed three-hourly precipitation fields and chemistry.

K. The ASTRAP model by J. D. Shannon (see Niemann and Young, 1981). It attempts to include diurnal and seasonal variations in many of the parameters. Following the suggestion that rain droplets can only take a limited amount of acidity into themselves, the model attempts to limit the uptake of sulphur.

L. The CAPITA Monte-Carlo model developed by the Center for Air Pollution Impact and Trend Analysis (see Niemann and Young, 1981). The model uses stochastic techniques to simulate horizontal diffusion and wet and dry deposition.

M. The SURE model by Hidy *et al.* (1976). It uses a small scale 80 × 80 km grid.

N. The ARL trajectory model by Pack *et al.* (1978). It uses the powerful trajectory determination techniques developed by Heffter at the Air Resources Laboratory to evaluate trajectories for any origin in the northern hemisphere. The plume is represented by a series of puffs.

O. The MEP-TRANS model by Weisman (see Niemann and Young, 1981). The trajectories are obtained using the sophisticated method of Sykes and Hatton (1976) based on sea-level pressures. The model includes an evolving mixing layer and carries SO_2 and NO_x.

P. The University of Michigan UMACID model (see Niemann and Young, 1981). It incorporates horizontal dispersion due to vertical wind shear. It uses radiosonde and pibal data to get mean advecting winds and mixing heights.

Q. The EURMAP model of W. Johnson *et al.* (1979). It advects a series of puffs from each source square using adjusted 850 mb winds. The puffs grow horizontally at a rate proportional to $t^{\frac{1}{2}}$. Concentrations and depositions are calculated for each grid square depending on the fraction of the square covered by the puff.

R. A sector analysis model by Fuller (1973). This is a receptor-orientated model. It is one of the few attempts to use a model to interpret the mass of invaluable operationally measured daily average concentrations and depositions in terms of their areas of origin.

S. The MESOS model developed by ApSimon (1979). The model was originally designed for the study of the transport of airborne radionuclides. It stimulates changing mixing depths and rainfall. Many parameters are made stability dependent and the stability is assessed from standard surface meteorological observation reports. A practical and sophisticated model which can be used operationally on a limited number of sources.

T. A partially developed stochastic model by Smith (1981). It suggests a method for simulating the sporadic nature of precipitation and its effect on long-range transport.

Statistical Lagrangian models of Type [A]

U. Fisher's (1978) model which uses statistics of winds and vertical dispersion categories. It recognizes regional variations in rainfall and their effect on wet deposition. The model includes Fisher's formulation of transport through wet and dry periods.

V. The National Radiological Protection Board's model for continuous releases, (Jones, 1981a). The model uses Fisher's formulation of wet and dry periods. All releases are assumed to occur in neutral stability conditions. Vertical dispersion is included and is modelled using Hosker's numerical formulae for the Pasquill σ_z-curves. The model is designed for radionuclide dispersion and therefore contains no air chemistry, but could be easily adapted.

W. A simple statistical model by Smith (1982). The model has been deliberately kept simple and yet open to modification so that different ideas can be tried out with minimum expense and computer rewriting. The most successful modification assumes a single wind rose for the whole area (the model has been applied to the European EMEP area) and a constant wind speed. Trajectories are assumed straight. Originally dry and wet deposition rates were assumed constant but better results are obtained recognizing wet and dry periods, probabilities of rain that depend on wind direction (but not position *per se*), and wet depositions which are *post facto* modified according to the magnitude of the annual rainfall. This is a quick versatile model producing deposition fields in good agreement with measured fields. Total deposition fields, calculated and observed, are shown in Figs. 7.7 and 7.8.

X. Fisher's first statistical model (1975). It uses observed wind roses and solves the vertical diffusion equation for SO_2 with constant eddy diffusivity and with a linear loss term to represent dry and wet deposition and the conversion of SO_2 to sulphate.

Y. Model by Bolin and Persson (1975). It is very similar to model X but has a more physically plausible eddy diffusivity profile which increases with height in the lower part of the mixing layer.

Z. Rodhe's models of 1972 and 1974. Emissions are represented by just four sources situated in the main industrial areas of Europe. The first model uses

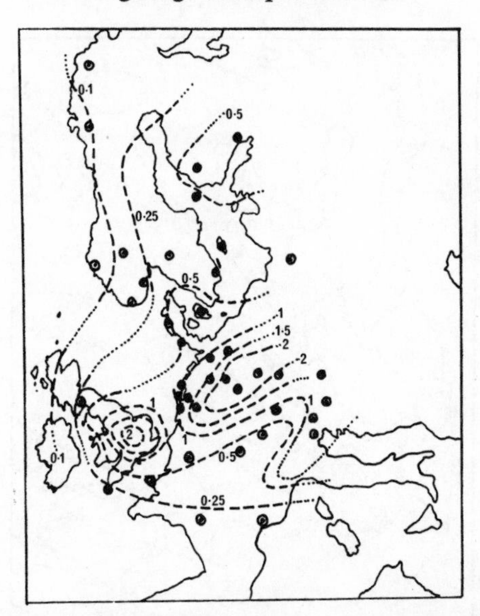

Fig. 7.7(a) – The dry deposition field, in units of g S $m^{-2}y^{-1}$, inferred from limited airborne SO_2 concentration data collected during 1978–79 assuming a uniform velocity of deposition of 0.8 cm s^{-1}. The degree of interpolation and extrapolation used to obtain the contours can be deduced from the spacing of the EMEP stations from which the data are obtained.

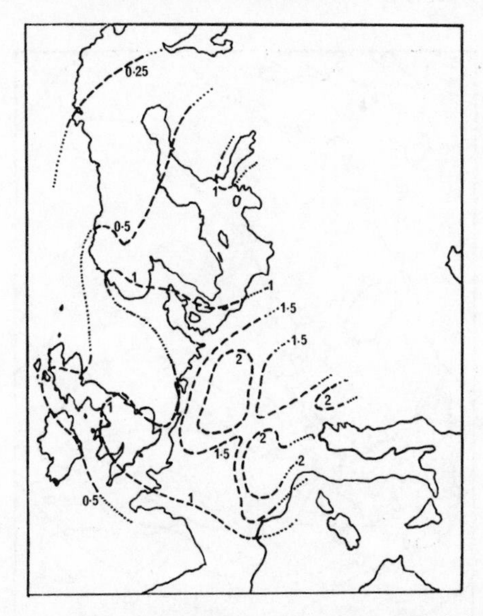

Fig. 7.7(b) – The wet deposition field, in units of g S $m^{-2}y^{-1}$, inferred from the analysis of precipitation at the same sites as in Fig. 7.7(a). Analysis techniques differed from one country to another at this stage in the Programme and these gave rise to systematic errors which have not been eliminated from the data used here, and may be the cause of some of the more obvious discrepancies between the observed and modelled deposition fields.

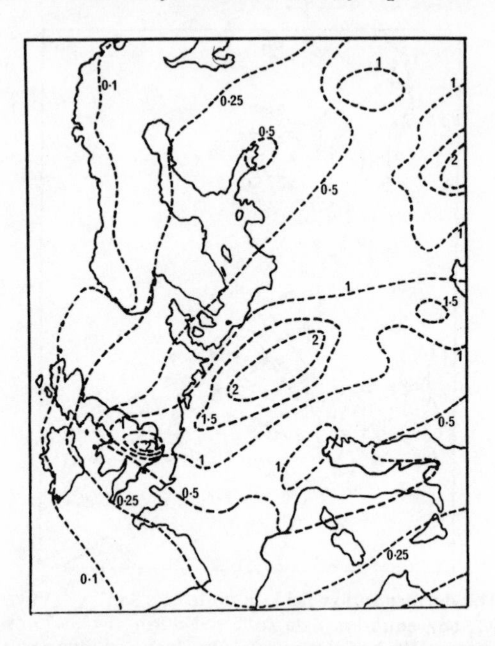

Fig. 7.8(a) – The dry deposition field resulting from Smith's (1982) simple model (model W) described in the text. The correlation between this field and the field given in Fig. 7.7(a) is 0.90.

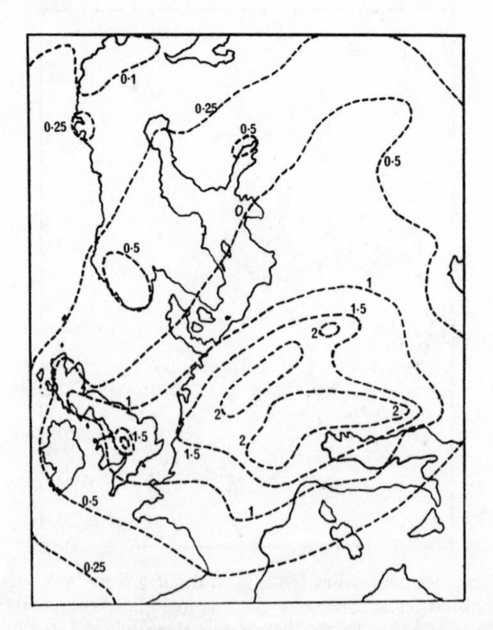

Fig. 7.8(b) – The wet deposition field from Smith's model. The correlation between this field and that given in Fig. 7.7(b) is 0.83.

850 mb wind frequencies and an exponential removal of sulphur with range. The second model introduces wind shear through the vertical by including stochastic particle-jumping from one level to another.

α OME-LRT model developed by Venkatram (see Niemann and Young, 1981). It has a single mean wind but has a superimposed large synoptic-scale diffusion. The model incorporates the concept of wet and dry regions. It is another extremely simple model and simulates the deposition fields surprisingly well.

β The RCDM-2 model of the University of Illinois, by Fay and Rosenweig (see Neimann and Young, 1981). It uses averaged winds and trajectories at 600 m for each source area. There is no vertical diffusion, but horizontal diffusion is simulated using *K*-theory. Obtains analytical solutions of coupled equations for SO_2 and sulphate.

Models suitable for application to the European situation may be tested against the vast amount of data collected at the monitoring stations consisting of daily averaged air concentrations of SO_2 and sulphate and depositions in rain of sulphate, as well as depositions of other related species. In general the model results compare poorly on a daily basis with little useful skill being evident, but on a seasonal or annual basis the comparison is generally good with correlations varying from about 0.7 to 0.9. Even here, a large part of this apparent success is fortuitous and arises from the inevitable link between the deposition fields and the emission fields, both in reality and in the results of any physically reasonable model, especially when the emissions show considerable large-scale geographical variation as they do in Europe.

Even with the use of wet and dry region models there is still a tendency for depositions at really long range (say in northern Scandinavia) to be somewhat underestimated. Nevertheless the overall picture is sufficiently good to give credence to the inter-country budgets of which Table 7.IX is typical. The Table is based on Eliassen's model (Model F) for the two years 1978/79. It should be remembered that these numbers are relevant to the potential long-term deposition effects but may have less relevance to ecological damage that may arise from short-term heavy episodic depositions.

Short releases

Model H referred to dispersion for relatively short releases ($\frac{1}{2}$–4 days) that might occur during a serious failure at a nuclear power plant for example. Provision of adequate emergency services for such a rare eventuality depends on assessments of the likely downwind areas that might be affected by the plume. This can only be done in probability terms. To assess these probabilities it is useful firstly to recognize that the plume width is the sum of two components: that due to turbulent spreading θ_t and that due to systematic changes in wind direction θ_w during the duration of release. Smith (1980) has analysed over two thousand

Table 7.IX – A budget of annual total depositions (g S $m^{-2}y^{-1}$) given by Eliassen's Lagrangian trajectory model (model F) when applied to 1978–79. The budget gives the estimated contributions to the average deposition in one country arising from other countries. Only a selection of countries is given, and these are ranked in order of deposition magnitude. It is interesting that Czechoslovakia receives over 12 times as much sulphur deposition as Norway. Note that the figures quoted for the USSR only refer to the area of that nation within the area of Eliassen's analysis (approximately given in Figs. 7.7 and 7.8)

Receiving country	Area (10^3 km^2)	Emitting country												
		Czech	DDR	Belg.	F.R.G.	Pol.	Neth.	U.K.	Fra.	U.S.S.R.	Nor.	Others	Undec.	Total
Czech.	128	4.5	1.8	0.1	1.0	0.9	0.1	0.3	0.4	0.1	0	2.1	0.8	12.2
D.D.R.	108	0.6	5.5	0.1	0.9	0.3	0.1	0.2	0.2	0	0	0.4	0.3	8.6
Belg.	30.5	0	0.1	2.6	0.9	0	0.2	0.7	1.1	0	0	0.3	0.4	6.3
F.R.G.	250	0.2	0.6	0.2	2.7	0.1	0.1	0.3	0.5	0	0	0.4	0.4	5.6
Pol.	313	0.5	0.8	0	0.1	2.2	0	0.1	0.1	0.1	0	0.8	0.3	5.1
Neth.	41	0.1	0.2	0.5	1.3	0	1.2	0.8	0.4	0	0	0.3	0.3	5.1
U.K.	244	0	0.1	0	0.1	0	0	3.3	0.1	0	0	0.1	0.4	4.2
Fra.	544	0	0	0.1	0.2	0	0	0.2	1.4	0	0	0.3	0.4	2.7
U.S.S.R.	3363	0.1	0.1	0	0.1	0.1	0	0	0	1.3	0	0.3	0.4	2.5
Nor.	324	0.03	0.08	0.01	0.07	0.04	0.01	0.15	0.04	0.03	0.07	0.14	0.27	0.94

geostrophic trajectories to determine the statistics of $\theta_w(x, T)$, the angle between the locations at which trajectories starting at different times cross circles of given radius x centred at the release point, after given time intervals T between the trajectories. The results of this study give the probability that the plume width is less than any given angle, averaged over all weather conditions. Simple formulae have been obtained to give the angles less than that experienced by 10, 50 and 90 per cent of the trajectories considered. The formulae and the lines are given on Fig. 7.9. Values of θ_w corresponding to other probabilities can be obtained from Fig. 7.10 which is applicable at a distance of 500 km from the source. It should be remembered that the analysis was done for trajectories evaluated over north-west Europe and the results should be used with caution in any other geographical region.

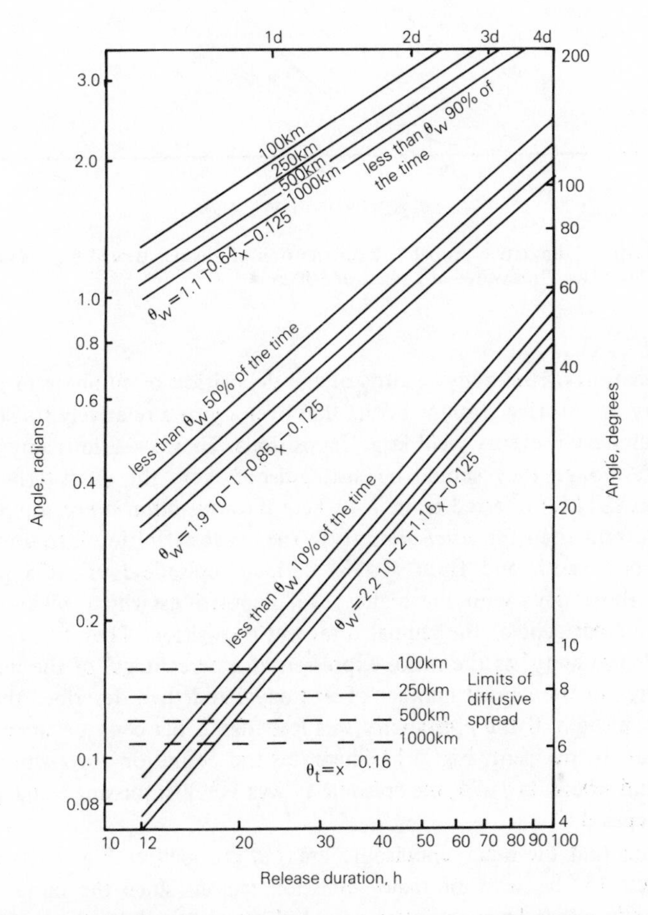

Fig. 7.9 – Plume width as a function of distance from the source and release duration for different probabilities.

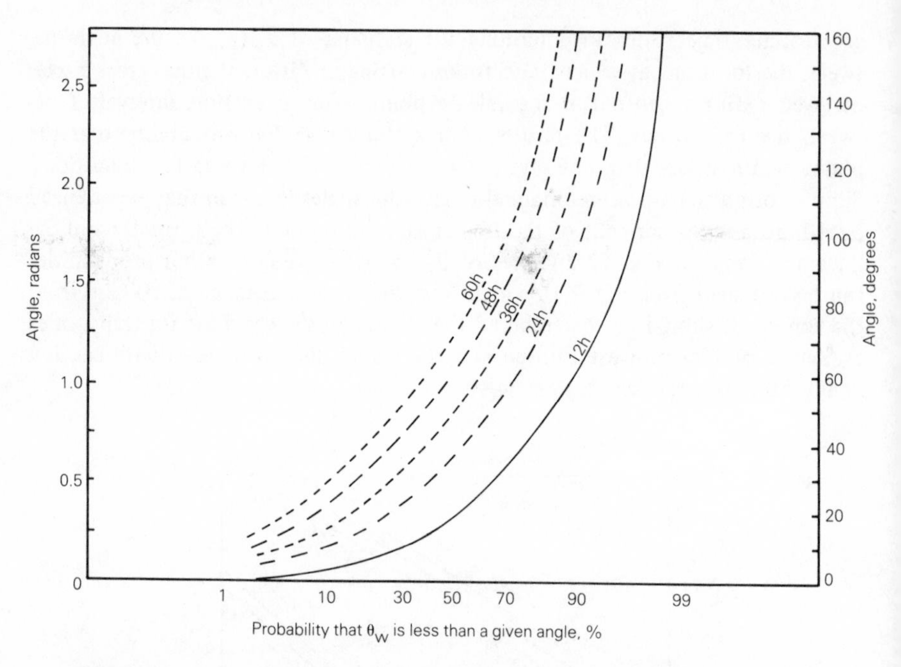

Fig. 7.10 – Cumulative probability distribution of angular spread θ_w for various sampling times T measured at a range of 500 km.

Episodes

An examination of the daily records of the deposition of sulphate in precipitation at any monitoring station reveals that over a year a relatively small fraction of all precipitation events yield large depositions. These occasions amy be called episodes (although they should be distinguished from the short-term episodic depositions (STED) referred to in 7.4 where the *concentration* of suplphate was more important than the *total qunatity*). The exact definition is to some extent arbitrary but Smith and Hunt (1978) defined 'episode-days' at a particular station as those days with the highest wet depositions which, when summed, make up 30 per cent of the annual total wet deposition. They further defined the term 'episodicity' as the ratio, expressed as a percentage, of the number of episode days to the annual number of wet days, and then described the station as 'highly episodic' if the episodicity was less than 5 per cent, or 'unepisodic' if greater than 10 per cent. Fig. 7.11 illustrates the definition for a site in south-east England where, in 1974, the episodicity was 5.3, just missing being described as 'highly episodic'.

It seems that the main 'unepisodic' areas in Europe (see Fig. 7.12) are often situated just to the west of major emission regions since the more frequent heavy rains in winds from the west are relatively clean whilst the less common lighter rainfalls in easterlies are much more polluted. The depositions reflecting the product of rainfall amount and sulphate concentration will therefore be rather uniform in magnitude and episodicity will be low. Highly episodic areas on the other hand are often remote from the largest source regions. Most precipitation is then relatively clean but occasionally air from a distant source region is involved and the resulting pollution may be drawn into frontal rain or thunderstorms giving very high depositions. Thunderstorms are often rather slow-moving and can draw into themselves vast quantities of boundary layer air, and pollution, as was discussed in 7.2. Boucher (1975) has shown areas within range of the main industrial areas of the world which are particularly subject to such storms. They include the southern Alps, Japan, the Transvaal and much of the mid-and eastern U.S.A. These areas are subject to at least twenty thunderstorm days per year.

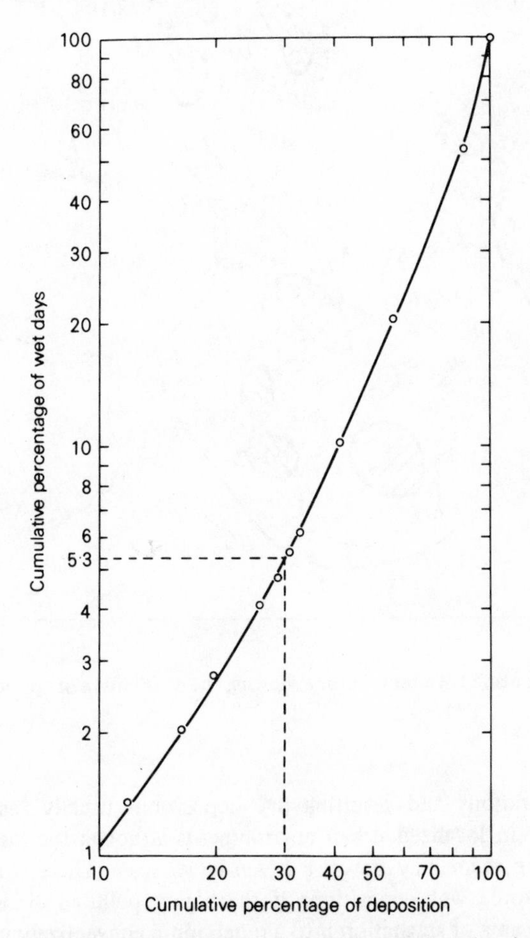

Fig. 7.11 – The cumulative percentage of deposition plotted against the cumulative percentage of wet days, plotted on logarithmic scales, at Cottered.

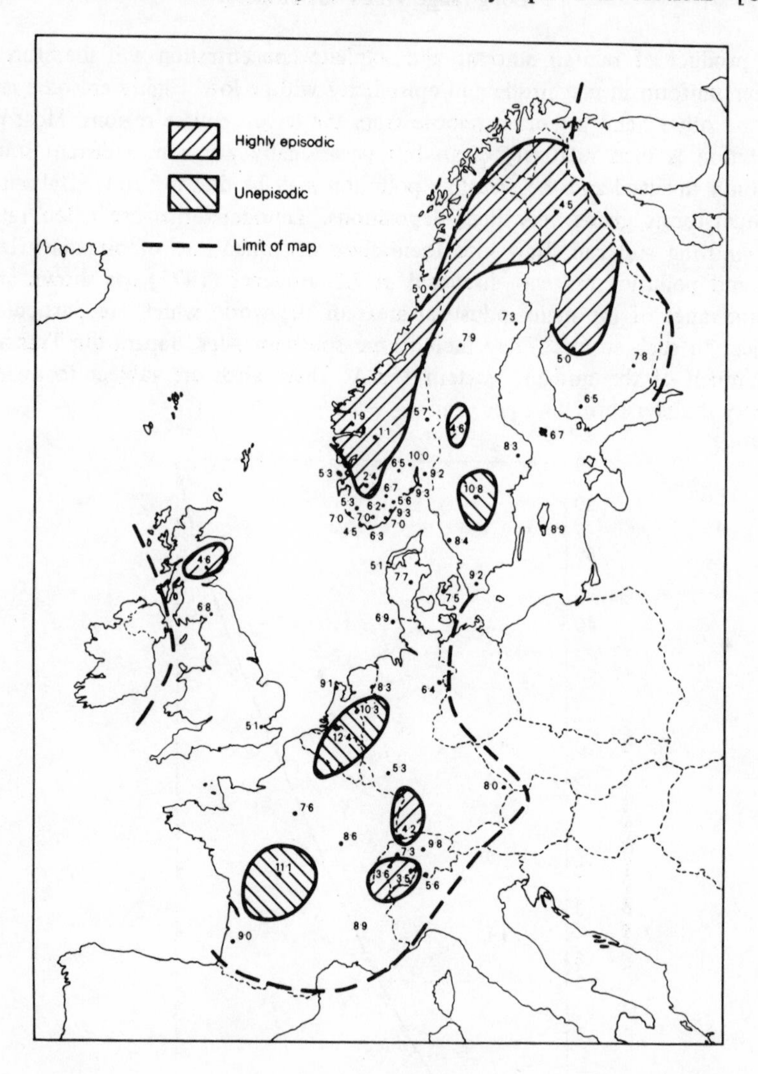

Fig. 7.12 – A map of Western Europe showing the distribution of episodicity for 1974.

Air concentrations and resulting dry depositions usually reach episodic proportions only in localized urban environments although the meteorological situations creating them may be on a much larger scale. These situations may also lead to episodic wet depositions if the highly polluted air is ultimately drawn out of the area of stagnation into a much more convectively unstable area or into the wet frontal zone of some depression passing round its perimeter.

For these reasons the formation of such areas of high pollution concentration is very relevant to this chapter.

In tropical regions, as industrialization proceeds and economic pressures tend to over-ride considerations of the environment, local episodes are becoming more common. Photochemical smogs generated by car exhaust fumes may be particularly troublesome in some of the larger cities further north like Los Angeles and Tokyo, owing to the high levels of incident solar radiation and the large number of vehicles. In both these cases surrounding hills accentuate the problem by reducing air mass ventilation.

In temperate latitudes urban episodes are usually associated with intense inversions at low level and with very light winds. These conditions usually take some days to form and are most frequently winter phenomena associated with moist cold surface air stagnating over the city, resulting in fog during the night and much of the day, being over-ridden by warm air associated with a weak slow-moving warm front which intensifies the inversion separating the two layers. The infamous 1952 London smog was of this type when SO_2 concentrations exceeded 2000 $\mu g\ m^{-3}$, ten times their current norm. The synoptic situation favourable to these kinds of episodes usually involves a blocking high pressure area. During November and December these blocking highs occur most frequently to the west of the British Isles, but occur also over Europe, the Soviet Union and Canada, and the eastern Pacific. Later in January and February the area of greatest frequency is over Scandinavia, eastern Europe and northern Siberia, the Gulf of Alaska and, less persistently, in the Bering Sea. Blocking highs are relatively uncommon in the U.S.A. and over Japan, although areas of very light winds are commonplace there.

References

Absalom, H. W. L., 1954, Meteorological aspects of smog, *Quart. J. R. Met. Soc.*, **80**, 261.
Angell, J. K., 1958, Lagrangian wind fluctuations at 300 mb derived from transosonde data, *J. Met.*, **15**, 522.
—— 1959, A climatological analysis of two years of routine transosonde flights from Japan, *Monthly Weather Review*, **87**, 427.
—— 1960, An analysis of operational 300 mb transosonde flights from Japan in 1957–1958, *J. Met.*, **17**, 20.
—— 1964, Measurements of Lagrangian and Eulerian properties of turbulence at a height of 2300 ft, *Quart. J. R. Met. Soc.*, **90**, 57.
Angell, J. K. and Pack, D. H., 1960, Analysis of some preliminary low-level constant level balloon (tetroon) flights, *Monthly Weather Review*, **88**, 235.
Anthes, R. A. and Warner, T. T., 1978, The development of hydrodynamic models suitable for air pollution and other mesometeorological studies, *Month. Weath. Rev.*, **106**, 1045.
Antonia, R. A., Chambers, A. J. and Satyaprakash, B. R., 1981, Kolmogorov constants for structure functions in turbulent shear flows, *Quart. J. R. Met. Soc.*, **107**, 579–589.
ApSimon, H. M., Goddard, A. and Wrigley, J. 1980, Estimating the possible transfrontier consequences of accidental releases: the MESOS model for long range atmospheric dispersal. Proceedings of a C.E.C. Seminar on 'Radioactive releases and their dispersion in the atmosphere following a hypothetical reactor accident' held at Risø, Denmark. C.E.C., Luxembourg, p. 819–842.
Arya, S. P. S. 1975, Geostrophic drag and heat transfer relations for the atmospheric boundary layer, *Quart, J. R. Met. Soc.*, **101**, 147–161.
Ball, F. K., 1958, Some observations of bent plumes, *Quart. J. R. Met. Soc.*, **84**, 61.
—— 1961, Viscous dissipation in the atmosphere, *J. Met.*, **18**, 333.

Ball, R. J. and Muschett, F. D., 1967, Air Pollution at Johnstown, Pennsylvania. Report of Department of Meteorology, College of Earth and Mineral Sciences, P.S.U.

Bannon, J. K., Dods, L. and Meade, P. J., 1962, Frequencies of various stabilities in the surface layer, *Meteorological Office Investigation Division Memorandum No. 88*.

Barad, M. L., 1958, Project Prairie Grass, a field program in diffusion, *Geophysical Research Paper No. 59,* Vols. I & II, G.R.D., A.F.C.R.C., Bedford, Mass.

—— 1959, Analysis of diffusion studies at O'Neill, *Atmospheric Diffusion and Air Pollution,* ed. F. N. Frenkiel and P. A. Sheppard, *Advances in Geophysics,* **6,** 389, Academic Press.

Barad, M. L. and Fuquay, J. J. (eds.), 1962, The Greem Glow Diffusion Program, *Geophys. Res. Paper No. 73,* Vols. 1 & 2.

—— 1962, Diffusion in shear flow, *J. Appl. Met.,* **1,** 257.

Barad, M. L. and Haugen, D. A., 1959, A preliminary eveluation of Sutton's hypothesis for diffusion from a continuous point source, *J. Met.,* **16,** 12.

Barad, M. L. and Shorr, B., 1954, Field studies of the diffusion of aerosols, *Am. Ind. Hyg. Assoc. Quart.,* **15(2),** 136–140.

Barrett, C. F., 1970, An experimental study by means of a fluorescent tracer of diffusion in two urban areas, *Warren Spring Laboratory Report LR 117 (AP)*.

Batchelor, G. K., 1946, Double velocity correlation function in turbulent motion, *Nature,* **158,** 883.

—— 1947, Kolmogoroff's theory of locally isotropic turbulence, *Proc. Camb. Phil. Soc.,* **43,** 533.

—— 1949, Diffusion in a field of homogeneous turbulence, I. Eulerian analysis, *Aust. J. Sci. Res.,* **2,** 437.

—— 1950, The application of the similarity theory of turbulence to atmospheric diffusion, *Quart. J. R. Met. Soc.,* **76,** 133.

—— 1952, Diffusion in a field of homogeneous turbulence, II. The relative motion of particles, *Proc. Camb. Phil. Soc.,* **48,** 345.

—— 1953, *The Theory of Homogeneous Turbulence,* Cambridge University Press.

—— 1959, Note on the diffusion from sources in a turbulent boundary layer (unpublished).

—— 1964, Diffusion from sources in a turbulent boundary layer, *Archiv. Mechaniki Stoswanej,* **3,** 661.

Batchelor, G. K. and Townsend, A. A., 1956, *Turbulent Diffusion, Surveys in Mechanics,* ed. G. K. Batchelor and R. M. Davies, 352. Cambridge University Press.

Beattie, J. R., 1961, An assessment of environmental hazards from fission product releases, *AHSB(5)R64,* U.K.A.E.A.

Benairie, M., 1971, Essai de provision synoptique de la pollution par l'acidité forte dans la région Rouennaise, *Atomos. Environ.*, **5**, 313–326.

Best, A. C., 1935, Transfer of heat and momentum in the lowest layers of the atmosphere, *Met. Office Geophys, Memoirs No. 65.*

—— 1954, Assessment of maximum concentration at ground level of gas from a heated elevated source, A paper of the Met. Res. Comm. (London), *M.R.P. No. 878.*

—— 1957, Maximum gas concentrations at ground-level from industrial chimneys, *J. Inst. Fuel,* **30**, 329.

Best, A. C. and Meade, P. J., 1956, The incidence of inversions over the United Kingdom, *Meteorological Office Investigation Division Memorandum No. 74.*

Bigg, E. K., Ayers, G. P. and Turvey, D. E. 1978, Measurement of the dispersion of a smoke plume at large distances from the source, *Atmos. Environ.*, **12**, 1815.

Blackadar, A. K., 1962, The vertical distribution of wind and turbulent exchange in a neutral atmosphere, *J. Geophys. Res.*, **67**, 3095–3102.

Blackall, R. M. and Gloster, J., 1981, Forecasting the airborne spread of foot and mouth disease, *Weather,* **36**, 162.

Blackman, R. B. and Tukey, J. W., 1958, The measurement of power spectra from the point of view of communications engineering. Part I. *The Bell System Technical Journal,* **37**, 185.

Bobileva, I. M., Zilitinkevitch, S. S. and Laikhtman, D. L., 1965, Turbulent exchange in the thermally-stratified planetary boundary layer of the atmosphere, *International Colloquium on Fine-scale Structure of the Atmosphere,* Academy of Sciences of the U.S.S.R., Moscow.

Bolin, B. and Persson, C., 1975, Regional dispersion and deposition of atmospheric pollutants with particular application to sulphur pollution over western Europe, *Tellus,* **27**, 281.

Booker, H. G., 1948, Some problems of radio-meteorology, *Quart. J. R. Met. Soc.*, **74**, 277.

Bosanquet, C. H., 1957, The rise of a hot waste gas plume, *J. Inst. Fuel,* **30**, 322.

Bosanquet, C. H., Carey, W. F. and Halton, E. M., 1950, Dust deposition from chimney stacks, *Proc. Inst. Mech. Eng.*, **162**, 355.

Bosanquet, C. H. and Pearson, J. L., 1936, The spread of smoke and gases from chimneys, *Trans. Faraday Soc.*, **32**, 1249.

Boucher, K., 1975, *Global Climate,* The English Universities Press Ltd.

Bradley, E. F., 1980, An experimental study of the profiles of wind speed, shearing stress and turbulence at the crest of a large hill, *Quart. J. R. Met. Soc.*, **106**, 101–123.

Bradley, E. F., Antonia, R. A. and Chambers, A. J., 1981, Turbulence Reynolds number and the turbulent kinetic energy balance in the atmospheric surface layer, *Boundary Layer Met.*, **21**, 2, 183.

Bradshaw, P., 1968, Calculation of boundary layer development using the turbulent energy equation, IV, *N.P.L. Aero Report 1271.*

Braham, R. R., Seely, B. K. and Crozier, W. D., 1952, A technique for tagging and tracing air parcels, *Trans. Amer. Geophys. Union,* **33**, 825.

Brier, G. W., 1950, The statistical theory of turbulence and the problem of diffusion in the atmosphere, *J. Met.,* **7**, 283.

Briggs, G. A., 1965, A plume rise model compared with observations, *J. Air Poll. Control Ass.,* **15**, 433.

—— 1969, *Plume Rise,* U.S. Atomic Energy Commission Div. Tech. Inf.

—— 1972, Chimney plumes in neutral and stable surroundings, *Atmos. Environ.,* **6**, 507–510.

—— 1973, Diffusion estimates for small emissions, in U.S. National Oceanic and Atmospheric Administration *E. R. L. Report ATDL-106*.

Briggs, G. A. and McDonald, K. R., 1978, Prairie Grass revisited: Optimum indicators of vertical spread, *Proceedings of 9th-NATO-CCMS Int. Tech. Symp. Air Pollution Modelling and its Application, Toronto*.

Brock, F. V., 1962, Analog computing techniques applied to atmospheric diffusion, continuous line source, *J. App. Met.,* **1**, 444.

Brooks, C. E. P. and Carruthers, N., 1953, *Handbook of Statistical Methods in Meteorology,* H.M S.O., London.

Brown, R. A., 1974, *Analytical Methods in Planetary Boundary-layer Modelling,* Adam Hilger, London.

Browning, K. A. and Ludlam, F. H., 1962, Airflow in convective storms, *Quart. J. R. Met. Soc.,* **88**, 117.

Browning, K. A. and Monk, G. A., 1982, A simple model for the synoptic analysis of cold fronts, *Quart. J. R. Met. Soc.,* **108**, 435.

Brunt, D., 1941, *Physical and Dynamical Meteorology,* Cambridge University Press.

Brutsaert, W. H., 1982, *Evaporation into the atmosphere,* D. Reidel Publishing Company, pp. 105 *et seq*.

Busch, N. E., Frizzola, J. A. and Singer, I. A., 1968, The micrometeorology of the turbulent flow field in the atmospheric surface boundary layer, *Acta Polytechnica Scandinavica Ph 59,* Copenhagen.

Busch, N. E. and Panofsky, H. A. 1968, Recent spectra of atmospheric turbulence, *Quart. J. R. Met. Soc.,* **94**, 132–148.

Businger, J. A., 1959, Data reduction technique, *Project prairie grass,* a field program in diffusion, *Geophysical Paper No. 59,* ed. D. A. Haughen, Vol. III, 29, G.R.D., A.F.C.R.C., Bedford, Mass.

—— 1966, Transfer of momentum and heat in the planetary boundary layer, *Proc. Symposium on Arctic Heat Budget and Atmospheric Circulation* (The Rand Corporation), 305–332.

—— 1973, Turbulent transfer in the atmospheric surface layer, *Workshop on Micrometeorology,* ed. D. A. Haugeni, American Meteorological Society.

Businger, J. A., Miyake, M., Dyer, A. J. and Bradley, E. F., 1967, On the direct determination of the turbulent heat flux near the ground, *J. App. Met.*, **6**, 1025–1032.

Businger, J. A. and Suomi, V. E., 1958, Variance spectra of the vertical wind component derived from observations with the sonic anemometer at O'Neill, Nebraska in 1953. *Archiv. f. Met. Geoph. un Biokl., A,* **10**, 415.

Businger, J. A., Wyngaard, J. C., Izumi, T. and Bradley, E. F., 1971, Flux-profile relationships in the atmospheric surface layer, *J.A.S.,* **28**, 181–189.

Calder, K. L., 1949, Eddy diffusion and evaporation in flow over aerodynamically smooth and rough surfaces: a treatment based on laboratory laws of turbulent flow with special reference to conditions in the lower atmosphere, *Quart. J. Mech. & Applied Math.*, **II**, 153.

—— 1949a, The criterion of turbulence in a fluid of variable density with particular reference to conditions in the atmosphere, *Quart. J. R. Met. Soc.,* **75**, 71–88.

—— 1952, Some recent British work on the problem of diffusion in the lower atmosphere, air pollution, *Proc. U.S. Tech. Conf. Air Poll.*, p. 787, McGraw-Hill, New York.

—— 1961, Atmospheric diffusion of particulate material considered as a boundary value problem, *J. Met.*, **18**, 413.

—— 1966, Concerning the similarity theory of Monin Obukhov, *Quart. J. R. Met. Soc.*, **92**, 141–146.

Caldwell, D. R., van Atta, C. W. and Helland, K. N., 1972, A laboratory study of the turbulent Ekman layer, *Geoph. Fluid Dynamics,* **3**, 125.

Carpenter, K. M., 1979, An experimental forecast using a non-hydrostatic mesoscale model, *Quart. J. R. Met. Soc.,* **105**, 629.

Carpenter, S. B., Frizzola, J. A., Smith, M. E., Leavitt, J. M. and Thomas, F. W., 1968, Report of full-scale study of plume rise at large electric generating stations, *J. Air Poll. Control Ass.*, **18**, 458.

Carras, J. N. and Williams, D. J., 1981, The long range dispersion of a plume from an isolated point source, *Atmos. Environ.,* **15**, 2205.

Carson, D. J., 1973, The development of a dry inversion-capped convectively unstable boundary layer, *Quart. J. R. Met. Soc.,* **99**, 450–467.

Caughey, S. J. 1982, Observed characteristics of the atmospheric boundary layer, *Atmospheric Turbulence and Air Pollution Modelling,* ed. F. T. M. Nieuwstadt and H. van Dop, D. Reidel Publishing Company, 107–158.

Caughey, S. J. and Kaimal, J. C., 1977, Vertical heat flux in the convective boundary layer, *Quart. J. R. Met. Soc.,* **103**, 811–815.

Caughey, S. J. and Palmer, S. G. 1979, Some aspect of turbulence structure through the depth of the convective boundary layer, *Quart. J. R. Met. Soc.,* **105**, 811–827.

Caughey, S. J., Wyngaard, J. C. and Kaimal, J. C., 1979, Turbulence in the evolving stable boundary layer, *J. Atmos. Sci.,* **6**, 1041–1052.

Chamberlain, A. C., 1953, Aspects of travel and deposition of aerosol and vapour clouds, *A.E.R.E., HP/R 1261,* H.M.S.O.

—— 1959, Deposition of iodine-131 in Northern England in October 1957, *Quart. J. R. Met. Soc.,* **85,** 350.

—— 1961, Aspects of the deposition of radioactive and other gases and particles, *Int. J. Air. Poll.,* **3.**

—— 1961a, Dispersion of activity from chimney stacks, *Atomic Energy Waste, its Nature, Use and Disposal,* ed. E. Glueckauf, Butterworth, 308.

—— 1966a, Transport of gases to and from grass and grass-like surfaces, *Proc. Roy, Soc.,* **A, 290,** 236.

—— 1966b, Transport of lycopodium spores and other small particles to rough surfaces, *Proc. Roy. Soc.,* **A, 296,** 45.

Chamberlain, A. C. and Chadwick, R. C., 1953, Deposition of airborne radioiodine vapour, *Nucleonics,* **II,** 22.

—— 1965, Transport of iodine from atmosphere to ground, *Tellus,* **18,** 226–237.

Charnock, H., 1951, Note on eddy diffusion in the atmosphere between one and two kilometres, *Quart. J. R. Met. Soc.,* **77,** 654.

Chatwin, P. C., 1968, The dispersion of a puff of passive contaminant in the constant stress region, *Quart. J. R. Met. Soc.,* **94,** 350–360.

Chatwin, P. C. and Sullivan, P. J., 1979, The basic structure of clouds of diffusing contaminant, *Mathematical modelling of Turbulent Diffusion in the Environment,* ed. C. J. Harris, Academic Press, 3–32.

Chaudhry, F. H. and Meroney, R. N., 1973, Similarity theory of diffusion and the observed vertical spread in the diabatic surface layer, *Boundary Layer Met.,* **3,** 405–415.

Christensen, O. and Prahm, L. P., 1976, A pseudospectral model for dispersion of atmospheric pollutants, *J. Appl. Meteorol.,* **15,** 1284.

Church, P. E., 1949, Dilution of waste stack gases in the atmosphere, *Ind. Eng. Chem.,* **41,** 2753.

Clarke, J. F. 1969, Nocturnal urban boundary layer over Cincinnati, Ohio, *M.W.R.,* **97,** 582–589.

Clarke, R. H., 1970, Observational studies in the atmospheric boundary layer, *Quart. J. R. Met. Soc.,* **96,** 91–114.

—— 1979, A model for short and medium range dispersion of radionuclides released to the atmosphere, The first Report of a Working Group on Atmospheric Dispersion, National Radiological Protection Baord, Harwell, U.K. NRPB-R91.

Committee on meteorological aspects of the effect of atomic radiation, 1956, report in *Science,* **124,** 105, U.S.A.

Committee on air pollution, 1954, *Report of Beaver Committee,* Cmd 9322, London, H.M.S.O.

Cooley, J. W. and Tukey, J. W., 1965, An algorithm for the machine calculation of complex Fourier series, *Maths. Comput.*, **19**, 297–301, Providence, R.I.

Corrsin, S., 1959, Progress report on some turbulent diffusion research. Atmospheric Diffusion and Air Pollution, *Advances in Geophysics*, **6**, 161.

—— 1963, Estimation of the relation between Eulerian and Lagrangian scales in large Reynold's number turbulence, *J. Atmos. Sci.*, **20**, 115.

Crabtree, J., 1959, The travel and diffusion of the radioactive material emitted during the Windscale accident, *Quart. J. R. Met. Soc.*, **85**, 362.

—— 1982, Studies of plume transport and dispersion over distances of travel up to several hundred kilometres, *Proc. 13th NATO/CCMS Conf. on Air Pollution Modelling and its Applications*, Plenum Press, New York.

Cramer, H. E., 1952, Preliminary results of a program for measuring the structure of turbulent flow near the ground. International symposium on atmospheric turbulence in the boundary layer. *Geophys. Res. Paper No. 19*, p. 187, G.R.D., Cambridge, Mass.

—— 1957, A practical method of estimating the dispersal of atmospheric contaminants, *Proceedings of the Conference on Applied Meteorology*, Am. Met. Soc.

—— 1959, Engineering estimates of atmospheric dispersal capacity, *Amer. Ind. Hyg. Ass. J.*, **20**, 183.

Cramer, H. E., Gill, G. C. and Record, F. A., 1957, Heated thermocouple anemometers and light bivanes, *Exploring the Atmosphere's First Mile*, ed. H. H. Lettau and B. Davidson, Vol. 1, p. 233, Pergamon Press.

Cramer, H. E., Record, F. A. and Vaughan, H. C., 1958, The study of the diffusion of gases in the lower atmosphere, *M.I.T. Department of Meteorology, Final Report under Contract No. AF 19(604)–1058*.

Crane, G. and Panofsky, H. A., 1976, A dispersion model for Los Angeles, *Third Symposium on Atmospheric Turbulence, Diffusion and Air Quality, Am. Met. Soc., 122–123.*

Crane, H. L. and Chilton, R. G., 1956, Measurements of atmospheric turbulence over a wide range of wavelengths for one meteorological condition, *NACA Technical Note 3702*.

Crawford, T. V., 1966, A computer program for calculating the atmospheric dispersion of large clouds, University of California Lawrence Livermore Laboratory, *Report UCRL/50179*.

Crozier, W. D. and Seely, B. K., 1955, Concentration distributions in aerosol plumes three to twenty-two miles from a point source, *Trans. Amer. Geophys. Union*, **36**, 42.

Csanady, G. T., 1961, Some observations on smoke plumes, *Int. J. Air Water Poll.*, **4**, 47–51.

—— 1963, Turbulent diffusion of heavy particles in the atmosphere, *J. Atmos. Sci.*, **20**, 201.

—— 1964, Turbulent diffusion in a stratified flow, *J. Atmos. Sci.*, **21**, 439.

—— 1967a, Concentration fluctuations in turbulent diffusion, *J. Atmos. Sci.*, **24**, 21–28.

—— 1967b, Variance of local concentration fluctuations, *Boundary Layers and Turbulence*, Physics of Fluid Supplement, 576–578.

—— 1972, Crosswind shear effects on atmospheric diffusion, *Atmos. Environ.*, **6**, 221–232.

—— 1973, *Turbulent Diffusion in the Environment*, D. Reidel Publishing Co.

Culkowski, W. M., 1967, Estimating the effect of buildings on plumes from short stacks, *Nuclear Safety*, **8**, 257–259.

Davidson, B. and Halitsky, J., 1958, A method of estimating the field of instantaneous ground concentration from tower bivane data. *J. Air Poll. Cont. Ass.*, **7**, 316.

Davies, P. O. A. L. and Moore, D. J., 1964, Experiments on the behaviour of effluent emitted from stacks at or near the roof level of tall reactor buildings, *Int. J. Air Wat. Poll.*, **8**, 515–523.

Deacon, E. L., 1949, Vertical diffusion in the lowest layers of the atmosphere, *Quart. J. R. Met Soc.*, **75**, 89.

Deardorff, J. W. 1970, Convective velocity and temperature scales for the unstable planetary boundary layer and for Rayleigh convection. *J. Atmos. Sci.*, **27**, 1211–1213.

—— 1973, Three-dimensional numerical modelling of the planetary boundary layer. Chap. 7, *Workshop on Micrometeorology*, ed. D. A. Haugen, Am. Met. Soc., Boston.

—— 1976, Clear and cloud-capped mixed layers: their numerical simulation, structure and growth and parameterization. *Seminar on the treatment of the boundary layer in numerical weather prediction*, European Centre for Medium Range Weather Forecasts, Reading, U.K.

Deardorff, J. W. and Willis, G. E., 1974, Physical modelling of diffusion in the mixed layer, *Second Symposium on Atmospheric Diffusion and Air Pollution*, Am. Met. Soc., 387–391.

Deardorff, J. W. and Willis, G. E., 1974, Computer and laboratory modelling of the vertical diffusion of non-buoyant particles in the mixed layer, *Advances in Geophysics*, Vol. 18B, Academic Press, 187–200.

Deardorff, J. W. and Willis, G. E., 1975, A parameterization of diffusion into the mixed layer, *J. Appl. Met.*, **14**, 1451–1458.

Department of Scientific and Industrial Research, 1945, *Atmospheric Pollution in Leicester, a Scientific Survey*, H.M.S.O.

Donaldson, C. du P., 1973, Construction of a dynamic model of the production of atmospheric turbulence and the dispersal of atmospheric pollutants, *Workshop on Micrometeorology*, ed. D. A. Haugen, Am. Met. Soc.

Doran, J. C., Horst, T. W. and Nickola, P. W. 1978a, Variations in measured values of lateral diffusion parameters, *J. App. Met.*, **17**, 825–831.

Doran, J. C., Horst, T. W. and Nickola, P. W., 1978b, Experimental observations of the dependence of lateral and vertical dispersion characteristics on source height, *Atmos. Environ.*, **12**, 2259–2264.

Drabløs, D. and Tollan, A. (eds.), 1980, Ecological impact of acid precipitation, SNSF-project, P.O. Box 61, 1432 As-NLH, Norway.

Draxler, R. R. 1976, Determination of atmospheric diffusion parameters, *Atmospheric Environment*, **10**, 99–105.

—— 1979, A summary of recent atmospheric diffusion experiments, *NOAA Technical Memorandum ERL ARL-78*, Dept. of Commerce, U.S.A.

Dumbauld, R. K., 1962, Meteorological tracer technique for atmospheric diffusion studies, *J. App. Met.*, **1**, 437.

Durbin, P. A. and Hunt, J. C. R., 1980, Dispersion from elevated sources in turbulent boundary layers, *J. Mecanique*, **19**, 679.

Durst, C. S., 1948, The fine structure of wind in the free air, *Quart. J. R. Met. Soc,,* **74**, 349.

Durst, C. S. and Davis, N. E., 1957, Accuracy of geostrophic trajectories, *Met. Mag.*, **86**, 138.

Durst, C. S., Crossley, A. F. and Davis, N. E., 1957, Horizontal diffusion in the atmosphere in the light of air trajectories, a paper of the Meterorological Research Committee (London), *M.R.P. No. 1058*.

—— 1959, Horizontal diffusion in the atmosphere as determined by geostrophic trajectories, *J. Fluid Mech.*, **6**, 401.

Dyer, A. J., 1967, The turbulent transport of heat and water vapour in an unstable atmosphere, *Quart. J. R. Met. Soc.*, **93**, 501–508.

—— 1974, A review of flux-profile relationships, *Boundary-layer Met.*, **7**, 363–372.

Dyer, A. J. and Bradley, E. F. 1982, An alternative analysis of flux-gradient relationships at the 1976 ITCE, *Boundary-layer Meteorol.*, **22**, 3–19.

Dyer, A. J. and Hicks, B. B., 1970, Flux-gradient relationships in the constant flux layer, *Quart. J. R. Met. Soc.*, **96**, 715–721.

Edinger, J. G., 1952, A technique for measuring the detailed structure of atmospheric flow, International Symposium on atmospheric turbulence in the boundary layer, *Geophys. Res. Paper No. 19*, G.R.D., Cambridge, Mass.

Egan, B. A., 1975, Turbulent diffusion in complex terrain, *Lectures on Air Pollution and Environmental Impact Analyses*, Am. Met. Soc., Boston, Mass., 112–135.

Egan, B. A. and Mahoney, J. R., 1972, Numerical modelling of advection and diffusion of urban area source pollutants, *J.A.M.*, **11**, 312–322.

Eggleton, A. E. J. and Thompson, N., 1961, Loss of fluorescent particles in atmospheric diffusion experiments by comparison with radioxenon tracer, *Nature*, **192**, 935.

Einstein, A., 1905, Uber die von molekularkinetischen Theorie der Wärme geforderte Bewegung von in ruhenden Flüssigkeiten suspendierten Teilchen, *Ann. Phys.*, **17**, 549.

Eliassen, A., 1978, The OECD study of long range transport of air pollutants: long range transport modelling, *Atmos. Environ.*, **12**, 479.

Eliassen, A; and Saltbones, J., 1982, Modelling of long range transport of sulphur over Europe: a two-year model run and some model experiments, *EMEP/MSC-W Report 1/82,* Norwegian Met. Instit., P.O. Box 320, Blindern, Oslo 3.

Elliott, W. P., 1961, The vertical diffusion of gas from a continuous source, *Int. J. Air & Water Poll.*, **4**, p. 33.

Eriksson, E., 1956, The chemical climate and saline soils in the arid zone, Australia – U.N.E.S.C.O. symp., *Arid Zone Climate,* Paper No. 45.

Etkes, P. W. and Brooks, C. F., 1918, Smoke as an indicator of gustiness and convection, *Monthly Weather Review,* **46**, 459, U.S. Weather Bureau.

Fackrell, J. E. and Robins, A. G., 1982a, The effects of source size on concentration fluctuations in plumes, *Boundary-layer Met.*, **22**, 335–350.

—— 1982b, Concentration fluctuations and fluxes in plumes from point sources in a turbulent boundary layer, *J. Fluid Mech.*, **117**, 1–26.

Fay, J. A. and Rosenzweig, J. J. 1980, An analytical diffusion model for long distance transport of air pollutants, *Atmos. Environ.*, **14**, 355.

Fisher, B. E. A., 1975, The long range transport of sulphur dioxide, *Atmos. Environ.*, **9**, 1063.

—— 1978, The calculation of long term sulphur deposition in Europe, *Atmos. Environ.*, **12**, 489.

Fortak, H. G., 1970, Numerical simulation of the temporal and spatial distribution of urban air pollution concentrations, *Symposium on Multiple Source Urban Diffusion Models,* U.S. Environmental Protection Agency.

Frenkiel, F. N., 1952a, On the statistical theory of turbulent diffusion, International symposium on atmospheric turbulence in the boundary layer. *Geophysical Research Paper, No. 19,* 415, G.R.D., Cambridge, Mass.

—— 1952b, Application of the statistical theory of turbulent diffusion to micrometeorology, *J. Met.*, **9**, 252.

Frenkiel, F. N. and Katz, I., 1956, Studies of small-scale turbulent diffusion in the atmosphere, *J. Met.*, **13**, 388.

Fuller, H. I., 1973, A study of sulphur deposition in the Skagerak area to evaluate the U.K. contribution in the period July 1972– June 1973. Report of the Institute of Petroleum, ESSO Research Centre, Abingdon, U.K.

Fuquay, J. J., Simpson, C. L. and Hinds, T. H. 1964, Prediction of environmental exposures from sources near the ground based on Hanford experimental data, *J. App. Met.*, **3**, 761–770.

Garland, J. A., Atkins, D. H. F., Readings, C. J., and Caughey, S. J., 1973, Deposition of gaseous sulphur dioxide to the ground, *Atmos. Environ.*, **8**, 75–79.

Gee, J. H. and Davies, D. R., 1963, A note on horizontal dispersion from an instantaneous ground source, *Quart. J. R. Met. Soc.*, **89**, 542.

—— 1964, A further note on horizontal dispersion from an instantaneous ground source, *Quart. J. R. Met. Soc.*, **90**, 478.

Georgii, H. W., 1969, Contribution to the sulphur budget based on SO_2 and sulphate measurements in the free atmosphere, Offenbach, a.M., D. Wetterd, *Ann. Met. Neue Folge*, Nr. 4, pp. 117–121.

Giblett, M. A. *et al.*, 1932, The structure of wind over level country, *Meteorological Office Geophysical Memoirs No. 54*.

Gifford, F. A., 1953, A study of low-level air trajectories at Oak Ridge, Tenn., *Monthly Weather Review*, **81**, 179, U.S. Weather Bureau.

—— 1955, A Simultaneous Lagrangian-Eulerian turbulence experiment, *Monthly Weather Review*, **83**, 293, U.S. Weather Bureau.

—— 1956, The relation between space and time correlations in the atmosphere, *J. Met.*, **13**, 289.

—— 1957a, Relative atmospheric diffusion of smoke puffs, *J. Met.*, **14**, 410.

—— 1957b, Further data on relative atmospheric diffusion, *J. Met.*, **14**, 475.

—— 1959, Statistical properties of a fluctuating plume dispersion model, *Atmospheric Diffusion and Air Pollution*, ed. F. N. Frenkiel and P. A. Sheppard, *Advances in Geophysics*, **6**, 117, Academic Press.

—— 1960, Peak to average concentration ratios according to a fluctuating plume dispersion model, *Int. J. Air Poll.*, **3**, 253.

—— 1961, Use of routine meteorological observations for estimating atmospheric dispersion, *Nucl. Safety*, **2**(4), 47–51.

—— 1962, Diffusion in the diabatic surface layer, *J. Geophy. Res.*, **67**, 3207.

—— 1975, Atmospheric dispersion models for environmental pollution applications, *Lectures on air pollution and environmental impact analyses*, Am. Met. Soc., 35–58.

—— 1977, Tropospheric relative diffusion observations, *J. Appl. Met.*, **16**, 311–313.

—— 1982a, Horizontal diffusion in the atmosphere: a Lagrangian-dynamical theory, *Atmos. Env.*, **16**, 505–512.

—— 1982b, Long-range plume dispersion: comparisons of the Mt. Isa data with theoretical and empirical formulas, *Atmos. Env.*, **16**, 883–886.

Gifford, F. A. and Hanna, S. R., 1971, Urban air pollution modelling, *2nd Int. Clean Air Congress of the Int. Union of Air Poll. Prevention Associations*, 1146–1151, Academic Press.

Godson, W. L. 1958, The diffusion of particulate matter from an elevated source, *Archiv. f. Met. Geoph. und Biokl.*, A, **10**, 305.

Gosline, C. A., 1952, Dispersion from short stacks, *Chemical Engineering Progress*, **48**, 165.

Grant, H. L. Stewart, R. W., and Moilliet, A. 1962, Turbulence spectra from a tidal channel, *J. Fluid Mech.*, **12**, 241–268.

Green, H. L. and Lane, W. R., 1957, *Particulate Clouds: Dusts, Smokes and Mists*, Spon, London.

Greenfield, S. M., 1957, Rain scavenging of radioactive particulate matter from the atmosphere, *J. Met.*, **14**, 115.

Gregory, P. H., 1945, The dispersion of airborne spores, *Trans. Brit. Myc. Soc.*, **28**, 26.

—— 1951, Deposition of airborne Lycopodium spores on cylinders, *Ann. App. Biol.*, **38**, 357.

Gryning, S., van Ulden, A. P. and Larsen, S. E. 1983, Dispersion from a continuous ground-level source investigated by a K model, *Quart. J. R. Met. Soc.*, **109**, 357–366.

Gurvic, A. S. and Yaglom, A. M., 1967, Breakdown of eddies and probability distributions for small-scale turbulence, *Boundary Layers and Turbulence*, Physics of Fluids Supplement.

Hage, K. D., 1961a, The influence of size distribution on the ground deposit of large particles emitted from an elevated source, *Int. J. Air Wat. Poll.*, **4**, 24.

—— 1961b, On the dispersion of large particles from a 15 m source in the atmosphere, *J. Met.*, **18**, 534–539.

—— 1964, Particle fallout and dispersion below 30 km in the atmosphere, *Final Report SC-DC-64-1463, Sandia Corp., Albuquerque, NM, U.S.A.*

Halitsky, J., 1968, Gas diffusion near buildings, *Meteorology and Atomic Energy*, 221–255. U.S. Atomic Energy Commission Div. Tech. Inf.

Hall, C. D., 1975, The simulation of particle motion in the atmosphere by a numerical random walk model, *Quart. J. R. Met. Soc.*, **101**, 235–244.

Hamilton, P. M., 1967, Plume height measurements at two power stations, *Atmos. Environ.*, **1**, 379–387.

—— 1969, The application of a pulsed-light rangefinder (lidar) to the study of chimney plumes, *Phil. Trans. Roy. Soc. Lond.*, A, **265**, 153–172.

Hanna, R. S., 1968, A method of estimating vertical eddy transport in the planetary boundary layer using characteristics of the vertical velocity spectrum, *J. Atmos. Sci.*, **25**, 1026.

—— 1978, A statistical diffusion model for use with variable wind fields *4th Symposium on Turbulent Diffusion and Air Pollution,* Am. Met. Soc., Boston.

—— 1979, Some statistics of Lagrangian and Eulerian wind fluctuations, *J Appl. Met.*, **18**, 518.

—— 1981a, Lagrangian and Eulerian time-scale relations in the daytime boundary layer, *J. Appl. Met.*, **20**, 242–249.

—— 1981b, Effects of release height on σ_y and σ_z in daytime conditions. Air Pollution Modelling and its Applications, *Proceedings of the 1980 NATO/CCMS Meeting,* ed C. de Wispelaere, Plenum Press, 337–356.

—— 1982, Applications in air pollution modelling, *Atmospheric Turbulence and Air Pollution Modelling,* ed. F. T. M. Nieuwstadt and H. van Dop, D. Reidel Publishing Co., 275–310.

Hanna, S. R. *et al.*, 1977, AMS Workshop on Stability Classification Schemes and Sigma Curves – Summary of Recommendations, *Bull. Am. Met. Soc.*, **58**, 1305–1309.

Hanna, S. R., Briggs, G. A. and Hosker, R. P., 1982, *Handbook on Atmospheric Diffusion,* Technical Information Center, U.S. Department of Energy.

Harriman, R. and Morrison, B., 1980, Ecology of acid streams draining forested and non-forested catchments in Scotland, *Ecological impact of acid precipitation,* ed. Drablφs, D. and Tollan, A., SNSF-project, 312.

Harvey, H. H., 1980, Widespread and diverse changes in the biota of North American lakes and rivers coincident with acidification, *Ecological impact of acid precipitation,* ed. Drablφs, D. and Tollan, A., SNSF-project, 93.

Haugen, D. A., 1959, Project Prairie Grass, A Field Programme in Diffusion, *Geographical Research Paper No. 59,* **Vol.** III, G.R.D.A.F.C., Bedford, Mass.

—— 1966, Some Lagrangian properties of turbulence deduced from atmospheric diffusion experiments, *J. App. Met.*, **5**, 647–652.

Haugen, D. A., Kaimal, J. C. and Bradley, E. F., 1971, An experimental study of Reynolds stress and heat flux in the atmospheric surface layer, *Quart. J. R. Met. Soc.*, **97**, 168–180.

Hawkins, J. E. and Nonhebel, G., 1955, Chimneys and the dispersal of smoke, *J. Inst. Fuel,* **28**, 530.

Hay, J. S. and Pasquill, F., 1957, Diffusion from a fixed source at a height of a few hundred feet in the atmosphere, *J. Fluid Mech.*, **2**, 299.

—— 1959, Diffusion from a continuous source in relation to the spectrum and scale of turbulence. *Atmospheric Diffusion and Air Pollution,* ed. F. N. Frenkiel and P. A. Sheppard, *Advances in Geophysics,* **6**, 345, Academic Press.

Hess, G. D. and Clarke, R. H., 1973, Time spectra and cross-spectra of kinetic energy in the planetary boundary layer, *Quart. J. R. Met. Soc.*, **99**, 130–153.

Hewson, E. W., 1945, The meteorological control of atmospheric pollution by heavy industry, *Quart. J. R. Met. Soc.*, **71**, 266.

Hicks, B. B., 1978, Some limitations of dimensional analysis and power laws. *Boundary-layer Meteorol.*, **14**, 567–569.

Hidy, G. M., Tong, E. Y. and Mueller, P. K., 1976, Design of the Sulfate Regional Experiment (SURE)), Electric Power Research Institute *Report EC-125*.

Hilst, G. R., 1957a, Observations of the diffusion and transport of stack effluents in stable atmospheres, Ph.D, thesis, University of Chicago.

—— 1957b, The dispersion of stack gases in stable atmospheres, *J. Air. Poll. Cont. Ass.*, **7**, 205.

—— 1970, Sensitivities of air quality prediction to input errors and uncertainties, *Symposium on Multiple Source Urban Diffusion Models,* U.S. Environmental Protection Agency.

Hilst, G. R. and Simpson, C. L., 1958, Observations of vertical diffusion rates in stable atmospheres, *J. Met.,* **15,** 125.

Hinds, W. T., 1970, Diffusion over coastal moutnains of southern California, *Atmos. Environ.,* **4,** 107–124.

Hoeber, H., 1982, KonTur, Convection and Turbulence Experiment, *Field Phase Report,* G.M.L. Wittenborn Sohne, 2000 Hamburg 13.

Hoffnagle, G. F., Smith, M. E., Crawford, T. V. and Lockhart, T. J. 1981, On-site meteorological instrumentation requirements to characterise diffusion from point sources – a Workshop, 15–17 Jan., 1980, Raleigh, N.C., *Bull. Amer. Met. Soc.,* **62,** 255–261.

Högström, U., 1964, An experimental study on atmospheric diffusion, *Tellus,* **16,** 205.

Holland, J. Z., 1953, A meteorological survey of the Oak Ridge Area, *U.S.A.E.C. Report ORO-99,* Tech. Inf. Ser., Oak Ridge, Tenn, U.S.A.

Holzworth, G. C., 1967, Mixing depths, wind speed and air pollution potential for selected locations in the U.S.A., *J. Appl. Met.,* **6,** 1039–1044.

Horst, T. W., 1977, A sruface depletion model for deposition from a Gaussian plume, *Atmos. Environ.,* **11,** 41–46.

—— 1978, Comments on 'A general Gaussian diffusion-deposition model for elevated point sources', *J. Appl. Met.,* **17,** 415–416.

—— 1979, Lagrangian similarity modelling of vertical diffusion from a ground-level source, *J. App. Met.,* **18,** 733–740.

—— 1979a, A review of Gaussian diffusion deposition models, *Atmospheric Sulfur Deposition,* Ann Arbor Science, 275–283.

Horst, T. W., Doran, J. C. and Nickola, P. W., 1979, Evaluation of empirical atmospheric diffusion data, *NUREG/CR-0798, PNL-2599,* Battelle Pacific Northwest Laboratories, Richland, Washington, U.S.A. and U.S. Nuclear Regulatory Commission.

Horst, T. W., van Ulden, A. P. and Nieuwstadt, F. T. M., 1980, Discussion of Nieuwstadt and van Ulden 1978 paper, *Atmos. Environ.,* **14,** 267–279.

Hosler, C. R., Pack, D. H. and Harris, T. B., 1959, *Meteorological Investigation of diffusion in a valley at Shippingport, Pennsylvania,* U.S. Dept. of Commerce, Weather Bureau.

Hunt, J. C. R. 1982, Diffusion in the stable boundary layer, *Atmospheric Turbulence and Air Pollution Modelling,* ed. F. T. M. Nieuwstadt and H. van Dop, D. Reidel Publishing Company, 231–274.

Hunt, J. C. R., Britter, R. E. and Puttock, J. S., 1979, Mathematical models of dispersion around buildings and hills, *Mathematical Modelling of Turbulent Diffusion in the Environment,* ed. C. J. Harris, Academic Press, 145–200.

Hunt, J. C. R. and Mulhearn, P. J., 1973, Turbulent dispersion from sources near two-dimensional obstacles, *J. Fluid Mech.*, **61**, 245–274.

Hunt, J. C. R., Puttock, J. S. and Snyder, W. H., 1979, Turbulent diffusion from a point source in stratified and neutral flows around a three-dimensional hill – Part I. Diffusion equation analysis, *Atmos. Environ.*, **13**, 1227–1239.

Hunt, J. C. R. and Weber, A. H. 1979, A Lagrangian statistical analysis of diffusion from a ground-level source in a turbulent boundary layer, *Quart. J. R. Met. Soc.*, **105**, 423–443.

Inoue, E., 1950, On the turbulent diffusion in the atmosphere (I), *J. Met. Soc., Japan*, **28**, 13.

—— 1951, On the turbulent diffusion in the atmosphere (II), *J. Met. Soc., Japan*, **29**, 32.

Irwin, J. S., 1979, Estimating plume dispersion – a recommended generalized scheme, *Am. Met. Soc. Fourth Symposium on Turbulence, Diffusion and Air Pollution*, 62–69.

Islitzer, N. F. 1961, Short-range atmospheric dispersion measurements from an elevated source, *J. Met.*, **18**, 443–450.

Islitzer, N. F. and Dumbauld, R. K., 1963, Atmospheric diffusion-deposition studies over flat terrain, *Int. J. Air Water Poll.*, **79**, 999.

Izumi, Y. and Caughey, S. J. 1976, Minnesota 1973 Atmospheric Boundary Layer Experiment Data Report, *Env. Res. Paper 547*, U.S.A.F. Cambridge Res. Lab.

Jarman, R. T. and de Turville, C. M., 1966, The screening power of visible smoke, *Int. J. Air Wat. Poll.*, **10**, 465–467.

Jarman, R. T. and de Turville, 1969, The visibility and length of chimney plumes, *Atmos. Environ.*, **3**, 257.

Jenkins, G. J., 1983, Meteorological Office, T.D.N. 142.

Jenkins, G. J. and Whitlock, J., 1983, Results of a dispersion study in the Sirhowy ridge-valley area of S. Wales. Met. Office, *Met. 0.14 Turbulence and Diffusion Note No. 141*.

Jenkins G. J., Mason, P. J., Moores, W. H. and Sykes, R. I., 1981, Measurements of the flow structure around Ailsa Craig, a steep, three-dimensional, isolated hill, *Quart. J. R. Met. Soc.*, **107**, 833–851.

Johnson, W. B., Wolf, D. E. and Mancuso, R. L. 1979, Long term regional patterns and transfrontier exchanges of airborne sulfur pollution in Europe, *Atmos. Environ.*, **12**, 511.

Jones, C. D., 1979, Statistics of the concentration fluctuations in short range atmospheric diffusion, *Mathematical Modelling of Turbulent Diffusion in the Environment*, ed. C. J. Harris, Academic Press, 277–295.

Jones, J. A., 1981a, The estimation of long range dispersion and deposition of continuous releases of radionuclides to the atmosphere, *The third report of a Working Group on Atmospheric Dispersion*, National Radiological Protection Board, Harwell, U.K., *NRPB-R123*.

—— 1981b, A model for long range atmospheric dispersion of radionuclides released over a short period, *The fourth report of a Working Group on Atmospheric Dispersion,* National Radiological Protection Board, Harwell, U.K., *NRPB-R124*.

Jones, J. I. P., 1963, A band-pass filter technique for recording atmospheric turbulence, *Brit. J. Appl. Phys.*, **14**, 95–101.

Jones, J. I. P. and Butler, H. E., 1958, The measurement of gustiness in the first few thousand feet of the atmosphere, *Quart. J. R. Met. Soc.,* **84**, 17.

Jones, J. I. P. and Pasquill, F., 1959, An experimental system for directly recording statistics of the intensity of atmospheric turbulence, *Quart. J. R. Met. Soc.,* **85**, 225.

Jones, R. A., 1957, A preliminary examination of the spectrum and scale of the vertical component at 2000 ft., a paper of the Meteorological Research Committee (London), *M.R.P. No. 1044*.

Kaimal, J. C., 1973, Turbulence spectra, length scales and structure parameters in the stable surface layer, *Boundary-layer Meteorol.*, **4**, 289–309.

Kaimal, J. C. and Haugen, D. A., 1967, Characteristics of vertical velocity fluctuations on a 430 metre tower, *Quart. J. R. Met. Soc.*, **93**, 305–317.

Kaimal, J. C., Wyngaard, J. C., Haugen, D. A., Coté, O. R., Izumi, Y., Caughey, S. J. and Readings, C. J. 1976, Turbulence structure in the convective boundary layer, *J. Atmos. Sci.*, **33**, 2152–2169.

Kaimal, J. C., Wyngaard, J. C., Izumi, Y. and Coté, O. R., 1972, Spectral characteristics of surface layer turbulence, *Quart. J. R. Met. Soc.,* **98**, 563–589.

Kampé de Fériet, M. J., 1939, Les fonctions aléatoires stationnaires et la théorie statistique de la turbulence homogène, *Ann. Soc. sci. Brux.*, **59**, 145.

Kazanskii, A. B. and Monin, A. S., 1957, The forms of smoke trails, *Izv. Akad. Nauk. U.S.S.R.* (Ser. Geofiz.), No. 8, 1020.

Kellogg, W. W., 1956, Diffusion of smoke in the stratosphere, *J. Met.*, **13**, 241.

Klug, W., 1965, Diabatic influence on turbulent wind fluctuations, *Quart. J. R. Met. Soc.,* **91**, 215.

—— 1968, Diffusion in the atmospheric surface layer: comparison of similarity theory with observation, *Quart. J. R. Met. Soc.,* **94**, 555–562.

Kolmogorov, A. N., 1941, The local structure of turbulence in incompressible viscous fluid for very large Reynolds numbers. *C. R. Acad. Sci. U.R.S.S.,* **30**, 301.

Lamb, R. G., 1978, A numerical simulation of dispersion from an elevated point source within a modelled convective planetary boundary layer, *Atmos. Environ.,* **12**, 1297–1304.

—— 1979, The effects of release height on material dispersion in the convective planetary boundary layer. Proceedings, *4th Symposium on Turbulence, Diffusion and Air Pollution,* Am. Met. Soc., 27–33.

—— 1982, Diffusion in the convective boundary layer, *Atmospheric Turbulence and Air Pollution Modelling,* ed. F. T. M. Nieuwstadt and H. van Dop, D. Reidel Publishing Co., 159–230.

Lange, R., 1978, ADPIC: a three-dimensional particle-in-cell model for the dispersal of atmospheric pollutants and its comparison to regional tracer studies, *J. Appl. Met.,* **17,** 320.

Larsen, R. I., 1971, A mathematical model for relating air quality measurements to air quality standards, *Publication AP-89,* Environmental Protection Agency, North Carolina.

Lawrence, E. N., 1962, Atmospheric pollution (sulphur dioxide) in hilly terrain, *Int. J. Air Water Poll.,* **6,** 5–26.

Leighton, P. A., Perkins, W. A., Grinnell, S. W. and Webster, F. X., 1965, The fluorescent particle atmospheric tracer, *J. Appl. Met.,* **4,** 334–348.

Lenschow, D. H., 1970, Airplane measurements of planetary boundary layer structure, *J. Appl. Met.,* **9,** 874–884.

Leonard, B. P., 1959, Long range cloud diffusion in the lower atmosphere, *J. Air Poll. Control Ass.,* **9,** 77.

Lettau, H., 1950, A re-examination of the 'Leipzig wind profile' considering some relations between wind and turbulence in the friction layer. *Tellus,* **2,** 125–129.

Lettau, H. H. and Davidson, B., 1957, *Exploring the Atmosphere's First Mile,* Vol. I, *Instrumentation and Data Evaluation,* Vol. 2, *Site Description and Data Tabulation,* Pergamon Press.

Lewellen, W. S. and Teske, M., 1973, Prediction of the Monin–Obukhov similarity functions from an invariant model of turbulence, *J. Atm. Sci.,* **30,** 1340–1345.

—— 1976, Second-order closure modelling of diffusion in the atmospheric boundary layer, *Boundary-layer Met.,* **10,** 69–90.

Ley, A. J., 1982, A random walk simulation of two-dimensional turbulent diffusion in the neutral surface layer, *Atmos. Environ.,* **16,** 2799–2808.

Lucas, D. H., 1958, The atmospheric pollution of cities, *Int. J. Air Poll.,* **1,** 71.

—— 1967, Application and evaluation of results of the Tilbury plume rise and dispersion experiment, *Atmos. Environ.,* **1,** 421–424.

Lucas, D. H., Moore, D. J. and Spurr, G., 1963, The rise of hot plumes from chimneys, *Int. J. Air Water Poll.,* 7, 473.

Ludwick, J. D., 1966, Xenon 133 as an atmospheric tracer, *J. Geophys. Res.,* **71,** 4743.

Lumley, J. L. and Panofsky, H. A., 1964, *The Structure of Atmospheric Turbulence,* John Wiley and Sons, New York.

Luna, R. E. and Church, H. W., 1972, A comparison of turbulence intensity and stability ratio measurements to Pasquill stability classes, *J. Appl. Met.,* **11,** 663–669.

MacCready, P. B., 1953, Atmospheric turbulence measurements and analysis, *J. Met.*, **10**, 325.

Manton, M. J., 1977, On the structure of convection, *Boundary Layer Meteorol.*, **12**, 491.

Marsh, K. J. and Withers, V. R., 1969, An experimental study of the dispersion of the emissions from chimneys in Reading – III. The investigation of dispersion calculations. *Atmos. Environ.*, **3**, 281–302.

Mason, B. J., 1957, *The Physics of Clouds*, O.U.P.

Mason, P. J. and Sykes, R. I., 1979, Separation effects in Ekman layer flow over ridges, *Quart. J. R. Met. Soc.*, **105**, 129.

—— 1981, On the influence of topography on plume dispersal, *Boundary-layer Meteorol.*, **21**, 137–157.

May, F. G., 1958, The washout by rain of lycopodium spores, *Quart. J. R. Met Soc.*, **84**, 451.

McBean, G. A., 1971, The variations of the statistics of wind, temperature and humidity fluctuations with stability, Boundary Layer Meteorology, **1**, 438–457.

McCormick, R. A., 1970, Meteorological aspects of air pollution in urban and industrial districts, *World Meteorological Organization Technical Note, No. 106*.

McCully, C. R., *et al.*, 1956, Scavenging action of rain in air-borne particulate matter, *Ind. Eng. Chem.*, **48**, 1512.

McDonald, J. E., 1963, Rain washout of partially wettable insoluble particles, *J. Geophys. Res.*, **68**, 4993.

McElroy, J. L. and Pooler, F., 1968, St. Louis dispersion study, Vol. II – *Analysis*, National Air Pollution Control Administration, Arlington, Va., Pub. No. AP-53.

Meade, P. J. and Pasquill, F., 1958, A study of the average distribution of pollution around Staythorpe, *Int. J. Air Poll.*, **1**, 60.

Meetham, A. R., 1950, Natural removal of pollution from the atmosphere, *Quart. J. R. Met. Soc.*, **76**, 359.

—— 1952, *Atmospheric Pollution*, Pergamon Press.

—— 1954, Natural removal of atmospheric pollution during fog, *Quart. J. R. Met. Soc.*, **80**, 96.

Meteorology and Atomic Energy, 1968, ed. D. H. Slade, U.S. Atomic Energy Commission, Division of Technical Information.

Mickelsen, W. R., 1955, An experimental comparison of the Lagrangian and Eulerian correlation coefficients in homogeneous isotropic turbulence, N.A.C.A. Washington, *Tech. Note No. 3570*.

Mistra, P. K., 1982, Dispersion of non-buoyant particles inside a convective boundary layer, *Atmos. Environ.*, **16**, 239–244.

Miyake, M., Stewart, R. W. and Burling, R. W., 1970, Spectra and co-spectra of turbulence over water, *Quart. J. R. Met. Soc.*, **96**, 138–143.

Monin, A. S., 1955, The equation of turbulent diffusion, *Dokl. Akad. Nauk.*, **105**, 256.

—— 1959, Smoke propagation in the surface layer of the atmosphere. *Atmospheric Diffusion and Air Pollution,* ed. F. N. Frenkiel and P. A. Sheppard, *Advances in Geophysics,* **6**, 331, Academic Press.

—— 1965, Structure of an atmospheric boundary layer, *Izv. ANSSSR Ser. Atm. and Oceanic Phys.,* **1**, 3, 258–265.

Monin, A. S. and Yaglom, A. M., 1965, Statistical hydromechanics, The Mechanics of Turbulence, Part 1, Trans-Joint Publication Research Service, Washington, *JPRS,* **37**, 763, 1966.

Monteith, J. L. and Szeicz, G., 1960, The carbon dioxide flux over a field of sugar beet, *Quart. J. R. Met. Soc.,* **86**, 205–214.

Moore, D. J., 1966, Physical aspects of plume models, *Atmos. Environ.,* **10**, 411–417.

—— 1967, Meteorological measurements on a 187 m tower, *Atmos. Environ.,* **1**, 367.

—— 1967, Variation of turbulence with height, *Atmos. Environ.,* **1**, 521–522.

—— 1969, The distributions of surface concentrations of sulphur dioxide emitted from tall chimneys, *Phil. Trans. Roy. Soc., London,* **A265**, 245.

—— 1974, Observed and calculated magnitudes and distances of maximum ground-level concentration of gaseous effluent material downwind of a tall stack, *Turbulent Diffusion in Environmental Pollution,* ed. F. N. Frenkiel and R. E. Munn, *Advances in Geophysics,* Vol. 18b, Academic Press, 201–221.

—— 1975, A simple boundary layer model for predicting time mean ground-level concentrations of material emitted from tall chimneys, *Proc. Instn. Mech. Engrs.,* **189**, 33–43.

—— 1976, Calculation of ground-level concentrations for different sampling periods and source locations, *Atmospheric Pollution,* Elsevier, Amsterdam.

—— 1980, Lectures on plume-rise, *Atmospheric Planetary Boundary Layer Physics,* ed. A. Longhetto, *Developments in Atmospheric Science,* Vol. 11, Elsevier, Amsterdam.

Mordukhovich, M. I. and Tsvang, L. R., 1966, Direct measurement of turbulent flows at two heights in the atmospheric ground layer, *Izv. Atm. and Oceanic Phys.,* **2**, 786.

Moses, H. and Carson, J. E. 1967, Stack design parameters influencing plume rise, Papers 67–84, *6th Annual Meeting Air Pollution Control Assoc.,* Cleveland, Ohio.

Munn, R. E. and Cole, A. F. W., 1967, Turbulence and diffusion in the wake of a building, *Atmos. Environ.,* **1**, 33–43.

Nappo, C. J., 1979, Relative and single particle diffusion estimates determined from smoke plume photographs, *Proc. Fourth Symposium on Turbulence, Diffusion and Air Pollution,* Am. Met. Soc., 45–47.

Nicholls, S. and Readings, C. J., 1981, Spectral characteristics of surface layer turbulence over the sea, *Quart. J. R. Met. Soc.*, **107**, 591–614.

Nicholls, S., LeMone, M. A. and Sommeria, G., 1982, The simulation of a fine-weather marine boundary layer in GATE using a three-dimensional model, *Quart. J. R. Met. Soc.*, **108**, 167–190.

Nickola, P. W., 1977, *The Hanford 67-Series: A volume of atmospheric field diffusion measurements*, PNL-2433, UC-11, U.S. Dept. of Energy and U.S. Nuclear Regulatory Commission.

Niemann, B. L. and Young, J. W. S., 1981, United States – Canada Memorandum of Intent on Transboundary Air Pollution, *Modelling Subgroup Report: Interim Working Paper*, 2–13.

Niemeyer, L. E., 1960, Forecasting air pollution potential, *Monthly Weather Review*, **88**, 88.

Niemeyer, L. E. and McCormick, R. A., 1968, Some results of multiple-tracer diffusion experiments at Cincinnati, *A.P.C.A. Journal*, **18**, 403.

Nieuwstadt, F. T. M., 1980a, Application of mixed-layer similarity to the observed dispersion from a ground-level source, *J. App. Met.*, **19**, 157–162.

Nieuwstadt, F. T. M., 1980b, An analytic solution of the time-dependent one-dimensional diffusion equation in the atmospheric boundary layer, *Atmos. Environ.*, **14**, 1361–1364.

Nieuwstadt, F. T. M. and van Ulden, A. P., 1978, A numerical study on the vertical dispersion of passive contaminants from a continuous source in the atmospheric surface layer, *Atmos. Environ.*, **12**, 2119–2124.

Nordlund, G., 1973, A particle-in-cell method for calculating long range transport of air pollutants, *Tech. Rep. No. 7*, Finnish Meteor. Instit.

Nordø, J., 1974, Quantitative estimates of long range transport of sulphur pollutants in Europe, *Ann. Met.*, **9**, 71.

Obukhov, A. M., 1941, Energy distribution in the spectrum of turbulent flow, *Izv. Akad. Nauk, Geogr. i Geofiz*, **5**, 453.

Obukhov, A. M. and Yaglom, A. M., 1959, On the micro-structure of atmospheric turbulence – a review of recent work in the U.S.S.R., *Quart. J. R. Met. Soc.*, **85**, 81.

Oden, S. 1968, The acidification of air and precipitation and its consequences on the natural environment, Swedish Nat. Sci. Res. Council, Ecology Committee, *Bul. 1:68* (in Swedish). Also: *Tr-1172*, Translation Consultants Ltd., Arlington, Va., U.S.A. (1968).

OECD, 1977, *The OECD Programme on Long-range Transport of Air Pollutants*, Paris.

Ogura, Y., 1952, The theory of turbulent diffusion in the atmosphere, *J. Met. Soc. Japan*, **30**, 23.

—— 1959, Diffusion from a continuous source in relation to a finite observation interval, *Atmospheric Diffusion and Air Pollution*, ed. F. N. Frenkiel and P. A. Sheppard, *Advances in Geophysics*, **6**, 149, Academic Press.

Oke, T. R., 1978, *Boundary Layer Climates,* Methuen, London.

Overcamp, T. J., 1976, A general Gaussian diffusion-deposition model for elevated point sources, *J Appl. Met.,* **15,** 1167–1171.

Owen, P. R. and Thompson, W. R. 1963, Heat transfer across rough surfaces, *J. Fluid Mech.,* **15,** 321.

Pack, D. H., Ferber, G. J., Heffter, J. L., Telegadas, K., Angell, J. K., Hoeker, W. H. and Machta, L., 1978, Meteorology of long-range transport, *Atmos. Environ.,* **12,** 425.

Panofsky, H. A., 1962, Scale analysis of atmospheric turbulence at 2 m, *Quart. J. R. Met. Soc.,* **88,** 57.

—— 1969, Budgets of turbulent fluctuations, *Radio Sci.,* **4,** 1385–1387.

Panofsky, H. A., Cramer, H. E. and Rao, V. R. K., 1958, The relation between Eulerian time and space spectra, *Quart. J. R. Met. Soc.,* **84,** 270.

Panofsky, H. A. and McCormick, R. A., 1952, The vertical momentum flux at Brookhaven at 109 meters, *Geophys. Res. Paper No. 19,* **219,** G.R.D., Cambridge, Mass.

—— 1954, Properties of spectra of atmospheric turbulence at 100 metres, *Quart. J. R. Met. Soc.,* **80,** 546.

Panofsky, H. A. and Prasad, B., 1965, Similarity theories and diffusion, *Int. J. Air Water Pollution,* **9,** 419–430.

—— 1967, The effect of meteorological factors on air pollution in a narrow valley, *J.A.M.,* **6,** 493.

Panofsky, H. A., Tennekes, H., Lenschow, D. H. and Wyngaard, J. C., 1977, The characteristics of turbulent velocity components in the surface layer under convective conditions, *Boundary-layer Meteorol.,* **11,** 355–361.

Pasquill, F., 1956, Meteorological research at Porton, *Nature,* **177,** 1148.

—— 1961, The estimation of the dispersion of windborne material, *Met. Mag.,* **90,** 33.

—— 1962, Some observed properties of medium-scale diffusion in the atmosphere, *Quart. J. R. Met. Soc.,* **88,** 70.

—— 1963, The statistics of turbulence in the lower part of the atmosphere, *Atmospheric Turbulence and its Relation to Aircraft,* H.M.S.O.

—— 1963a, The determination of eddy diffusivity from measurements of turbulent energy, *Quart. J. R. Met. Soc.,* **89,** 95.

—— 1966, Lagrangian similarity and vertical diffusion from a source at ground level, *Quart. J. R. Met. Soc.,* **92,** 185.

—— 1967, The vertical component of atmospheric turbulence at heights up to 1200 metres, *Atmos. Env.,* **1,** 441–450.

—— 1968, Some outstanding issues in the theory and practice of estimating diffusion from sources, Sandia Laboratories, *Symposium on the Theory and Measurement of Atmospheric Turbulence and Diffusion in the Planetary Boundary Layer,* pp. 17–30.

—— 1969, The influence of the turning of wind with height on crosswind diffusion, *Phil Trans. Roy. Soc., Lond.*, A, **265**, 173.

—— 1970, Prediction of diffusion over an urban area – current practice and future prospects, *Proceedings of Symposium on Multiple Source Urban Diffusion Models,* U.S. Environmental Protection Agency.

—— 1971, Atmospheric dispersion of pollution, *Quart. J. R. Met. Soc.,* **97,** 369–395.

—— 1974, *Atmospheric Diffusion* (2nd edn.), Ellis Horwood Ltd., Chichester, p. 228.

—— 1975, Some topics relating to modelling of dispersion in the boundary layer, *U.S.E.P.A. Res. Rep. EPA-650/4-75-015.*

—— 1978, Dispersion from individual sources, *Air Quality Meteorology and Atmospheric Ozone,* ed. A. L. Morris and R. C. Barras, American Society for Testing and Materials, Publication 653, 235–261. (Additional details in *Atmospheric Dispersion Parameters in Plume Modelling,* U.S. Environmental Protection Agency, EPA 600/4-78-021.)

Pasquill, F. and Butler, H. E., 1964, A note on determining the scale of turbulence, *Quart. J. R. Met. Soc.,* **90,** 79.

Pearson, H. J., Puttock, J. S. and Hunt, J. C. R., 1983, A statistical model of fluid element motions and vertical diffusion in a homogeneous stratified turbulent flow, *J. Fluid Mech.,* in press.

Pedgley, D., 1982, *Windborne Pests and Diseases,* Ellis Horwood Ltd., Chichester, ISBN 0-85312-312-8, and Halsted Press, New York, ISBN 0-470-27516-2.

Pemberton, C. S., 1961, Scavenging action of rain on non-wettable particulate matter suspended in the atmosphere, *Int. J. Air Poll.,* **3.**

Perkins, W. A., Leighton, P. A., Grinnell, S. W. and Webster, F. X., 1952, A fluorescent atmospheric tracer technique for mesometeorological research, *Proc. 2nd National Air Pollution Symposium,* Pasadena, U.S.A.

Peterson, E. W., 1969, Modification of mean flow and turbulent energy by a change in surface roughness under conditions of neutral stability, *Quart. J. R. Met. Soc.,* **95,** 561–575.

Philip, J. R., 1967, Relation between Eulerian and Lagrangian statistics, *Boundary Layers and Turbulence,* Physics of Fluid Supplement, p. 69.

Phillips, P. and Panofsky, H. A., 1982, A re-examination of lateral dispersion from continuous sources, *Atmos. Environ.,* **16,** 1851–1860.

Pond, S., Stewart, R. W. and Burling, R. W., 1963, Turbulence spectra in the wind over waves, *J. Atmos. Sci.,* **20,** 319–324.

Prahm, L. P. and Christensen, O., 1977, Long-range transmission of pollutants simulated by the 2-dimensional pseudospectral dispersion model, *J. Appl. Meteorol.,* **16,** 896.

Prandtl, L., 1952, *The Essentials of Fluid Dynamics,* Blackie, London.

Pressman, A., 1981, see: *Report of the Expert Meeting on the Assessment of the Meteorological Aspects of the First Phase of EMEP,* W.M.O. EMEP.

Priestley, C. H. B., 1956, A working theory of the bent-over plume of hot gas, *Quart. J. R. Met. Soc.,* **82,** 165.

—— 1959, *Turbulent Transfer in the Lower Atmosphere,* University of Chicago Press.

Priestley, C. H. B. and Swinbank, W. C., 1947, Vertical transport of heat by turbulence in the atmosphere, *Proc. Roy. Soc., Lond.,* A, **189,** 543–561.

Pruitt, W. O., Morgan, D. L., and Lourence, F. J., 1973, Momentum and mass transfers in the surface boundary layer, *Quart. J. R. Met. Soc.,* **99,** 370–386.

Puttock, J. S. and Hunt, J. C. R., 1979, Turbulent diffusion from sources near obstacles with separated wakes, Pt. I: An eddy diffusivity model, *Atmos. Environ.,* **13,** 1.

Randerson, D., 1979, Review panel on sigma computations, *Bull. Am. Met. Soc.,* **60,** 682–683.

Rauch, H., 1964 Zür Schornstein – Uberhohung, *Beitr. Phys. Atmos.,* **37,** 132–158, trans. *U.S.A.E.C. Report ORNL-tr-1029,* Oak Ridge National Laboratory.

Rayment, R., 1970, Introduction to the fast fourier transform (FFT) in the production of spectra, *Met. Mag.,* **99,** 261–269.

—— 1973, An observational study of the vertical profiles of the high frequency fluctuations of the wind in the atmospheric boundary layer, *Boundary-Layer Meteorology,* **3,** 284–300.

Readings, C. J. and Rayment, R., 1969, The high-frequency fluctuation of the wind in the first kilometre of the atmosphere, *Radio Sci.,* **4,** 1127–1131.

Record, F. A. and Cramer, H. E., 1958, Preliminary analysis of Project Prairie Grass diffusion measurements, *J. Air Poll. Cont. Ass.,* **8,** 240.

Record, F. A., Bubenick, D. V. and Kindya, R. J., 1982, *Acid Rain Information Book,* Noyes Data Corp., New Jersey.

Reid, J. D., 1979, Markov chain simulations of vertical dispersion in the neutral surface layer for surface and elevated releases, *Boundary-layer Meteorol.,* **16,** 3–22.

Reiquam, H., 1970, An atmospheric transport and accumulation model for air-sheds, *Atmos. Environ.,* **4,** 233–247.

Richardson, L. F., 1920, The supply of energy from and to atmospheric eddies, *Proc. Roy. Soc.,* **A, 97,** 354–373.

—— 1926, Atmospheric diffusion shown on a distance-neighbour graph, *Proc. Roy. Soc.,* **A, 110,** 709.

Richardson, L. F. and Proctor, D., 1925, Diffusion over distances ranging from 3 km to 86 km, *Memoirs of the R. Met. Soc.,* Vol. 1, No. 1.

Roberts, O. F. T., 1923, The theoretical scattering of smoke in a turbulent atmosphere, *Proc. Roy. Soc.,* A, **104,** 640.

Robins, A. G. and Fackrel, J. E., 1979, Continuous plumes – their structure and prediction, *Mathematical Modelling of Turbulent Diffusion in the Environment,* ed. C. J. Harris, Academic Press.

Rodhe, H., 1972, A study of the sulfur budget for the atmosphere over Northern Europe, *Tellus,* **24,** 128.

—— 1974, Some aspects of the use of air trajectories for the computation of large-scale dispersion and fall-out patterns, *Advances in Geophysics,* Vol. 18B, 95.

Rodhe, H. and Grandell, J., 1972, On the removal time of aerosol particles from the atmosphere by precipitation scavenging, *Tellus,* **24,** 442.

Rounds, W., 1955, Solutions of the two-dimensional diffusion equations, *Trans. Amer. Geoph. Union,* **36,** 395.

Saffman, P. G., 1962, The effect of wind shear on horizontal spread from an instantaneous ground source, *Quart. J. R. Met. Soc.,* **88,** 382.

—— 1963, An approximate calculation of the Lagrangian autocorrelation coefficient for stationary homogeneous turbulence, *Applied Science Res.,* **11,** 245.

Sakagami, J., 1974, On the vertical concentration distribution of the atmospheric diffusion, *Natural Science Report of the Ochanomizu University, Tokyo,* **25,** 69–77.

Sawford, B. L., 1982, Comparison of some different approximations in the statistical theory of relative dispersion, *Quart. J. R. Met. Soc.,* **108,** 191–206.

Schmidt, W., 1925, Der Massenaustrausch in freier Luft und verwandte Erscheinungen, *Probleme der Kosmischen Physik,* Hamburg, Verlag von Henri Grand.

Scorer, R. S., 1958, *Natural Aerodynamics,* Pergamon Press, London.

—— 1959a, The behaviour of chimney plumes, *Int. Journ. Air Poll.,* **1,** 198.

—— 1959b, The rise of bent-over hot plumes, *Atmospheric Diffusion and Air Pollution,* ed. F. N. Frenkiel and P. A. Sheppard, *Advances in Geophysics,* **6,** 399, Academic Press.

—— 1968, *Air Pollution,* Pergamon Press.

—— 1978, *Environmental Aerodynamics,* Ellis Horwood Ltd., Chichester, ISBN 0-85312-094-3, and Halsted Press, New York, ISBN 0-470-99270-0.

Scrase, F. J., 1930, Some characteristics of eddy motion in the atmosphere, *Meteorological Office Geophysical Memoirs No. 52.*

Seaman, N. L. and Anthes, R. A., 1981, A mesoscale semi-implicit numerical model, *Quart. J. R. Met. Soc.,* **107,** 167.

Sheih, C. M., Tennekes, H. and Lumley, J. L., 1971, Airborne hot-wire measurements of the small-scale structure of atmospheric turbulence, *Physics of Fluids,* **14,** 201–215.

Sheppard, P. A., 1947, The aerodynamic drag of the earth's surface and the value of von Karman's constant in the lower atmosphere, *Proc. Roy. Soc.*, A, **188**, 208.

Sherman, C. A., 1978, A mass-consistent model for wind fields over complex terrain, *J. Appl. Met.*, **17**, 312–319.

Singer, I. A. and Smith, M. E., 1953, Relation of gustiness to other meteorological parameters, *J. Met.*, **10**, 121.

—— 1966, Atmospheric dispersion at Brookhaven National Laboratory, *Int. J. Air Water Poll.*, **10**, 125–135.

Sklarew, R. C., Fabrick, A. J. and Prager, J. E., 1971, A particle-in-cell method for numerical solution of the atmospheric diffusion equation and applications to air pollution problems. *Systems, Science & Software*, California, Report 3SR-844 Vol. 1, under U.S. Environ. Protection Agency contract no. 68-02-0006.

Slade, D. H., Editor, 1968, *Meteorology and Atomic Energy*, U.S. Atomic Energy Commission, Div. Tech. Inf.

Slawson, P. R., 1966, Observations of plume rise from a large industrial stack, *Research Report 1*, U.S.A.E.C. Report NYO-3685-7, University of Waterloo, Dept. of Mech. Eng.

Smith, D. B., 1968, Tracer study in an urban valley, *J. Air Poll. Cont. Ass.*, **18**, 600.

Smith, E. J. and Heffernan, K. J., 1956, The decay of the ice-nucleating properties of silver iodide released from a mountain top, *Quart. J. R. Met. Soc.*, **82**, 301.

Smith, F. B., 1957a, The diffusion of smoke from a continuous elevated point-source into a turbulent atmosphere, *J. Fluid Mech.*, **2**, 49.

—— 1957b, Convection-diffusion processes below a stable layer, a paper of the Meteorological Research Committee (London), *M.R.P. No. 1048*.

—— 1957c, Convection-diffusion processes below a stable layer – Part II, a paper of the Meteorological Research Committee (London), *M.R.P. No. 1073*.

—— 1959, The turbulent spread of a falling cluster, *Atmospheric Diffusion and Air Pollution*, ed. F. N. Frenkiel and P. A. Sheppard, *Advances in Geophysics*, **6**, 193–210, Academic Press.

—— 1961, An analysis of vertical wind-fluctuations at heights between 500 and 5000 feet, *Quart. J. R. Met. Soc.*, **87**, 180–193.

—— 1962, The problem of deposition in atmospheric diffusion of particulate matter, *J. Atmos. Sci.*, **19**, 429.

—— 1962a, The effect of sampling and averaging on the spectrum of turbulence, *Quart. J. R. Met. Soc.*, **88**, 177–180.

—— 1965, The role of wind shear in horizontal diffusion of ambient particles, *Quart. J. R. Met. Soc.*, **91**, 318–329.

—— 1968, Conditioned particle motion in a homogeneous turbulent field, *Atmos. Environ.*, **2**, 491–508.

—— 1973, A scheme for estimating the vertical dispersion of a plume from a source near ground-level, Met. Office, *Turbulence and Diffusion Note No. 40*.

—— 1973a, The basic equations for finite-velocity one-dimensional random-walk diffusion, Met. Office, *Turbulence and Diffusion Note No. 33*.

—— 1978, A comparison of analytic solutions and Lagrangian similarity solutions of the diffusion equation in the neutral surface layer of the atmosphere, *Meteorological Office Turbulence and Diffusion Note, T.D.N. No. 96*.

—— 1980, The influence of meteorological factors on radioactive dosages and depositions following an accidental release, *Proc. of CEC Seminar on Radioactive Releases and their Dispersion in the Atmosphere following a Hypothetical Reactor Accident,* Risø, Denmark, p. 22.

—— 1980, Probability prediction of the wet deposition of airborne pollution, Proc. 11th NATO/CCMS Conf. on Air Pollution Modelling and its Applications, Plenum Press, New Jersey.

—— 1981, The significance of wet and dry synoptic regions on long-range transport of pollution and its deposition, *Atmos. Environ.*, **15**, 863.

—— 1982b, The integral equation of diffusion, *Proc. 13th NATO/CCMS Conf. on Air Pollution Modelling and its Applications,* Plenum Press, New York.

—— 1982c, A review of the European EMEP Programme on the long-range transport of pollution, and some ideas on how to treat wet deposition, *Scientific Lecture to the 33rd Session of the WMO Executive Committee,* Geneva, June, 1981, WMO Report (in press).

—— 1983, Discussion on 'Horizontal diffusion on the Atmosphere: A Lagrangian-dynamical theory' by F. A. Gifford, *Atmos. Environ.*, **17**, 194–197.

Smith, F. B. and Abbott, P. F., 1961, Statistics of lateral gustiness at 16 metres above ground, *Quart. J. R. Met. Soc.*, **87**, 549.

Smith, F. B. and Blackall, R. M. 1979, The application of field experimental data to the parameterisation of the dispersion of plumes from ground level and elevated sources, *Mathematical Modelling of Turbulent Diffusion in the Environment,* ed. C. J. Harris, Academic Press, 201–236.

Smith, F. B. and Hay, J. S., 1961, The expansion of clusters of particles in the atmosphere, *Quart. J. R. Met. Soc.*, **87**, 82.

Smith, F. B. and Hunt, R. D. 1978, Meteorological aspects of the transport of pollution over long distances, *Atmos. Environ.*, **12**, 461.

Smith, F. B. and Hunt, R. D., 1979, The dispersion of sulphur pollutants over western Europe, *Phil. Trans. R. Soc. Lond. A.*, **290**, *523*.

Smith, F. D. and Jeffrey, G. H., 1971, The prediction of high concentrations of sulphur dioxide in London and Manchester air, Meteorological Office, *Met. O. 14 Turbulence and Diffusion Note No. 19*.

—— 1975, Airborne transport of sulphur dioxide from the United Kingdom, *Atmos. Environ.*, **9**, 643.

Smith, M. E., 1951, The forecasting of micrometeorological variables, *Met. Monogr.*, **1**, 50, Am. Met. Soc.

Smith, M. E. (ed.), 1968, *Recommended Guide for the Prediction of the Dispersion of Airborne Effluents,* Am. Soc. of Mech. Engineers, New York.

Smith, M. E. and Singer, I. A., 1966, An improved method of estimating concentrations and related phenomena from a point source emission, *J. Appl. Met.*, **5**, 631–639.

Smith, R. A., 1872, *Air and Rain, the Beginnings of a Chemical Climatology,* Longmans, Green, London.

Snyder, W. H. and Lumley, J. L., 1971, Some measurements of particle velocity autocorrelation functions in turbulent flow. *J. Fluid Mech.*, **48**, 41–71.

Start, G. E., Ricks, N. R. and Dickson, C. R., 1976, Effluent dilutions over mountainous terrain, *Third Symposium on Atmospheric Turbulence, Diffusion and Air Quality,* Am. Met. Soc.

Stümke, H., 1963, Suggestions for an empirical formula for chimney elevation, *Staub,* **23**, 549–556.

Stewart, N. G., Gale, H. J. and Crooks, R. N., 1954 and 1958, The atmospheric diffusion of gases discharged from the chimney of the Harwell Pile (Bepo). *A.E.R.E. HP/R 1452,* H.M.S.O. (in shortened version *Int. J. Air. Poll.*, **1**, 87, 1958).

Stewart, R. E. and Csanady, G. T., 1967, Deposition of heavy particles from a continuous elevated source, *Proc. 1st Canadian Conference on Micrometeorology,* Part II, 395.

Stewart, R. W., 1969, Turbulence and waves in a stratified atmosphere, *Radio Science,* **4**, 1269–1278.

Stratmann, H., 1956, Investigation of sulphur dioxide emission from a bituminous coal-fired power station with very high chimneys. *Mitt. Var. Grosskessel-besitzer,* **40**, 49.

Summers, P. W., 1967, An urban heat island model, its role in air pollution problems with applications to Montreal (abstract only in *Proc. 1st Canadian Conference on Micrometeorology*).

Sutton, O. G., 1932, A theory of eddy diffusion in the atmosphere, *Proc. Roy. Soc.*, A, **135**, 143.

—— 1934, Wind structure and evaporation in a turbulent atmosphere, *Proc. Roy. Soc.*, A, **146**, 701.

—— 1947a, The problem of diffusion in the lower atmosphere, *Quart. J. R. Met. Soc.*, **73**, 257.

—— 1947b, The theoretical distribution of airborne pollution from factory chimneys, *Quart. J. R. Met. Soc.*, **73**, 426.

—— 1953, *Micrometeorology,* McGraw-Hill, New York.

Sutton, W. G. L., 1943, On the equation of diffusion in a turbulent medium, *Proc. Roy. Soc.*, A, **182**, 48.

Swinbank, W. C. and Dyer, A. J., 1967, An experimental study in micrometeorology, *Quart. J. R. Met. Soc.*, **93**, 494–500.

Sykes, R. I. and Hatton, L., 1976, Computation of horizontal trajectories based on the surface geostrophic wind, *Atmos. Environ.*, **10**, 925.

Tank, W. G., 1957, The use of large-scale parameters in small-scale diffusion studies, *Bull. Am. Met. Soc.*, **38**, 6.

Tapp, M. C. and White, P. W., 1976, A non-hydrostatic mesoscale model, *Quart. J. R. Met. Soc.*, **102**, 277.

Taylor, G. I., 1915, Eddy motion in the atmosphere, *Phil. Trans. Roy. Soc.*, A, **215**, 1.

—— 1921, Diffusion by continuous movements, *Proc. London Math. Soc.*, Ser. 2, **20**, 196.

—— 1927, Turbulence, *Quart. J. R. Met. Soc.*, **53**, 201.

—— 1935, Statistical theory of turbulence, Pts 1–4, *Proc. Roy. Soc.*, A, **151**, 421.

—— 1938, The spectrum of turbulence, *Proc. Roy. Soc.*, A, **164**, 476.

—— 1953, Dispersion of soluble matter in solvent flowing slowly through a tube, *Proc. Roy. Soc., Lond.*, **A, 219**, 186.

—— 1959, The Present Position in the Theory of Turbulent Diffusion, *Atmospheric Diffusion and Air Pollution,* ed. F. N. Frenkiel and P. A. Sheppard, *Advances in Geophysics,* **6**, 101, Academic Press.

Taylor, R. J., 1955, Some observations of wind velocity autocorrelations in the lowest layers of the atmosphere, *Aust. J. Phys.*, **8**, 535.

Tennekes, H., 1970, Free convection in the turbulent Ekman layer of the atmosphere, *J. Atmos. Sc.*, **27**, 1027–1034.

—— 1973, The logarithmic wind profile, *J. Atmos. Sci.*, **30**, 234–238.

—— 1979, The exponential Lagrangian correlation function and turbulent diffusion in the inertial subrange, *Atmos. Environ.*, **13**, 11, 1565–1568.

—— 1982, Similarity relations, scaling laws and spectral dynamics, *Atmospheric Turbulence and Air Pollution Modelling,* ed. F. T. M. Nieuwstadt and H. van Dop, D. Reidel Publishing Co., 37–68.

Thomas, D. M. C., 1964, Comment on 'Turbulent diffusion of heavy particles in the atmosphere', *J. App. Sci.*, **21**, 322.

Thomas, F. W., Carpenter, S. B. and Colbraugh, W. C., 1969, Plume rise estimates for electricity generating stations, *Phil. Trans. Roy. Soc., Lond.*, A, **265**, 221.

Thomas, M. D., Hill, G. R. and Abersold, J. N., 1949, Dispersion of gases from tall stacks, *Ind. and En. Chemistry,* **41**, 2409.

Thomas, P. W., Hubschmann, W., Konig, L. A., Schuttelkopf, H., Vogt, S. and Winter, M., 1976a, Experimental determination of the atmospheric dispersion parameters over rough terrain, Part 1, Measurements at the Karlsruhe

Nuclear Research Centre, KFK2285, Central Division for Nuclear Research, M.B.H. Karlsruhe, Federal Republic of Germany.

Thomas, P. W. and Nestor, K., 1976b, Experimental determination of the atmospheric dispersion parameters over rough terrain, Part 2, Evaluation of measurements, KFK 2286, Central Division of Nuclear Research, M.B.H. Karlsruhe, Federal Republic of Germany.

Thompson, N., 1965, Short-range vertical diffusion in stable conditions, *Quart. J. R. Met. Soc.*, **91**, 175.

—— 1966, The estimation of vertical diffusion over medium distances of travel, *Quart. J. R. Met. Soc.*, **92**, 270–276.

Thompson, R., 1971, Numeric calculation of turbulent diffusion, *Quart. J. R. Met. Soc.*, **97**, 93–98.

Thomson, D. J. and Ley, A. J., 1982, A random walk dispersion model, applicable to diabatic conditions, unpublished Meteorological Office Note TDN 138.

Thorpe, A. J., 1981, Thunderstorm dynamics – a challenge to the physicist, *Weather*, **36**, 108.

Tukey, J. W., 1950, The sampling theory of power spectrum estimates. *Symposium on Application of Autocorrelation Analysis to Physical Problems,* Woods Hole, Mass., Office of Naval Research, Washington, D.C.

Turner, D. B., 1964, A diffusion model for an urban area, *J. App. Met.*, **3**, 83.

Turner, D. B., 1970, Workbook of atmospheric dispersion estimates, *Office of Air Programs Pub. No. AP-26*, Environmental Protection Agency, U.S.A.

Tyldesley, J. B., 1967, Contribution to discussion on short-range vertical diffusion in stable conditions, *Quart. J. R. Met. Soc.*, **93**, 383–384.

Tyldesley, J. B. and Wallington, C. E., 1965, The effect of wind shear and vertical diffusion on horizontal dispersion, *Quart. J. R. Met. Soc.*, **91**, 158–174.

Van der Hoven, I., 1957, Power spectrum of horizontal wind speed in the frequency range from 0.0007 to 900 cycles per hour, *J. Met.*, **14**, 160.

van Ulden, A. P., 1978, Simple estimates for vertical diffusion from sources near the ground, *Atmos. Environ.*, **12**, 2125–2129.

Venkatram, A., 1980, A scheme to incorporate scavenging processes in statistical long-range transport models. Unpublished note.

Venkatram, A., 1980a, Dispersion from an elevated source in the convective boundary layer, *Atmos. Environ.*, **14**, 1–10.

Venkatram, A. and Vet, R., 1981, Modelling of dispersion from tall stacks, *Atmos Environ.*, **15**, 1531–1538.

Vinnichenko, N. K., Pinus, N. Z. and Shur, G. N. 1965, Some results of the experimental turbulence investigations in the troposphere, *International colloquium on fine-scale structure of the atmosphere, Moscow*.

Vogt, K. J., 1977a, A new system of release height dependent diffusion parameters for the Gaussian plume model, *Fourth International Clean Air Congress, 1977, Tokyo, Japan*.

—— 1977b, Empirical Investigations of the diffusion of waste air plumes in the atmosphere, *Nucl. Tech.*, **34**, 43–57.

Vogt, K. J., Geiss, H., Nordsieck, H., Polster, G. and Roohloff, F., 1973, Investigations on the propagation of contaminated air trails in the atmosphere, *Jul-998-ST,* Julich Nuclear Research Institute, Julich, Federal Republic of Germany.

Vogt, K. J., Geiss, H., and Polster, G., 1978, New sets of diffusion parameters resulting from tracer experiments with 50 and 100 m release height, *Proc. Ninth Internat. Tech. Meeting on Air Pollution Modelling and its Applications, No. 103, NATO/CCMS,* 221–239.

Walker, E. R., 1965, A particulate diffusion experiment, *J.A.M.*, **4**, 614.

Wamser, H. and Muller, 1977, On the spectral scale of wind fluctuations within and above the surface layer, *Quart. J. R. Met. Soc.*, **103**, 721–730.

Webb, E. K., 1955, Autocorrelations and energy spectra of atmospheric turbulence, *Tech. Paper No. 5,* Melbourne, C.S.I.R.O., Div. Met. Phys.

Webb, E. K., 1970, Profile relationships: the log-linear range and extension to strong stability, *Quart. J. R. Met. Soc.*, **96**, 67–90.

Wieringa, J., 1980, A revaluation of the Kansas mast influence on measurements of stress and cup anemometer overspeeding, *Boundary-layer Meteorol.*, **18**, 411–430.

Wilkins, E. M., 1958, Observations on the separation of pairs of neutral balloons and applications to atmospheric diffusion theory, *J. Met.*, **15**, 324.

Wilkins, E. T., 1954, Air pollution aspects of the London fog of December 1952, *Quart. J. R. Met. Soc.*, **80**, 267.

Willis, G. E. and Deardorff, J. W., 1976, A laboratory model of diffusion into the convective planetary boundary layer, *Quart. J. R. Met. Soc.*, **102**, 427–446.

Willis, G. E. and Deardorff, J. E., 1978, A laboratory study of dispersion from an elevated source within a modelled convective planetary boundary layer, *Atmos. Environ.*, **12**, 1305–1311.

Wilson, J. D., Thurtell, G. D. and Kidd, G. E., 1981, Numerical simulation of particle trajectories in inhomogeneous turbulence, II: Systems with variable turbulent velocity scale. *Boundary-layer meteorol.*, **21**, 423.

World Meteorological Organization, 1958, Turbulent diffusion in the atmosphere, *WMO-No. 77 TP,* 31.

World Meteorological Organization Report, 1981, Environmental Pollution Monitoring Programme; Report of the Expert Meeting on the Assessment of the Meteorological Aspects of the First Phase of EMEP, held at Shinfield Park, U.K. W.M.O., Geneva.

Wrigley, J., ApSimon, H. M. and Goddard, A. J. H., 1979, Meteorological data and the MESOS model for the long-range transport of atmospheric pollutants, *WMO Symposium Proceedings, Sofia, Oct. 1979, WMO Report No. 538.*

Wyngaard, J. C. and Coté, O. R., 1971, The budgets of turbulent kinetic energy and temperature variance in the atmospheric surface layer, *J. Atmos. Sci.*, **28**, 190–201.

Wyngaard, J. C., Coté, O. R. and Izumi, Y., 1971, Local free convection, similarity and the budgets of shear stress and heat flux, *J. Atmos. Sci.*, **28**, 1171–1182.

Yaglom, A. M., 1972, Turbulent diffusion in the surface layer of the atmosphere, *Isv. Atmos. Oceanic Phys.*, **8**, 333–340.

—— 1977, Comments on wind and temperatures flux-profile relationships, *Boundary-layer Meteorol.*, **11**, 89–102.

Yudine, M. I., 1959, Physical considerations on heavy-particle diffusion. *Atmospheric Diffusion and Air Pollution*, ed. F. N. Frenkiel and P. A. Sheppard, *Advances in Geophysics*, **6**, 185, Academic Press.

Zeman, P. and Lumley, J. L., 1976, Turbulence and diffusion modelling in buoyancy driven mixed layers, *Third Symposium on Atmospheric Turbulence, Diffusion and Air Quality*, Am. Met. Soc., 38–45.

Zilitinkevich, S. S., Laikhtman, D. L. and Monin, A. S. 1967, Dynamics of the atmospheric boundary layer, Isvestiya, Academy of Sciences, U.S.S.R., *Atmospheric and Oceanic Physics*, **3**, 297–333.

Index

acceleration of fluid particle, 138
of growth of cluster, 153
acid rain, 378f
advecting wind in boundary layer, 362, 383
aerodynamic resistence, 256
Ailsa Craig Experiment, 298, 370
air motion, horizontal equations of, 39
air pollution:
distribution in urban areas, 287f
effects of topography on, 293
forecasting levels of, 303, 353
from individual stacks, 270f
long range transport of, 378f
prediction from meteorological data, 303, 351
prediction from historical data on pollution, 353
uncertainties in prediction of, 357
aliasing, 34
aluminium leached from soil, 380
animal-virus dispersion model, 372
area source solutions, 343f
Austausch coefficient, 89
autocorrelation, autocovariance, 21
averaging time: 24
effect on measurements of turbulence, 24,77
significance in treatments of dispersion, 120

ballistic model for heavy particle diffusion, 249
balloons:
neutral, 84, 226, 293, 374
tethered, use in measuring atmospheric turbulence, 67f
use in dispersion studies on large scale, 361
band pass filter, for growth of clusters, 156
bias velocity, 139
bidirectional vane, 14, 190
blocking highs, 399
boundary layer:
backing of winds, 36
breakdown processes, 35
depth and its evolution, 36, 48
general features of, 13, 35
in convective cinditions, 37, 49
in stable conditions, 38 ,48
longitudinal rolls in, 37
quasi-neutral conditions, 37
similarity theory of surface stress layer, 41f
treatment of layer as a whole, 46
box model of dispersion, 346f
Bremen Survey, 302
Brent Knoll Experiment, 298
Brunt-Vaisala frequency, 39
buildings, effect on dispersion, 283f
buoyancy, effect on:
eddy diffusivity, 91
rise of effluent plumes, 234f, 275
turbulent kinetic energy, 49, 91
buoyancy forces, 37, 49, 234

canyons, 295
capping inversion, 35
cavity region, 284
cellular motions, 37
chemical transformation of SO_2, 379
chimney plumes:
characteristic forms of, 312
dispersion of, 265
effect of turbulence on rise, 275
effect of wind speed on rise, 372
formula for rise of, 335
growth induced by buoyant rise, 265f, 334

chimney plumes – continued
observed rise of, 237
observed rise compared with theory, 239f
observed rise in stratified air, 242f
penetration of inversions, 244
theory of rise of, 233
clusters, 149f
collection efficiency, 257f
collision efficiency, 257, 259
concentration of windborne material:
calculation of, from routine meteorological data, 335f
calculation of, in case of limited vertical spread, 328
crosswind integrated data tested against theory, 209f
depletion by deposition, 260
frequency distribution of, in urban pollution, 302
from experimental sources, 180
from individual chimneys, 270, 280
instantaneous and average aspects, 278f, 311, 320
long-term value of, 280f
maximum from elevated source, 268, 272
variations in, 278, 300
complex terrain, 296
concentration profile in surface layer, 148
conditioned release, 121f, 144
coning, 311
conjugate power law, 91, 97
constant level balloons, 361
continuous line source, 108
convective boundary layers, 50
convective diffusion, 102, 112
convective velocity scale, 72, 112, 276, 314
conveyor-belt flow, 372
core-bulk structure, 154
Coriolis force, 36, 39
correlation of eddy velocities:
auto, 22
correlograms, 116, 124, 127
longitudinal, 57
space, 21
transverse, 57
covariance, 21
counter-gradient flux, 102
crosswind-spread:
definition of, 185
effect of sampling time, 77
estimates of, in relation to stability, 309f
from wind shear, 169f, 376
of smoke plumes and puffs, 375
of tracer material, 190f
on large scale, 229, 375
crosswind-spread – continued
practical formulae for, 322, 324f
relation between overall width and standard deviation, 188
relation with turbulence, 193, 198f, 199
test of Hay-Pasquill adaption of Taylor's theory, 200
test of mixed layer similarity, 203
variation with distance or time, 195f, 375
when vertical diffusion extends above surface layer, 330
cumulus clouds, organised, 366

degrees of freedom in spectral estimates, 33
deposition, effect of concentration on, 260f
deposition of:
gases, 254
heavy particles, 246f
iodine-131, 377
radioactive material, 377
small particles, 251f
spores, 251
deposition velocity, 158, 252, 384
depths of boundary layer, 36
diffusion equation, 94f
diffusion:
coefficient of (*see* eddy diffusivity)
data on (*see* concentration, crosswind spread, vertical spread)
effect of inertia of particles on, 164
effect of intensity of turbulence on, 198f, 248, 272, 286, 291, 314
effect of scale of turbulence on, 315
effect of topography on, 293f, 369f
effect of wind shear on, 169f, 376
Fickian, 89
hyperbolic equation of, 100
methods of observing, 179f, 220
of cluster or puff, 15, 149f
of effluent from chimneys, 265
of falling particles, 246f, 249
of radioactive material, 377
of spores, 251
parabolic equation of, 94f
processes, 15
random walk, 133f
simple processes of, 89
statistical theory of, 115f
urban effects on, 287, 298f
diffusion formulae:
for clusters, 156, 230
for elevated source, 268
for long range, 331
for short range, 331

diffusion formulae – continued
general functional forms of, 321
limitations of, 357
practical systems, 335f
recommended for short range, 319f
direction of travel of windborne material, 362, 383
dispersion (*see* diffusion)
dissipation of turbulent energy, 52, 60, 65, 315
profile of, in boundary layer, 66
relation with shearing stress and heat flux, 65
distance neighbour function, 149
dosage, 185
downdraughts, 284
downslope flow, 368
downwash, 285
drag, aerodynamic, 39, 256
drag coefficient, 40
drainage winds, 293
dry adiabatic lapse rate, 42, 310
dry deposition, 379

ecological damage, 378f
eddy diffusivity:
concept of, 41, 88f, 315
effect of thermal stratification on, 45
explicit forms for vertical transfer, 92
for momentum transfer in neutral sheared flow, 42
in relation to dissipation of turbulent energy, 66
in relation to wind profile, 42, 46, 92
in relation to turbulence intensity and scale, 92, 327
limiting forms of, on similarity theory, 66
magnitude above surface layer, 47
eddy energy, creation and dissipation of, 49
eddy flux, 89
eddy viscosity (*see* eddy diffusivity)
effective source height of hot plumes, 272
effluent (*see* chimney plumes)
efflux velocity, 284
Ekman layer, 13
Ekman instabilities, 37
elevated source:
formulae for concentration from, 333
maximum concentration from, 270f
observations of dispersion from, 215f, 270f
EMEP study of long-range transport, 378
emission-deposition budgets, 383, 394
emissions of sulphur dioxide, 380
energy dissipation rate, 53, 62, 65, 66
energy spectrum, 22f
episodes, 396
episode-days, 396
episodicity, 396
episodic depositions, 368, 379
Eulerian reference system, 20f
evaporation, 98
effect on buoyancy, 38
evolving boundary layer, Carsons's formula, 49
expanding clusters, puffs, instantaneous releases, 149f

falling particle diffusion, 158f
fanning, 311
fast Fourier transforms, 34f
Fickian diffusion, 89
filtering, 30
fish-kill, 378
fluctuations:
of concentration, 165f, 278, 311
of the wind, 67f
fluctuating plume model, 167
fluorescent particle tracers, 182, 374
flux Richardson number, 50, 65
foot and mouth disease dispersion model, 372
Fourier transforms for correlogram and spectrum, 22, 33
frequency, 22f
friction velocity, 41, 314
nomograms for, 316, 318
Friedman-Keller equation, 90
Froude number, 285, 297, 370
frozen eddy hypothesis, 21, 23, 56, 81
fumigation, 312

Gaussian (normal or Fickian) distribution, 110
plume model, 320
geostrophic-drag coefficient, 40
departure derivation of K(z), 47
trajectories, 361
wind, 40
Gifford's equation for conditioned releases, 122
gradient transfer theory, 88f
tests of, 208f
gravity waves, 39
grid-cell systems, 349
ground-level sources, 96
gustiness (*see* fluctuations, intensity of turbulence)

harmonic analysis, 30
heat flux, 49f, 314
 in representation of stability, 41, 103, 314
 nomograms for, 316, 318
Heaviside operational methods, 99, 159
heavy particles, 246f
 ballistic model for, 249
hills, effect on plumes, 296, 369f
homogeneous turbulence, 29
horizontal mesoscale eddies, 371
hot plumes (*see also* chimney plumes), 233f
 effect of sampling, 279
 effect of wind speed, 272
 induced growth, 265
 long term concentrations, 280f
 surface concentrations, 268f
 peak to average ratio, 278
hyperbolic equation of diffusion, 100

image sources, 268
industrial stacks, distribution of effluent, 264f
inertia of particles, 160, 164
inertial sub-range of eddies, 60f
instantaneous releases at ground level, 107
integral equation of diffusion:
 applied to continuous sources, 141f
 applied to clusters, 157
integral length and time scales, 22
 (*see also* time scales)
intensity of turbulence:
 data for atmosphere, 67f
 effect of sampling duration on, 76
 horizontal components of, 72
 practical significance in relation to dispersion, 314
 relation with scale and dissipation, 62f
 variation with height, 71
 vertical component of, 70
 (*see also* crosswind spread, diffusion, eddy diffusivity, vertical spread)
intermittant turbulence, 38
internal boundary layers, 287
inversion of temperature:
 at elevation, 35
 at surface, 310
 at top of boundary layer, 35
 penetration by plumes, 236, 243
ion-tracer technique, 154
isotropic turbulence, theory of locally, 59
isotropy, 57

JASIN-Experiment data, 70

Kansas-Experiment data, 67f
Kármán's constant, 42, 43
Kolmogorov constant, 63
 scale, 61
kurtosis, 143

Lagrangian:
 system, 21f, 80f
 spectrum, 83
 effect of spectral form on diffusion, 117, 124
 Lagrangian-Eulerian scale ratio, 127
 Hay and Pasquill's form of similarity with Eulerian form, 127
 indirect estimates from variation of plume width with distance, 201f
 observations of spectra, 84f
 time-scale, 116, 161
 theory of relation with Eulerian spectra, 81f
land breezes, 367
Langevin equation, 122, 138
latent heat flux, 38
lateral (*see* crosswind):
 correlation, 57
 spread, 290
 spread in urban areas, 290
lee eddies, 37
lee vortex behind hills, 370
Leicester study, 299
lengthscale of turbulence, 67f
leptokurticity, 139, 143
lidar, use with plumes, 265
line source:
 equation of diffusion, 96
 formulae for distribution from, 96, 321
 similarity treatment, 108f
 (*see also* vertical spread)
linear growth stage of clusters, 156
local isotropy, 59
lofting, 312
log-normal distribution, 303
logarithmic wind profile, 42, 48, 90
London Study by Lucas, 303
London 1952 smog, 307, 399
long range transport of pollution, 378f
long range transport models:
 complex, 382
 general, 380f
 parametric, 379, 382
 stochastic, 389
longitudinal correlation, 57
looping of plumes, 242, 311
loss processes from boundary layer, 381

Markov equation, 133
Markovian correlation, 133f

mean deviation:
of particles, 105, 186
in relation to standard deviation, 186
mesoscale models, 372
mesoscale motions, 360
Minnesota Experiment data, 67f
mixed Eulerian-Lagrangian correlations, 154
mixing layer and depth thereof, 13, 49, 317
nomogram for, 319
formula for, 317
mixing length, 42, 89
molecular diffusion, 154
moments, method of, 170
Monin-Obukhov functions, 41, 43
analytical forms of, 45f
Monin-Obukhov length, 41, 103, 314
monitoring surveys, 18
Mt. Isa smelter plume data, 375

non-Markovian velocity regression, 137
normal distribution, 110, 320
numerical modelling of concentration from area sources, 349f

OECD Experiment on long-range transport of pollution, 378
oxidation rates of sulphur dioxide, 379

particles, 158f, 246f, 249
particle-in-cell models, 387
peak concentrations, 278
periodogram, 30
photochemical oxidant smogs, 399
platykurtosis, 143
plume of windborne material:
fluctuations of, 165f, 278, 311
general properties of, 311
instantaneous properties of, 220
plume rise:
general, 234f
observations on, 237f
in stratified air, 242f
effect of wind speed on, 272
effect of turbulence on, 275
point source:
influence of finite duration of release or of sampling, 127, 278f, 311, 320
equation of diffusion, 95
similarity treatment, 108f
(*see also* crosswind spread)
pollution (*see* air pollution)
potential flow around hills, 296f
potential temperature, 42, 310
prewhitening of time series, 34
probability distribution:
of eddy velocities, 76, 141, 143
of pollutant concentrations, 157, 303
of tracer concentrations, 186f
pseudo-diffusion velocity, 383
pseudo-spectral model, 386
puffs, 149f

radioactive material, windborne travel of:
iodine-131, 377
radio-xenon, 187
Windscale accident, 377
rainout of industrial pollutants, 381, 384
rainout of atomic debris, 259
random walk modelling, 133f
effect of turbulent-energy gradient on, 134, 141
Reading Survey, 301, 304
reciprocal theorem, 98
recirculation zone, 284, 296, 369
regression, linear and multiple, 136
relative diffusion of particles as in a cluster, 151f, 220f
from observations of smoke puffs, 221f
from observations of balloons, 226
from tracer sampling, 226f
testing of observations against F. B. Smith's theory, 230f
Reynold's analogy, 93
Reynold's number, 285
Reynold's stress, 90
Richardson's law for effect of scale on diffusion, 149f
Richardson number, 38, 50f, 300, 314
rolls within boundary layer, 37, 76
Rossby number similarity, 316
roughness (surface roughness):
effect on diffusion, 338
effect on turbulence, 316
length, 42, 106, 287, 314, 317

sampling of windborne material, techniques, 181f
sampling duration, 24, 77
effect on diffusion and concentration, 278f
effect on spectrum and variance of turbulence, 24f, 76f
satellite pictures, 366
scale of turbulence, 22, 83, 85, 91, 116, 122, 126f, 315, 361
in relation to low frequency spectral density, 23
relation between Lagrangian and Eulerian, 81f
variation with height, 68

sea breeze circulations, 366
second-order closure techniques, 102f
sedimentation, 158f, 246
sensible heat flux, 38, 49, 314
separation of flow near hills, 284, 296, 369
shearing stress, 39, 314
relation with turbulent components, 70f
relation with wind profile, 41f, 90
short releases of hazardous material, 393
similarity theory, Monin-Obukhov, 40
Kolmogorov, 59
Rossby, 316
second-order closure, 102f
similarity treatments:
of diffusion, 103f, 139
of intensity and spectra of turbulence, 70f
of vertical profiles of wind, etc, 41
of thermal stratification effect on diffusion, 110f
tested against observations of vertical diffusion from surface releases, 208f
skewness of vertical velocity distribution, 76
smog, 307
smoke:
opacity theory, 221
visual observation of diffusion of plumes, 180
of puffs, 180, 221
soap bubble studies of diffusion, 84
source depletion models, 260
spectral density, definition, 23
in inertial sub-range, 63, 65
spectral scale represented by wavelength at peak, 54, 67f, 80, 315
spectral gap, 80
spectral smoothing, 33
spectral window, in relation to turbulence, 30
spectrum of turbulence, 23
analytical derivation of, 29f
application of cosine-transform method in practice, 33
data for horizontal components, 74f
data for vertical component, 74f
dependence on sampling duration and averaging time, 24f
empirical forms of, 73f
Eulerian, 124
gap in, 80
inertial subrange form, 61f
Lagrangian, 117, 124
large-scale form, 360
longitudinal and transverse forms, 57
one-dimensional form, 57
spectrum of turbulence – continued
relation between Lagrangian and Eulerian forms, 83
relation between space and time forms, 55
significance of one-dimensional nature of observations, 24
three-dimensional form, 58
speed of travel of windborne material, 362, 383
speed-up factors of wind over hills, 369
spores, deposition of, 251
stack parameter, 235
Stanton number, 254
stability, 49f, 311
stability categories, 336, 340
standard deviation of wind fluctuations:
in the vertical, 21, 70f
in the horizontal, 71f
stationarity, 22, 29
stationary quality, lack of in atmospheric turbulence, 30
statistical description of turbulence, 20f
statistical theory of dispersion, 115f
adaption for release of finite duration, 127
including effect of wind shear on horizontal dispersion, 169f
in homogeneous turbulence, 115f
in stable conditions, 132
in surface layer, 130f
Stokes law, 165
storms, effect on boundary layer, 368
effect on depositions, 397
stratosphere, 363
stress (*see* shearing stress)
structure function, 21, 61
super-geostrophic jets, 39
suppression of turbulence, 51
surface absorption, 246
surface depletion models, 263
surface impaction, 246
surface layer diffusion, 103f, 145
surface resistance for perfect sink, 256
surface Rossby number, 47
surface roughness (*see* roughness), 106, 287, 314
surface stress layer, 40, 105
sulphur dioxide:
budget over large regions or between countries, 393
deposition of, 256, 384
from industrial stacks, 268f
within cities, 298f

Taylor's equation for any stochastic variable, 27

Taylor's equation – continued
for diffusion, 115
Taylor's hypothesis for relation between time and space correlation, 21f, 55
terminal velocity of particles, 158
effect on deposition, 249, 253
effect on diffusion, 158f, 163f
effect on washout by rain, 256
tilted plume model, 249
tetroons, 84, 361
thermals, 38
thermals, effect on diffusion, 99
thermal stratification of the atmosphere, effect on:
crosswind spread, 192f, 203, 309f, 330
eddy diffusivity, 41f
turbulence, 70f
wind direction trace, 313
wind profile, 43f
vertical spread, 110, 204f, 267f, 311, 326f
tilted plume model for heavy particles, 249
time of travel, effective weighting function to spectrum, 120
time-scale of spectra, 85, 116, 122, 361
ratio of Lagrangian to Eulerian, 83
topographical influences, 293f, 369f
topographically induced rainfall, 380
touch-down distance of plumes, 276
trace correlogram, 156
tracer techniques, 181
trajectories:
deduced from geostrophic winds, 181
long range, 361
of balloons, 181
of radioactive material, 377
transverse correlation, 57
turbulence:
early history of measurement, 14
(*see* intensity of, spectrum of)
turbulent energy equation, 53
generation, 49, 91
Turner's stability scheme, 304

urban circulations, 369
urban diffusion, 287, 298f
urban heat fluxes, 292
urban heat island, 291, 298
urban heat plume, 292
urban pollution, 298f
urban surface roughness, 289

valley dispersion, 294
valley, trajectories within, 293
variance of velocity fluctuations, 21, 71f
vertical component of turbulence, 21, 70f
vertical spread from surface releases:
at medium range, 208
at short range, 207
effect of intensity of turbulence, 208
in relation to theory, 208f
performance of practical formulae over extended range, 331f
shape of distribution, 205f
theoretical treatments of, 88f (Chapter 3)
working estimates for use in diffusion and pollution calculations, 335f
von Kármán's constant, 42, 43
vorticity within boundary layer, 37

wakes behind buildings, 284
Wangara Experiment, 360
washout by rain:
of particles, 256f
of soluble gases and vapour, 260, 381, 384
washout coefficient, 257
wavenumber, 23
weighting function applied to spectra:
for sampling duration and averaging, 26
for cluster size, 155
wet and dry periods and regions, 385
wet deposition, 381, 384
windborne infection, 372
wind fluctuations (*see* intensity of turbulence)
wind shear:
effect on horizontal diffusion, 169f, 376
effect of thermal stratification on, 41
role in maintenance of turbulence, 49f
theoretical estimates of effect on horizontal dispersion, 178
wind speed:
effect on concentration from continuous sources, 313f
effect on maximum concentration from elevated source, 272f
effect on rise of plume, 237f, 272
in relation to stability categories and thermal stratification, 335f
vertical profile near the ground, 42
wind tunnel modelling, 369
Windscale accident, 377

zinc cadmium sulphide tracer, 294